Cytochromes P450

Cytochromes P450: Drug Metabolism, Bioactivation and Biodiversity 2.0

Editors

**Patrick M. Dansette
Danièle Werck-Reichhart**

MDPI • Basel • Beijing • Wuhan • Barcelona • Belgrade • Manchester • Tokyo • Cluj • Tianjin

Editors
Patrick M. Dansette
Université de Paris
France

Danièle Werck-Reichhart
University of Strasbourg
France

Editorial Office
MDPI
St. Alban-Anlage 66
4052 Basel, Switzerland

This is a reprint of articles from the Special Issue published online in the open access journal *International Journal of Molecular Sciences* (ISSN 1422-0067) (available at: https://www.mdpi.com/journal/ijms/special_issues/p450_2.0).

For citation purposes, cite each article independently as indicated on the article page online and as indicated below:

LastName, A.A.; LastName, B.B.; LastName, C.C. Article Title. *Journal Name* **Year**, *Volume Number*, Page Range.

ISBN 978-3-0365-0256-4 (Hbk)
ISBN 978-3-0365-0257-1 (PDF)

Cover image courtesy of Danièle Werck-Reichhart.

© 2021 by the authors. Articles in this book are Open Access and distributed under the Creative Commons Attribution (CC BY) license, which allows users to download, copy and build upon published articles, as long as the author and publisher are properly credited, which ensures maximum dissemination and a wider impact of our publications.
The book as a whole is distributed by MDPI under the terms and conditions of the Creative Commons license CC BY-NC-ND.

Contents

About the Editors vii

Preface to "Cytochromes P450: Drug Metabolism, Bioactivation and Biodiversity 2.0" ix

Lydia Benkaidali, François André, Gautier Moroy, Bahoueddine Tangour, François Maurel and Michel Petitjean
Four Major Channels Detected in the Cytochrome P450 3A4: A Step toward Understanding Its Multispecificity
Reprinted from: *Int. J. Mol. Sci.* **2019**, *20*, 987, doi:10.3390/ijms20040987 1

Irina F. Sevrioukova
Structural Insights into the Interaction of Cytochrome P450 3A4 with Suicide Substrates: Mibefradil, Azamulin and 6',7'-Dihydroxybergamottin
Reprinted from: *Int. J. Mol. Sci.* **2019**, *20*, 4245, doi:10.3390/ijms20174245 23

Ghulam Mustafa, Prajwal P. Nandekar, Neil J. Bruce and Rebecca C. Wade
Differing Membrane Interactions of Two Highly Similar Drug-Metabolizing Cytochrome P450 Isoforms: CYP 2C9 and CYP 2C19
Reprinted from: *Int. J. Mol. Sci.* **2019**, *20*, 4328, doi:10.3390/ijms20184328 35

William R. Arnold, Susan Zelasko, Daryl D. Meling, Kimberly Sam and Aditi Das
Polymorphisms of CYP2C8 Alter First-Electron Transfer Kinetics and Increase Catalytic Uncoupling
Reprinted from: *Int. J. Mol. Sci.* **2019**, *20*, 4626, doi:10.3390/ijms20184626 59

Diana Campelo, Francisco Esteves, Bernardo Brito Palma, Bruno Costa Gomes, José Rueff, Thomas Lautier, Philippe Urban, Gilles Truan and Michel Kranendonk
Probing the Role of the Hinge Segment of Cytochrome P450 Oxidoreductase in the Interaction with Cytochrome P450
Reprinted from: *Int. J. Mol. Sci.* **2018**, *19*, 3914, doi:10.3390/ijms19123914 75

Olubadewa A. Fatunde and Sherry-Ann Brown
The Role of CYP450 Drug Metabolism in Precision Cardio-Oncology
Reprinted from: *Int. J. Mol. Sci.* **2020**, *21*, 604, doi:10.3390/ijms21020604 91

Yazun Bashir Jarrar and Su-Jun Lee
Molecular Functionality of Cytochrome P450 4 (CYP4) Genetic Polymorphisms and Their Clinical Implications
Reprinted from: *Int. J. Mol. Sci.* **2019**, *20*, 4274, doi:10.3390/ijms20174274 117

Kelli Gerth, Sunitha Kodidela, Madeline Mahon, Sanjana Haque, Neha Verma and Santosh Kumar
Circulating Extracellular Vesicles Containing Xenobiotic Metabolizing CYP Enzymes and Their Potential Roles in Extrahepatic Cells Via Cell–Cell Interactions
Reprinted from: *Int. J. Mol. Sci.* **2019**, *20*, 6178, doi:10.3390/ijms20246178 139

Bongumusa Comfort Mthethwa, Wanping Chen, Mathula Lancelot Ngwenya, Abidemi Paul Kappo, Puleng Rosinah Syed, Rajshekhar Karpoormath, Jae-Hyuk Yu, David R. Nelson and Khajamohiddin Syed
Comparative Analyses of Cytochrome P450s and Those Associated with Secondary Metabolism in *Bacillus* Species
Reprinted from: *Int. J. Mol. Sci.* **2018**, *19*, 3623, doi:10.3390/ijms19113623 157

Puleng Rosinah Syed, Wanping Chen, David R. Nelson, Abidemi Paul Kappo, Jae-Hyuk Yu, Rajshekhar Karpoormath and Khajamohiddin Syed
Cytochrome P450 Monooxygenase CYP139 Family Involved in the Synthesis of Secondary Metabolites in 824 Mycobacterial Species
Reprinted from: *Int. J. Mol. Sci.* **2019**, *20*, 2690, doi:10.3390/ijms20112690 . . . . **171**

Olufunmilayo Olukemi Akapo, Tiara Padayachee, Wanping Chen, Abidemi Paul Kappo, Jae-Hyuk Yu, David R. Nelson and Khajamohiddin Syed
Distribution and Diversity of Cytochrome P450 Monooxygenases in the Fungal Class *Tremellomycetes*
Reprinted from: *Int. J. Mol. Sci.* **2019**, *20*, 2889, doi:10.3390/ijms20122889 . . . . **185**

Makhosazana Jabulile Khumalo, Nomfundo Nzuza, Tiara Padayachee, Wanping Chen, Jae-Hyuk Yu, David R. Nelson and Khajamohiddin Syed
Comprehensive Analyses of Cytochrome P450 Monooxygenases and Secondary Metabolite Biosynthetic Gene Clusters in *Cyanobacteria*
Reprinted from: *Int. J. Mol. Sci.* **2020**, *21*, 656, doi:10.3390/ijms21020656 . . . . **199**

Fanele Cabangile Mnguni, Tiara Padayachee, Wanping Chen, Dominik Gront, Jae-Hyuk Yu, David R. Nelson and Khajamohiddin Syed
More P450s Are Involved in Secondary Metabolite Biosynthesis in *Streptomyces* Compared to *Bacillus*, *Cyanobacteria*, and *Mycobacterium*
Reprinted from: *Int. J. Mol. Sci.* **2020**, *21*, 4814, doi:10.3390/ijms21134814 . . . . **215**

Nokwanda Samantha Ngcobo, Zinhle Edith Chiliza, Wanping Chen, Jae-Hyuk Yu, David R. Nelson, Jack A. Tuszynski, Jordane Preto and Khajamohiddin Syed
Comparative Analysis, Structural Insights, and Substrate/Drug Interaction of CYP128A1 in *Mycobacterium tuberculosis*
Reprinted from: *Int. J. Mol. Sci.* **2020**, *21*, 4816, doi:10.3390/ijms21144816 . . . . **235**

Xue-Gui Wang, Yan-Wei Ruan, Chang-Wei Gong, Xin Xiang, Xiang Xu, Yu-Ming Zhang and Li-Tao Shen
Transcriptome Analysis of *Sogatella furcifera* (Homoptera: Delphacidae) in Response to Sulfoxaflor and Functional Verification of Resistance-Related P450 Genes
Reprinted from: *Int. J. Mol. Sci.* **2019**, *20*, 4573, doi:10.3390/ijms20184573 . . . . **253**

About the Editors

Patrick M. Dansette is Director of Research Emeritus at CNRS UMR 8601, University of Paris. He is 75 and has 54 years of research experience. He is a chemical engineer (ESCIL, CPE-Lyon) and obtained his PhD in biochemistry in ORSAY (1972). Since his work at NIH with JW Daly and Don Jerina and after 1977 with D. Mansuy, he has been involved in the study of xenobiotic-metabolizing enzymes and reactive metabolites (arene oxides, thiophene sulfoxides, sulfenic acids, and other electrophiles). In 1987, with Ph. Beaune, he first demonstrated that auto-antibodies in drug-induced hepatitis (tienilic acid, TA) are directed against a cytochrome P450 (CYP2C9) and that TA is a mechanism-based inactivator. He also showed that ticlopidine is a mechanism-based inhibitor of CYP2C19. Together with EF Johnson, he has studied the active site topology of several CYP 2Cs using a family of substrates and inhibitors, NMR relaxation methods, computer modeling, and crystallization. He demonstrated the pharmacological activation pathway of antithrombotic agents Ticlopidine, Clopidogrel and Prasugrel via a sulfenic acid-reactive intermediate and extended the studies on sulfenic acid intermediates in drug metabolism. He has also studied the metabolism of ferrocifens, ferrocenyl analogs of tamoxifen, as anticancer compounds. Finally, with Jean-Luc Boucher, he studied the metabolism of dithiolethiones and their involvement in H_2S production.

Danièle Werck-Reichhart is Emeritus Research Director at the CNRS Institute of Plant Molecular Biology, University of Strasbourg. She has published more than 120 papers related to plant cytochrome P450 enzymes and performed pioneering work in the field in connection with phenolic and terpenoid plant metabolism, as well as herbicide detoxification. She is an elected member of EMBO.

Preface to "Cytochromes P450: Drug Metabolism, Bioactivation and Biodiversity 2.0"

Nearly 70 years ago, R.T. Williams and B.B. Brodie developed the concept of drug metabolism and described the types of reactions and mechanisms by which the body facilitates drug excretion. A decade later, a protein that absorbs at 450 nm in the presence of carbon monoxide was independently discovered by Klingenberg and Garfinkel. Five years later, Omura and Sato identified this protein as cytochrome P450 (P450 or CYP). Numerous P450 investigations in over 100,000 papers have established the enzyme's prominent role in drug metabolism and endobiotic biosynthesis. They have also identified different types of P450s with diverse ranges of substrate specificity and functions. These enzymes are essential for life and are widely distributed among archaea, prokaryotes, and eukaryotes.

The success of and interest in the first Special Issue, titled "Cytochromes P450: Drug Metabolism and Bioactivation", spawned a second issue to include additional topics, titled "Cytochrome P450: Drug Metabolism and Bioactivation and Biodiversity". This book comprises 12 original papers and 3 reviews that were published in this Special Issue and recommended by the *IJMS* Editor. As Guest Editors, we present the work described below.

The first five papers examine human P450s and the human P450 reductase.

The first paper investigates access to the substrate of CYP3A4, the major human liver P450. The authors used *in silico* modeling to analyze access channels for the substrate and product of CYP3A4. Calculations were performed with version 2 of the CCCPP software, which was developed for this project. The article provides a detailed description of these channels, together with associated quantitative data.

The second paper reports the use of X-ray crystallography to explain the mechanism-based inactivation of CYP3A4 by three suicide substrates: mibefradil, an antihypertensive drug that was quickly withdrawn from the market; a semi-synthetic antibiotic azamulin; and a natural furanocoumarin, 6,7-dihydroxy-bergamottin. These findings can increase our understanding of suicide substrate binding and inhibitory mechanisms and can be used to improve the predictions of binding, metabolic sites, and inhibitory/inactivation potential of newly developed drugs.

The third paper investigates the membrane interaction of two P450 enzymes of the 2C family, CYP2C9 and CYP2C19, the structures of which have been solved in their truncated forms (without the N-terminal transmembrane helix) by X-ray crystallography. The aim of this work was to model the missing transmembrane helix and to study its interaction with the phospholipid layer. The authors provided a mechanistic interpretation of experimentally observed effects of mutagenesis on substrate selectivity.

The fourth paper examines electron transfer between NADPH–cytochrome P450 reductase (CPR) and CYP2C8. The study used *in vitro* and fast kinetic methods to study electron transfer and hydrogen peroxide production for three polymorphic forms of CYP2C8, namely, *1, *2, and *3. Anaerobic stopped-flow measurements revealed that the kinetics of the first electron transfer in these genetic variants were altered compared with those of CYP2C8*1 (wildtype), suggesting that electron transfer from CPR is disfavored.

The fifth paper presents the interactions between CPR and cytochrome P450. CPR is the unique redox partner of microsomal cytochrome P450s (CYPs). CPR exists in a dynamic conformational equilibrium between open and closed conformations during electron transfer (ET). This study used *in silico* and *in vitro* approaches to probe the effects of a specific mutation in the hinge segment of CPR on electron transfer with three different human P450 isoforms. The investigation showed that CPR has a highly flexible hinge that results in a conformational distribution of open CPR conformers that can accommodate ET interactions with a variety of redox partners.

The next three papers review the clinical implications of cytochrome P450s as possible clinical biomarkers.

The first paper in this series reviews the role of cytochrome P450 (CYP450) enzymes in the field of cardio-oncology. The paper highlights the importance of cardiac medications in preventive cardio-oncology for high-risk patients or in the management of cardiotoxicities during or following cancer treatment. Common interactions between anticancer and cardiovascular drugs are reviewed. This work emphasizes that metabolic differences between drugs can lead to unpredictable bioavailability, which can drive inter-individual variability in drug disposition and cardiovascular toxicity.

The clinical implications of cytochrome P450 family 4 are reviewed in the next paper. This P450 family is responsible for the metabolism of fatty acids, xenobiotics, therapeutic drugs, and signaling molecules, including eicosanoids, leukotrienes, and prostanoids. Genetic polymorphisms within this P450 family have been associated with a range of human diseases; for example, genetic variants of the CYP4A11 and 4F2 genes have been associated with cardiovascular diseases. The risk of cancer is increased with mutations in CYP4B1, CYP4Z1, and other CYP4 genes that generate 20-hydroxyeicosatetraenoic acid (20-HETE). CYP4V2 gene variants are associated with ocular disease, while those of CYP4F22 are linked to skin disease, and CYP4F3B is associated with dysfunctions in the inflammatory response.

The focus of the final paper in this series is on CYPs within circulating extracellular vesicles (EVs). Recent studies have revealed an abundance of several CYPs in plasma EVs and other cell-derived EVs. This review covers the abundance of CYPs in plasma EVs and EVs derived from CYP-expressing cells, as well as the potential role of EV CYPs in cell–cell communication and their application with respect to novel biomarkers and therapeutic interventions. Moreover, studies have demonstrated that CYP-containing EVs can cause xenobiotic-induced toxicity via cell–cell interactions. Ultimately, mitigating CYP-mediated toxicity will require an understanding of the mechanism by which CYPs are loaded in EVs, as well as their circulation via plasma and their role in extrahepatic cells.

The next six papers are studies from K. Syed's group, who analyzed microorganism genomes to identify cytochrome P450s, collectively known as the CYPome.

The first paper from K. Syed's group describes the CYPome of bacterial species in the *Bacillus* genus that produce secondary metabolites. Genetic studies on these bacteria recently revealed the presence of secondary metabolite biosynthetic gene clusters (BGCs). *In silico* analysis of 128 species within the *Bacillus* genus identified 507 P450s divided into 13 families and 28 subfamilies. A large number of P450 genes were found within secondary metabolite BGCs and are associated with specific P450 families.

The CYPomes of the *Mycobacteria* (M.) genus, which includes species that are involved in tuberculosis (*M. tuberculosis*) and leprosy (*M. leprae*), are described in the next paper from K. Syed's group. This paper shows that the cytochrome P450 139 A (CYP139A) subfamily is present in 894 species in three mycobacterial groups: *M. tuberculosis* complex (850 species), *Mycobacterium avium* complex (34 species), and non-tuberculosis mycobacteria (10 species). Biosynthetic gene cluster analyses suggest that 92% of CYP139A members produce different secondary metabolites

K. Syed's group examined CYPs of the fungal world in their third paper. The paper analyzes the subphylum Agaricomycotina within the dimorphic fungus class Tremellomycetes. Analysis of the CYPome revealed 203 CYPs (excluding 16 pseudo-CYPs) in 23 species of Tremellomycetes; the identified proteins are grouped into 38 CYP families and 72 subfamilies. Of the identified CYPs, 23 CYP families are new, and 3 CYP families (CYP5139, CYP51, and CYP61) are conserved across 23 species within the fungal class Tremellomycetes. This paper focuses on the CYP51 family in particular because mutations lead to resistance to fungicides such as fluconazole.

The focus of the fourth paper from K. Syed's group is on the cyanobacterial CYPome. Cyanobacteria are the oldest known photosynthetic organisms and responsible for the oxygenation of the Earth's atmosphere. Analysis of the genomes of 114 cyanobacterial species revealed 341 P450s from 88 species within 36 families and 79 subfamilies. In total, 770 secondary metabolite BGCs were found in 103 cyanobacterial species. Comparative analyses were performed with other bacterial species in the *Bacillus*, *Streptomyces*, and *Mycobacteria* genera. Compared with cyanobacteria, the species in these genera had fewer P450s and BGCs and smaller P450 fractions within a given BGC.

The CYPomes of *Streptomyces* and secondary metabolites of bacteria from *Bacillus*, *Cyanobacterium*, and *Mycobacterium* are compared in their fifth paper. Bacteria in the *Streptomyces* genus have a higher number of P450s than species from the *Bacillus* genus and Cyanobacteria. The average number of secondary metabolite BGCs and the number of P450s within BGCs were found to be higher in bacteria from the *Streptomyces* genus than the other examined bacterial species. This result corroborates the superior capacity of *Streptomyces* bacteria to generate diverse secondary metabolites. The CYP107 family was found consistently in bacterial species of the *Streptomyces* and *Bacillus* genera, implying a central role in secondary metabolite synthesis.

Using *in silico* approaches, the penultimate paper provided by K. Syed's group investigates CYP128 in bacterial species from the *Mycobacterium* genus, which is best known for the tuberculosis-causing *Mycobacterium* (M.) *tuberculosis*. Genomic analysis of the bacteria revealed a large number of CYP128s that fall into six categories. The paper also examines the interrelationships and patterns of CYP128 genes and biosynthetic gene clusters (BGCs) in different mycobacterial species. The authors report different features in the CYP128 gene distribution, subfamily patterns, and characteristics of the secondary metabolite biosynthetic gene clusters (BGCs) between the *M. tuberculosis* complex (MTBC) and other mycobacterial species. In all MTBC species (except one), CYP128 P450s belong to subfamily A, whereas subfamily B is predominant in four other mycobacterial species. Of CYP128 P450s, 78% are in BGCs with CYP124A1 or with both CYP124A1 and CYP121A1. The CYP128 family ranks fifth in conservation among species. Unique amino acid patterns are present in the EXXR and CXG motifs. Molecular dynamic simulation studies indicate that CYP128A1 binds to MK9 with a higher affinity compared with the azole drugs analyzed. This study provides a comprehensive comparative analysis and structural insights into CYP128A1 in *M. tuberculosis*.

Finally, the last paper explores cytochrome P450 in the white-back planthopper, *Sogatellas* (S.) *furcifera*, an insect and major rice pest in China and in several other rice-growing countries of Asia. The aim of the study was to explore key genes related to the development of resistance to the insecticide sulfoxaflor in *S. furcifera* and to verify their functions. The paper reports the predicted interactions between CYP6FD1 and CYP4FD2 and sulfoxaflor. It also predicts that CYP6FD1 will have higher metabolic activity with respect to sulfoxaflor.

Patrick M. Dansette, Danièle Werck-Reichhart
Editors

Article

Four Major Channels Detected in the Cytochrome P450 3A4: A Step toward Understanding Its Multispecificity

Lydia Benkaidali [1], François André [2], Gautier Moroy [3,†], Bahoueddine Tangour [4], François Maurel [5] and Michel Petitjean [3,6,*,†]

1. Laboratoire de Biochimie Théorique, CNRS UPR 9080, Institut de Biologie Physico-Chimique, 75005 Paris, France; lydia.benkaidali@gmail.com
2. CEA/I2BC, CNRS UMR 9198, Université Paris Saclay, 91190 Gif-sur-Yvette, France; francois.andre@i2bc.paris-saclay.fr
3. MTi, INSERM UMR-S 973, Université Paris Diderot, 75013 Paris, France; gautier.moroy@univ-paris-diderot.fr
4. Unité de Recherche de Modélisation en Sciences Fondamentales et Didactique, BP244, Université de Tunis El Manar, 2092 Tunis, Tunisie; bahoueddine.tangour@ipeiem.utm.tn
5. ITODYS, CNRS UMR 7086, Université Paris Diderot, 75013 Paris, France; francois.maurel@univ-paris-diderot.fr
6. E-pôle de Génoinformatique, CNRS UMR 7592, Institut Jacques Monod, 75013 Paris, France
* Correspondence: petitjean.chiral@gmail.com
† Current address: CMPLI, INSERM U1133 (BFA, CNRS UMR 8251), Université Paris Diderot, 75013 Paris, France.

Received: 18 January 2019; Accepted: 20 February 2019; Published: 25 February 2019

Abstract: We computed the network of channels of the 3A4 isoform of the cytochrome P450 (CYP) on the basis of 16 crystal structures extracted from the Protein Data Bank (PDB). The calculations were performed with version 2 of the CCCPP software that we developed for this research project. We identified the minimal cost paths (MCPs) output by CCCPP as probable ways to access to the buried active site. The algorithm of calculation of the MCPs is presented in this paper, with its original method of visualization of the channels. We found that these MCPs constitute four major channels in CYP3A4. Among the many channels proposed by Cojocaru et al. in 2007, we found that only four of them open in 3A4. We provide a refined description of these channels together with associated quantitative data.

Keywords: cytochromes P450; CYP3A4; active site access channels; cavities boundaries; minimal cost paths

1. Introduction

The cytochromes P450 (CYP) constitute the largest superfamily of hemoproteins, which have been studied since the late 1940s (see [1,2] for an historical survey). With the emergence of genomic data and the quickly growing number of P450 sequences, the superfamily has been phylogenetically classified in families, subfamilies and individuals, respectively, denoted by the first identification number, the letter following it and the second identification number (e.g., in human 1A2, 3A4, etc.). More than 18,000 P450s sequences in all living kingdoms were recognized in 2013 [3], but, since then, the number of known sequences keeps growing and can be estimated to be higher than 300,000 taking into account the current plant genomic projects [4]. CYPs are found in many bacteria, plants and animals. It is estimated that, in human or mammal metabolism, 75% of drug transformation reactions involve catalysis by P450s [5,6]. The secondary and tertiary structures of the CYPs have largely been conserved throughout evolution [7]. The crystal structures of a large number of CYPs, in both the free and the substrate-bound forms, have been solved. The core is highly conserved within the structural fold, and the heminic active site of the CYP is buried inside the enzyme [7].

The human genome encodes 57 CYP isoforms, that play a major role in the biotransformation of drugs, pesticides or many other chemicals, and in the metabolism of endogenous compounds such as steroids and vitamins [8,9]. The detoxification reaction mediated by these CYPs can yield reactive intermediates which can damage DNA, as well as lipids and proteins [10], while the alteration of their activity often leads to serious diseases [11]. Tables of substrates of human isoforms are available [6,12–14]. Despite several attempts to predict substrates of CYPs [15–19], no clear prediction rule is known. It is known that lipophilicity can play a crucial role [20] and a summary of substrates/selectivity rules is proposed [21]. However, there is an urgent need to improve the accuracy, interpretability and confidence of the computation models used in drug discovery process (see [19] and references therein).

In this paper, we consider the human isoform 3A4. It lies in the human liver and is estimated to contribute to the phase I metabolism of roughly half of the drugs on the market [22,23]. The other isoforms accounting for more than 90% of the oxidation of drugs are 1A2, 2A6, 2C9, and 2D6 [24]. Although most of the CYPs have a binding allosteric site for their substrates which reversibly accommodates one molecule of substrate at a time, 3A4 isoform can accommodate more than one molecule in its binding site at the same time [25]. Recent advances about CYP3A4 show limited information on the pathways to the heminic site [26,27].

P450s catalyze an oxidation where the substrate binds in the active site on the distal side of the heme. Although the oxidation step has been investigated for a long [28–31], the ingress and egress of the compounds to and from the active site remain unclear. Structural flexibility is essential to allow chemical compounds to get in and out the active site, and it was shown that it correlates with substrate preferences for several CYPs, including for 3A4 [32].

Few biophysical and biochemical approaches have been proposed by wet biology teams to experimentally address the role of ligand access channels, as reviewed in [33], but never for CYP3A4. To our knowledge, only one article presents clear-cut results suggesting that a ligand diffuses through a given channel and not another one [34]. The authors mutated selected residues in one of the channels (double mutant Y309C/S360C) to introduce cross-linking by disulfide bond that resulted in one channel closure. They could then measure the kinetics of metabolism of two different substrates (benzphetamine and 7-EFC, i.e., 7-ethoxy-trifluoro-coumarine), and show that the double mutant exhibited unchanged activity for benzphetamine (98% of wild type activity), while it dropped to 19% compared to wild-type activity for 7-EFC, indicating that the two substrates do not cross the same channel to access the active site. This experiment necessitates choosing carefully the two residues to mutate, and obtaining an active form of the recombinant enzyme, which is never obvious.

Molecular dynamics based studies of the channels of several CYPs were performed [32,35–55]. Some of them were applied to 3A4 [32,42–45,48,51,53,55]. However, these studies do not lead to a consensus on the number and type of channels for a given isoform. Due to the prohibitively long time scale required to observe opening or closing of channels, we preferred to use a rapid geometric method to identify through which paths are travelling the compounds. Identifying and characterizing the access channels and their lining amino acids is not a trivial task because the channels can dynamically open/close in response to water or ligand passage and enzyme breathing motions [56].

In this context, we computed cavities and channels with CCCPP, which takes in account both the size and the shape of the ligands through a cylindrical model of the ligands, proved to be more realistic than the spherical model used almost everywhere in the literature [57]. The effect of the ligands conformational flexibility on their shape was taken in account in preliminary studies [58,59]. Then, we found that the 3A4 isoform has three major conformations while only two conformations are considered in the literature [60].

For the present study, we built version 2 of CCCPP to perform a refined analysis of the channels. Unless otherwise stated, further mentions of CCCPP refer to its enhanced version. The secondary structure of the members of the P450 family is described by a nomenclature defined by Poulos et al. [61] (see also Figure 1 in [62] and Figure 2 in [63]), which is widely used in the P450 scientific community and that we use throughout this paper. The major description of the channels [64] is based on a geometric method in terms of the secondary structures at which there is an egress of the computed channels. These channels were computed with the software CAVER [65], and gave rise to a channel nomenclature which is still in use [64]. However, there are dozens of other cavities and channels calculation softwares (see [57] for a review). They give a variety of results due to the diversity of output data structures. Thus, it is rather difficult to compare these results. For example, the CAVER based nomenclature established in [64] from all CYP crystal structures available in the Protein Data Bank (PDB) in March 2006 (total: 143 PDB files, 192 chains, 26 CYPs) indicates 14 channels, while CYP3A4 alone (PDB code 1TQN, 1 chain, included in the 2006 study) gave rise in 2012 to 21 channels with MOLEonline 2.0 [66], one of the successors of CAVER. In fact, only three channels were attributed to 3A4 in the 2006 study. Such discrepancies are not unusual. They appeared also between the P450$_{cam}$ (CYP101) channels calculated with CAVER and MOLEonline and the ones computed in [67], and the situation remains unclear despite the help of several molecular dynamics simulations [35,37]. The experimentalists are left with dozens of published software packages and they have to face to a huge of potentially contradictory results about the channels they are looking for: Which channels should be retained? Easy and rapid comparisons are needed. Giving the name of the secondary structure at which there is a channel egress does not suffice to describe the channels. For a given CYP chain, most of the channels have common parts. Thus, in our opinion, the network of channels should be described with the help of graph theory tools, in terms of paths along nodes and edges, as done in the present study. To compare these networks for different input CYPs, it is better to give a full description of the channels in terms of protein heavy atoms and residues, not only at the egress locations of the channels, but also all along the channels. These functionalities were unavailable in the original version of CCCPP described in [57]. Thus, no more visual examination of the secondary structures is needed to locate the egress of the channels, as it was needed with CAVER. Moreover, the lists of atoms and residues are returned by CCCPP, plus the data structure defining the boundary of each channel. This latter functionality was also available in the version 1 of CCCPP. Throughout this paper, channels named 1, 2a, 2b, etc., refer to the nomenclature of Cojocaru et al. [64] based on the secondary structures elements at the protein surface where the channels emerge.

2. Methods

2.1. The Standard Approach: Terminology

The channels in proteins were calculated with the CCCPP software (binaries and documentation available at http://petitjeanmichel.free.fr/itoweb.petitjean.freeware.html). The first part of the method implemented in CCCPP is described in [57]. For clarity, we summarize it as follows. The smallest convex domain enclosing the heavy atoms of the protein is a polyhedron partitioned in non overlapping tetrahedral cells with atoms at their vertices (Delaunay triangulation). Two adjacent cells are separated by a triangle with atoms at its vertices, acting as a door between two tetrahedral rooms, which let or not the ligand pass through to travel from one cell to its neighbor. Having flagged all triangular doors with their status, open or closed, it is easy to exhibit the protein shape and its concavities: the protein shape is modelized by the set of tetrahedral cells interconnected by triangles, which can not be passed by the ligand, although the other cells are part of the concavities. Thus, it can be seen whether or not the ligand is sterically allowed to travel from the exterior of the protein to the location of the active site.

It is emphasized that the concavities (or channels) available to the ligand depend on which ligand is considered, and by no way constitute a universal network of concavities (or channels). That should not be shocking: e.g., the space available in the protein to a small molecule such as water cannot be identical to the space available to a large ligand such as cyclosporin or erythromycin.

We also emphasize that the usual terminology dealing with voids inside proteins does not yet make consensus: channels, concavities, pores, pockets, etc. Here, we call channels the concavities linking the exterior of the protein to its buried active site. In the case of a protein with an active site at its surface, we would say that the concavity is a pocket, while surface concavities without any active site are also often called pockets. A concavity throughout the protein and linking its exterior at two places can be called a pore, without reference to any active site. We insist that these intuitive definitions are introduced for clarity but are not intended to be mathematically rigorous.

However, our data structure is rigorously defined and can be handled with graph theory tools. The facial graph was defined as follows: each tetrahedral cell is a node of this graph, and each triangle between two adjacent tetrahedra (i.e., two nodes) is an edge of the graph linking these two nodes if and only if the ligand can pass through this triangle. In general, the facial graph is not connected: it has several components. Any component linking the exterior of the protein to the active site is called a channel. Each ligand has a smallest size (thickness) denoted by CV (critical value) [57]. There is a largest CV for which at least one access channel to the active site exists: it is called the limiting CV, and is denoted CV_{lim}. Above this value, it is declared that the ligand cannot access to the active site due to sterical constraints. The reader is referred to the original paper [57] for advanced technical details.

2.2. The Improved Approach: Minimal Cost Paths

The new part of CCCPP that we developed in the framework of the present study is presented below. The full CCCPP software is publicly available on a repository located at http://petitjeanmichel.free.fr/itoweb.petitjean.freeware.html.

It appeared that the channels of the CYPs have large parts at the protein surface and that the main channel to the active site is a funnel which permits several potential pathways for the ligand. To find preferential trajectories for the ligand, we defined a minimal cost path, denoted MCP, as follows. To each edge of the facial graph is associated the cost CV/CV_{max}, where CV is the critical value of the current ligand, and CV_{max} is the maximal critical value which would allow a hypothetical ligand to pass through the triangle associated to this edge. This cost is in the interval (0,1). The smaller is the cost, easier is the passage. In the facial graph defined in Section 2.1 we can seek for the MCP among all possible paths linking the exterior of the protein to the active site. This is performed with the algorithm of Dijkstra [68]. To detect further potential pathways of interest, all edges of the current MCP are removed, then Dijkstra's algorithm is applied again, and so on until no new MCP can be found.

Each MCP is an ordered sequence of triangles, but it is also an ordered sequence of tetrahedra. Discarding if it is a channel or a MCP inside a channel, a set of tetrahedra has a volume, which is the sum of the volumes of the tetrahedra. It also has a boundary, which is the set of the triangular faces through which the ligand cannot pass. Thus, it has a surface, which is the sum of the surfaces of these latter triangles. The MCPs are clusterized. Each cluster defines a trajectory: it has surrounding atoms, residues and secondary structures [60]. Here, these trajectories correspond to channels, in the sense of [64].

2.3. The Two Modes of Visualization of the Channels and Pathways

These two modes of visualization are exemplified in Figure 1.

The first mode of visualization of the channels relies on the facial graph of the channels, or parts of this facial graph. It is done by generating a molecular file such that each tetrahedron is a virtual atom located at the barycenter of its four surrounding protein atoms, and the edge connecting two tetrahedra is a bond between their two respective associated virtual atoms.

The second mode of visualization applies mainly to pathways in channels. MCPs can be visualized by generating a molecular file containing the edges of the tetrahedral cells as bonds linking protein atoms. It is pointed out that these bonds originate from the triangulation of the protein, and as such in general they are not chemical bonds between protein atoms: this is just a functionality of CCCPP.

All figures displayed in this paper were generated with the help of PyMOL^{TM} (The PyMOL Molecular Graphics System, Version 1.2r3pre, Schrödinger, LLC, https://pymol.org/). Some of these figures are based on a mix of the two modes of visualization with appropriate clipping planes, sometimes together with the heme and the ligand.

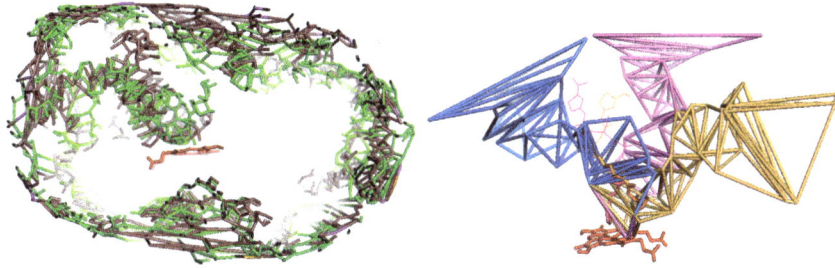

Figure 1. The two basic modes of visualization of CCCPP (images from [69]). The target atom is the iron of the heme group (in red). (**Left**) Superposition of the networks of channels of two complexes of CYP3A4, PDB codes 1TQN and 2V0M, respectively, in green and in brown, computed at CV_{lim} 6 Å and 7 Å. The edges are those of the facial graph of the pockets and channels: They show the location of the voids in the CYP (it is why most of them lie at the surface of the CYP). (**Right**) The channels 2a (in brown), 2f (in purple) and S (in blue) computed by CCCPP in the complex 4K9U of CYP3A4, a,t respectively, CV = 5.75 Å, 6.25 Å and 6.75 Å. The edges are those of the tetrahedra:They show the boundaries of the channels (they are inside the CYP).

3. Results and Discussion

3.1. The Main Channel

A funnel shaped channel appears in the apo form 1TQN of CYP3A4 (Figure 2a). It is located in the deformable area of channel 2 (as named by Cojocaru et al. [64]) and lets a wide opening at the neighbor of the active site. It lies between the following secondary structures: β_1 sheet, A-anchor, B-C block and β_4 sheet (in 3A4, it is the C-terminal loop, as denoted further in the text; size slightly depending on the conformations of a given isoform), and F-G block (helix-loop-helix), very flexible due to channel opening (Figure 2b). The B-C block is defined to extend from the beginning of the B helix to the end of the C helix, and the F-G block to extend from the beginning of the F helix to the end of the G helix. The F-G block region offers highly variable amino acids sequences and structures for different CYPs, and thus can be important for substrate specificity through making contacts with the substrate. These regions are bordered by putative SRSs (Substrate Recognition Site: see [70]). The flexibility of the F-F' loop let us define three conformations of CYP3A4 [60]: the closed one, labeled C, and the opened conformations, labeled O1 or O2 depending, respectively, if block 1 or block 2 opens. Molecular dynamics simulations of cytochrome P450$_{cam}$ showed that substrates and products could egress from the active site via pathways in the vicinity of the three routes identified by TMP (thermal motion pathway) analysis, but with pathway 2 being energetically favored over pathways 1 and 3 [36,37]. Analysis of the simulations of cytochromes P450$_{cam}$, P450-BM3 and P450$_{eryF}$, showed that egress trajectories in the region of channel 2 could be clustered into subclasses, named 2a, 2b, etc., according to the secondary structure elements lining the ligand pathway as it emerges from the protein surface. Although the overall fold of the CYPs is well conserved, the length and secondary structure of the B-C and F-G blocks vary considerably inside the P450s family. The flexibility of these 2 blocks is the main source of conformational change for a given isoform [32].

The facial graph (see Section 2.1) of this main channel is connected, i.e., there is only one channel there: see Figure 2a,b. Knowing the volumes of the tetrahedra issued from the triangulation, we computed that about 88% of the void appears to be surface pockets and only 12% of the void is located at the distal face of the heme (Figure 2a,b). Some ligands are in surface binding pockets, such as the progesterone [71] (PDB code 1W0F; see also Figure 4 in [57]). It could be considered that surface pockets are concavities which are not part of the protein domain, but there is no consensus in the literature about the definition of the protein domain and of the pockets. The binding cavity has a large volume, reported to range from 1173 to 1332 $Å^3$, increasing up to 2000 $Å^3$ when binding large substrates such as ketoconazole and erythromycin [72]. The whole channel 2 computed by CCCPP for CV = 6 Å (see the definition of CV in Section 2.1), which includes the binding cavity, the path accessing to it and the mouth of the channel, has a total volume of 42,400 $Å^3$ and a bounding surface of 41,800 $Å^2$. These values are large because they include the contribution of the mouth of the channel, which is a large cavity lying at the protein surface. It is pointed out that the status of such a surface cavity is unclear because, while it is inside the convex hull of the protein, it is difficult to decide if it is indeed a part of the protein domain or if it is a void region exterior to the non convex shaped protein.

The holo form 2V0M of CYP3A4 contains two ketoconazoles molecules. This antifungal molecule is bulky (van der Waals volume: 450 $Å^3$). The funnel appears for a maximal ligand size CV_{lim} = 7 Å (see the definition of CV_{lim} in Section 2.1), and leads to observe two pathways, each corresponding to a subchannel and containing one ketoconazole. These two subchannels are in block 1 and block 2 [60]. In the apo form, only block 1 appears to let pass the first ketoconazole (Figure 2a). In the holo form, block 2 is opened at CV = 5.75 Å because its access is obstructed by a bottleneck of CV < 6 Å (see Figure 2c). At this CV, block 2 appears (Figure 2a).

Several crystallographic structures of CYP3A4 exist, in several conformations corresponding to different states: interaction with one or two ligands, or none. These differences are related to conformational changes, which can be correlated to the differences between the channels, in function of the ligand. To accommodate two molecules or a large molecule, the protein undergoes a significant conformational change, especially in the F-G region and around the Phenyle cluster, which contains eight phenylalanines residues (57, 108, 213, 215, 219, 220, 241 and 304). Positioning of the I-helix and the C-terminal loop are also altered. The apo form 1TQN is flagged as closed subtype 1 [64,73]. It is such that the F-F' loop binds the ligand in 2V0M (Figure 3c), and is flagged as open subtype 2 [60,64,73]. It is bound to two ketoconazoles, a known inhibitor of CYP3A4. Ketoconazole has an imidazole group (highly polar) and it has nine rotatable bonds and thus it is highly flexible [59]. In 2V0M there are two co-crystallized ketoconazoles (respective thicknesses of 5.19 Å and 6.52 Å, measured as indicated in [57]).

The first one has its imidazole group near the heme: the N3 atom of the imidazole ring binds the iron atom of the heme [74]. The dichlorinated aromatic ring is in the active site and the main skeleton is in the part of channel 2a common with channel 2f (channels 2a and 2f separate at Thr224). The bottleneck in channel 2a is at Phe108 and Thr224. Compared to the apo structure, an enlargement of channel 2a of 1 Å is needed to accept a large ligand (Figures 3c and 4a,b).

The second ketoconazole is inside channel 2f with an orientation opposite to the one of the first ketoconazole, and with the acetyl group near the active site and the dichlorinated aromatic ring (more hyrophobic) near the entry of the channel (Figure 5a,b). The second ketoconazole is inside channel 2f but it has the opposite orientation, with the acetyl group near the active site and the dichlorinated aromatic ring (more hyrophobic) near the entry of the channel (Figure 5a,b). There is a bottleneck in channel 2f, surrounded by four residues (Tyr53, His54, Phe215, and Leu216), three of them being aromatic and acting as gating residues: see Figure 3b. Phe215 and Leu216 are in F-F' loop after conformational change of the apo structure where Phe215 obstructed the opening of channel 2f. Tyr53 and His54 are on anchored helix A in the membrane without significant change from the apo structure, and are thus assumed to have no role in building the bottleneck. Thus, this latter would be due only to a move of the F-F' loop. Phe215 moves in the holo structure toward the mouth of channel 2f and catches the hydrophobic ligand at the membrane/water interface: this is a difference with the first molecule, which enters in channel 2a through the membrane, thus explaining the opposite orientations of the two molecules. The movement of F-F' loop let Arg212 go to the interior of the protein at a location at the opposite of the active site: Arg212 is no longer able to interact with the ligand.

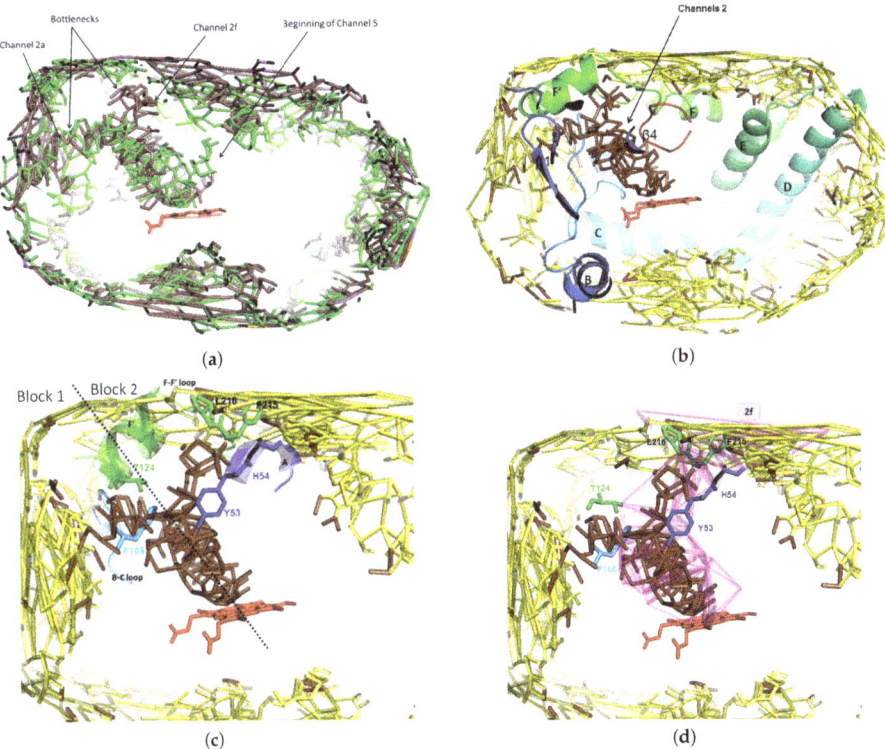

Figure 2. Four representations of the main channel through its facial graph. The target atom is the iron of the heme group (in red). (**a**) Facial graph of the main channel to the active site of CYP3A4 computed from 2V0M (in brown, for CV_{lim} = 7 Å), superposed to the facial graph of this channel computed from 1TQN (in green, for CV_{lim} = 6 Å). The two facial graphs show the way in channel 2a: the one of 2V0M is the beginning of channel 2f with a bottleneck at its entry (not connected to the exterior); the two facial graphs show also the beginning of channels S. (**b**) The funnel shaped channel leading to the active site, computed from 2V0M, appearing at CV_{lim} = 7 Å (in brown) superposed on the one computed at CV = 7.25 Å (in yellow); the part of the facial graph of channels 2 lies within β_1 sheet and C-terminal loop and B-C and F-G loops; only surface pockets were visible for $CV > CV_{lim}$ (in yellow), thus the evidence of a bottleneck at the entry of the two pathways. (**c**) Zoom of (**b**); bottleneck toward 2a: Phe108 (in B-C loop) and Thr224 (in helix F'); bottleneck toward 2f: Tyr53 and His54 (both in helix A), Phe215 and Leu216 (both in F-F' loop); the motions of these residues induce the opening/closing of the bottlenecks; the residues constituting the bottleneck are in green and blue sticks; the dashed line separates blocks 1 and 2. (**d**) Zoom of (**b**), superposed on channel 2f computed at CV = 5.75 Å; six residues border the bottlenecks of channel 2f: four are located at the entry (in helix A and FF' loop), Phe108 in the common part of 2a and 2f, and Thr224 where 2a and 2f separate.

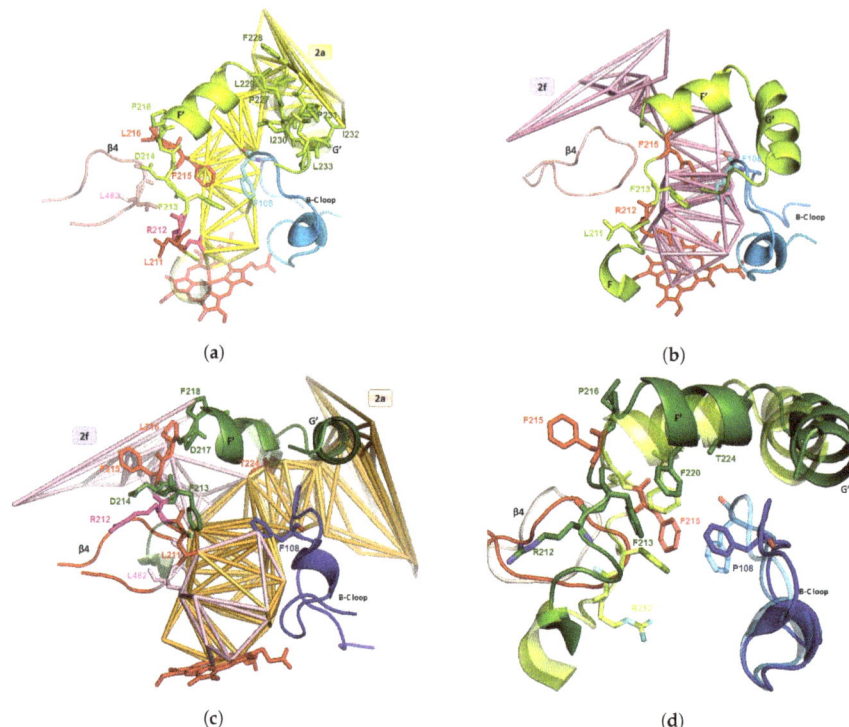

Figure 3. Apo forms are in light colors, holo forms are in dark colors. Each holo form refers to 2V0M, with two ketoconazoles (see Figure 5). (**a**) Channel 2a in 1TQN with some of its bounding residues in the F-F′ and B-C loops; the mouth of the channel is lined by G′ helix hydrophobic residues; the gating residues Phe213 and Phe215 of the F-F′ loop and Phe108 of the B-C loop are bounded by channel 2a in closed form. (**b**) Channel 2f of 2V0M, superposed on 1TQN; shows the steric obstruction of lining residues of F-F′ and B-C loops (Arg212 and gating residues Phe213, Phe215 and Phe108) and of helix F′ to open channel 2f. (**c**) Channels 2a and 2f in 2V0M with some bounding residues; the common part of channels 2a and 2f is lined by F-F′ and B-C loops and by C-terminal loop; the mouth of 2a is lined by helix G′ helix and by the mouth of 2f by F-F′ loop; 2a and 2f are separated at exit by hydroxy-Thr224 in helix F′ at the hydroxy group; the common part is in orange and the separated parts are in light rose; the gating residues Phe 213 and Phe215 in loop F-F′ and Phe108 in B-C loop borders 2a and 2f in the open form of CYP3A4. (**d**) 1TQN superposed on 2V0M; the secondary structures are in cartoon and the residues are in stick (gating ones: Leu216, Arg212); 1TQN is transparent, showing 3A4 conformation before entrance of the two ketoconazoles; indicate what could be the moves of B-C and F-G blocks and of C-terminal loop, involved in the opening of access channels 2a and 2f.

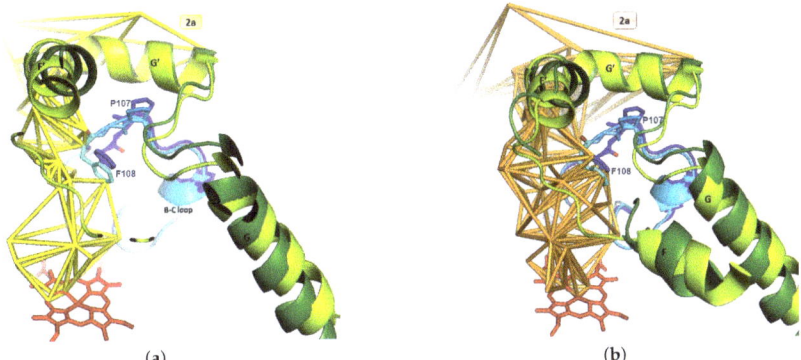

Figure 4. Superposition of CYP3A4 1TQN and 1V0M at blocks F-G and B-C, which bound channel 2a. For clarity, a piece of F-F' loop was removed. Apo forms are in light colors, holo forms are in dark colors. Each holo form refers to 2V0M, with two ketoconazoles (see Figure 5). (**a**) Channel 2a of 1TQN (CV = 6 Å) superposed on 2V0M. (**b**) Channel 2a of 2V0M (CV = 7 Å) superposed on 1TQN.

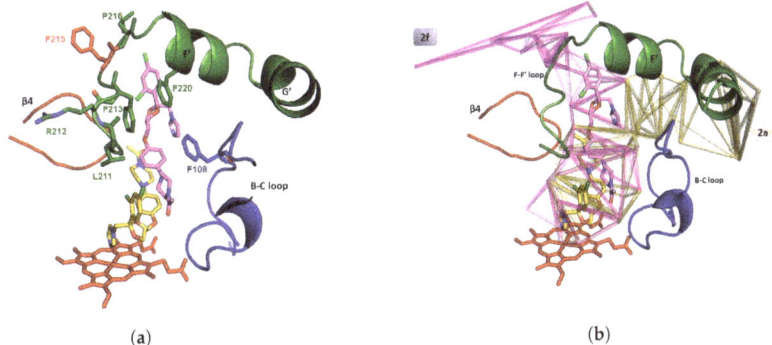

Figure 5. (**a**) The two ketoconazoles of 2V0M inside channels 2a and 2f, within the secondary structures bounding these channels (these latter are not delineated). The first ketoconazole, in yellow, enters through channel 2a. The second ketoconazole, in purple, enters through channel 2f. (**b**) Idem, with delineation of channels 2a and 2f by the boundaries of the trajectories (residues are not displayed).

3.2. Access Channels and Narrow Channels

3.2.1. Channels 2a and 2f

We know that the access to 3A4 active site is through channel 2a, computed from the apo structure with CV = 6 Å. This channel is opened in the apo form, and it enlarges to accept a first ketoconazole ligand at CV = 7 Å between the F-G and the B-C loops and the β_1 sheet. Channel 2a is the first found by CCCPP, and it is the biggest one for this conformation of CYP3A4. It contains the first ketoconazole, bound to the heme. This channel is assumed to be the first pathway found by the ketoconazole. Pathway 2a, in which the ligand passes between F-G loop, B-C loop and the β_1 sheet, appeared to be the most likely route for substrate access and product egress in previous study on bacterial proteins [75]. In the bacterial P450 structures, 2a is opened in most of the structures that have at least one channel opened. Simulations of product egress from CYP101 indicate that this route would be used by the product as well as the substrate [36]. The MCP (see Section 2.2) corresponding to channel 2a has a volume of 613 Å^3 and a bounding surface of 273 Å^2 in the apo form, and it has a volume of 774 Å^3 and a bounding surface of 896 Å^2 in the holo form. This volume is small because it corresponds to the path accessing to the binding site rather than to the full binding cavity. It should be compared to the volume of convex hull of the ketoconazole, which is estimated to be in the interval 300–340 Å^3, depending on the conformation. The surface boundary is small because the path is not a closed one: it is bounded only at some places while it opens on the rest of the channel at other places. Such quantitative results got by CCCPP cannot be obtained by other channel calculation programs.

Then, other pathways are needed to accept more ligands. In the holo form, a second channel is found: it is channel 2f, in block 2, located between F-G block and C-terminal loop, and containing the second ketoconazole. It is identified as distinct channel and it lies between channel 2a and the solvent channel S. In our study of the 16 crystallographic structures, we observe that the channel 2f opens for CV in the range 5.75–8.75 Å, depending on the ligand size. In 2V0M, channels 2a and 2f have a common part and separate at exit in helix F' at hydroxy group of Thr224. We found that the second ketoconazole of structure 2V0M is oriented from the heme in the direction of channel 2f. At CV_{lim} = 7 Å, channel 2f appears as a dead end, not connected to the exterior (Figure 2a). It has a total volume of 49,200 Å^3 and a bounding surface of 48,100 Å^2. The MCP in channel 2f has a volume of 851 Å^3 and a bounding surface of 360 Å^2 in the holo form. A bottleneck at CV < 6 Å obstructs the connection to the exterior of the protein. An opening at CV = 5.75 Å and a slight channel flexibility should permit the second ligand to enter (CV = 6.52 Å). Channels 2a and 2f are bordered by SRS-2, lying above the active site cavity.

Channels 2a and 2f are hydrophobic. They have a common part then separate near the hydroxy group of Thr224, on helix F' (see Figure 2c). Channel 2a is bordered by SRS-1, SRS-2 and SRS-3, and channel 2f is bordered by SRS-2 and SRS-6. Both channels 2a and 2f open at the C-terminal region of the helix F and in the F-F' loop, forming SRS-2 (near the active site cavity). The other part of the entry of channel 2a lies at the N-terminal region of the helix G and in the G-G' loop, forming SRS-3 (see Figure 5a). The B-C loop/B' helix (bordering channel 2a) was identified as a putative SRS-1 [70], and involved in substrate binding. We found with CCCPP that the opening of channel 2a is apolar for 1TQN, located at His28, Phe228, Ile232 and Val235. It is more polar for 2V0M, located at Lys35, Phe46, Phe228, Val235, Asp380, and Lys390. For both 1TQN and 2V0M, the entry of 2a is bordered on one of its sides by the hydrophobic helix G' (residues 230–237; see Figure 6a,b), which is anchored in the membrane (see Figures 3 and 6a,c). The following residues of channel 2a have an orientation common to the apo form 1TQN and to the holo form 2V0M: Pro227, Phe228, Leu229, Ile230, Pro231, Ile232, Leu233, Glu234, Val235, and Leu236. Channel 2a enlarges of ΔCV~1 Å and channel 2f opens to accept two ketoconazoles in 2V0M. The superposition of 1TQN and 2V0M shows that the two B-C loops are superposed due to a move of B-C in 2V0M at Phe108, Gly109, and Pro110, at the exterior of the funnel area, so that it permits the entrance of the ligand in 2a and 2f. These two channels are bordered by par the B-C loop (see Figure 3a,b). In the apo form, Phe108 obstructs the channel. Its move enlarges the common space admitting the two ketoconazoles. In 2V0M, R106 (in B-C loop) and E374 are lining channel 2a. Tyr53 and His54 are on the anchored helix A, without change between the apo form 1TQN and the holo form 2V0M. These few flexible two residues are not involved in the move of the bottleneck of channel 2f. This move is thus due to the move of F-F' loop. The two bottleneck residues are at the mouth of 2f: Phe215 and Leu216, after the conformational change of F-F' where Phe215 obstructed the opening of 2f in the apo form 1TQN (Figure 2b).

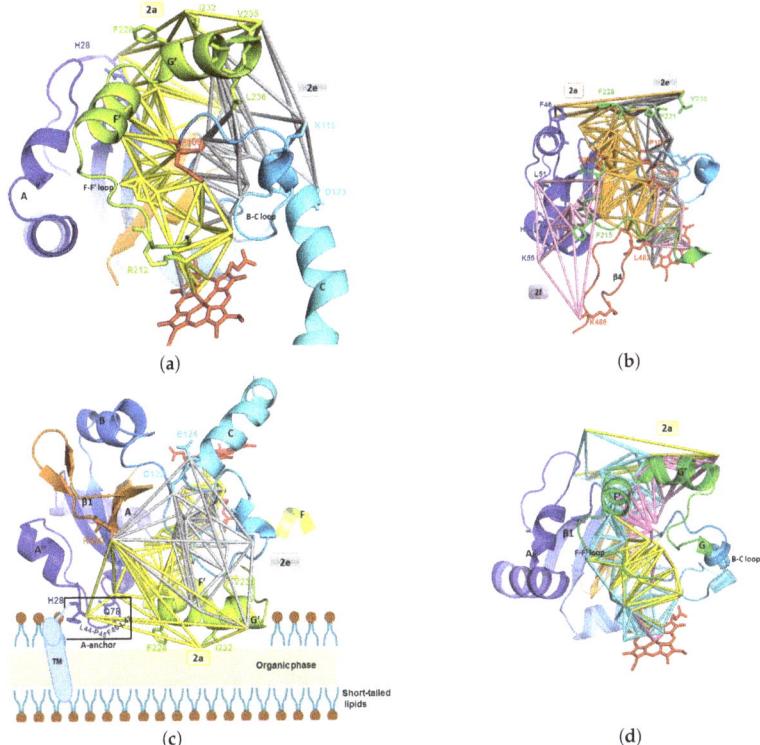

Figure 6. (**a**) Channels 2a and 2e within lining secondary structures of 1TQN (β_1 sheet, F-G and B-C blocks): the two mouths are neighboring, then the channels separate at Pro107 and Phe108 in B-C loop. (**b**) Channels 2a, 2f and 2e of 2V0M enclosed by β_1 sheet, F-G and B-C blocks, and C-terminal loop; channels 2a and 2e are separated at Pro107 and Phe108 in B-C loop. Channels 2a and 2f are separated by Thr224 in helix F'. (**c**) Same as (**a**), rotated: the bottom part of this figure is a part of Figure 1 in [27] (©American Chemical Society, https://pubs.acs.org/doi/abs/10.1021%2Fja4003525). Membrane-bound form of CYP3A4 shows the location of the mouths of 2a and 2e in 1TQN and highlights residues inserting into the membrane and interacting with it via the hydrophobic side chains of helices F' and G' and of A-anchor. A-anchor has a fixed position on A'A" loop, and is formed by residues His28, Leu44, Pro45, Phe46, Leu47 and Gln78: it is labeled in dark blue and boxed. Channel 2a leads to the membrane; channel 2e leads to cytosol; the membrane is viewed on the side of the globular domain; the transmembrane helix is visible; and the location of the organic phase and the short-tailed lipids are represented. (**d**) Channel 2a represented by three MCPs, respectively, in purple, yellow and cyan, surrounded by secondary structures. The three MCPs have a common part in channel 2a.

F-F' loop is hydrophobic and has three charged residues: Arg212, Asp214 and Asp217. Its move may act as a closing/opening valve, at Leu211, Arg212, Phe213, and Phe215 (Figure 2a,c). F-F' loop in a closed conformation is exposed to block 1. Its residues Arg212, Phe213 and Phe215 border the body of the hydrophobic channel 2a in the apo form 1TQN. In this latter, only one polar residue, Arg212 (oriented toward the active site: Figure 2a) borders channel 2a. Arg212 was suggested to regulate solvation of the active site [61], and was reported to assist the binding of ligands to the active site [76,77]. Phe215 is in F-F' loop, bordering channel 2a in conformation C (Figure 2a). Phe215 is involved in the substrate orientation within CYPs active sites and in catalytic mechanisms of these substrates [78]. In the 2V0M, F-F' loop is distorted to open channel 2f. The orientation of the F-F' loop residues follow either the same direction (K208, K209, L210, snf D214), the opposite direction (Leu211,

Arg212, Phe215, and Leu216), or are oriented in the same sense but shifted (valves: Phe213, Asp217, and Pro218). The opening of channel 2f is at Leu51, His54, Lys55, Phe215, Pro218 and Lys486. Helix F' borders channel 2f with apolar residues Leu221, Phe220 and Thr224. In the holo form, an opening occurs via the move of the F-F' loop such that Phe215 (bordering 2a in the apo form), is pushed toward block 2. This push of Phe215 toward the surface is useful: (i) to suppress the steric hindrance to free space to open channel 2f and enlarge the cavity near the part common to 2a and 2f; and (ii) to border the entry of channel 2f, as Leu216 goes to the mouth of 2f at the water interface of the membrane/cytosol to catch the second hydrophobic molecule (see Figure 2c). The same phenomenon is observed in the structure 4K9U, which welcomes two molecules of GS5 (a ritonavir analog [79]), with Log(P) = 3.06. This phenomenon is observed neither for 4K9T, which welcomes one GS4 molecule (another ritonavir analog [79]), nor for 2J0D, which welcomes the bulky erythromycin molecule. For these two more hydrophilic molecules (Log(P) = 2.22 and 2.60, respectively), Phe215 is deformed without being located on the mouth of the computed channel 2f. Pro218 remains at the surface and borders the mouth of channel 2f: that lets the mouth of channel 2f more hydrophobic, thus it helps to catch the hydrophobic ligand. Asp217 is at the entry of block 2 but is not exposed at the surface, i.e., it does not border the mouth of 2f, at the opposite of Pro218, which is hydrophobic and borders the mouth of 2f. The residues rearrangements occur with the move of Leu211 (inside the protein in the apo form), to become (in 3A4-ketoconazole complex) oriented in the active site, bordering the common part of channels 2a and 2f, near the active site, at the opposite of Arg212 which is pushed on the other side of channels 2a and 2f (Figures 2c and 3a). This orientation can be correlated with the fact that substituting Leu211 by a phenylalanine residue affects the homotropic cooperativity in the binding with testosterone [80]. The presence of Leu211 compensates the hydrophobicity of channels 2a and 2f after departure of Phe215. Other significant changes are a shift of 4 Å of the C-terminal loop and a deformation of the of the crevice of helix I around Tyr307 (the C_α is shifted at more than 2 Å) [81]. C-terminal loop becomes closer to the common body of the two channels 2a and 2f delineated by Gly481 and Leu482 near the active site (Figure 2a,c).

3.2.2. Detection of Narrow Channels 2e and S

The narrow channel 2e appears at CV = 5.75 Å in 1TQN and at 5.5 Å in 2V0M. It appears with a total volume in the apo form of 44,500 Å^3 (bounding surface: 44,100 Å^2) and a volume in the holo form of 51,000 Å^3 (bounding surface: 48,200 Å^2). It is a rather common channel observed in twelve different crystallized P450 isoforms [64]. This channel 2e was observed in the 16 crystallized 3A4 structures (available in the pdb data bank) considered this study (see Table 1 in Section 3.3). It has been observed that the channel 2e opens for CV in the range of CV between 5.5 Å and 6 Å, without any relationship with the fact that the protein is bound or not. Channel 2e appears for all conformations of CYP3A4 (C, O1, O2). It threads through the B-C loop (Figure 6), and its opening could depend on the length and of the flexibility of this loop. In the stage of the reaction where these structures of CYP3A4 were crystallized, channel 2e is opened in this range of $\Delta CV = 0.5$ Å, not containing ligand. Channel 2e is detected as a secondary egress route for substrates or products in molecular dynamics simulations of P450s [64]. It is also observed as a secondary exit pathway in simulations of P450eryF [75] and CYP2C5 [40]. It is difficult to conclude about the role of channel 2e as an egress route as long as there is no known structure of complexes with oxidation products. MD (Molecular Dynamics) studies give information on the dynamics of ligand tunnels (opening/closure), but do not involve simulation of the process of ligand egress. MD and SMD (Steering Molecular Dynamics) studies mainly focused on ligand preferred exit tunnel. It was found that channel 2e is an exit one for the hydroxylated product of diazepam in 1TQN, and that channel S is an exit one for 6-hydroxytestosterone [44,45]. That suggests the enlargement of the small opened channel 2e for ligand exit. As shown in Section 1, a controlled hydration of the substrate bound P450 active site is extremely important for catalysis. A solvent filled channel from bulk solvent to heme let water circulate as a water pump. It is likely that its function is related to active site hydration, although it may also have a role in proton transfer (furthermore channel 2e opens on the cytosol). It was suggested that a water channel exists between B-C loop and the C-terminal part of helix B [72]. Channel 2e is displayed with channels 2a and 2f in Figure 7.

In the apo form, channel 2e exits at Lys115, Asp123, Glu124, Pro231, Val235 and Lys390, and channel S exits at Phe22, Pro23, Val235, Asp380 and Lys390. Channels 2a and 2e have a common part near the active site, then separate (Figure 6a,b). Channel 2e and S exit in the cytosol. The residues of the mouth of channel 2e are rather polar, while those of 2a and 2f are apolar (Figure 6c). For the 16 complexes considered in this study, the ligands were located in channels 2a, 2f and S, but none was found in channel 2e. It suggests that channel 2e could be an exit channel, either for oxidized products or for water or protons. Channel 2e may also be for dropping water out of access channels to free space for ligands.

In the apo and holo forms, channel 2e is opened simultaneously with channel 2a, 2f and S, suggesting that the access channels may alternatively open by a F-G move away from the B-C loop, without affecting the 2e channel. This is in agreement with a study on the structure of CYP2C5 [82].

A contiguous water channel from the bulk solvent to the active site is possible [83]. Channel S, i.e., the solvent channel, is detected at CV = 5 Å in the apo form 1TQN. Then, it was computed at CV = 5.75 Å (slightly larger), as a host channel for one of the two GS5 molecules (a ritonavir analog) in the 4K9U complex [79]. It is flagged as important in several isoforms in [64,84,85]. It faces the cytosol and contains a charged gating residue which could lead the product out of the CYP3A4 active site [44]. MD simulations of expulsion of temazepam and 6β-hydroxy testosterone out of CYP3A4 were done: channel S has the largest opening for these two products, so that it may be an exit way for the substrate [44]. Channel S was proposed as a route for controlling water access and egress to the active site for water based on its observation in P450-BM3 [84]. It may also be used for substrate egress [84,86]. It was proposed as the main gateway to the active site of the human 2D6 [86].

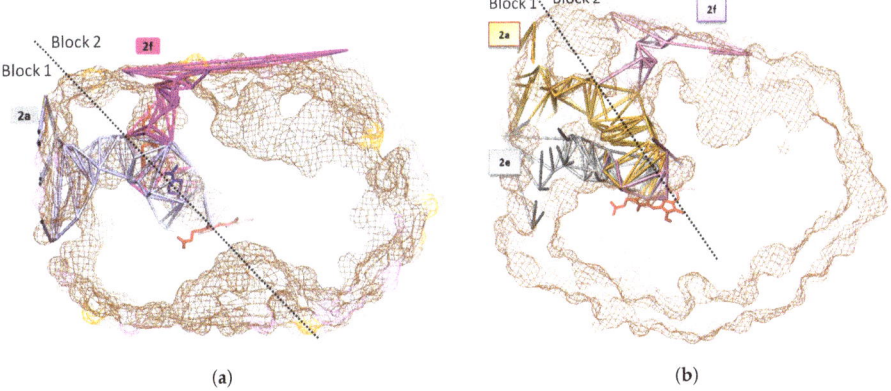

Figure 7. Superposition of the channels of 2V0M. The dashed lines separate blocks 1 and 2. (**a**) Channels 2a and 2f computed at CV = 5.75 Å. (**b**) Channels 2a, 2f and 2e, computed at CV = 5.50 Å.

3.2.3. Channels Opening/Closing and Interactions with the Lipid Bilayer

In a recent paper [87], it is stated that access channels to the buried active site control substrate specificity in CYP1A P450 enzymes. Then, in a recent review [33], it is suggested that the network of channels is involved in the control of the P450 enzymes substrate specificity for all P450 family. The diversity and the deformability of the channels could explain the diversity of its substrates. We emit the assumption that the specificity of the CYPs relies on what happens at the entry of the channels and that, due to sterical constraints, the orientation of the ligand does not change until it reaches the active site. It was suggested that the channels are often gated by aromatic residues all along them [73]. It was also suggested from the analysis of crystal structures that aromatic residues can form a network of gates, which regulates cooperatively the opening and closing of different tunnels [88]. Except Tyr53 for channel 2f, the following gating residues are phenylalanines:

- channel 2a: 57, 108, 213, 215, 228, and 304.
- channel 2f: 46, 57, 108, 213, 215, 220 (on helix F'), 226, and 304.
- channel 2e: 108, 213, 215, 228, and 304.
- channel S: 108, 213, 220, and 304.

Mammalian CYPs are generally membrane-bound proteins [89,90]. The mechanisms of substrate access and product egress from the mammalian membrane bound P450s may differ from those in the soluble bacterial P450s studied before [53]. The microsomal CYPs are anchored in the membrane by an N-terminal transmembrane alpha-helix and there is evidence that their globular domain dips into the membrane [91]. A membrane bound model of human CYP3A4 provides the structure of the protein membrane complexes consistent with most experimental data [51]. Membrane binding of the globular domain in CYP3A4 significantly reshapes the protein at the membrane interface, where most channels open, inducing conformational changes relevant to access tunnels [27].

The CYPs substrates are rather hydrophobic and in the case of membrane bound P450s they are expected to come from the lipid bilayer. As the products of the P450 catalyzed reactions are more hydrophilic, they may be released into the aqueous environment or the polar headgroup region rather than back into the lipid bilayer. The multiplicity of channels suggests possibilities for ligand channelling to and from the P450s sitting in or on the membrane. The P450 protein topology favors channel formation on the distal side of the heme. The proximal side is the likely reductase binding site and corresponds to the smallest channels found by CCCPP.

Helices F' and G' do not completely insert into the membrane, with helix G' establishing a closer contact to the membrane than helix F'. CYP3A4 is anchored into the membrane helix G', which partitions mainly within the headgroup region [27]. The mouth of channel 2f is bordered by helix F' at the water/membrane interface. It was assumed from modeling studies that channel 2f opens at arrival of the molecule [92,93]. Channel opening was observed as a consequence of ligand-induced conformational changes [94]. The mouths of channels 2f that we computed have hydrophobic residues in F-F' loop: Phe213, Phe215, Leu216 and Pro218; thus, the interaction with the channel mouth is facilitated for hydrophobic molecules.

Given the dynamic nature of membrane-anchored CYPs, the precise positions of channels may change dynamically over time [93]. We retained for our calculations the positioning used in [27].

The dynamic motions of the protein can cause the opening of channels not seen in the crystal structures as well as changes in the relative dimensions of the channels [36,40,75]. Even though these motions cannot be seen dynamically in the crystal structures, the location of the channels in these latter, supported by their capacity of ligands, provides a useful basis for exploring ligand access and egress routes, particularly when the snapshots from different crystal structures are considered together.

The mammalian CYPs are characterized by a subdivision of their larger F-G region in F' et G' helix [51]. Insertion of F'-G' helix-loop of CYP3A4 in the membrane moves the β domain towards the heme plane, allowing channel 2a to open, whereas this opening does not occur in soluble bacterial P450s. In these latter, the beta domain plane is farther from the heme plane: the opening occurs between F-G loop and B helix [26], at the level of the opening between F-F' loop and the C-terminal loop in 3A4, corresponding to the channel 2f. The opening of block 2 (due to the move of F-F' loop), which characterizes 3A4, offers a more diverse exterior environment for the compound than the prokaryotic CYPs which offers only solvent exterior channel 2a as environment [36,75].

3.3. Characteristics of the Four Major Channels

For convenience, we summarize these characteristics in Table 2. The four major channels contains substrate recognition sites (SRS). We also summarize in Table 1 the characteristics of the channels computed by CCCPP for the crystallographic structures of CYP3A4 considered in the present study.

We recall that CV is the value of the channel bottleneck (see Section 2). When the critical thickness CV of a ligand exceeds CV_{lim}, the bulky part of the ligand is not in the bottleneck: the passage of the ligand requires flexibility. We also recall that the topology of the channels may be constituted by several MCPs having one or several common parts. This was observed for 2J0D and 4K9T, but for clarity in these cases we provide only the CV values of the main paths in Table 1. We show the four channels in Figure 8 and we summarize the location of the four channels as follows:

In block 1:

- 2a: main channel, apolar mouth (helices F' and G'), opening in the membrane, hydrophobic ligands.

In block 2:

- 2f: alternate channel, mouth opening in transmembrane region, more hydrophilic ligands.
- S: alternate channel, may be an egress channel, mouth opening on cytosol, polar environment, accepting less hydrophobic molecules.

Neither in block 1 nor in block 2:

- 2e: may be an egress channel, mouth opening on cytosol, apparait at CV_{min} ~5.5 Å (narrow), exists in all structures in the three conformations, does not contain a molecule.

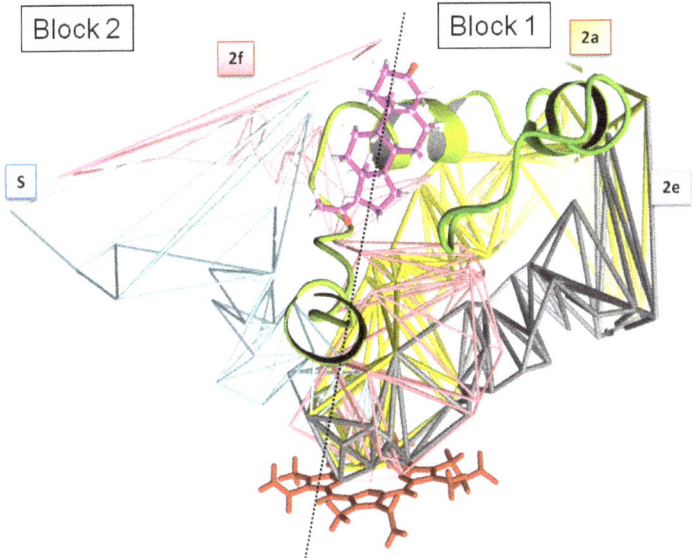

Figure 8. Summary of the four channels of CYP3A4. The heme is shown at the bottom of the channels. The F-F' loop is in green. The progesterone molecule (in purple) lies in a surface pocket (see also Figure 4 in [57]). The dashed line separates blocks 1 and 2.

Table 1. The four channels of CYP3A4 and its conformational states (C, O1, O2). The first channel appears at CV_{lim} and the next channels appear at $CV < CV_{lim}$. [a] Critical thickness of the ligand in Å computed as in [57]. When there are two ligands, two CV values are reported. [b] Cf: conformation (C, O1, O2), according to the authors of [60,69]. [c] See Section 2. [d] Third channel and eventually fourth channel. For 4K9U, the same CV was observed for the third and the fourth channels. [e] Ritonavir analog: see [79]. [f] Other ritonavir analog: see [95]. [g] The topology is constituted by several MCPs having common parts (see Section 2.2).

PDB Code	Ligand	CV [a]	Cf [b]	First Channel	CV_{lim} [c]	Second Channel	CV [c]	Next [d] Channels	CV [c,d]
1TQN			C	2a	6.00	2e	5.75	S	5.00
1W0E			C	2a	5.75	2e	5.75		
1W0F	Progesterone	4.10	C	2a	6.00	2e	5.75		
1W0G	Metyrapone	4.19	C	2a	6.50	2e	5.50		
3UA1	Bromoergocryptin	6.06	C	2a	6.75	2e	5.50		
4K9V	GS6 [e]	5.53	C	2a	6.50	2e	6.00		
3NXU	Ritonavir	7.37	O1	2a	6.25	2e	5.75		
4I4G	GS2 [f]	6.54	O1	2a	6.00	2e	5.75		
4I4H	GS3 [f]	6.72	O1	2a	6.25	2e	6.00		
4K9W	GS7 [e]	6.47	O1	2a	6.00	2e	5.50		
4K9X	GS8 [e]	4.77	O1	2a	6.00	2e	5.75		
2J0D	Erythromycine	7.16	O2	2f	8.75	2f [g]	6.75	2e	5.50
2V0M	Ketoconazole	5.19, 6.52	O2	2a	7.00	2f	5.75	2e	5.50
4K9T	GS4 [e]	5.96, 6.37	O2	2f	6.75	2f [g]	5.75	2e	5.50
4K9U	GS5 [e]	5.84, 7.89	O2	S	6.75	2f	6.25	2e, 2a	5.75 [d]

Table 2. Location of the four major access channels of CYP3A4. [a] According to the authors of [60,69]: C (closed conformation), O1 (conformation opened, in block 1), O2 (conformation opened, in block 2; opening of O2 is larger than opening of O1). [b] The triangulations of the channels performed with CCCPP can be described with lengthy technical details that we consider to be non essential in this paper, such as the lining atoms and residues: see these latter in Table 3.10 in [69]. [c] Substrate recognition sites (SRS), according to the authors of [70]. [d] Depending on the ligand size, channel 2f can offer a larger opening than channel 2a.

Channel	Block	Conformation [a]	Exit Location In the CYP [b]	Exit Location Outside the CYP	SRS [c]
2a	1	C, O1, O2	Between F-G and B-C loops and β_1 sheet	Plasma membrane (PM)	1, 2, 3
2f [d]	2	O2	Between helix F/F-G loop and C-terminal loop	Interface PM/cytosol	2, 6
S	2	O2	Between helices E,F,I and C-terminal loop	Cytosol	4, 6
2e		C, O1, O2	Through B-C loop	Cytosol	1

4. Conclusions

Our analysis, carried out with CCCPP program on several crystal structures of CYP3A4, enabled identifying relevant ingress/egress channels, with their lining heavy atoms and residues. Our calculations support the hypothesis of channels 2a and 2f as major channels of substrate/product egress in CYP3A4, plus two secondary ones, 2e and S. We propose potential pathways of the ligands, inside these channels, toward the distal face of the heme together with information on the movements of the residues associated to the opening/closing of the channels.

Our analysis suggests that block 1 anchored in the membrane opens at ligand entrance and channel 2a is the only ingress/egress channel. Then, in the case of one large ligand or two ligands, block 2 opens. Smaller channels 2e and S could be involved in the egress of the metabolites, either by enlarging or by use for circulation of water or dioxygen. We did not consider proximal channels smaller than distal channels.

Channels 2a and 2f are occupied by ligands that are not yet oxidized. Either channel 2a enlarges to accept the ligand, or a new path (channel 2f) opens due to the nature and/or the large size of the ligand [60]. Residues obstructing the channel create a bottleneck. These changes are influenced by the location of the channels with respect to the protein topology and the protein's overall global motion [37,38]. Channel 2e exits near the cytosol (see Figure 6 in [27]). Channel 2a is opened in the apo structure, although channel 2f is not. At input of the ligand, channel 2a enlarges until reaching a critical value, above which other paths are needed to accept the entrance of more ligands. When the second ligand enters in channel 2f, its orientation is the opposite of the one of the first ligand, thus fitting the weaker hydrophobicity of channel 2f which starts at the cytosol/membrane interface. These movements are located at flexible secondary structures such as B-C and F-F' loops and C-terminal loop [64]. Molecular dynamics simulations could help to know if channel 2f is closed in the apo form then opens to accept the ligand, or if it is already opened. The F-G block acts as a multi-hinged lid on the distal side of the protein and many channels border it, permitting an opening at the membrane's interface or on the cytosol.

The major channel lining residues mentioned in our study, which we suggested to have a role in channel opening, were proved to be involved in ligand binding, in the activity or cooperativity of CYP3A4 and to be key residues governing allosteric processes in P450 catalyzed substrate oxidations [92]. Another important structural element is the B-C loop, which borders channels 2a and 2e.

Although molecular dynamics simulations were able to exhibit channels not visible by a rough examination of crystallographic structures [94], it was possible to detect such channels by geometric methods because the ligands were indeed present there. A major use of our results could be to provide pertinent starting points for molecular dynamics computations to observe the opening and closing of the channels.

Author Contributions: The main results came from the doctoral work of L.B., realized under the co-supervision of F.M. and M.P., with the collaboration of F.A. and B.T.; L.B. and M.P. wrote the original draft; M.P. produced the software CCCPP; and F.A. and G.M. critically reviewed the manuscript. All authors approved the final version of the text.

Funding: The early stages of this work were funded by the IBC company (Integrative BioComputing, Rennes, France). The final part of this work was funded by the Maghreb edition of the programme L'Oréal-UNESCO For Women in Science, which awarded L.B. in October 2016.

Acknowledgments: L.B. gratefully acknowledges Pridi Siregar, founder and manager of the IBC company, for supplying financial support.

Conflicts of Interest: The authors declare no conflict of interest. The funders had no role in the design of the study, in the collection, analyses, or interpretation of data, in the writing of the manuscript, or in the decision to publish the results.

Abbreviations

The following abbreviations are used in this manuscript:

CCCPP	Computing Cavities, Channels, Pores and Pockets (software)
CV	Critical Value
CYP	Cytochrome P450
MCP	Minimal Cost Path
MD	Molecular Dynamics
PM	Plasma Membrane
RAMD	Random Acceleration Molecular Dynamics
SMD	Steering Molecular Dynamics
SRS	Substrate Recognition Site

References

1. Estabrook, R.W. A passion for P450s (remembrances of the early history of research on cytochrome P450). *Drug. Metab. Disp.* **2003**, *31*, 1461–1473. [CrossRef] [PubMed]
2. Omura, T. Recollection of the early years of the research on cytochrome P450. *Proc. Jpn. Acad. Ser. B* **2011**, *87*, 617–640. [CrossRef]
3. Guengerich, F.P. New trends in cytochrome P450 research at the half-century mark. *J. Biol. Chem.* **2013**, *288*, 17063–17064. [CrossRef] [PubMed]
4. Nelson, D.R. Cytochrome P450 diversity in the tree of life. *Biochim. Biophys. Acta Proteins Proteom.* **2018**, *1866*, 141–154. [CrossRef] [PubMed]
5. Guengerich, F.P.; Rendic, S. Update information on drug metabolism systems—2009, Part I. *Curr. Drug. Metab.* **2010**, *11*, 1–3. [CrossRef] [PubMed]
6. Rendic, S.; Guengerich, F.P. Survey of human oxidoreductases and cytochrome P450 enzymes involved in the metabolism of xenobiotic and natural chemicals. *Chem. Res. Toxicol.* **2015**, *28*, 38–42. [CrossRef] [PubMed]
7. Johnson, E.F.; Stout, C.D. Structural diversity of eukaryotic membrane cytochrome P450s. *J. Biol. Chem.* **2013**, *288*, 17082–17090. [CrossRef] [PubMed]
8. Guengerich, F.P. Cytochrome P450 and chemical toxicology. *Chem. Res. Toxicol.* **2008**, *21*, 70–83. [CrossRef] [PubMed]
9. Guengerich, F.P.; Waterman, M.R.; Egli, M. Recent structural insights into cytochrome P450 function. *Trends Pharmacol. Sci.* **2016**, *37*, 625–640. [CrossRef] [PubMed]
10. Nebert, D.W.; Dalton, T.P. The role of cytochrome P450 enzymes in endogenous signalling pathways and environmental carcinogenesis. *Nat. Rev. Cancer* **2006**, *6*, 947–960. [CrossRef] [PubMed]
11. Pikuleva, I.A.; Waterman, M.R. Cytochromes P450: Roles in diseases. *J. Biol. Chem.* **2013**, *288*, 17091–17098. [CrossRef] [PubMed]
12. Rendic, S. Summary of information on human CYP enzymes: Human P450 metabolism data. *Drug Metab. Rev.* **2002**, *34*, 83–448. [CrossRef] [PubMed]
13. Brown, C.M.; Reisfeld, B.; Mayeno, A.M. Cytochromes P450: A structure-based summary of biotransformations using representative substrates. *Drug Metab. Rev.* **2008**, *40*, 1–100. [CrossRef] [PubMed]
14. Roy, K.; Roy, P.P. QSAR of cytochrome inhibitors. *Expert Opin. Drug Metab. Toxicol.* **2009**, *5*, 1245–1266. [CrossRef] [PubMed]
15. Kriegl, J.M.; Arnhold, T.; Beck, B.; Fox, T. A support vector machine approach to classify human cytochrome P450 3A4 inhibitors. *J. Comput. Aided Mol. Des.* **2005**, *19*, 189–201. [CrossRef] [PubMed]
16. Terfloth, L.; Bienfait, B.; Gasteiger, J. Ligand-based models for the isoform specificity of cytochrome P450 3A4, 2D6, and 2C9 substrates. *J. Chem. Inf. Model.* **2007**, *47*, 1688–1701. [CrossRef] [PubMed]
17. Stjernschantz, E.; Vermeulen, N.P.E.; Oostenbrink, C. Computational prediction of drug binding and rationalisation of selectivity towards cytochromes P450. *Expert Opin. Drug Metab. Toxicol.* **2008**, *4*, 513–527. [CrossRef] [PubMed]
18. Sridhar, J.; Foroozesh, M.; Klein Stevens, C.L. QSAR models of cytochrome P450 enzyme 1A2 inhibitors using CoMFA, CoMSIA and HQSAR. *SAR QSAR Environ. Res.* **2011**, *22*, 681–697. [CrossRef] [PubMed]
19. Mishra, N.K. Computational modeling of P450s for toxicity prediction. *Expert Opin. Drug Metab. Toxicol.* **2011**, *7*, 1211–1231. [CrossRef] [PubMed]
20. Lewis, D.F.V.; Jacobs, M.N.; Dickins, M. Compound lipophilicity for substrate binding to human P450s in drug metabolism. *Drug Disc. Today* **2004**, *9*, 530–537. [CrossRef]
21. Smith, D.A.; Ackland, M.J.; Jones, B.C. Properties of cytochrome P450 isoenzymes and their substrates. Part 2: Properties of cytochrome P450 substrates. *Drug Disc. Today* **1997**, *2*, 479–486. [CrossRef]
22. Wrighton, S.A.; Schuetz, E.G.; Thummel, K.E.; Shen, D.D.; Korzekwa, K.R.; Watkins, P.B. The human CYP3A subfamily: Practical considerations. *Drug Metab. Rev.* **2000**, *32*, 339–361. [CrossRef] [PubMed]
23. Guengerich, F.P. Human cytochrome P450 enzymes. In *Cytochrome P450, Structure, Mechanism, and Biochemistry*, 3rd ed.; Ortiz de Montellano, P.R., Ed.; Kluwer/Plenum: New York, NY, USA, 2005; Chapter 10, Section 6.20.3, pp. 425–426, ISBN 0-306-48324-6.
24. Guengerich, F.P.; Parikh, A.; Yun, C.H.; Kim, D.; Nakamura, K.; Notley, L.M.; Gillam, E.M.J. What makes P450s work? Searches for answers with known and new P450s. *Drug Metab. Rev.* **2000**, *32*, 267–281. [CrossRef] [PubMed]

25. Boobis, A.; Watelet, J.B.; Whomsley, R.; Strolin Benedetti, M.; Demoly, P.; Tipton, K. Drug interactions. *Drug Metab. Rev.* **2009**, *41*, 486–527. [CrossRef] [PubMed]
26. Sevrioukova, I.F.; Poulos, T.L. Understanding the mechanism of cytochrome P450 3A4: Recent advances and remaining problems. *Dalton Trans.* **2013**, *42*, 3116–3126. [CrossRef] [PubMed]
27. Baylon, J.L.; Lenov, I.L.; Sligar, S.G.; Tajkhorshid, E. Characterizing the membrane-bound state of cytochrome P450 3A4: Structure, depth of insertion, and orientation. *J. Am. Chem. Soc.* **2013**, *135*, 8542–8551. [CrossRef] [PubMed]
28. White, R.E.; Coon, M.J. Oxygen activation by cytochrome P-450. *Ann. Rev. Biochem.* **1980**, *49*, 315–356. [CrossRef] [PubMed]
29. Sligar, S.G.; Gelb, M.H.; Heimbrook, D.C. Bio-organic chemistry and cytochrome P-450-dependent catalysis. *Xenobiotica* **1984**, *14*, 63–86. [CrossRef] [PubMed]
30. Guengerich, F.P. Cytochrome P450 oxidations in the generation of reactive electrophiles: Epoxidation and related reactions. *Arch. Biochem. Biophys.* **2003**, *409*, 59–71. [CrossRef]
31. Isin, E.M.; Guengerich, F.P. Substrate binding to cytochromes P450. *Anal. Bioanal. Chem.* **2008**, *392*, 1019–1030. [CrossRef] [PubMed]
32. Skopalik, J.; Anzenbacher, P.; Otyepka, M. Flexibility of human cytochromes P450: Molecular dynamics reveals differences between CYPs 3A4, 2C9, and 2A6, which correlate with their substrate preferences. *J. Phys. Chem. B* **2008**, *112*, 8165–8173. [CrossRef] [PubMed]
33. Urban, P.; Lautier, T.; Pompon, D.; Truan, G. Ligand access channels in cytochrome P450 enzymes: A review. *Int. J. Mol. Sci.* **2018**, *19*, 1617. [CrossRef] [PubMed]
34. Zhang, H.; Kenaan, C.; Hamdane, D.; Hoa, G.H.; Hollenberg, P.F. Effect of conformational dynamics on substrate recognition and specificity as probed by the introduction of a de novo disulfide bond into cytochrome P450 2B1. *J. Biol. Chem.* **2009**, *284*, 25678–25686. [CrossRef] [PubMed]
35. Lüdemann, S.K.; Carugo, O.; Wade, R.C. Substrate access to cytochrome P450cam: A comparison of a thermal motion pathway analysis with molecular dynamics simulation data. *J. Mol. Model.* **1997**, *3*, 369–374. [CrossRef]
36. Lüdemann, S.K.; Lounnas, V.; Wade, R.C. How do substrates enter and products exit the buried active site of cytochrome P450cam? 1. Random expulsion molecular dynamics investigation of ligand access channels and mechanisms. *J. Mol. Biol.* **2000**, *303*, 797–811. [CrossRef] [PubMed]
37. Lüdemann, S.K.; Lounnas, V.; Wade, R.C. How do substrates enter and products exit the buried active site of cytochrome P450cam? 2. Steered molecular dynamics and adiabatic mapping of substrate pathways. *J. Mol. Biol.* **2000**, *303*, 813–830. [CrossRef] [PubMed]
38. Wade, R.C.; Winn, P.J.; Schlichting, I.; Sudarko. A survey of active site access channels in cytochromes P450. *J. Inorg. Biochem.* **2004**, *98*, 1175–1182. [CrossRef] [PubMed]
39. Li, W.; Liu, H.; Scott, E.E.; Gräter, F.; Halpert, J.R.; Luo, X.; Shen, J.; Jiang, H. Possible pathway(s) of testosterone egress from the active site of cytochrome P450 2B1: A steered molecular dynamics simulation. *Drug Metab. Dispos.* **2005**, *33*, 910–919. [CrossRef] [PubMed]
40. Schleinkofer, K.; Sudarko; Winn, P.J.; Lüdeman, S.K.; Wade, R.C. Do mammalian cytochrome P450s show multiple ligand access pathways and ligand channelling? *EMBO Rep.* **2005**, *6*, 584–589. [CrossRef] [PubMed]
41. Seifert, A.; Tatzel, S.; Schmid, R.D.; Pleiss, J. Multiple molecular dynamics simulations of human p450 monooxygenase CYP2C9: The molecular basis of substrate binding and regioselectivity toward warfarin. *Proteins Struct. Funct. Bioinf.* **2006**, *64*, 147–155. [CrossRef] [PubMed]
42. Li, W.; Liu, H.; Luo, X.; Zhu, W.; Tang, Y.; Halpert, J.R.; Jiang, H. Possible pathway(s) of metyrapone egress from the active site of cytochrome P450 3A4: A molecular dynamics simulation. *Drug Metab. Dispos.* **2007**, *35*, 689–696. [CrossRef] [PubMed]
43. Rydberg, P.; Rod, T.H.; Olsen, L.; Ryde, U. Dynamics of water molecules in the active-site cavity of human cytochromes P450. *J. Phys. Chem. B* **2007**, *111*, 5445–5457. [CrossRef] [PubMed]
44. Fishelovitch, D.; Shaik, S.; Wolfson, H.J.; Nussinov, R. Theoretical characterization of substrate access/exit channels in the human cytochrome P450 3A4 enzyme: Involvement of phenylalanine residues in the gating mechanism. *J. Phys. Chem. B* **2009**, *113*, 13018–13025. [CrossRef] [PubMed]
45. Krishnamoorthy, N.; Gajendrarao, P.; Thangapandian, S.; Lee, Y.; Lee, K.W. Probing possible egress channels for multiple ligands in human CYP3A4: A molecular modeling study. *J. Mol. Model.* **2010**, *16*, 607–614. [CrossRef] [PubMed]

46. Berka, K.; Hendrychová, T.; Anzenbacher, P.; Otyepka, M. Membrane position of ibuprofen agrees with suggested access path entrance to cytochrome P450 2C9 active site. *J. Phys. Chem. A* **2011**, *115*, 11248–11255. [CrossRef] [PubMed]
47. Cojocaru, V.; Balali-Mood, K.; Sansom, M.S.P.; Wade, R.C. Structure and dynamics of the membrane-bound cytochrome P450 2C9. *PLoS Comput. Biol.* **2011**, *7*, e1002152. [CrossRef] [PubMed]
48. Hendrychová, T.; Anzenbacherová, E.; Hudeček, J.; Skopalík, J.; Lange, R.; Hildebrandt, P.; Otyepka, M.; Anzenbacher, P. Flexibility of human cytochrome P450 enzymes: Molecular dynamics and spectroscopy reveal important function-related variations. *Biochim. Biophys. Acta* **2011**, *1814*, 58–68. [CrossRef] [PubMed]
49. Li, W.; Shen, J.; Liu, G.; Tang, Y.; Hoshino, T. Exploring coumarin egress channels in human cytochrome P450 2A6 by random acceleration and steered molecular dynamics simulations. *Proteins Struct. Funct. Bioinf.* **2011**, *79*, 271–281. [CrossRef] [PubMed]
50. Cojocaru, V.; Winn, P.J.; Wade, R.C. Multiple, ligand-dependent routes from the active site of cytochrome P450 2C9. *Curr. Drug Metab.* **2012**, *13*, 143–154. [CrossRef] [PubMed]
51. Denisov, I.G.; Shih, A.Y.; Sligar, S.G. Structural differences between soluble and membrane bound cytochrome P450s. *J. Inorg. Biochem.* **2012**, *108*, 150–158. [CrossRef] [PubMed]
52. Shen, Z.; Cheng, F.; Xu, Y.; Fu, J.; Xiao, W.; Shen, J.; Liu, G.; Li, W.; Tang, Y. Investigation of indazole unbinding pathways in CYP2E1 by molecular dynamics simulations. *PLoS ONE* **2012**, *7*, e33500. [CrossRef]
53. Berka, K.; Paloncýová, M.; Anzenbacher, P.; Otyepka, M. Behavior of human cytochromes P450 on lipid membranes. *J. Phys. Chem. B* **2013**, *117*, 11556–11564. [CrossRef] [PubMed]
54. Perić-Hassler, L.; Stjernschantz, E.; Oostenbrink, C.; Geerke, D.P. CYP 2D6 binding affinity predictions using multiple ligand and protein conformations. *Int. J. Mol. Sci.* **2013**, *14*, 24514–24530. [CrossRef] [PubMed]
55. Paloncýová, M.; Navrátilová, V.; Berka, K.; Laio, A.; Otyepka, M. Role of enzyme flexibility in ligand access and egress to active site: Bias-exchange metadynamics study of 1,3,7-trimethyluric acid in cytochrome P450 3A4. *J. Chem. Theory Comput.* **2016**, *12*, 2101–2109. [CrossRef] [PubMed]
56. Hendrychova, T.; Berka, K.; Navratilova, V.; Anzenbacher, P.; Otyepka, M. Dynamics and hydration of the active sites of mammalian cytochromes P450 probed by molecular dynamics simulations. *Curr. Drug Metab.* **2012**, *13*, 177–189. [CrossRef] [PubMed]
57. Benkaidali, L.; André, F.; Maouche, B.; Siregar, P.; Benyettou, M.; Maurel, F.; Petitjean, M. Computing cavities, channels, pores and pockets in proteins from non spherical ligands models. *Bioinformatics* **2014**, *30*, 792–800. [CrossRef] [PubMed]
58. Meslamani, J.E.; André, F.; Petitjean, M. Assessing the geometric diversity of cytochrome P450 ligand conformers by hierarchical clustering with a stop criterion. *J. Chem. Inf. Model.* **2009**, *49*, 330–337. [CrossRef] [PubMed]
59. Benkaidali, L.; Mansouri, K.; Tuffery, P.; André, F.; Petitjean, M. How well is conformational space covered? In *Chemical Information and Computational Challenges in the 21st Century*; Putz, M., Ed.; Nova Science: New York, NY, USA, 2012; Chapter 13, pp. 299–313, ISBN 978-1-61209-712-1.
60. Benkaidali, L.; André, F.; Moroy, G.; Tangour, B.; Maurel, F.; Petitjean, M. The cytochrome P450 3A4 has three major conformations: new clues to drug recognition by this promiscuous enzyme. *Mol. Inf.* **2017**, *36*. [CrossRef] [PubMed]
61. Poulos, T.L.; Finzel, B.C.; Howard, A.J. High-resolution crystal structure of cytochrome P450cam. *J. Mol. Biol.* **1987**, *195*, 687–700. [CrossRef]
62. Graham, S.E.; Peterson, J.A. How similar are P450s and what can their differences teach us? *Archiv. Biochem. Biophys.* **1999**, *369*, 24–29. [CrossRef] [PubMed]
63. Sirim, D.; Widmann, M.; Wagner, F.; Pleis, J. Prediction and analysis of the modular structure of cytochrome P450 monooxygenases. *BMC Struct. Biol.* **2010**, *10*, 34. [CrossRef] [PubMed]
64. Cojocaru, V.; Winn, P.J.; Wade, R.C. The ins and outs of cytochrome P450s. *Biochim. Biophys. Acta* **2007**, *1770*, 390–401. [CrossRef] [PubMed]
65. Petřek, M.; Otyepka, M.; Banáš, P.; Košinová, P.; Koča, J.; Damborský, J. CAVER: A new tool to explore routes from protein clefts, pockets and cavities. *BMC Bioinform.* **2006**, *7*, 316. [CrossRef] [PubMed]
66. Berka, K.; Hanák, O.; Sehnal, D.; Banáš, P.; Navrátilová, V.; Jaiswal, D.; Ionescu, C.M.; Vařeková, R.S.; Koča, J.; Otyepka, M. MOLEonline 2.0: Interactive web-based analysis of biomacromolecular channels. *Nucleic Acids Res.* **2012**, *40*, W222–W227. [CrossRef] [PubMed]

67. Jankowski, C.K.; Chiasson, J.B.; Dako, E.; Doucet, K.; Surette, M.E.; André, F.; Delaforge, M. Modeling of cytochrome P450 (Cyt P450, CYP) channels. *Spectroscopy* **2011**, *25*, 63–87. [CrossRef]
68. Dijkstra, E.W. A note on two problems in connexion with graphs. *Numer. Math.* **1959**, *1*, 269–271. [CrossRef]
69. Benkaidali, L. Etude et Applications de Nouveaux Modèles Géométriques des Canaux D'accès au Site Actif de Certains Cytochromes P450 Humains par des Ligands Volumineux. Ph.D. Thesis, University Pierre et Marie Curie, Paris, France, 15 September 2016. Available online: https://hal.archives-ouvertes.fr/tel-01483643 (accessed on 11 October 2018).
70. Gotoh, O. Substrate recognition sites in cytochrome P450 family 2 (CYP2) proteins inferred from comparative analyses of amino acid and coding nucleotide sequences. *J. Biol. Chem.* **1992**, *267*, 83–90. [PubMed]
71. Williams, P.A.; Cosme, J.; Vinković, D.M.; Ward, A.; Angove, H.C.; Day, P.J.; Vonrhein, C.; Tickle, I.J.; Jhoti, H. Crystal structures of human cytochrome P450 3A4 bound to metyrapone and progesterone. *Science* **2004**, *305*, 683–686. [CrossRef] [PubMed]
72. Mustafa, G.; Yu, X.; Wade, R.C. Structure and dynamics of human drug-metabolizing cytochrome P450 enzymes. In *Drug Metabolism Prediction*; Kirchmair, J., Ed.; Wiley: New York, NY, USA, 2014; Volume 63, Section 4, pp. 77–102, ISBN 978-3-527-33566-4.
73. Yu, X.; Cojocaru, V.; Wade, R.C. Conformational diversity and ligand tunnels of mammalian cytochrome P450s. *Biotechnol. Appl. Biochem.* **2013**, *60*, 134–145. [CrossRef] [PubMed]
74. Davis, J.L.; Papich, M.G.; Heit, M.C. Antifungal and antiviral drugs. In *Veterinary Pharmacology and Therapeutics*, 9th ed.; Riviere, J.E., Papich, M.G., Eds.; Wiley-Blackwell: New York, NY, USA, 2009; Chapter 39, Section 9, pp. 1019–1020, ISBN 978-0-8138-2061-3.
75. Winn, P.J.; Lüdemann, S.K.; Gauges, R.; Lounnas, V.; Wade, R.C. Comparison of the dynamics of substrate access channels in three cytochrome P450s reveals different opening mechanisms and a novel functional role for a buried arginine. *Proc. Natl. Acad. Sci. USA* **2002**, *99*, 5361–5366. [CrossRef] [PubMed]
76. Sevrioukova, I.F.; Poulos, T.L. Interaction of human cytochrome P4503A4 with ritonavir analogs. *Arch. Biochem. Biophys.* **2012**, *520*, 108–116. [CrossRef] [PubMed]
77. Sevrioukova, I.F.; Poulos, T.L. Structural and mechanistic insights into the interaction of cytochrome P4503A4 with bromoergocryptine, a type I ligand. *J. Biol. Chem.* **2012**, *287*, 3510–3517. [CrossRef] [PubMed]
78. Moore, C.D.; Shahrokh, K.; Sontum, S.F.; Cheatham, T.E., III; Yost, G.S. Improved cytochrome P450 3A4 molecular models accurately predict the Phe215 requirement for raloxifene dehydrogenation selectivity. *Biochemistry* **2010**, *49*, 9011–9019. [CrossRef] [PubMed]
79. Sevrioukova, I.F.; Poulos, T.L. Dissecting cytochrome P450 3A4-ligand interactions using ritonavir analogues. *Biochemistry* **2013**, *52*, 4474–4481. [CrossRef] [PubMed]
80. Harlow, G.R.; Halpert, J.R. Analysis of human cytochrome P450 3A4 cooperativity: Construction and characterization of a site-directed mutant that displays hyperbolic steroid hydroxylation kinetics. *Proc. Natl. Acad. Sci. USA* **1998**, *95*, 6636–6641. [CrossRef] [PubMed]
81. Ekroos, M.; Sjögren, T. Structural basis for ligand promiscuity in cytochrome P450 3A4. *Proc. Natl. Acad. Sci. USA* **2006**, *103*, 13682–13687. [CrossRef] [PubMed]
82. Lee, D.S.; Yamada, A.; Sugimoto, H.; Matsunaga, I.; Ogura, H.; Ichihara, K.; Adachi, S.; Park, S.Y.; Shiro, Y. Substrate recognition and molecular mechanism of fatty acid hydroxylation by cytochrome P450 from Bacillus subtilis. Crystallographic, spectroscopic, and mutational studies. *J. Biol. Chem.* **2003**, *278*, 9761–9767. [CrossRef] [PubMed]
83. Oprea, T.; Hummer, G.; García, A.E. Identification of a functional water channel in cytochrome P450 enzymes. *Proc. Natl. Acad. Sci. USA* **1997**, *94*, 2133–2138. [CrossRef] [PubMed]
84. Haines, D.C.; Tomchick, D.R.; Machiusm, M.; Peterson, J.A. Pivotal role of water in the mechanism of P450BM-3. *Biochemistry* **2001**, *40*, 13456–13465. [CrossRef] [PubMed]
85. Wester, M.R.; Johnson, E.F.; Marques-Soares, C.; Dijols, S.; Dansette, P.M.; Mansuy, D.; Stout, C.D. Structure of mammalian cytochrome P450 2C5 complexed with diclofenac at 2.1 Å resolution: Evidence for an induced fit model of substrate binding. *Biochemistry* **2003**, *42*, 9335–9345. [CrossRef] [PubMed]
86. Rowland, P.; Blaney, F.E.; Smyth, M.G.; Jones, J.J.; Leydon, V.R.; Oxbrow, A.K.; Lewis, C.J.; Tennant, M.G.; Modi, S.; Eggleston, D.S.; et al. Crystal structure of human cytochrome P450 2D6. *J. Biol. Chem.* **2006**, *281*, 7614–7622. [CrossRef] [PubMed]
87. Urban, P.; Truan, G.; Pompon, D. Access channels to the buried active site control substrate specificity in CYP1A P450 enzymes. *Biochim. Biophys. Acta* **2015**, *1850*, 696–707. [CrossRef] [PubMed]

88. Zawaira, A.; Coulson, L.; Gallotta, M.; Karimanzira, O.; Blackburn, J. On the deduction and analysis of singlet and two-state gating-models from the static structures of mammalian CYP450. *J. Struct. Biol.* **2011**, *173*, 282–293. [CrossRef] [PubMed]
89. Black, S.D. Membrane topology of the mammalian P450 cytochromes. *FASEB J.* **1992**, *6*, 680–685. [CrossRef] [PubMed]
90. Headlam, M.J.; Wilce, M.C.; Tuckey, R.C. The F-G loop region of cytochrome P450scc (CYP11A1) interacts with the phospholipid membrane. *Biochim. Biophys. Acta* **2003**, *1617*, 96–108; Erratum in **2004**, *1661*, 219. [CrossRef]
91. Black, S.D.; Martin, S.T.; Smith, C.A. Membrane topology of liver microsomal cytochrome P450 2B4 determined via monoclonal antibodies directed to the halt-transfer signal. *Biochemistry* **1994**, *33*, 6945–6951. [CrossRef] [PubMed]
92. Hlavica, P. Challenges in assignment of allosteric effects in cytochrome P450-catalyzed substrate oxidations to structural dynamics in the hemoprotein architecture. *J. Inorg. Biochem.* **2017**, *67*, 100–115. [CrossRef] [PubMed]
93. Šrejber, M.; Navrátilová, V.; Paloncýová, M.; Bazgier, V.; Berka, K.; Anzenbacher, P.; Otyepka, M. Membrane-attached mammalian cytochromes P450: An overview of the membrane's effects on structure, drug binding, and interactions with redox partners. *J. Inorg. Biochem.* **2018**, *183*, 117–136. [CrossRef] [PubMed]
94. Kingsley, L.J.; Lill, M.A. Including ligand-induced protein flexibility into protein tunnel prediction. *J. Comput. Chem.* **2014**, *35*, 1748–1756. [CrossRef] [PubMed]
95. Sevrioukova, I.F.; Poulos, T.L. Pyridine-substituted desoxyritonavir is a more potent inhibitor of cytochrome P450 3A4 than ritonavir. *J. Med. Chem.* **2013**, *56*, 3733–3741. [CrossRef] [PubMed]

© 2019 by the authors. Licensee MDPI, Basel, Switzerland. This article is an open access article distributed under the terms and conditions of the Creative Commons Attribution (CC BY) license (http://creativecommons.org/licenses/by/4.0/).

Article

Structural Insights into the Interaction of Cytochrome P450 3A4 with Suicide Substrates: Mibefradil, Azamulin and 6′,7′-Dihydroxybergamottin

Irina F. Sevrioukova

Department of Molecular Biology and Biochemistry, University of California, Irvine, CA 92697-3900, USA; sevrioui@uci.edu

Received: 9 August 2019; Accepted: 29 August 2019; Published: 30 August 2019

Abstract: Human cytochrome P450 3A4 (CYP3A4) is the most important drug-metabolizing enzyme. Some drugs and natural compounds can act as suicide (mechanism-based) inactivators of CYP3A4, leading to unanticipated drug-drug interactions, toxicity and therapeutic failures. Despite significant clinical and toxicological implications, the mechanism-based inactivation remains incompletely understood. This study provides the first direct insights into the interaction of CYP3A4 with three suicide substrates: mibefradil, an antihypertensive drug quickly withdrawn from the market; a semi-synthetic antibiotic azamulin; and a natural furanocoumarin, 6′,7′-dihydroxybergamottin. Novel structural findings help better understand the suicide substrate binding and inhibitory mechanism, and can be used to improve the predictability of the binding ability, metabolic sites and inhibitory/inactivation potential of newly developed drugs and other chemicals relevant to public health.

Keywords: CYP3A4; mechanism-based inhibitor; crystal structure

1. Introduction

Human cytochrome P450 3A4 (CYP3A4) oxidizes over 50% of administered drugs [1], along with natural compounds, some of which can act as inhibitors of CYP3A4 [2]. The mechanism-based (suicide) inhibition (MBI) of CYP3A4 is the most common mechanism that could lead to clinically significant drug-drug interactions (DDIs), toxicity and therapeutic failures [3]. MBI is characterized by NADPH-, time- and concentration-dependent enzyme inactivation due to the formation and covalent attachment of a reactive metabolite(s) to the heme and/or apoprotein. The early elimination of the MBI/DDI potential in drug candidates is crucial [4] but highly challenging due to poor predictability of CYP3A4-ligand interactions. This results from high promiscuity [5] and conformational flexibility of CYP3A4 [6–8] and very limited structural information on the substrate association modes. To date, only three crystal structures of CYP3A4 with drug substrates bound in the active site have been reported [6,7,9] and none with suicide inactivators.

To fill this knowledge gap, this study investigated the binding manner of mibefradil, azamulin, bergamottin and 6′,7′-dihydroxybergamottin (DHB) (Figure 1), which are known to act as potent mechanism-based inhibitors of CYP3A4. Mibefradil (or Posicor) is a benzimidazoyl-substituted tetraline, designed as a long-acting T-type calcium channel blocker for the treatment of chronic hypertension. Mibefradil interacts with multiple cytochrome P450 isoforms, but was withdrawn from the market mainly due to its high inhibitory potency for CYP3A4, leading to numerous life-threatening DDIs [10]. Azamulin is a semi-synthetic antibiotic and derivative of pleuromutilin, which failed stage I clinical trials due to poor bioavailability. Azamulin was recently identified as an effective and highly selective inhibitor of CYP3A4 and is currently recommended for use in reaction phenotyping studies instead of less specific ketoconazole [11,12]. Bergamottin and its derivative DHB, on the other hand, are natural furanocoumarins found most abundantly in grapefruits. In addition to several health-promoting effects [13], both compounds mediate food-drug interactions primarily through inhibition of CYP3A4 [14]. Here, the results of spectral investigations are reported, as well as crystal structures of CYP3A4 bound to mibefradil, azamulin and DHB. Our findings help better understand the suicide substrate binding and inhibitory mechanism, and could improve the computational tools for modeling and prediction of the CYP3A4-ligand interactions, which is vital for designing safer and more effective drugs.

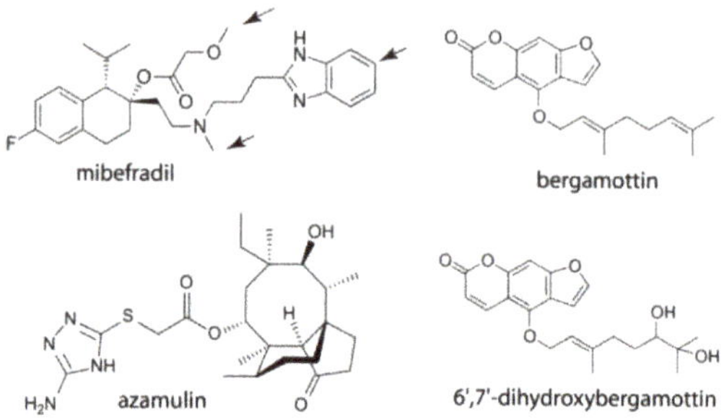

Figure 1. Chemical structures of the investigated compounds. Arrows indicate known sites of metabolism.

2. Results and Discussion

2.1. Interaction of CYP3A4 with Mibefradil

Equilibrium titrations of the full-length CYP3A4 were conducted to determine and relate the binding affinity and dissociability of the investigated compounds to their association mode. Mibefradil is a type I ligand that causes a blue shift in the Soret band (Figure 2A). The spectral dissociation constant (K_s) derived from the titration plot (left inset in Figure 2A) was 3.3 µM, which is ~5-fold higher than the previously reported value [15], possibly due to differences in the CYP3A4 form. Since CYP3A4 precipitates during prolonged dialysis, the dissociation ability was assessed by titrating the ligand-bound protein with ritonavir, a high-affinity inhibitor that easily displaces other type I substrates, such as bromocryptine and midazolam (Figure S1). As seen in Figure 2B, ritonavir could fully replace mibefradil, but its binding affinity was 74-fold lower than for the ligand-free CYP3A4: K_s^{RIT} of 1.4 µM vs. 0.019 µM [16], respectively. In contrast, only a 3.5-fold decrease in K_s^{RIT} was observed for the bromocryptine- and midazolam-bound CYP3A4 (Table 1). Thus, mibefradil interacts with CYP3A4 stronger and/or is better protected and has a lower ability to dissociate.

Table 1. Parameters for the ligand binding to CYP3A4.

Compound	K_s [a] µM	High-Spin Shift	K_s^{RIT} [b] µM
mibefradil	3.3 ± 0.4	65%	1.42 ± 0.03
azamulin	1.7 ± 0.3	100%	0.88 ± 0.09
bergamottin	3.0 ± 0.3 [c] (n = 1.8)	48%	0.035 ± 0.008
DHB	0.22 ± 0.03 [d]	<3%	0.032 ± 0.002
	10.3 ± 1.4 [d]		
bromocryptine	0.43 ± 0.08	90%	0.070 ± 0.005
midazolam	24 ± 3	85%	0.071 ± 0.006
none			0.019 ± 0.002

[a] Spectral dissociation constant for ligand-free CYP3A4. [b] Binding affinity of ritonavir for ligand-bound/free CYP3A4. [a,b] Values represent an average of three measurements with the standard error. [c] S_{50}; n is a Hill coefficient. [d] Dissociation constants for two binding sites.

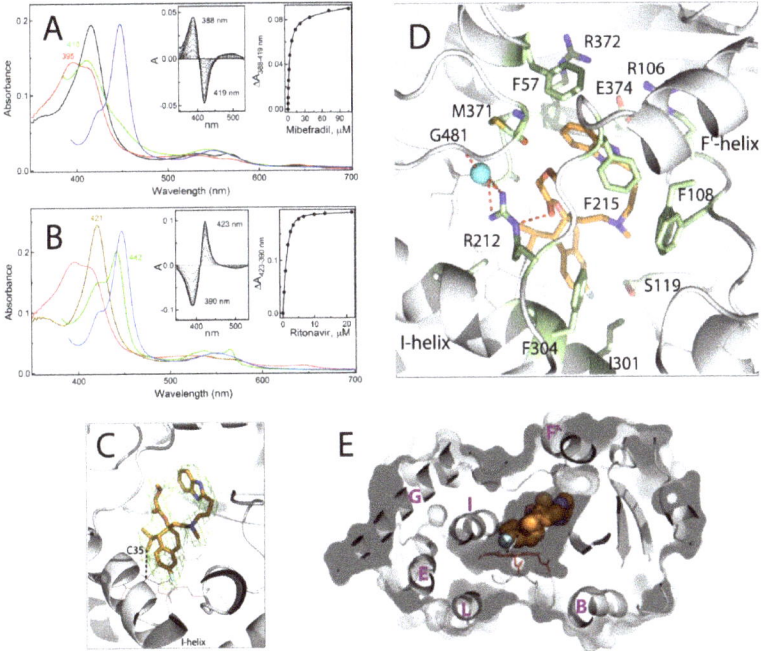

Figure 2. Spectral and structural properties of the CYP3A4-mibefradil complex. (**A,B**) Spectral changes observed during equilibrium titrations of CYP3A4 with mibefradil and upon displacement of mibefradil with ritonavir, respectively. In panel (**A**), the spectrum of ligand-free CYP3A4 is in black. In panel (**B**), the spectrum of ritonavir-bound CYP3A4 is in brown. In both panels, the spectra of the CYP3A4-mibefradil complex and its ferrous and ferrous CO-bound forms are in red, green and blue, respectively. In the competitive displacement experiment (panel **B**), the concentration of mibefradil was 120 µM. The left and right insets are the difference spectra and titration plots with hyperbolic fittings, respectively. The derived K_s values are given in Table 1. (**C**) The active site of CYP3A4 bound to mibefradil (shown in orange sticks; PDB ID 6O09). Green mesh is a polder omit electron density map contoured at 3σ level. Simulated annealing omit map for mibefradil is shown in Figure S2A. The labeled C35 atom of mibefradil is the closest to the heme iron (~3.8 Å away). (**D**) Interaction of mibefradil with surrounding residues (shown in green sticks and labeled). Red dotted lines are H-bonds. Cyan sphere is a water molecule. (**E**) A slice through the CYP3A4 molecule showing how well mibefradil (in space-filling representation) fits into the active site cavity. The visible helices are labeled.

2.1.1. Crystal Structure of the CYP3A4-Mibefradil Complex

Mibefradil inhibits microsomal and recombinant CYP3A4 with IC_{50} of 0.3–2 µM [17], but the chemical nature of its reactive metabolite(s) is still unknown. A previous attempt to elucidate the molecular basis for MBI led to a conclusion that the inactivation of CYP3A4 proceeds through heme destruction rather than covalent modification of the heme or apoprotein [15]. To identify the potential oxidation sites that could lead to bioactivation, CYP3A4 was co-crystallized with mibefradil. The crystal structure was determined to 2.25 Å resolution (Table S1) and contained one drug molecule in the active site (Figure 2C–E). The tetraline moiety is 3.3–3.6 Å above the heme plane, with the propyl C35 atom being the closest to the iron (~3.8 Å away; Figure 2C). The fluorophenyl portion is parallel to the I-helix and partially inserts into a hydrophobic pocket formed by F304, A305 and I301. The benzimidazole moiety provides additional hydrophobic and aromatic interactions with F57, F215 and M371 (Figure 2D). The methoxyacetate functionality, in turn, is H-bonded via the carbonyl oxygen to the R212 guanidine group. This polar interaction is part of the H-bonding network that links the F-F′-loop to the protein core and likely contributes to the inhibitory potency, as the derivatives lacking the methoxyacetate group are weaker inhibitors of CYP3A4 than mibefradil [15,18]. The ligand binding mode is further stabilized by multiple van der Waals interactions.

As seen from Figures 2E and S3A, mibefradil binds compactly and fits into the catalytic cavity without triggering any notable conformational change. However, the compact binding mode may limit its motional freedom. This limitation and the fact that CYP3A4 remains in a resting conformation, tightly locked through the R212-mediated contacts, could explain the low dissociability of mibefradil.

2.1.2. Possible Inhibitory Mechanism of Mibefradil

CYP3A4 metabolizes mibefradil via methoxyacetate and tertiary amine demethylation and hydroxylation of the benzimidazole ring (indicated in Figure 1) rather than the oxidation of the closest to the iron propyl group [10]. Thus, the re-entry or reorientation of mibefradil would be required to allow an access to the primary, non-inhibitory sites of metabolism. Whether the crystallographic binding mode could lead to products capable of escaping the active site is yet to be proven. Even so, based on our and earlier findings [10,15,18], it is plausible to suggest that the crystal structure represents an inhibitory complex, where mibefradil could decrease the CYP3A4 activity in two ways: (i) through formation of a slowly dissociable complex that would prevent other substrate molecules from reaching the catalytic center; and (ii) by producing an alkyl radical intermediate upon a hydrogen abstraction from the C35 atom, which could attack and destroy the heme. The formation of a highly reactive free radical would explain why the C35 oxidation product was not observed.

2.2. Interaction of CYP3A4 with Azamulin

Azamulin is also a medium affinity type I ligand (K_s of 1.7 µM; Figure 3A) which, unlike mibefradil, causes a complete high-spin shift in CYP3A4. The observed absorbance changes suggest that azamulin docks with the bulky pleuromutilin head on, as the heme ligation to the amino-triazolyl nitrogen would lead to a red shift in the Soret band (type II spectral perturbations). Despite the higher affinity for CYP3A4, azamulin was displaced by ritonavir more easily than mibefradil but less easily than other type I ligands (Figure 3B; Table 1). This implies that azamulin is a slowly dissociable ligand that forms fewer/weaker contacts and/or is less protected in the active site than mibefradil.

It should be noted also that K_s^{RIT} serves as an estimate for the displaced ligand dissociation constant. The fact that the K_s^{RIT} and K_s values derived for mibefradil and azamulin are very close (only 2-fold difference; Table 1) suggests that in the competitive displacement experiments the substrate dissociation is a limiting step, which is only weakly affected by ritonavir.

Crystal Structure of CYP3A4 Bound to Azamulin

The crystal structure of the CYP3A4-azamulin complex was determined to 2.52 Å resolution (Table S1) and contains one drug molecule bound to the active site in an extended conformation (Figure 3C). As the spectral data predicted, the pleuromutilin functionality is placed near the heme. The complex could be productive, because two carbon atoms of the hexane ring, C08 and C09, are close enough for oxidation: 4.6 Å and 4.3 Å from the iron, respectively. The thioacetyl linker is stretched above the I-helix and, as a result, the amino-triazolyl end-group points upward rather than toward the substrate channel (Figure 3C,D). This conformation is stabilized by multiple van der Waals contacts and two hydrogen bonds, formed between the hydroxyl group of pleuromutilin's eight-membered ring and S119 side chain, and between the amino-triazolyl nitrogen and E308 carboxyl (Figure 3E). Importantly, due to steric clashing with the amino-triazolyl moiety, the F-F'-loop (residues 210–217) becomes disordered, leaving the end-portion of azamulin partially solvent exposed (Figure S4). This suboptimal binding mode could explain why azamulin is displaced by ritonavir more easily than mibefradil.

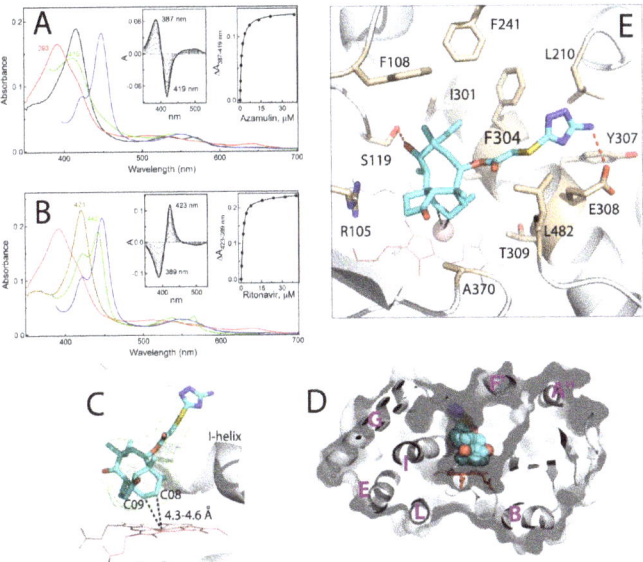

Figure 3. Spectral and structural properties of the CYP3A4-azamulin complex. (**A,B**) Spectral changes observed during equilibrium titrations of CYP3A4 with azamulin and upon displacement of azamulin with ritonavir, respectively. In panel (**A**), the spectrum of ligand-free CYP3A4 is in black. In panel (**B**), the spectrum of ritonavir-bound CYP3A4 is in brown. In both panels, the spectra of the CYP3A4-azamulin complex and its ferrous and ferrous CO-bound forms are in red, green and blue, respectively. In the competitive displacement experiment (panel **B**), the concentration of azamulin was 60 µM. The left and right insets are the difference spectra and titration plots with hyperbolic fittings, respectively. The derived K_s values are listed in Table 1. (**C**) The binding mode of azamulin (in cyan; PDB ID 6OOA). Green mesh is a polder omit electron density map contoured at 3σ level. Simulated annealing omit map for azamulin is shown in Figure S2B. The closest to the iron C08 and C09 atoms are indicated. (**D**) A slice through the CYP3A4 molecule showing that azamulin (in space-filling representation) extends over the I-helix rather than along the substrate channel. The visible helices are labeled. (**E**) Interaction of azamulin with surrounding residues (shown in beige sticks and labeled). The H-bonds are depicted as red dotted lines.

Azamulin was identified as a potent and highly selective competitive and mechanism-based inhibitor of microsomal and recombinant CYP3A4 (IC_{50} of 0.03–0.24 µM) [11]. The pleuromutilin group is thought to undergo metabolic activation [11], but the reactive intermediates are yet to be identified. The crystal structure corroborates the notion that the pleuromutilin is required for MBI, because this functionality is the closest to the heme, mediates the majority of protein-ligand interactions and orients suitably for oxidation. It needs to be tested though whether the C08/9 oxidation products could be reactive or they would have to dissociate and rebind in a distinct orientation to enable bioactivation at other sites. In any case, considering the bulkiness, high hydrophobicity and low dissociability of azamulin, it is plausible to suggest that this compound could effectively inhibit CYP3A4 by crowding/blocking the active site as well.

2.3. Interaction of CYP3A4 with Bergamottin and DHB

Bergamottin, but not DHB, causes a partial high-spin shift in CYP3A4 (Figure 4A,B). An increase in DHB concentration leads to a small decrease rather than a shift in the Soret band, meaning that DHB could approach and alter the heme environment without changing the coordination state. Another notable difference was in the shape of the titration curves: sigmoidal for bergamottin and hyperbolic for DHB (right insets in Figure 4A,B). The S_{50} and n values (substrate concentration at half-saturation and the Hill coefficient, respectively) derived from the sigmoidal plot suggest that bergamottin binds to CYP3A4 cooperatively and with affinity comparable to those of mibefradil and azamulin (Table 1). The DHB titration curve, in turn, was best fit to a two-site binding hyperbolic equation, indicating that (i) two DHB molecules can simultaneously bind to CYP3A4, and (ii) the occupation of the high affinity site (K_s of 0.22 µM) leads to very small changes in the Soret band (~17% of the total absorbance decrease). Another notable feature is the incomplete reduction of bergamottin- and DHB-bound CYP3A4 with sodium dithionite, as evident from the red-shifted absorbance maxima of their ferrous forms: 413 and 416 nm, respectively, versus 410 nm for mibefradil- and azamulin-bound CYP3A4 (compare green spectra in Figures 2A, 3A and 4A,B). Thus, both furanocoumarins seem to limit an access of the reductant to the heme iron. Since the reaction of CYP3A4 with sodium dithionite was carried out under aerobic conditions and was the slowest for the DHB-bound form, the lower than expected 450 nm absorption of the ferrous DHB/CO-bound species (blue spectrum in Figure 4B) could be the consequence of both the incomplete reduction and oxidative damage of the heme. Despite the markedly distinct binding manner, both bergamottin and DHB could be easily displaced by ritonavir (Figure S5), whose binding affinity was only mildly affected (<2-fold decrease; Table 1).

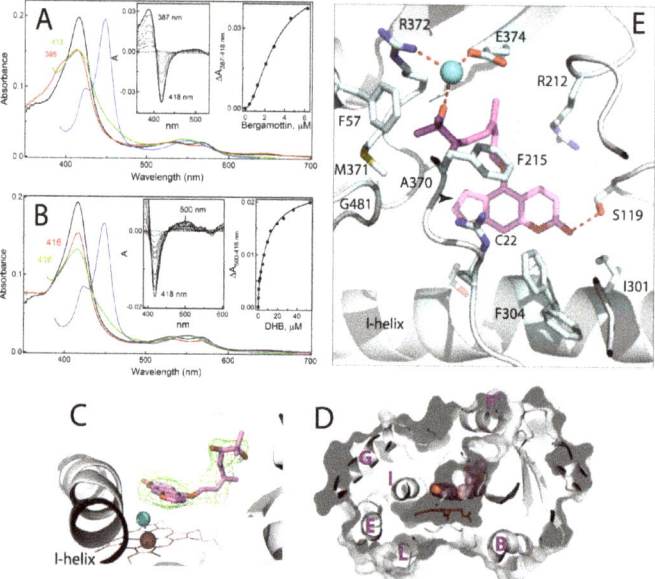

Figure 4. Interaction of CYP3A4 with bergamottin and DHB. (**A,B**) Spectral changes observed during equilibrium titration of CYP3A4 with bergamottin and 6′,7′-dihydroxybergamottin (DHB), respectively. The spectra of ligand-free CYP3A4 are in black. The spectra of the CYP3A4-substrate complexes and their ferrous and ferrous CO-bound forms are in red, green and blue, respectively. The left and right insets are the difference spectra and titration plots with sigmoidal and hyperbolic fittings, respectively. The derived K_s values are listed in Table 1. (**C**) The binding mode of DHB (in magenta; PDB ID 6OOB). Green mesh is a polder omit electron density map contoured at 3σ level. Simulated annealing omit map for DHB is shown in Figure S2C. Cyan sphere is the heme-bound water molecule. (**D**) A slice through the CYP3A4 molecule showing how DHB (in space-filling representation) orients in the active site cavity. The visible helices are labeled. (**E**) Interaction of DHB with surrounding residues (shown in cyan sticks and labeled). Red dotted lines are H-bonds. Cyan sphere is a water molecule. The furan double bond, a possible bioactivation site, is indicated by an arrow. The C22 atom of psoralen's phenyl ring, the closest to the heme iron, is labeled.

2.3.1. Crystal Structure of the CYP3A4-DHB Complex

We attempted to co-crystallize CYP3A4 with both bergamottin and DHB but succeeded in obtaining only the DHB-bound crystals. The latter structure was solved to 2.2 Å resolution (Table S1) and contains one DHB molecule in the active site (Figure 4C–E). The psoralen ring of DHB is placed 3.6–5.2 Å above the heme (~15° tilt) and parallel to the I-helix, with the carbonyl oxygen H-bonded to the S119 hydroxyl group. As the spectral data predicted, the heme iron remains hexa-coordinate but the water ligand shifts toward the I-helix, altering the Fe-O bond perpendicularity by ~7° (Figure 4C). This subtle perturbation in the heme ligation explains why only small spectral changes could be detected during the equilibrium titrations.

The dihydroxygeranyl group curls along the substrate channel and anchors to its wall through the water-mediated H-bond, linking 7'-OH to the R372 and E374 side chains. F57, F215, A370 and M371 create a hydrophobic environment for the aliphatic portion of the geranyl moiety, whereas R212 and F304 flank the psoralen ring. The F304 side group adopts two alternative conformers: one pointing away and another toward the active site, as in ligand-free CYP3A4 (Figure 4E). Since the DHB site is fully occupied, this conformational heterogeneity is likely caused by steric hindrance with the psoralen ring. Considering the largely different affinities for the two binding sites (Table 1), it is reasonable to conclude that DHB occupies the high affinity site. However, the crystallographic binding mode is non-productive, because the nearest to the iron C22 atom of psoralen's phenyl ring (indicated in Figure 4E) is too far for oxidation (>5Å).

2.3.2. Possible Mechanism for DHB Bioactivation

Bergamottin and DHB mediate food-drug interactions primarily through reversible and mechanism-based inhibition of CYP3A4 (IC$_{50}$ of 2–23 µM) [14,19,20]. The structure-activity studies on natural furanocoumarins showed that a plain tricyclic ring containing the furan moiety is strictly required for inhibiting CYP3A4 activity [21]. This led to a suggestion that bioactivation of bergamottin and DHB proceeds via epoxidation and opening of the furan ring [22,23]. In the crystal structure, however, the furan double bond (indicated in Figure 4E) is too remote and cannot approach the catalytic center without rotation of the psoralen ring, which might be difficult to achieve due to close proximity of the heme and I-helix.

To reconcile our and the previous results [22,23], we propose the following mechanism (Figure 5). Having a more hydrophobic aliphatic chain, bergamottin preferably enters the active site with the geranyl being the closest to the heme iron (step 1), as observed in the 6DWM structure of CYP1A1 [24]. This binding mode is productive and leads to formation of DHB and singly hydroxylated products (step 2). Being more hydrophilic, the oxidation products would dissociate and reenter the active site with the psoralen group approaching the heme (steps 3–4). This orientation could enable association of the second DHB molecule, which may alter the conformation of DHB bound to the high-affinity site (step 5). Occupation of both binding sites would enhance the inhibitory potential, as furanocoumarin dimers are known to inhibit CYP3A4 stronger than DHB [21,25,26]. Nonetheless, the higher-affinity ligands could fully displace DHB and, thus, its metabolic activation would be a prerequisite for the potent inhibition. Among multiple DHB orientations, only some would be suitable for the enzymatic bioactivation of the furan ring (steps 6–7). Alternatively, since CYP3A4 catalysis is highly uncoupled [27], the furan double bond could be oxidized by a by-product, hydrogen peroxide, even when DHB adopts the crystallographic binding mode (step 7a). Upon epoxidation and opening of the furan ring, the radical intermediate could attack the heme (step 8). The epoxide product, in turn, could diffuse out of the active site and modify the apoprotein [22]. This could lead to alterations in conformational dynamics and structural integrity of CYP3A4 and its accelerated degradation [28].

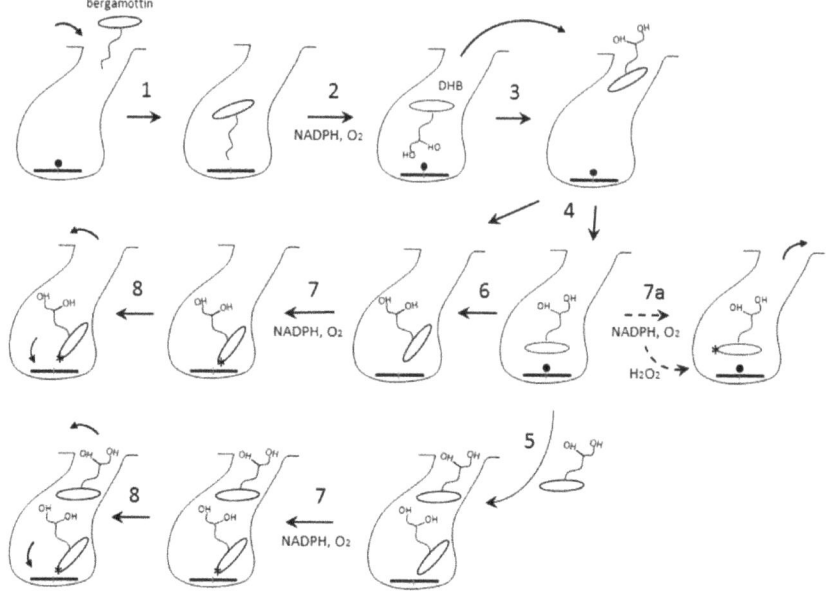

Figure 5. Possible DHB binding and bioactivation mechanism. The parent compound, bergamottin, preferably binds to CYP3A4 with the hydrophobic geranyl group entering the substrate channel (step 1). This binding mode is productive and leads to formation of DHB, which dissociates and re-enters the active site with the psoralen group approaching the heme iron (steps 2–3). DHB could have multiple orientations and two binding sites (steps 4–6). In some binding modes, psoralen's furan ring will be close enough to the iron to allow enzymatic bioactivation (step 7). Alternatively, the furan double bond can be oxidized by a by-product, hydrogen peroxide (step 7a). The reactive metabolite(s) could attack and destroy the heme or escape the active site and modify the apoprotein (step 8), leading to alterations in conformational dynamics and structural integrity of CYP3A4. The bioactivation site is indicated by an asterisk.

In summary, this study provided the first direct insights into the interaction of human drug-metabolizing CYP3A4 with three structurally diverse suicide substrates: mibefradil, azamulin and DHB. Only minimal conformational adjustments were needed to accommodate these compounds which, instead, were molded or stretched for a better fit and optimization of protein-ligand contacts. In addition to S119 and R212, frequently involved in the ligand binding, three more polar residues, E308, R372 and E372, were identified as important for the formation/stabilization of the substrate-bound complexes. The CYP3A4-DHB structure identified the high-affinity area where furanocoumarins and other small hydrophobic molecules bearing few polar groups could preferably dock. The binding manner of mibefradil and azamulin, on the other hand, suggested that these compounds could exert their inhibitory action not only upon bioactivation of the newly identified or other oxidation sites, but also by forming slowly dissociable complexes that disallow other substrates to access the catalytic site. Together, the spectral and structural data help better understand the suicide substrate binding and inhibitory mechanism, enable more accurate mapping of the CYP3A4 active site, and can be used to improve computational tools for the prediction of the binding ability, metabolic sites and MBI/DDI potential of newly developed medicines and other chemicals relevant to human health.

3. Materials and Methods

Mibefradil was obtained from Tocris Bioscience (Minneapolis, MN, USA) azamulin and DHB from Cayman Chemical (Ann Arbor, MI, USA), and bergamottin from Sigma-Aldrich (St. Louis, MO, USA).

3.1. Protein Expression and Purification

The codon-optimized full-length and Δ3-22 human CYP3A4 were produced as reported previously [29] and used for the spectral and structural studies, respectively.

3.2. Spectral Binding Titrations

Ligand binding to CYP3A4 was monitored in a Cary 300 spectrophotometer at ambient temperatures in 0.1 M phosphate buffer, pH 7.4, containing 20% glycerol and 1 mM dithiothreitol. The investigated compounds were dissolved in dimethyl sulfoxide (DMSO) to 0.2–20 mM concentration and added to a 1.5–2 µM protein solution in small aliquots, with the final solvent concentration <2%. After the addition of each aliquot, the reaction mixture was permitted to stand until no further absorbance changes could be detected, usually less than 20 min. At the end of titrations, the spectra of the ferrous CO-adduct were recorded to ensure that there was no CYP3A4 inactivation during lengthy measurements. Difference spectra were recorded in a separate experiment, where glycerol was omitted and equal amounts of DMSO were added to the reference cuvette to correct for the solvent-induced spectral perturbations. The spectral dissociation constant (K_s) was derived from the hyperbolic, sigmoidal or quadratic fits to the titration curves using IgorPro (WaveMetrics, Inc., Portland, OR, USA). To assess the ligands' dissociability, CYP3A4 was saturated with the substrate and then titrated with ritonavir (0.2–5 mM solution in DMSO) to determine how occupation of the active site affects its dissociation constant (K_s^{RIT}). Each titration experiment was repeated three times. The average K_s values and standard errors are given in Table 1. The high-spin content was estimated based on the absorbance spectra of ligand-free (100% low-spin) and azamulin-bound CYP3A4 (100% high-spin conversion).

3.3. Determination of the X-ray Structures

Δ3-22 CYP3A4 was co-crystallized with the investigated compounds at room temperature using a sitting drop vapor diffusion method. The protein (50–70 mg/mL or 0.9–1.25 mM in 100 mM potassium phosphate buffer, pH 7.4, 20% glycerol and 2 mM dithiothreitol) was incubated with a 5-fold ligand excess and centrifuged to remove the precipitate (<5% protein loss). The supernatant (0.4–0.5 µL) was mixed with 0.4–0.6 µL of the crystallization solution containing 6–10% polyethylene glycol 3350 and either 70 mM malate, 50 mM succinate, or 80 mM ammonium citrate tribasic, pH 7.0, for mibefradil-, azamulin- and DHB-bound CYP3A4, respectively. After harvesting, crystals were cryoprotected with Paratone-N and frozen in liquid nitrogen. The X-ray diffraction data were collected at the Stanford Synchrotron Radiation Lightsource beamline 7-1 and the Advanced Light Source beamline 5.0.2. The high-resolution cutoffs were chosen based on the CC1/2 value [30]. Crystal structures were solved by molecular replacement with PHASER [31], using the 5VCC structure as a search model. The ligands were built with eLBOW [32] and manually fit into the density with COOT [33]. The initial models were rebuilt and refined with COOT and PHENIX [32]. The polder and simulated annealing omit electron density maps were calculated with PHENIX. Data collection and refinement statistics are summarized in Table S1. The atomic coordinates and structure factors for mibefradil-, azamulin- and DHB-bound CYP3A4 were deposited to the Protein Data Bank with the ID codes 6OO9, 6OOA and 6OOB, respectively.

Supplementary Materials: Supplementary materials can be found at http://www.mdpi.com/1422-0067/20/17/4245/s1 and include five figures, showing spectral changes observed upon displacement of CYP3A4-bound bromocryptine, midazolam, bergamottin and DHB by ritonavir, simulated annealing omit electron density maps, superposition of the ligand-free and mibefradil-, DHB- and azamulin-bound structures of CYP3A4, and a table with the X-ray data collection and refinement statistics.

Funding: This work was funded by the National Institutes of Health grant ES025767.

Acknowledgments: This study involves research carried out at the Stanford Synchrotron Radiation Lightsource and the Advanced Light Source. Use of the Stanford Synchrotron Radiation Lightsource, SLAC National Accelerator Laboratory, is supported by the U.S. Department of Energy, Office of Science, Office of Basic Energy Sciences under Contract No. DE-AC02-76SF00515. The SSRL Structural Molecular Biology Program is supported by the DOE Office of Biological and Environmental Research, and by the National Institutes of Health, National Institute of General Medical Sciences (including P41GM103393). The Advanced Light Source is supported by the Director, Office of Science, Office of Basic Energy Sciences, of the U.S. Department of Energy under Contract No. DE-AC02-05CH11231.

Conflicts of Interest: The author declares no competing financial interest.

Abbreviations

CYP3A4	cytochrome P450 3A4
DDI	drug-drug interaction
DHB	6′,7′-dihydroxybergamottin
MBI	mechanism-based inactivation

References

1. Guengerich, F.P. Cytochrome P-450 3A4: Regulation and role in drug metabolism. *Annu. Rev. Pharmacol. Toxicol.* **1999**, *39*, 1–17. [CrossRef] [PubMed]
2. Zhou, S.F. Drugs behave as substrates, inhibitors and inducers of human cytochrome P450 3A4. *Curr. Drug Metab.* **2008**, *9*, 310–322. [CrossRef] [PubMed]
3. Zhou, S.; Yung Chan, S.; Cher Goh, B.; Chan, E.; Duan, W.; Huang, M.; McLeod, H.L. Mechanism-based inhibition of cytochrome P450 3A4 by therapeutic drugs. *Clin. Pharmacokinet.* **2005**, *44*, 279–304. [CrossRef] [PubMed]
4. Zhou, S.F. Potential strategies for minimizing mechanism-based inhibition of cytochrome P450 3A4. *Curr. Pharm. Des.* **2008**, *14*, 990–1000. [CrossRef] [PubMed]
5. Li, A.P.; Kaminski, D.L.; Rasmussen, A. Substrates of human hepatic cytochrome P450 3A4. *Toxicology* **1995**, *104*, 1–8. [CrossRef]
6. Ekroos, M.; Sjogren, T. Structural basis for ligand promiscuity in cytochrome P450 3A4. *Proc. Natl. Acad. Sci. USA* **2006**, *103*, 13682–13687. [CrossRef] [PubMed]
7. Sevrioukova, I.F.; Poulos, T.L. Structural basis for regiospecific midazolam oxidation by human cytochrome P450 3A4. *Proc. Natl. Acad. Sci. USA* **2017**, *114*, 486–491. [CrossRef] [PubMed]
8. Skopalik, J.; Anzenbacher, P.; Otyepka, M. Flexibility of human cytochromes P450: Molecular dynamics reveals differences between CYPs 3A4, 2C9, and 2A6, which correlate with their substrate preferences. *J. Phys. Chem. B.* **2008**, *112*, 8165–8173. [CrossRef]
9. Sevrioukova, I.F.; Poulos, T.L. Structural and mechanistic insights into the interaction of cytochrome P4503A4 with bromoergocryptine, a type I ligand. *J. Biol. Chem.* **2012**, *287*, 3510–3517. [CrossRef]
10. Welker, H.A.; Wiltshire, H.; Bullingham, R. Clinical pharmacokinetics of mibefradil. *Clin. Pharmacokinet.* **1998**, *35*, 405–423. [CrossRef]
11. Stresser, D.M.; Broudy, M.I.; Ho, T.; Cargill, C.E.; Blanchard, A.P.; Sharma, R.; Dandeneau, A.A.; Goodwin, J.J.; Turner, S.D.; Erve, J.C.; et al. Highly selective inhibition of human CYP3Aa in vitro by azamulin and evidence that inhibition is irreversible. *Drug Metab. Dispos.* **2004**, *32*, 105–112. [CrossRef] [PubMed]
12. Khojasteh, S.C.; Prabhu, S.; Kenny, J.R.; Halladay, J.S.; Lu, A.Y. Chemical inhibitors of cytochrome P450 isoforms in human liver microsomes: A re-evaluation of P450 isoform selectivity. *Eur. J. Drug Metab. Pharmacokinet.* **2011**, *36*, 1–16. [CrossRef] [PubMed]
13. Hung, W.L.; Suh, J.H.; Wang, Y. Chemistry and health effects of furanocoumarins in grapefruit. *J. Food Drug Anal.* **2017**, *25*, 71–83. [CrossRef] [PubMed]
14. He, K.; Iyer, K.R.; Hayes, R.N.; Sinz, M.W.; Woolf, T.F.; Hollenberg, P.F. Inactivation of cytochrome P450 3A4 by bergamottin, a component of grapefruit juice. *Chem. Res. Toxicol.* **1998**, *11*, 252–259. [CrossRef] [PubMed]
15. Foti, R.S.; Rock, D.A.; Pearson, J.T.; Wahlstrom, J.L.; Wienkers, L.C. Mechanism-based inactivation of cytochrome P450 3A4 by mibefradil through heme destruction. *Drug Metab. Dispos.* **2011**, *39*, 1188–1195. [CrossRef] [PubMed]

16. Samuels, E.R.; Sevrioukova, I. Structure-activity relationships of rationally designed ritonavir analogs: Impact of side-group stereochemistry, head-group spacing, and backbone composition on the interaction with CYP3A4. *Biochemistry* **2019**, *58*, 2077–2087. [CrossRef] [PubMed]
17. Prueksaritanont, T.; Ma, B.; Tang, C.; Meng, Y.; Assang, C.; Lu, P.; Reider, P.J.; Lin, J.H.; Baillie, T.A. Metabolic interactions between mibefradil and HMG-CoA reductase inhibitors: An in vitro investigation with human liver preparations. *Br. J. Clin. Pharmacol.* **1999**, *47*, 291–298. [CrossRef] [PubMed]
18. Bui, P.H.; Quesada, A.; Handforth, A.; Hankinson, O. The mibefradil derivative NNC55-0396, a specific T-type calcium channel antagonist, exhibits less CYP3A4 inhibition than mibefradil. *Drug Metab. Dispos.* **2008**, *36*, 1291–1299. [CrossRef]
19. Tassaneeyakul, W.; Guo, L.Q.; Fukuda, K.; Ohta, T.; Yamazoe, Y. Inhibition selectivity of grapefruit juice components on human cytochromes P450. *Arch. Biochem. Biophys.* **2000**, *378*, 356–363. [CrossRef]
20. Edwards, D.J.; Bellevue, F.H., 3rd; Woster, P.M. Identification of 6',7'-dihydroxybergamottin, a cytochrome P450 inhibitor, in grapefruit juice. *Drug Metab. Dispos.* **1996**, *24*, 1287–1290.
21. Guo, L.Q.; Taniguchi, M.; Xiao, Y.Q.; Baba, K.; Ohta, T.; Yamazoe, Y. Inhibitory effect of natural furanocoumarins on human microsomal cytochrome P450 3A activity. *Jpn. J. Pharmacol.* **2000**, *82*, 122–129. [CrossRef] [PubMed]
22. Lin, H.L.; Kenaan, C.; Hollenberg, P.F. Identification of the residue in human CYP3A4 that is covalently modified by bergamottin and the reactive intermediate that contributes to the grapefruit juice effect. *Drug Metab. Dispos.* **2012**, *40*, 998–1006. [CrossRef] [PubMed]
23. Lin, H.L.; Kent, U.M.; Hollenberg, P.F. The grapefruit juice effect is not limited to cytochrome P450 (P450) 3A4: Evidence for bergamottin-dependent inactivation, heme destruction, and covalent binding to protein in P450s 2B6 and 3A5. *J. Pharmacol. Exp. Ther.* **2005**, *313*, 154–164. [CrossRef] [PubMed]
24. Bart, A.G.; Scott, E.E. Structures of human cytochrome P450 1A1 with bergamottin and erlotinib reveal active-site modifications for binding of diverse ligands. *J. Biol. Chem.* **2018**, *293*, 19201–19210. [CrossRef] [PubMed]
25. Row, E.; Brown, S.A.; Stachulski, A.V.; Lennard, M.S. Development of novel furanocoumarin dimers as potent and selective inhibitors of CYP3A4. *Drug Metab. Dispos.* **2006**, *34*, 324–330. [CrossRef]
26. Oda, K.; Yamaguchi, Y.; Yoshimura, T.; Wada, K.; Nishizono, N. Synthetic models related to furanocoumarin-CYP 3A4 interactions. comparison of furanocoumarin, coumarin, and benzofuran dimers as potent inhibitors of CYP3A4 activity. *Chem. Pharm. Bull.* **2007**, *55*, 1419–1421. [CrossRef] [PubMed]
27. Grinkova, Y.V.; Denisov, I.G.; McLean, M.A.; Sligar, S.G. Oxidase uncoupling in heme monooxygenases: Human cytochrome P450 CYP3A4 in Nanodiscs. *Biochem. Biophys. Res. Commun.* **2013**, *430*, 1223–1227. [CrossRef]
28. Paine, M.F.; Criss, A.B.; Watkins, P.B. Two major grapefruit juice components differ in time to onset of intestinal CYP3A4 inhibition. *J. Pharmacol. Exp. Ther.* **2005**, *312*, 1151–1160. [CrossRef]
29. Sevrioukova, I.F. High-level production and properties of the cysteine-depleted cytochrome P450 3A4. *Biochemistry* **2017**, *56*, 3058–3067. [CrossRef]
30. Karplus, P.A.; Diederichs, K. Linking crystallographic model and data quality. *Science* **2012**, *336*, 1030–1033. [CrossRef]
31. McCoy, A.J.; Grosse-Kunstleve, R.W.; Adams, P.D.; Winn, M.D.; Storoni, L.C.; Read, R.J. Phaser crystallographic software. *J. Appl. Crystallogr.* **2007**, *40*, 658–674. [CrossRef] [PubMed]
32. Adams, P.D.; Afonine, P.V.; Bunkoczi, G.; Chen, V.B.; Davis, I.W.; Echols, N.; Headd, J.J.; Hung, L.W.; Kapral, G.J.; Grosse-Kunstleve, R.W.; et al. PHENIX: A comprehensive Python-based system for macromolecular structure solution. *Acta Crystallogr. Sect. D* **2010**, *66*, 213–321. [CrossRef] [PubMed]
33. Emsley, P.; Lohkamp, B.; Scott, W.G.; Cowtan, K. Features and development of Coot. *Acta Crystallogr. Sect. D* **2010**, *66*, 486–501. [CrossRef] [PubMed]

© 2019 by the author. Licensee MDPI, Basel, Switzerland. This article is an open access article distributed under the terms and conditions of the Creative Commons Attribution (CC BY) license (http://creativecommons.org/licenses/by/4.0/).

Article

Differing Membrane Interactions of Two Highly Similar Drug-Metabolizing Cytochrome P450 Isoforms: CYP 2C9 and CYP 2C19

Ghulam Mustafa [1,2], Prajwal P. Nandekar [1,2], Neil J. Bruce [1] and Rebecca C. Wade [1,2,3,*]

1. Molecular and Cellular Modeling Group, Heidelberg Institute for Theoretical Studies (HITS), 69118 Heidelberg, Germany
2. Zentrum für Molekulare Biologie der Universität Heidelberg, DKFZ-ZMBH Alliance, 69120 Heidelberg, Germany
3. Interdisciplinary Center for Scientific Computing (IWR), Heidelberg University, 69120 Heidelberg, Germany
* Correspondence: rebecca.wade@h-its.org; Tel.: +49-6221-533247

Received: 6 August 2019; Accepted: 1 September 2019; Published: 4 September 2019

Abstract: The human cytochrome P450 (CYP) 2C9 and 2C19 enzymes are two highly similar isoforms with key roles in drug metabolism. They are anchored to the endoplasmic reticulum membrane by their N-terminal transmembrane helix and interactions of their cytoplasmic globular domain with the membrane. However, their crystal structures were determined after N-terminal truncation and mutating residues in the globular domain that contact the membrane. Therefore, the CYP-membrane interactions are not structurally well-characterized and their dynamics and the influence of membrane interactions on CYP function are not well understood. We describe herein the modeling and simulation of CYP 2C9 and CYP 2C19 in a phospholipid bilayer. The simulations revealed that, despite high sequence conservation, the small sequence and structural differences between the two isoforms altered the interactions and orientations of the CYPs in the membrane bilayer. We identified residues (including K72, P73, and I99 in CYP 2C9 and E72, R73, and H99 in CYP 2C19) at the protein-membrane interface that contribute not only to the differing orientations adopted by the two isoforms in the membrane, but also to their differing substrate specificities by affecting the substrate access tunnels. Our findings provide a mechanistic interpretation of experimentally observed effects of mutagenesis on substrate selectivity.

Keywords: cytochrome P450; isoform; membrane protein; protein-membrane interactions; enzyme substrate specificity; mutagenesis; molecular dynamics simulation

---

1. Introduction

Human cytochrome P450 (CYP) enzymes play important roles in the metabolism of drugs, steroids, fatty acids, and xenobiotics. CYPs also catalyze the conversion of some prodrugs into active drugs. Only about a dozen human CYPs metabolize 70–80% of all drugs, and these mainly belong to families CYP1, CYP2, and CYP3, and their subfamilies [1]. The human CYP2C subfamily contributes significantly to the hepatic clearance of many drugs. Although the members of the subfamily exhibit about 70% sequence similarity, they have unique substrate specificity profiles [2]. The human CYP2C subfamily consists of four isoforms: *CYP2C8*, *CYP2C18*, *CYP2C9*, and *CYP2C19* [3]. CYP 2C9 is the most highly expressed CYP protein after CYP3A4 and it is responsible for the metabolism of over 12.8% of drugs, with its substrates being mostly weak acids, such as non-steroidal anti-inflammatory drugs (NSAID) [1]. CYP 2C19 has a 10-fold lower expression level than CYP 2C9, but contributes to the metabolism of 6.8% of drugs [1], although without the specificity for acidic drugs of CYP 2C9. Nevertheless, the polymorphism of *CYP2C19* can dramatically affect drug treatments. For example, it has been observed in the treatment of *Helicobacter pylori* infections with proton-pump inhibitors that are substrates of CYP 2C19, such as omeprazole, that the therapeutic efficiency is improved in patients with a poorly metabolizing *CYP2C19* genotype due to slower drug clearance [4]. Furthermore, *CYP2C19* is important for the enzymatic activation of the antiplatelet agent, clopidogrel, to its active thiol metabolite [5,6], and loss of function in the common *CYP2C19*2* allele, which has a splicing variant leading to truncation of the protein, results in poor response to clopidogrel [7]. On the other hand, *CYP2C9* polymorphism results in reduced affinity for cytochrome P450 reductase (*CYP2C9*2*) and altered substrate specificity (*CYP2C9*3*) [8].

CYP 2C9 and CYP 2C19 have distinct substrate specificities, despite having high sequence conservation with 91.2% sequence identity (see sequence alignment in Figure 1). Crystal structures of the globular domains of the proteins have been resolved by X-ray crystallography after truncation to remove the N-terminal transmembrane (TM) domain and flexible linker sequences, as well as mutation to introduce terminal expression tags (see Figure 1). Only one crystal structure of CYP 2C19 has been resolved (Protein Data Bank (PDB) identifier 4GQS) [9], whereas a number of crystal structures of CYP 2C9 in various liganded and mutated states have been determined (currently 11 PDB entries) [10–17]. The crystal structures show that CYP 2C19 differs from CYP 2C9 at two residues in the active site: L208 and L362 in CYP 2C9 are substituted by V208 and I362 in CYP 2C19 [9,11]. Other differences are seen on the outer surface of the globular domain. The three-dimensional fold of CYP 2C19 is closer to the structure of CYP 2C8 (PDB 2NNI) [12], which shares 78% sequence identity with CYP 2C19, than to the structures of CYP 2C9 (PDB 1R9O) [11] or CYP 2C9m7 (PDB 1OG2, 1OG5) [10], despite their higher sequence identity (91.2%). The latter structures were resolved after making seven substitutions (K206E, I215V, C216Y, S220P, P221A, I222L, and I223L) in the F'–G' loop region of CYP 2C9 for the purpose of crystallization, as this part of the protein is hydrophobic and interacts with the membrane [10]. The structure of CYP 2C9m7 differs from that of CYP 2C9 in the B–C loop, which is highly flexible in the 1R9O structure, and the conformations of the F' and G' helices, which are missing in the 1R9O structure. The F'–G' region shows high structural variation amongst the crystal structures of CYP 2C9; in structures in which the protein has the wild-type sequence, the F'–G' region is either missing (e.g., in PDB 1R9O) [11], has an extended loop conformation and a small G' helix (PDB 5W0C) [17], or has an F' helix followed by a loop in the G' region interacting with a peripherally bound ligand [15]. CYP 2C9m7 also differs in the position of the sidechain of R108, which points out of the binding cavity in the CYP 2C9m7 (1OG2) structure and inside in the CYP 2C9 (1R9O) structure. The structure of CYP 2C19 shows a more than 3.0 Å Cα atom deviation from both the CYP 2C9 (1R9O) and CYP 2C9m7 (1OG2) structures on the outer surface entrance region of the protein responsible for substrate access and selectivity. The main differences are observed in helices F, F', G', and G and their turns, the turn in β-hairpin 1, and the B–C loop region.

The sequence differences outside the CYP active site binding cavity may be responsible for the differential selection of drugs entering the binding pocket [18–21]. Indeed, differences in the use of the access tunnels have been suggested by mutagenesis studies on CYP 2C9/2C19 chimeras [18–21] and simulations of the globular domain of CYP 2C9 [22]. For example, CYP 2C19 selectively hydroxylates omeprazole and S-mephenytoin, whereas CYP 2C9 has little activity against these substrates. However, substitution of residues outside the binding site (I99H, S220, and P221T) at the entrance to tunnel 2b (using the nomenclature of Cojocaru et al. [23]) increased the omeprazole 5-hydroxylase activity of CYP 2C9 to a level similar to CYP 2C19 [18]. On the other hand, the E72K substitution in CYP 2C19 was shown to decrease its enzymatic metabolic activity against three tricyclic antidepressant (TCA) CYP 2C19 substrates, amitriptyline, imipramine, and dothiepin, whereas the K72E mutation in CYP 2C9 increased its metabolic activity against these compounds [21]. Most of these differences are found in the substrate recognition sites (SRS) identified by Zawaira et al. [24] (see Figure 1). Since most residues that differ between CYP 2C9 and CYP 2C19 are found in these SRS regions, we hypothesized that the sequence differences in the SRS regions and, thereby, the conformational differences observed between the two CYPs, can contribute to different protein–membrane interactions which, in turn, can lead to differences in the substrate access tunnels to the binding cavity and the product release tunnels

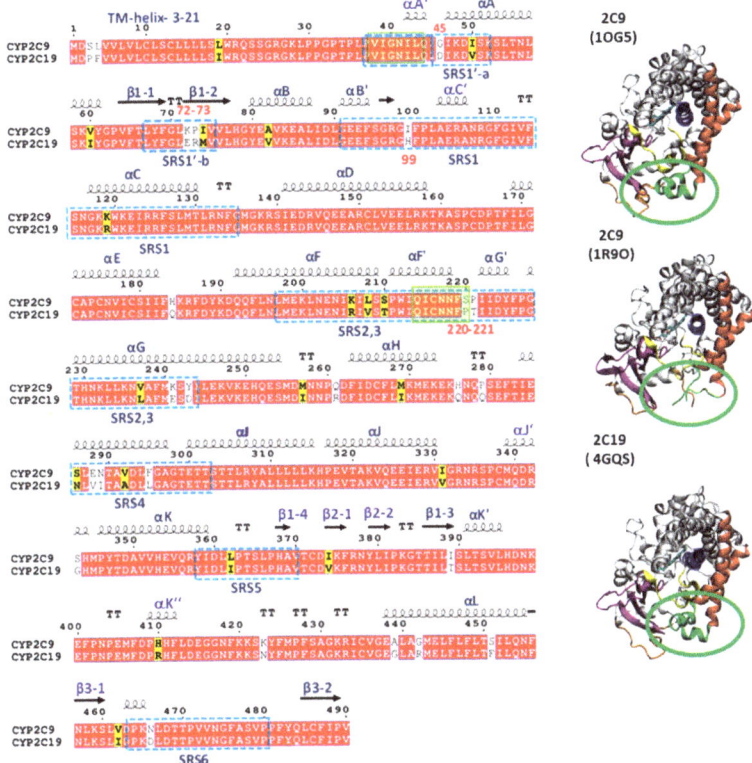

Figure 1. (Left) Sequence alignment of CYP 2C9 and CYP 2C19. Identical residues are shown with a red background, similar residues with a yellow background, and differing residues with a white background. The secondary structure in the crystal structure of CYP 2C19 (PDB 4GQS) is indicated by arrows for β-strands, springs for α-helices, and 'TT' for turns; long loops are unmarked. The substrate recognition sites (SRS) are shown by blue dashed line boxes. The residues in the globular domain differing at the membrane interface are highlighted by red numbers. The missing regions in the crystal structure (PDB 1R9O) of the globular domain of CYP 2C9 are shown by transparent green boxes. (Right) Cartoon representations of the crystal structures of CYP 2C9m7 (PDB 1OG5), CYP 2C9 (PDB 1R9O), and CYP 2C19 (PDB 4GQS), showing the structural differences in the F'–G' region highlighted by the green rings, the heme in stick representation, and key secondary structure elements colored as follows: β-strand regions in magenta, the B–C loop in yellow, the F and G helices in red, the F'–G' helices/loop in green, the I-helix in blue, and the linker in orange. The active site is lined by the heme and the I-helix.

To investigate the effect of sequence differences between CYP 2C9 and CYP 2C19 on the protein–membrane interactions and the orientation of the protein globular domain in the membrane, we applied our optimized multiscale modeling and simulation protocol [25] to model and simulate the two proteins in a 1-palmitoyl-2-oleoyl-sn-glycero-3-phosphocholine (POPC) bilayer. We have previously applied a similar procedure to simulate CYP 2C9 in a POPC bilayer [22]. For each system, this protocol starts with building a model of the full protein in a POPC bilayer based on the crystal structure of the globular domain (see Figure 2), and then proceeds with optimizing the system to reach a converged arrangement by coarse-grained (CG) and all-atom (AA) molecular dynamics (MD) simulations (see Materials and Methods (Section 4) for details). We compared the behavior of the two proteins in the simulations and compared our results with previously reported experimental and computational data.

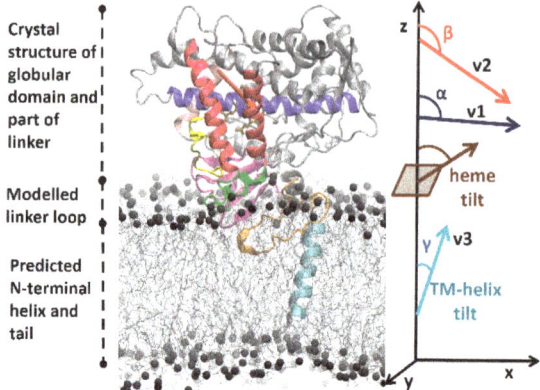

Figure 2. Initial model of human cytochrome P450 (CYP) 2C9 showing its three domains and the initial information on which modeling and simulation of its arrangement in the phospholipid bilayer was based. The crystal structure (PDB 1R9O) of the globular domain (residues 50–490) and part of the linker region (residues 37–49) are shown in cartoon representation. Secondary structure predictions indicate the length of the N-terminal transmembrane (TM)-helix (cyan, residues 3–21). These two components are connected by a modeled linker loop of unknown conformation (orange, residues 22–36). The flexible C-terminal tail (residues 491–492) was not included in the model. The F′–G′ helices (residues 210–220) were not observed in the structure and were modeled from the crystal structure of CYP 2C19 (PDB 4GQS). Important secondary structure elements in the globular domain are colored as follows: β-strand region in magenta, F and G helices in red, I-helix in blue, B–C loop in yellow, and F′–G′ helices in green. The heme is shown in brown stick representation. Experimentally, it is known that the globular domain interacts with the bilayer (shown in grey line representation with grey spheres representing the phosphorous atoms) and, during the coarse-grained (CG) simulations, it approached and dipped into the bilayer. The heme tilt angle and the angles α and β defining its orientation in the bilayer are shown on the right, along with the definition of the TM-helix tilt angle (γ), and the vectors (v1 along the I-helix , v2 shown by the red arrow from the C to the F helix, and v3 along the TM-helix) computed to define these angles; the definitions of these angles are given in the Materials and Methods section.

2. Results and Discussion

2.1. CG Simulations Show That the CYP 2C9 and CYP 2C19 Globular Domains Adopt Distinct Orientations in the Membrane Bilayer

The coarse-grained (CG) simulations carried out are listed in Table S1. The trajectories were analyzed to assess convergence of the orientation of the globular domain and its interactions with the membrane. Different starting structures, different lengths of the flexible linker region, and different simulation parameters were tested to ensure that reliable CYP-membrane interactions and globular domain orientations were obtained. The converged positions and orientations from all sets of simulations of CG systems are given in Table 1.

Simulations of the full wild-type sequences of the two isoforms (systems S1 and S2) show different orientations from each other in the membrane. For the full-length wild-type CYP 2C9, the angles α and β (see Figure 2 and Materials and Methods for definition of these angles) ranged from 89 to 91° and 112 to 118°, respectively. For the full-length wild-type CYP 2C19, the α and β angles ranged from 97 to 100° and 134 to 137°, respectively. The axial distance of the center of mass (CoM) of the globular domain from the membrane CoM was 42–43 ± 2 Å for CYP 2C9 and 47–48 ± 2 Å for CYP 2C19. The different orientations of the two isoforms were classified into three different classes, A, A/B, and B. When the β angle was below 125°, the orientation was categorized in class A, from 125° to 130° into class A/B, and above 130° into class B. All wild-type CYP 2C9 CG systems (S1–S3) converged to class A, and all wild-type CYP 2C19 CG systems (S1–S3) converged to class B. The angles and distance values were plotted for all the CG systems in Figure S1. Thus, the two isoforms adopt distinct positions and orientations of their globular domains in the phospholipid bilayer in the CG simulations.

Table 1. Results of CG simulations of CYP 2C9 and CYP 2C19 for the full length (S1, S2) and truncated (S3) wild-type proteins (based on crystal structures 1R9O and 4GQS, respectively), various models of full length CYP 2C9 (M1–M4, based on structures Model 1–4), and the mutant chimeras, mt2C9 and mt2C19. The systems simulated are listed in Table S1. The mean and standard deviation values are given for parameters defining the position of the globular domain with respect to the membrane: angles α and β (defining the orientation, as shown in Figure 2 and class (A, A/B, or B)) and the axial distances of the center of mass (CoM) of the bilayer from the CoM of the linker, the F'–G' region, and the globular domain, respectively. The simulations each had an average duration of 10 µs, and the parameters were computed for snapshots from the last 9 µs collected at intervals of 1 ns.

CYP Systems	Angles (°)		Distances (Å)			No. of Simulations	Class
	α	β	Linker	F'–G'	Globular		
2C9:S1	89 ± 7	112 ± 7	25 ± 4	28 ± 4	43 ± 2	10	A
2C9:S2	91 ± 8	118 ± 14	26 ± 4	27 ± 4	42 ± 2	5	A
2C9:S3 [1]	92 ± 7	109 ± 7	-	28 ± 2	46 ± 2	1	A
2C9:M1	94 ± 6	119 ± 8	20 ± 2	25 ± 2	46 ± 2	6	A
2C9:M2	92 ± 8	120 ± 12	19 ± 2	25 ± 2	46 ± 2	6	A
2C9:M3	95 ± 6	138 ± 6	22 ± 2	24 ± 2	48 ± 2	6	B
2C9:M4	85 ± 9	106 ± 9	25 ± 2	26 ± 3	47 ± 2	5	A
mt2C9	98 ± 7	129 ± 10	27 ± 4	29 ± 4	44 ± 2	5	A/B
C19:S1	100 ± 7	137 ± 10	21 ± 2	26 ± 2	47 ± 2	10	B
2C19:S2	97 ± 8	137 ± 12	20 ± 2	25 ± 2	48 ± 2	5	B
2C19:S3 [1]	106 ± 5	133 ± 6	-	27 ± 2	46 ± 2	1	B
mt2C19	95 ± 8	127 ± 13	19 ± 2	25 ± 3	46 ± 2	5	A/B

[1] Systems containing only the CYP globular domain, residues 47–490 for both proteins, without the TM and linker regions.

2.2. The Globular Domain Converges to the Same Orientations and Positions with Respect to the Membrane in CG Simulations of Full Length and of N-Terminally Truncated Protein

In order to ensure that the initial modelled structures did not bias the results, two separate simulations of CYP 2C9 and CYP 2C19 were performed using the N-terminally truncated form with only the globular domain (S3). In these simulations, the globular domain could explore more configurations before reaching a stable orientation. For both isoforms, the final orientations of the globular domains in the phospholipid bilayer were stabilized by insertion of the F'–G' helices in the bilayer. Snapshots from these CG simulations are shown in Figure 3A,B. The region of the F'–G' helices is one of the hydrophobic regions in CYPs that keep the globular domain anchored in the membrane, even after truncation of the TM-helix. In the simulations of truncated CYP 2C9, the orientation of the globular domain converged in 3.5 µs with a sharp decrease in the CoM distance of the F'–G' helices, shown by the arrow in Figure 3D. The orientation of the CYP 2C19 globular domain converged quickly in 200 ns and remained stable throughout the simulation (Figure 3B,F). In both CYPs, after the F'–G' helices developed contacts with the membrane, no further change was observed in the orientation of the globular domain. For both isoforms, the converged orientation of the globular domain was the same in the simulations of the globular domain only (S3) and of the full-length protein (S1, S2) (Table 1 and Figure 3C,E). The difference in the orientations of the two CYP isoforms in the membrane was maintained in the simulations of both the full length and truncated forms of the proteins. However, in the simulations of the truncated proteins (S3), the CoM distance was the same for both proteins

(46 ± 2 Å). Overall, the CYP 2C9 CG simulations showed final orientations differing from CYP 2C19 despite using the same simulation parameters, water models, and protein components, whether full length or globular domain only.

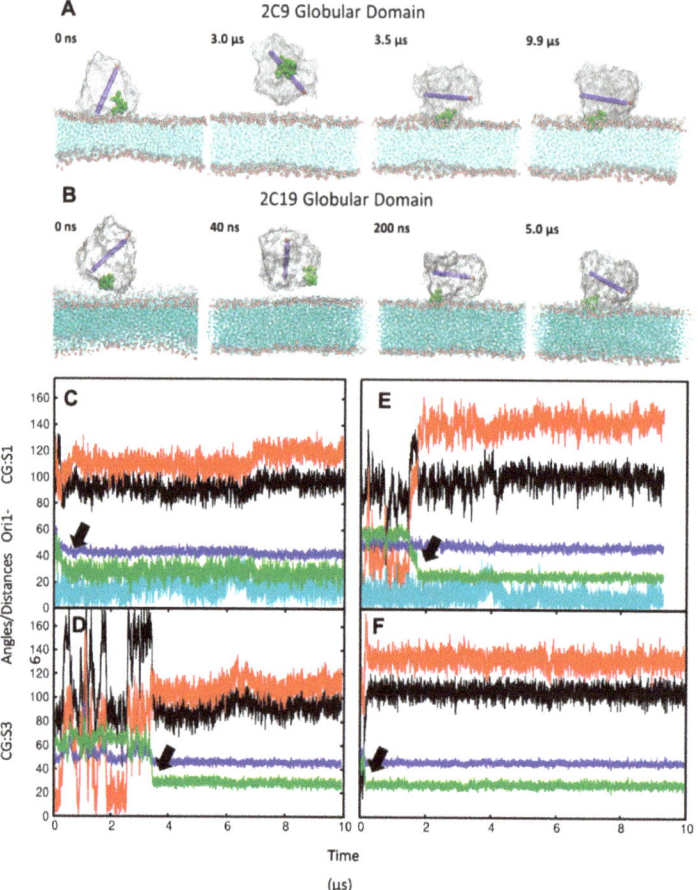

Figure 3. (**A,B**) Snapshots from CG simulations of globular domains (S3) of (**A**) CYP 2C9 and (**B**) CYP2 C19 showing exploration of different orientations followed by convergence to the same orientation as observed for CG simulations of the full-length wild-type proteins. The globular domain is shown with a silver surface representation, with the F'–G' helices shown as green VDW spheres and the I-helix (residues 286–316) shown as a blue cylinder with an arrow and a red sphere at the C-terminal end. The 1-palmitoyl-2-oleoyl-sn-glycero-3-phosphocholine (POPC) bilayer is shown in cyan, with the phosphate atoms in the head groups shown as red spheres. (**C–F**) Convergence of the orientation and position of the globular domain during CG simulations of CYP 2C9 (**C,D**) and CYP 2C19 (**E,F**). The angles (°) and distance (Å) values vs time (μs) are shown for selected trajectories from CG systems: (**C,E**) S1 (full-length proteins); (**D,F**) S3 (globular domain only). The angles α (black), β (red), and the TM-helix tilt angle (cyan) (defined in Figure 2 and Materials and Methods) are shown along with the axial distances of the bilayer CoM to the CoM of the globular domain (blue) and the F'–G' helices (green). The thick black arrows point to the decrease in the distance of the F'–G' helices from the membrane center, which is coincident with convergence to stable orientations.

2.3. Structural Differences in the Interfacial Residues Affect the Protein-Membrane Interactions and Globular Domain Orientations in CG Simulations

To test the effect of the initial structure of the globular domain on its positioning in the membrane, we next performed CG simulations starting with four different models of CYP 2C9 (systems: M1–M4) (see Table S1). These full-length models of CYP 2C9 differed slightly in the side chain conformations over the whole globular domain, and more significantly in the membrane-interacting regions due to the different templates used for modeling the protein (see Appendix A). We focused on CYP 2C9 for these tests because the F'–G' helices were missing in the crystal structure used (PDB 1R9O), and this region shows high structural variability in the crystal structures of CYP 2C9. Models M1 and M2 were built by employing two different strategies to use the template information from the structures of both CYP 2C9 and CYP 2C19, whereas M3 was built using the CYP 2C19 structure as a template, and M4 was built using a previous model [22] for the CYP 2C9 F'–G' helices. Important differences were observed in the conformations of β-hairpin 1, the B–C loop, and the F'–G' helices, which are critical for developing CYP-membrane interactions and thereby influence the final orientation of the globular domain in the membrane. The differences in the orientations in the CG simulations of the different models of CYP 2C9 are shown in Table 1, Table S2, and Figure S2.

In all these CG simulations (M1–M4), the distance of the CoM of the globular domain to the CoM of the lipid bilayer increased from 43 ± 2 to 46–48 ± 2 Å. The new CoM distance value for CYP 2C9 was the same as observed for CYP 2C19. The angles α and β increased in the CG simulations using modeled structures compared to those starting with the crystal structure of CYP 2C9 (S1, S2). In the CG simulations of M3, for which the CYP 2C19 crystal structure was the template, 50% of the simulations (three out of six trajectories) showed higher angles (class B), resembling the orientation of CYP 2C19 in the membrane (Table S2, Figure S2). The increased angle and distance values were attributed to the initial conformations of the globular domain, due to the selection of the modeling template. The CG simulations of the four CYP 2C9 models (M1–M4) indicated that it is not only the primary sequence but also the initial conformational differences in the linker, B–C loop, and F'–G' helix regions that influence the final positioning of the CYP globular domain in the membrane. However, it should be borne in mind that the different conformations come from templates with different sequences, and that conformational preferences are dependent on sequence.

2.4. CG Simulations of Chimeric Mutant Models Show Sequence Differences in the Interfacial Residues Affect the Protein-Membrane Interactions and Globular Domain Orientations

To identify the residues contributing to the differing position of the globular domain in the membrane, structures of chimeric mutants of CYP 2C9 (mt2C9) and CYP 2C19 (mt2C19) were prepared by swapping residues mainly at the membrane interface in the linker (G46D), β-strand 1–2 (K72E and P73R), the B–C loop (I99H), and the F'–G' helices (S220P and P221T) (see Table S1 and Figure 1). In CG simulations of mt2C9, two out of five trajectories converged to a CYP 2C19-like orientation (class B), one converged to an intermediate orientation (class A/B), and two converged to the same orientation as wild-type CYP 2C9 (class A). In CG simulations of mt2C19, two out of five trajectories converged to a CYP 2C9-like orientation (class A), one to an intermediate orientation (class A/B), and two retained wild-type CYP 2C19 orientations (class B), as shown in Table S3 and Figure S3. On average, the orientation changed to the intermediate class A/B for both mt2C9 and mt2C19 (see Table 1). The membrane insertion depths of the globular domain and F'–G' region of the chimeras showed small shifts towards the depths of the other protein.

2.5. All-Atom MD Simulations Based on the CG Simulations Result in Stable Systems for Both Isoforms.

We next performed all-atom (AA) MD simulations to obtain refined atomic-level interactions between the membrane and the proteins. Two different initial configurations obtained from the CG simulations (S1 and S2) were used for AA simulations of each isoform. In each case, two simulations (SIM1, SIM2) were performed starting with different initial velocities assigned from the CG:S1 configuration, one for the apo from and one for a ligand-bound form, and one simulation (SIM3) was performed starting from the CG:S2 configuration.

For both proteins, the structure of the globular domain was well preserved in these simulations, as indicated by the Cα-atom root mean squared deviation (RMSD) of the globular domain with respect to the initial energy minimized structure (see Figure S4). The structures stabilized with an RMSD of about 2–2.5 Å during the simulations. Furthermore, comparison of computed and crystallographic B-factors showed that the regions of the globular domain with large fluctuations were regions with high crystallographic B-factors (see Figure S5). This was the case for both isoforms, even though the crystallographic B-factors indicated somewhat different flexible regions in the crystal structures (see Figure S6). In CYP 2C19, the HI loop, the meander region (consisting β-bulge region and the K' helix region), and β-strand 3 (residues 460–475) had higher B-factors (see Figures S5 and S6). The presence of the membrane did not restrict the flexibility of the membrane interacting regions, such as the linker preceding the A-helix, the B–C loop, and the F'–G' region, and strands β1-1, β1-2, β2-1, and β2-2. Indeed, partial unwinding of the G'-helix was observed in the simulations of CYP 2C19 (2C19:SIM1). The high linker flexibility contributed to higher fluctuations in the transmembrane helix angle, γ, in this particular simulation (see Table 2).

Table 2. Positioning of the CYP globular domain with respect to the membrane in the AA MD simulations of CYP 2C9 and CYP 2C19. Values of the mean and standard deviations of angles (as defined in Figure 2) and distances characterizing the positioning of the globular and TM domains were calculated for the last 50 ns of all-atom simulations, with different starting configurations from CG simulations and with different initial velocities assigned. The orientations of the globular domain were assigned to class A, A/B, or B.

AA MD Simulation	Residues		Angles (°)				Globular Domain-Bilayer Distance (Å)	Class	Time (ns)
	TM-Helix	Flexible Linker	α	β	γ	Heme-tilt			
2C9:CG:S1	3–21	22–36	91.9	111.9	17.6	30.2	45.0	A	10,000 [2]
2C9:SIM1	3–21		74.8 ± 4.3	119.9 ± 4.5	11.9 ± 5.3	43.2 ± 4.8	45.5 ± 1.5	A	216.88
2C9:SIM2 [1]	3–21		95.9 ± 4.4	123.3 ± 4.9	13.5 ± 4.1	39.8 ± 4.9	48.3 ± 2.3	A	156.1
2C9:CG:S2	3–21	26–38	90.5	111.8	13.9	33.8	42.3	A	10,000 [2]
2C9:SIM3	3–21		86.6 ± 4.1	126.8 ± 3.0	5.9 ± 2.7	40.1 ± 5.5	44.2 ± 0.9	A/B	50.6
2C19:CG:S1	2–23	26–38	99.6	135.3	13.0	52.4	46.7	B	10,000 [2]
2C19:SIM1	2–23	–	106.3 ± 4.2	148.6 ± 5.1	25.4 ± 7.8	60.5 ± 4.5	46.2 ± 2.6	B	108.4
2C19:SIM2 [1]	2–23	–	97.0 ± 5.0	140.1 ± 4.2	25.3 ± 4.6	58.1 ± 5.3	45.8 ± 1.6	B	113.4
2C19:CG:S2	3–21	22–36	99.5	133.3	10.2	45.9	50.3	B	10,000 [2]
2C19:SIM3	3–21	–	94.9 ± 4.8	135.8 ± 6.6	17.2 ± 4.1	55.4 ± 6.5	46.0 ± 1.6	B	95.2

[1] Simulated with a ligand in the active site (all other simulations were for the apoproteins). For CYP 2C9, the ligand was flurbiprofen from the crystal structure, and for CYP 2C19, it was the inhibitor from the crystal structure (Protein Data Bank chemical component 0XV). [2] CG simulations were run for an average of 10 μs. The angles were computed for the representative structure that was selected for starting AA MD simulations.

2.6. All-Atom MD Simulations Confirm Differences in the Positioning of the Globular Domain on the Membrane Between CYP 2C9 and CYP 2C19

The positioning of the globular domain with respect to the membrane was analyzed by calculating the heme-tilt angle in addition to the angles and distances computed for the CG simulations. The computed values are given in Table 2. The normalized angle and distance distribution plots characterizing the position of the globular domain observed in the simulations are shown in Figure 4. The simulations of CYP 2C9 showed some readjustment in the orientations from the starting configurations in which the β angle increased from about 112° to 120–127°, corresponding to remaining in class A in two cases (for the apoprotein and for the holoprotein with a substrate, the drug flurbiprofen, bound in the active site) and transitioning to class A/B in one case (SIM3), in which the globular domain structure was slightly less stable (see Figures S4 and S5). Concomitantly, the heme-tilt angle increased from 30–34° to 40–43°. The average axial distance of the globular domain CoM from the membrane CoM increased by 0.5 to 3 Å during the simulations. Compared to previous simulations conducted with the GAFF lipid force field [22], in which the globular domain CoM distance from the bilayer center decreased from 39.5 ± 2.5 Å (from CG simulations) to 34.1 ± 1.0 Å during AA simulations, the globular domain was less immersed in the membrane, facilitating the slightly lower observed values of the α and β angles. These differences can be attributed in large part to differences in both the protein and the lipid force field used. Our studies have shown that the force field used in the current work produces results in better agreement with experimental data for simulations of CYP-membrane systems, including excellent agreement with linear dichroism measurements of the heme-tilt angle of CYPs in Nanodiscs [25–27].

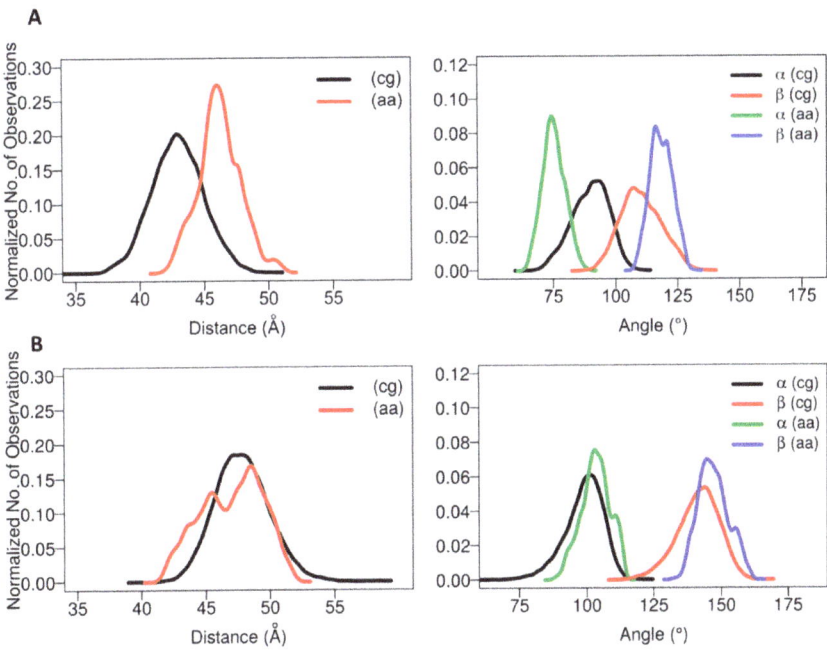

Figure 4. Plots of the distance and angle distributions defining the positioning of the globular domain with respect to the membrane in CG and AA MD simulations of CYP 2C9 (**A**) and CYP 2C19 (**B**). The globular domain of CYP 2C9 tended to be more immersed in the membrane than CYP 2C19 and the two adopted distinct orientations: class A (angle β < 125°) for CYP 2C9, and class B (angle β > 130°) for CYP 2C19.

The starting structures of the two AA models of CYP 2C19 in the membrane (for the apoprotein and for the holoprotein with the inhibitor 0XV bound in the active site) varied slightly in the heme-tilt angle: 52° in the structure from CG:S1 and 46° in the structure from CG:S2. The distance of the CoM of the globular domain to the bilayer center was also different in the two starting structures: 47 Å in CG:S1 and 50 Å in CG:S2. During all three AA simulations, the globular domain fluctuated around the starting position. The angles ranged between 95 ± 5° to 106 ± 4° for α and 136 ± 7° to 149 ± 5° for β (corresponding to class B), with an increase during the simulations of the heme-tilt angle to 55 ± 7° to 61 ± 5°. The axial distance of the CoM of the globular domain from the bilayer center stayed constant at about 46 ± 2 Å in the AA simulations, independently of whether a ligand was present (SIM2) or the TM-helix was truncated (SIM3).

In summary, the differences in the orientations and interactions of the two proteins (CYP 2C9 and CYP 2C19) observed in the CG simulations were maintained and, in some cases, became more pronounced during the AA simulations. Comparison of simulations of the apo- and holo- forms of the proteins indicate that the orientation of the proteins in the membrane was not significantly affected by the presence of a compound in the active site.

2.7. Key Residues Contribute to Differences in the Membrane-Protein Interactions of CYP 2C9 and CYP 2C19

From the CG and AA MD simulations, we found that despite high sequence conservation (92% sequence identity), the two isoforms of the CYP2C subfamily, CYP 2C9 and CYP 2C19, maintained differences in the interactions, orientations, and degree of insertion in the membrane of the globular domain. The most important differences in residues were found in the substrate recognition sites (SRSs). For example, the SRS1′a, SRS1′b, SRS1 and SRS2,3 regions defined by Zawaira et al. [24] covered residue differences in the linker region (G46D), between strands β1-1 and β1-2 (72–73 KP-ER), in the B–C loop (I99H) and between the F′ and G′ helices (220–221 SP-PT) (see Figure 1).

We therefore analyzed the trajectories to differentiate residues interacting with the membrane head or tail regions, calculated by defining a 5 Å distance cutoff between protein and lipid head group (phosphate atoms) and hydrophobic tail, separately. The % contact time or occupancy of these residues with the membrane components in the trajectories is shown in Figure 5A. In the simulations, the interactions of the globular domains of both isoforms with the membrane were mainly developed through strands β1-1 and β1-2 (residues 64–74), and the F′–G′ region (residues 210–226) (see Figure 5A,B). However, CYP 2C9 showed further interactions with the membrane through the A-helix (residues 50–60), the B–C loop (residues 95–110) and the C-terminal β-sheet 2 (residues 370–385). The peripheral interactions developed by CYP 2C9 were similar to the hydrophobic surface identified in CYP 2C5 (residues 30–45, 60–69 after the A-helix, 376–379 in β-strand 2-1 and the F′–G′ helices) [28]. These secondary interactions, with either lipid tail or head regions, were established by the SRS regions that showed primary sequence differences in the two CYP isoforms studied here. Therefore, differences in the above-mentioned SRS regions can be crucial for CYP-membrane interactions and the orientation of the globular domain, and may lead to distinct substrate specificity.

We extended our sequence comparison to include the four main human CYP2C family members, CYP 2C8, 2C9, 2C18, and 2C19, and rabbit CYP 2C5 to examine the residues differing at the interface region (see Figure 5C). Sequence comparison showed that only CYP 2C19 had a positively charged residue, R73, at this position, as all other CYP2C members had P73. Furthermore, only CYP 2C9 had a positively charged residue at position 72 (K72) when compared to the other CYPs; this residue could play a role as a selectivity filter for attracting the acidic substrates preferred by CYP 2C9 and repelling basic compounds. In the F′–G′ region, the S220 and P221 residues were only found in CYP 2C9, whereas the polar residue T221 was present only in CYP 2C19. Substitution of S220P and P221A in the CYP 2C9m7 structure (PDB 1OG2) shifted the position of P221 to that of P220, as in all other CYP2C members, which resulted in a turn between the F′–G′ helices and further stabilized the G′-helix. Experimentally, it has been observed that the substitution of CYP 2C9 residues I99H, S220P, and P221T enhanced omeprazole 5′-hydroxylation activity of CYP 2C9 [18]. In another experimental study, E72K substitution in CYP 2C19 decreased the metabolic activity, whereas the K72E substitution in CYP 2C9 increased the binding affinity of tricyclic antidepressant (TCA) drugs such as imipramine [21]. Together, the analysis of primary sequence, protein-membrane orientation and interactions, and experimental findings, supports the role of different residues in SRS regions in determining the distinct orientations of the two CYPs, leading to differences in substrate access and selectivity.

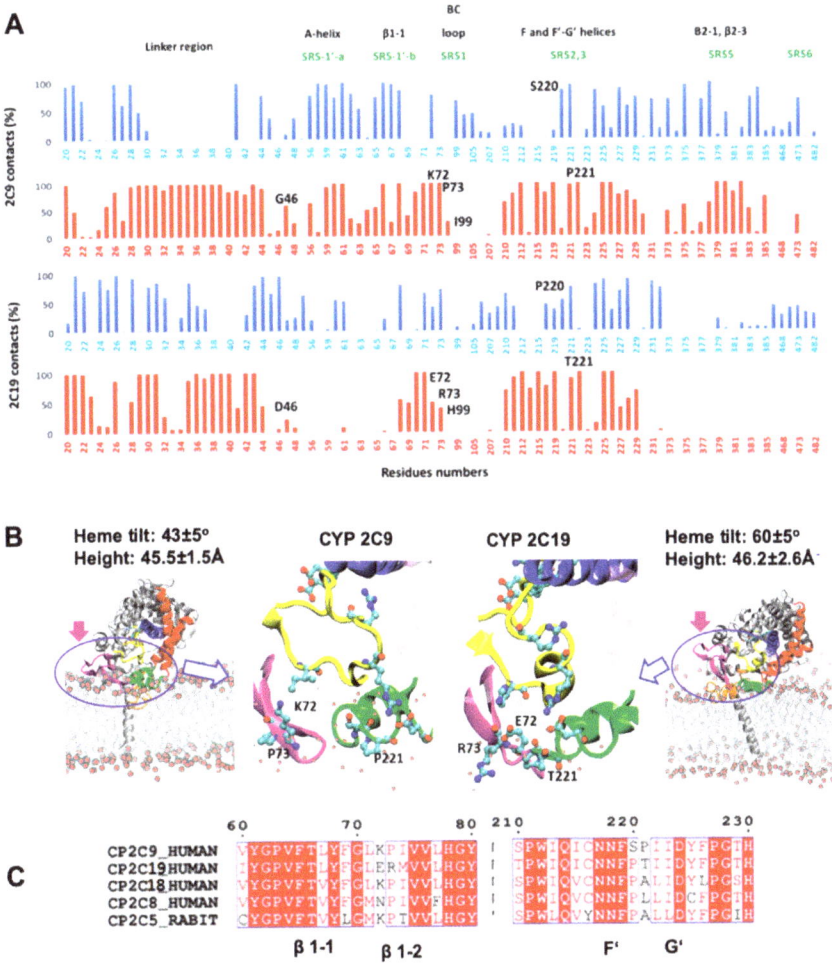

Figure 5. Differences in the arrangements of the CYP 2C9 and CYP 2C19 residues at the membrane interface. (**A**) Residues in contact with the lipid head group (blue) and tail region (red) during the AA MD simulations of CYP 2C9 (above) and CYP 2C19 (below). The percentage of snapshots in which a contact was present is shown on the y-axis, and the residues interacting with the membrane are given on the x-axis. The secondary structures and substrate recognition sites are shown on the top. The residues differing in the interacting regions between CYP 2C9 and CYP 2C19 are labeled. (**B**) For CYP 2C9 (left) and CYP 2C19 (right), the last frames from AA MD simulations of the apo form (SIM1) are shown for the full system and for the membrane interface region. The protein is shown in cartoon representation with selected side chains in ball-and-stick representation colored by atom type with cyan carbons. The linker is shown in orange, β-sheets 1 and 2 in magenta, the B–C loop in yellow, the F and G helices in red, the F' and G' helices in green, the central I-helix in blue, and the heme and key residues in cyan licorice representation. The POPC bilayer is shown in grey line representation with phosphate atoms as red spheres. The magenta arrows indicate differences in β-sheet 2 (residues 370–380). (**C**) Part of a sequence alignment of human CYP2C subfamily members (2C9, 2C19, 2C18, 2C8) and rabbit CYP 2C5. The conserved residues are shown in red for similar residues or with a red background and in white for identical residues. The residues outside the blue boxes differed amongst the aligned CYPs.

2.8. Structural Differences Result in Different Membrane-Protein Interactions

Structural analysis of CYP 2C9 and CYP 2C19 (PDB 1R90 and 4GQS, respectively) revealed different conformations of β-sheets 1 and 2 and the B–C loop (highlighted by rings in Figure 5B and which were observed in simulations of apo and ligand-bound proteins (Figure S7). During simulations, strands β1-1 and β1-2 remained inserted in the membrane or interacted with lipid head groups in CYP 2C9, whereas these strands made far fewer contacts in CYP 2C19. The conformation and orientation of strands β1-1 and β1-2 favored the interaction of β-sheet 2 (residues 370–385) with the membrane head groups in CYP 2C9, whereas this interaction was almost completely absent in CYP2 C19 (Figure 5A).

β-sheet 1 in CYP 2C9 differed in sequence (residues K72–P73) from CYP 2C19, (residues E72–R73), as well as in conformation (see Figure 5C). The turn between strands β1-1 and β1-2 in CYP 2C9 pointed away from the globular domain towards the membrane surface (see Figure 5B). K72 in CYP 2C9 pointed towards the binding pocket and, during the simulations, its positively charged ε-amino group transiently formed a hydrogen bond with S220 in the F′–G′ helices and with the phosphate head groups of lipid molecules (see Figure 6C). The interaction with the lipid head groups resulted in the further insertion of the β1-hairpin residues into the membrane (Figure 6). K72 has been suggested to play an important role in the selection of anionic substrates in CYP 2C9, and is positioned along pathway 2b (Figure 6, a description of ligand pathways is given by Cojocaru et al. [29]) for ligand entrance into the binding pocket from the membrane [21]; it is replaced by E72 in CYP 2C19. Besides K72, the presence of P73 in CYP 2C9 favored interactions with the hydrophobic interior of the membrane (also seen in the % occupancy contact plot, Figure 5A). Thus, the K72 and P73 residues of the β-hairpin (between the β1-1 and β1-2 strands) in CYP 2C9 could be determinants of the difference in the orientation of the globular domain in the membrane compared to CYP 2C19.

In the simulations of CYP 2C19, the β1-1 and β1-2 strands showed fewer interactions with the membrane surface, which could be attributed partly to its charged residues E72 and R73 and the conformational differences observed in the crystal structure. In various studies on membrane–protein interactions, it has been found that arginine has a propensity to stay in the lipid head group region [30–33]. In CYP 2C19, the R73 sidechain pointed towards the membrane and, thereby, appeared to restrict insertion of the β-strands in the membrane. Together, the differences in the interactions with the membrane resulted in greater tilting of the distal side of the globular domain towards the membrane, resulting in higher β angles and higher heme-tilt angles, for CYP 2C19 than for CYP 2C9 (see Figure 5B).

An important difference between CYP 2C9 and CYP 2C19 was seen in the B–C loop, which differed in only one residue, residue 99 (I in CYP 2C9, H99 in CYP 2C19). The B–C loop in CYP 2C9 was highly mobile compared to CYP 2C19, in which a B′–C′ helical conformation was present. The B–C loop also differed in the side chain conformations of R105 and R108 in the two CYPs. R105 in CYP 2C19 pointed outward and showed electrostatic interactions with D224 in the G′ helix. In the CYP 2C9 crystal structure (PDB 1R9O), R105 had a different conformation and no interactions were reported with D224 due to the missing G′-helix. However, after modeling of the F′–G′ helices and simulations, similar interactions between R105 and D224 were observed in most simulations.

From the CG simulations, we identified a role for the F′–G′ helices in stabilizing the interactions and orientation of the globular domain in the membrane. The AA simulations showed differences between the two CYPs in the F′–G′ helices (S220P and P221T). P221 in CYP 2C9 was located on the outer surface of the G′-helix, which is in direct contact with the membrane and thereby favored insertion of P221 into the lipid tail region, whereas T221 at the same position in CYP 2C19 made slightly less contact with the bilayer.

Together, these results imply that not only sequence differences but also conformational differences in the regions involved in membrane-protein interactions contribute to the differences in the orientations adopted by the two isoforms in the membrane.

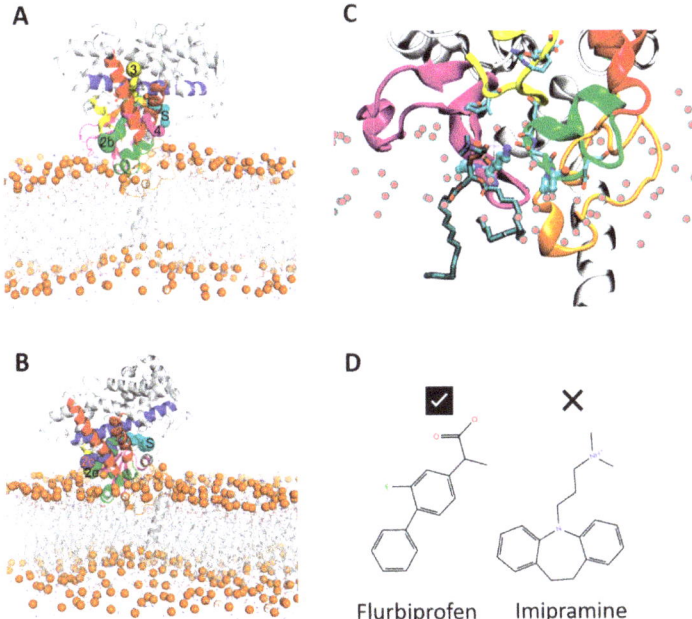

Figure 6. Initial (**A**) and final (**B**) snapshots of the AA MD simulation (SIM1) of the apo form of CYP 2C9, showing tunnels accessible to a water molecule probe between the active site and the protein surface. Tunnel 2b (green) connects the active site and the membrane and is present in both snapshots, as is tunnel S (cyan). Tunnel 3 (yellow) is present in the initial snapshot and tunnel 2c (blue) is present in the final snapshot. The protein and bilayer are shown with the same color scheme as in Figure 5. (**C**) Close-up view of the entrance to the 2b tunnel showing how the phosphate group of a phospholipid molecule (shown in stick representation colored by atom type with cyan carbons) makes a hydrogen bond with the amino group of K72 (all other lipid molecules are represented by spheres for the phorphorous atoms only; the protein is represented and colored as in Figure 5). This interaction is important for pulling the phospholipid molecule somewhat out of the membrane towards the tunnel to the active site. This motion leads to partial opening of the β-sheet and the F'–G' regions; further opening would be required for a substrate molecule to access the active site. (**D**) K72 may interact analogously with acidic substrates, such as flurbiprofen, a drug that is a substrate of CYP 2C9 (left), and may repel basic substrates such as the tricyclic antidepressant (TCA) drug, imipramine, which is a substrate of CYP 2C19 (right).

2.9. Comparison with Experiments and Previous Simulations

There are no experimental data characterizing full length CYPs and their interactions with the membrane in atomic detail. However, various experiments have been performed to study the membrane topology of CYPs and their interactions beyond the N-terminal transmembrane helix. Engineered CYP 2C9 without an N-terminal helix remained membrane-associated through the catalytic domain, as seen by atomic force microscopy (AFM) [34]. The height of the catalytic domain above the membrane was reported as 35 ± 9 Å using atomic force microscopy [35], consistent with our simulations. The binding orientation and height were also reasonably consistent with site-directed antibody-antipeptide studies [36] and the surface hydrophobicity pattern in the crystal structure of mammalian CYP 2C5 [28], the first mammalian CYP to have its 3D structure determined.

The insertion depth of the catalytic domain in the membrane was studied by tryptophan fluorescence scanning of CYP 2C2, which suggested that L36 at the start of the A'-helix, F69 at the end of the β1-1 strand, and L380 in the β2-2 strand were inserted in the lipid bilayer, whereas Y225 in the F′–G′ region is in or near the phospholipid head groups [37]. Primary sequence analysis of CYP 2C2, CYP 2C9, and CYP 2C19 showed that the CYP 2C2 residues identified by tryptophan fluorescence scanning are conserved in all three CYPs. Similar interactions were observed in our simulations of CYP 2C9 where residues L36 and L380 showed interactions with the lipid tail region (100% occupancy), while residue F69 interacted with the tail region for 40% of the simulation time (see Figure 5A). In CYP 2C19, residue L36 was buried in the membrane, while residue F69 showed interactions with the membrane tail region for 49% of the simulation time. Due to the difference in the orientation of the globular domain, strands β2-1 and β2-2 did not interact with the membrane in CYP 2C19, and therefore residue L380 remained outside the membrane. In both CYPs, the F′–G′ helices formed a strong anchoring point and residue Y225 remained buried in the membrane.

The rearrangement of the linker region in the two CYPs during the simulations to expose polar residues in the flexible region next to the TM-helix (see Figure 5A) is consistent with the observed cytoplasmic accessibility of rat CYP 2B2 in rough microsomes to an antibody raised against residues 24–38 [36]. The linker region consists of a patch of polar residues, including several positively charged residues (22–30), a hydrophobic proline-rich patch (residues 30–40), and a patch of polar residues (40–49). The linker orientation and interactions in the two CYPs, the distribution of amino acid residues in the lipid bilayer and their propensity to reside in the lipid head or tail region matched well with the hydrophobicity scale for amino acids determined by various experiments and MD simulation studies [38,39]. During simulations, the linker remained highly mobile and changed conformation to keep polar residues in the polar patches outside the membrane core, including polar residues and charged residues that were buried in the hydrophobic core of the bilayer in the initial structure.

The orientations of CYP 2C9 in the membrane in the current study using the LIPID14 forcefield matched well with previously our published work on CYP 2C9 [22] (see Table 3 for heme-tilt angles). MD simulations of CYP 2C9 in a 1,2-dioleoyl-sn-glycero-3-phosphocholine (DOPC) bilayer by Berka et al. [40,41] resulted in orientations with a higher heme-tilt angle and greater burial of the globular domain in the bilayer, with more ligand pathways leading from the heme into the bilayer. The authors used a different initial crystal structure of CYP 2C9 (PDB 1OG2), and a different procedure to model and generate starting orientations. The Berger united atom forcefield for lipids was used, which could also contribute to different membrane-protein interactions and orientations. A bilayer of DOPC is slightly thinner than a bilayer of POPC (by 0.3 Å) and has a larger area per lipid (by 4 Å^2) [42]. These small differences might lead to slightly more tilting of the TM-helix in DOPC than POPC, but we would expect the dipping of the CYP globular domain into the bilayer to be similar for DOPC and POPC, as they have the same head group. Interestingly, however, Berka et al. observed that the globular domain of CYP 2C9 was immersed in a depression in the bilayer, surrounded by phospholipid head groups [41].

Table 3. Comparison of the computed heme-tilt angle from AA MD simulations of CYP 2C9 and CYP 2C19 with previously reported values. The heme-tilt angle is the angle between the heme plane and the membrane normal (see Materials and Methods and Figure 2).

Source	Reference	Lipid/Force Field	Protein PDB ID	Heme-Tilt Angle (°)
		CYP 2C9		
MD Simulation	Current Study	POPC/LIPID14	1R9O	40–43 ± 5
	[22]	POPC/GAFF lipid	1R9O (1)	44 ± 4
			1R9O (2)	41 ± 4
	[40]	DOPC/Berger	1OG2	55 ± 5
	[41]	DOPC/Berger	1OG2	61 ± 4
OPM Database	[43]	DOPC/OPM	1R9O	59.8
			1OG5	71.9
		CYP 2C19		
MD Simulation	Current Study	POPC/LIPID14	4GQS	55–61 ± 5
OPM Database	[43]	DOPC/OPM	4GQS	74.0

The predicted orientation of CYPs in a DOPC bilayer has been reported in the OPM (orientation of proteins in membranes) database (https://opm.phar.umich.edu/). The heme-tilt angle of the orientation of CYP 2C19 (PDB: 4GQS) reported in the OPM database is 74°, notably higher than that observed in our MD simulations (see Table 3). A similar orientation with a heme-tilt angle of 72° is given in the OPM database for the CYP 2C9 crystal structure with PDB ID 1OG5, whereas the 1R9O structure has an orientation with a heme-tilt angle of 60°. These discrepancies may have arisen because the OPM uses the crystal structure as is to predict the protein orientation in the membrane, and therefore the orientation may be affected by the missing (PDB 1R9O) or mutated (PDB 1OG5) residues in the F'–G' loop region, or the lack of flexible linker and TM-helix residues.

3. Concluding Discussion

The two isoforms of the CYP2C subfamily CYP 2C9 and CYP 2C19 exhibit ~92% sequence identity, yet they show distinct substrate specificity. Since mammalian CYPs are anchored in the endoplasmic reticulum membrane by an N-terminal helix and secondary contacts from the catalytic domain, differences in the sequence and 3D structure in the membrane-interacting region in the catalytic domain can lead to different membrane-protein interactions. As it has been hypothesized that lipophilic substrates enter into the binding pockets of CYPs from the membrane core, determining the orientation of CYPs in the membrane can provide insights into differences in the opening of ligand entrance tunnels to the membrane, substrate specificity, and the mechanism of drug selectivity. Here, we have used a multiscale simulation methodology to understand the differences in primary sequence and 3D structure of two CYPs and their impact on their interactions and orientations in the membrane.

The results of multiple CG and AA simulations showed consistency and a clear tendency for the two CYPs to adopt different orientations and positions with respect to the membrane bilayer. The orientations adopted by the globular domains of the two CYPs were classified into classes A, B, or A/B (intermediate orientation). CYP 2C9 mainly adopted a class A orientation, which has lower α, β and heme-tilt angles, and CYP 2C19 adopted a class B orientation. The class B orientation is similar to the orientations observed in simulations with the same bilayer and force field for CYP 3A4 and N-terminal mutants of CYP 17A1 and CYP 19A1, with measured heme-tilt angles of about 60° [27]. CYP 2C9 thus appears to be unusual in this set of CYPs (simulated under the same conditions) in adopting an orientation with a lower heme-tilt angle. Berka et al. simulated the six major drug-metabolizing CYPs (which vary much more in sequence than the CYP 2C9/CYP 2C19 pair studied here) in a DOPC bilayer [41]. The computed heme-tilt angle varied over the six proteins between 56 ± 5° and 72 ± 6°, with CYP 2C9 having a heme-tilt angle of 61°: at the lower end of this range, although higher than found in our simulations. The tendency to a low heme-tilt angle for CYP 2C9 was also observed by Cojocaru et al. [22], with a slightly lower heme-tilt angle when an F–G loop was present instead of the F'–G' helices. It is reasonable to expect that ligand passage may involve unwinding of the F'–G' helices into a loop extending further into the membrane that can open up the entrance to the ligand access route from the membrane. However, we do not expect this change in conformation to result in an increase in the heme-tilt angle.

The difference in the sequence and in conformations in the SRSs near the membrane interface resulted in different orientations and insertion depths in the membrane of the globular domains of the two CYPs. A mutational swap of the key residues differing at the membrane interface in CYP 2C9 (or 2C19) in the CG simulations resulted in similar orientations (about 50% of simulation results) to the wild-type CYP 2C19 (or 2C9) orientation. Therefore, we concluded that altering a few key residues that differ in the linker, the β1-1 strand, and the F'–G' region through mutation or a change in conformation can significantly influence the orientation and interactions of CYP-membrane systems. McDougle et al. showed that a double mutant in the F–G loop of CYP 2J2 lowered membrane insertion in MD simulations and tryptophan fluorescence studies [44]. Notably, despite the importance of the F'–G' region, we found that mutating residues in the F–G loop alone was not sufficient to switch the orientation of CYP 2C9 to that of CYP 2C19 or vice versa. The mutation of additional residues in the β1-1 and 1-2 (K72E and P73R) and B–C loop (I99H) was necessary.

The orientation of the CYP globular domain in a membrane may be affected by simulation parameters such as the force field used and the procedure used to model and sample the conformational space of the simulated systems. We have shown previously that the force field and procedure employed here can give good agreement with experiments [25,27]. However, other factors relevant in vivo may affect the orientation of the CYP globular domain, such as homodimerization [37], heterodimerization, or the binding of redox partner proteins. Indeed, the orientational preferences of the different CYPs may affect their ability to present the proximal binding face for effective electron transfer from cytochrome P450 reductase, or to engage in CYP oligomers. The orientation of the protein also affects the access of substrates from the membrane to the active site. We have here observed the opening of tunnels to a water probe. Further exploration would require simulation of the passage of substrate molecules by standard or enhanced sampling approaches [26,29,45]. Furthermore, allosteric ligands may affect protein orientation and substrate tunnel opening, e.g., in CYP 3A4, the allosteric ligand binds at the protein-membrane interface [46].

The membrane composition may also have an influence on protein positioning in the membrane. We have here simulated the proteins in a pure POPC bilayer. Phosphatidylcholine is the main lipid component of the mammalian ER membrane [47], and POPC bilayers are often used as a simple ER mimic in in vitro studies. For example, we previously compared the heme-tilt angle computed from simulations of three different CYPs in a POPC bilayer with that measured in experiments done on these CYPs in a Nanodisc containing a POPC bilayer and found very good agreement [27]. However, the ER membrane in fact contains a variety of glycerophospholipids, as well as cholesterol and ceramide. This more heterogeneous membrane composition may affect protein positioning and dynamics as well as ligand entrance to the active site. Indeed, Navratilova et al. [48] found that the orientation of CYP 3A4 changed as the cholesterol content of a DOPC bilayer was changed, with the heme-tilt angle increasing with increasing cholesterol. The addition of cholesterol also altered the substrate access tunnel opening patterns due to interactions of the protein with the cholesterol and the ordering and thickening of the membrane due to cholesterol. Molecular simulation studies with more realistic membrane compositions, such as that employed by Park et al. in their simulation of CYP 19A1 to mimic the rat liver endoplasmic reticulum membrane [49], will be necessary to fully understand the structural and dynamics interplay between the CYP proteins, substrates, and the membrane.

In conclusion, our MD simulations demonstrate that small sequence changes at key positions can result in distinct orientations of CYP proteins in a phospholipid bilayer. These differences affect substrate access tunnels to the active site from the membrane. The differences observed for CYP 2C9 and CYP 2C19 are consistent with their differences in substrate selectivity, providing further evidence that substrate selectivity is governed by the residues lining the substrate access route as well as those in the active site.

4. Materials and Methods

Preparation of structures of full-length models of CYP 2C9 and CYP 2C19—Our simulations of CYP 2C9 (Uniprot id P11712) were based on the crystal structure of CYP 2C9 (PDB 1R9O, in the Protein Data Bank http://www.rcsb.org), resolved at 2.0 Å resolution in complex with flurbiprofen. This structure of the globular domain of CYP 2C9 was chosen because it was the highest resolution structure available that was determined with the wild-type sequence (apart from removal of the N-terminal residues 1–25 and addition of terminal expression tags). The structure had missing residues in the linker region (residues 38–42) and in the F′–G′ region (residues 214–220). The crystal structure of CYP 2C19 (Uniprot id P33261) (PDB 4GQS: chain A) in complex with the inhibitor (2-methyl-1-benzofuran-3-yl) (4-hydroxy-3,5-dimethylphenyl) methanone (Protein Data Bank chemical component 0XV) was used. It was resolved at 2.87 Å, after truncating residues 1–28 from the N-terminus and adding expression tags [9]. The missing residues in the linker and globular domain of CYP 2C9 were similar to those in

CYP 2C19 as shown in Figure 1, where the sequence alignment and secondary structure were generated using the online tool ESPript3.0 (http://espript.ibcp.fr/ESPript/ESPript) [50]. Therefore, the crystal structure of CYP 2C19 was used as a template for modeling the missing linker residues and F′–G′ residues in CYP 2C9. The TM-helix (residues 3–21) and missing linker (residues 22–25) of CYP 2C9 were modeled similarly to Cojocaru et al. [22], who modeled and simulated CYP 2C9 in a POPC membrane starting from the crystal structure (PDB 1R9O). The TM-helix of CYP 2C19 was predicted to span from residues 4–20 or 3–22 by the online server Psipred (http://bioinf.cs.ucl.ac.uk/psipred/), which uses the membrane protein structure and topology (MEMSAT3) software and transmembrane protein topology prediction using support vector machines (SVM-MEMSAT) software [51]. The PredictProtein server (https://www.predictprotein.org/) [52] suggested an N-terminal alpha-helical conformation spanning residues 2–23, and this assignment was used for the simulations of CYP 2C19, along with additional simulations with assignment of the TM-helix to residues 3–21 for consistency with the simulations of CYP 2C9. The final models of each protein consisted of the crystal structure of the globular domain with modeled missing regions (we have referred to these below as systems with the letter S). For each of the proteins, 10 different starting orientations of the globular domains above the membrane were generated by changing the dihedral angles in the linker regions to generate a diverse set of initial structures with the CYP globular domain positioned to ensure that it was outside the membrane bilayer when the protein is immersed in a bilayer (see below). These structures were used for the construction of CG models.

Preparation of additional models of CYP 2C9—Additional CG systems of CYP 2C9 were prepared by using four modeled structures. Since the crystal structure (PDB 1R9O) of CYP 2C9 lacks the F′–G′ helices (or F–G loop), different modeling approaches with different template structures were used to assemble four structures of full-length CYP 2C9 (see Appendix A). The CG systems (M1–M4) prepared with these modeled structures of CYP 2C9 have been designated by the letter M for "models", which differ from the CG systems indicated by S, for which crystal structures were used (with modeled missing regions only) (Table S1).

Modeling of chimeric mutants of CYP 2C9 and CYP 2C19—The residues at the protein-membrane interface differing between CYP 2C9 and CYP 2C19 were substituted to create chimeric CYP 2C9/2C19 structures. The residues of CYP 2C9 substituted by CYP 2C19 residues were in the linker (G46D), β-strand 1–2 (K72E and P73R), B–C loop (I99H), and F′–G′ helices (S220P and P221T). The corresponding substitutions were also made in CYP 2C19. Five different orientations of the wild-type all-atom models (S1) were selected to make the substitution mutations, while keeping the initial orientations of the globular domain of the mutant and wild-type structures the same. These modeled mutants are referred to as mt2C9 and mt2C19.

Preparation of coarse-grained systems—The MARTINI CG forcefield was used for CG simulations. A similar procedure was used to generate CG models of CYP 2C9 and CYP 2C19 in a POPC bilayer in water, as described in our previous work [25]. All-atom protein models were converted to MARTINI CG models using the martinize.py script (http://cgmartini.nl) and the TM-helix was immersed in a pre-equilibrated modeled CG POPC lipid bilayer consisting of 594 POPC molecules. The MARTINI version 2.2 forcefield with the standard non-polarizable water model (NPW) was used [53]. The elastic network model was used to apply additional harmonic restraints with an elastic force constant of 500 kJ·mol^{-1}·nm^{-2} and a distance cut-off of 5 to 9 Å to preserve the secondary and tertiary structure of the protein during simulation. The secondary structure information was provided in a DSSP file obtained from the DSSP server (www.cmbi.ru.nl/dssp.html).

The effects of differing linker flexibility on the final orientations of CYP 2C9 and CYP 2C19 were checked by defining two different flexible linker regions: residues 22–36 and residues 26–38. The linker was kept flexible by removing the restraints on specified residues in the elastic network. CG systems consisting of the globular domain only (S3), residues 47–490, were prepared for CYP 2C9 and CYP 2C19 to allow an unbiased conformational search of the protein orientation and to evaluate convergence of the orientations in the membrane. The CG systems were solvated using the MARTINI standard water model (NPW) (S1–S3). Tests were also performed for the MARTINI polarizable water (PW) model (S4–S5) (see SI, Table 1).

Coarse-grained simulations—After preparation of the CG models, several different CG simulations were performed with the Gromacs software [54]. The MARTINI standard water model (NPW) was used and the non-bonded interactions were treated with a reaction field (RF) for Coulomb interactions, and the cut-off distance for these and for van der Waals' interactions was set to 1.1 nm. We also tested the polarizable water (PW) model with electrostatic and van der Waals interactions calculated by the Shift method (Gromacs 4.5.5), PME and a cut-off, or RF and a cut-off (using Gromacs 5.0.4), as in Mustafa et al. [25]. Similar positioning of the globular domain with respect to the membrane was obtained as that in simulations with the NPW model, but simulation times to achieve convergence were much longer with the PW model, and convergence was not always achieved.

Each simulation started with a short steepest-descent energy minimization until the maximum force on a CG particle was less than 10 kJ·mol^{-1}·nm^{-1}. A 40 ns equilibration simulation at a constant temperature of 310 K and pressure of 1 atm was performed in the NPT ensemble, using velocity rescale (v-rescale) and the Berendsen procedure for pressure coupling before switching to the Parrinello–Rahman barostat method for production simulations of 12–20 µs. A coupling constant of 12 ps was used to maintain semi-isotropic pressure coupling with a compressibility of 3.0×10^{-5}. A time step of 20 fs was applied.

Convergence of coarse-grained simulations—The CG simulations were considered converged when no further significant changes in the orientations of the CYP globular domains above the membrane were observed. The orientation and position of the CYP globular domain was specified by the angles and distances defined previously [22,25,55] (see Figure 2). The angles were computed by defining the following vectors: v1, from the center of mass (CoM) of the backbone particles/atoms of the first four residues to the CoM of the last four residues of the I-helix; v2, from the CoM of the first four residues of the C-helix to the CoM of the last four residues of the F-helix; v3, the vector between the CoMs of the first and last four residues of the TM-helix; and the z-axis perpendicular to the membrane. The angle α was then defined as the angle between v1 and the z-axis and angle β was defined as the angle between v2 and the z-axis. Angles α and β define the orientation of the globular domain above the lipid membrane. Similarly, the TM-helix tilt angle (γ) in the lipid membrane was defined as the angle between v3 and the z-axis. The axial distances of the CoM of the globular domain (residues 50–490), the linker region (residues 22–49), and the F'–G' helices (residues 210–220) to the CoM of the lipid bilayer were monitored during the trajectories.

Back conversion from CG to AA models—For each system, representative frames from the converged parts of each set of CG production runs were selected for back-conversion to an all-atom model. The representative frame was chosen to have angle and distance values within 1% of their mean value over the converged parts of the production runs [55]. The back-conversion of the POPC bilayer was performed as described in Cojocaru et al. [22], whereas the protein back-conversion was done using scripts backward.py and initram.sh, available at the MARTINI website (http://cgmartini.nl) [56]. In the absence of the heme cofactor in the CG model, conformational changes in the side chains of the heme-binding pocket residues were observed. Therefore, the globular domain (residues 50–490) from the crystal structure was superimposed on the back-mapped structure and used in subsequent AA simulations. The AA model of the globular domain contained the heme-cofactor. If there was a co-crystallized ligand in the crystal structure, it was also reincorporated in the model. The TM-helix and flexible linker region obtained from the back-conversion procedure were then connected to the globular domain, resulting in a full-length all-atom model. Finally, the all-atom model of the CYP was placed into the all-atom model of the POPC bilayer to obtain a complete CYP-membrane complex.

All-atom molecular dynamics simulations of CYP 2C9 and CYP 2C19—AA MD simulations were performed with two different starting orientations of the CYPs in the membrane for each CYP. Different orientations were obtained for each CYP from two different CG simulation systems, S1 and S2. AA forcefields AMBER ff14SB [57] and LIPID14 [58] were used for the protein residues and for the

POPC lipids, respectively. The heme parameters were provided by D. Harris with the partial atomic charges derived from DFT calculations [59]. The ionic concentration was maintained at 150 mM using Na$^+$ and Cl$^-$ ions in a periodic box of TIP3P [60] water molecules. The same procedure for AA MD simulation was used as described by Cojocaru et al. [22]. The simulation protocol began with energy minimization with a decreasing harmonic force constant of 1000 to 0 kcal/mol.Å^2 on non-hydrogen atoms of the protein and lipid residues, as described in Reference [22]. The system was then equilibrated using NAMD 2.10 [61] in a constant surface area, pressure, and temperature (NPAT) ensemble, for 1.5 ns with a gradual decrease in the harmonic restraints from 100 to 0 kcal/mol.Å^2 on non-hydrogen atoms of the protein and lipid residues. The equilibration simulations in the NPAT ensemble were extended to 10 ns without harmonic restraints with a 1 fs integration time, keeping water molecules rigid. During subsequent production simulations, all bonds were kept rigid and the time step was increased to 2 fs. Anisotropic pressure coupling was applied, in which the cell fluctuates independently in the x, y, and z cell dimensions.

Control calculations were also performed with the GAFF lipid force field which was used in previous work [22]. The GAFF lipid forcefield requires surface tension to maintain the structural properties of the membrane bilayer, whereas the LIPID14 parameters are optimized for use without application of surface tension. We also assessed semi-isotropic pressure coupling for the simulations with the LIPID14 force field. The results show that the alternative simulation parameters gave the same class of orientation of the globular domain in the bilayer as obtained with LIPID14 and anisotropic pressure coupling. We have previously found that the combination of AMBERff14SB and LIPID14 gives better agreement with experiment than the GAFF force field and that it results in heme-tilt angles for CYPs in bilayers in excellent agreement with linear dichroism data for CYPs in Nanodiscs [27].

The orientation and position of the CYP globular domain in the AA MD simulations was characterized by computing the same angles and distances as for the CG MD simulations. In addition, the heme-tilt angle, the angle between the heme plane defined by the four nitrogen atoms coordinating the iron and the z-axis, was computed (see Figure 2). VMD (www.ks.uiuc.edu/Research/vmd/) [62] was used for the analysis and to generate the molecular graphics figures. Tunnels accessible to a water molecule probe were computed using the MOLEonline webserver (mole.upol.cz/) [63] with default parameters.

Supplementary Materials: Supplementary materials can be found at http://www.mdpi.com/1422-0067/20/18/4328/s1. Supplementary information is provided as a separate document containing Tables S1–S3 and Figures S1–S7, as well as coordinates from AA MD simulations.

Author Contributions: Conceptualization, G.M. and R.C.W.; methodology, G.M., P.P.N., N.J.B., R.C.W.; validation, G.M.; formal analysis, G.M., and P.P.N; investigation, G.M.; resources, R.C.W.; data curation, G.M.; writing—original draft preparation, G.M.; writing—review and editing, G.M. and R.C.W.; visualization, G.M. and P.P.N.; supervision, R.C.W.; project administration, R.C.W.; funding acquisition, G.M., P.P.N., R.C.W.

Funding: This research was funded by Klaus Tschira Foundation and the German Academic Exchange Service (DAAD) (scholarships to G.M., P.P.N.). The authors acknowledge support for computing resources from the state of Baden-Württemberg through bwHPC and the German Research Foundation (DFG) through grant INST 35/1134-1 FUGG and for use of the Hazel Hen Cray XC40 at the high-performance computing center Stuttgart, Germany (HLRS; Project Dynathor). The APC was funded by the Deutsche Forschungsgemeinschaft within the funding programme Open Access Publishing, by the Baden-Württemberg Ministry of Science, Research and the Arts and by Ruprecht-Karls-Universität Heidelberg.

Acknowledgments: G.M. gratefully acknowledges the support of the PhD program of the Institute of Pharmacy and Molecular Biotechnology, Heidelberg University. We thank Stefan Richter for assistance in optimizing software performance.

Conflicts of Interest: The authors declare no conflict of interest. The funders had no role in the design of the study; in the collection, analyses, or interpretation of data; in the writing of the manuscript, or in the decision to publish the results.

Abbreviations

AA	all-atom
CG	coarse-grained
CoM	center of mass
CYP	cytochrome P450
DOPC	1,2-dioleoyl-sn-glycero-3-phosphocholine
MD	molecular dynamics
POPC	1-palmitoyl-2-oleoyl-sn-glycero-3-phosphocholine
PDB	Protein Data Bank
RMSD	root mean squared deviation
SRS	substrate recognition site
TCA	tricyclic antidepressant
TM	Transmembrane

Appendix A

Preparation of additional models of CYP 2C9

Four additional models of CYP 2C9 were built to investigate sensitivity of the CG simulations to the initial structures used.

CYP 2C9 model 1: The CYP 2C9 model was generated using the CYP 2C9 crystal structure (PDB 1R9O) as a template. For modeling the missing residues in the linker region (residues 38–42) and in the F′–G′ helices region (residues 214–220), CYP 2C19 was used as a template only for these missing regions. The corresponding CG system is referred to as M1.

CYP 2C9 model 2: Another model of CYP 2C9 was generated using the complete 1R9O and 4GQS crystal structures as templates, which generated intermediate side chain conformations between the CYP 2C9 and CYP 2C19 structures. The corresponding CG system is referred to as M2.

CYP 2C9 model 3: Modeling of CYP 2C9 was also performed by using only a single template of the CYP 2C19 crystal structure (PDB 4GQS). This model of CYP 2C9 resembles CYP 2C19 in side chain conformations and the corresponding CG system is called M3.

CYP 2C9 model 4: Modeled structure of CYP 2C9 (residues 26–490) with the F′–G′ helices taken from previous studies by Cojocaru et al. [22]. The details of the modeling procedure are given in that paper. The missing TM-helix was modeled as for the other models. The corresponding CG system is referred to as M4.

References

1. Zanger, U.M.; Schwab, M. Cytochrome P450 enzymes in drug metabolism: Regulation of gene expression, enzyme activities, and impact of genetic variation. *Pharm. Ther.* **2013**, *138*, 103–141. [CrossRef] [PubMed]
2. Lewis, D.F.V.; Watson, E.; Lake, B.G. Evolution of the cytochrome P450 superfamily: Sequence alignments and pharmacogenetics. *Mutat. Res.—Rev. Mutat. Res.* **1998**, *410*, 245–270. [CrossRef]
3. Goldstein, J.A.; de Morais, S.M.F. Biochemistry and molecular biology of the human CYP2C subfamily. *Pharmacogenetics* **1994**, *4*, 285–300. [CrossRef] [PubMed]
4. Furuta, T.; Ohashi, K.; Kamata, T.; Takashima, M.; Kosuge, K.; Kawasaki, T.; Hanai, H.; Kubota, T.; Ishizaki, T.; Kaneko, E. Effect of genetic differences in omeprazole metabolism on cure rates for Helicobacter pylori infection and peptic ulcer. *Ann. Intern. Med.* **1998**, *129*, 1027–1030. [CrossRef] [PubMed]
5. Dansette, P.M.; Rosi, J.; Bertho, G.; Mansuy, D. Cytochromes P450 catalyze both steps of the major pathway of clopidogrel bioactivation, whereas paraoxonase catalyzes the formation of a minor thiol metabolite isomer. *Chem. Res. Toxicol.* **2012**, *25*, 348–356. [CrossRef] [PubMed]
6. Gong, I.Y.; Crown, N.; Suen, C.M.; Schwarz, U.I.; Dresser, G.K.; Knauer, M.J.; Sugiyama, D.; Degorter, M.K.; Woolsey, S.; Tirona, R.G.; et al. Clarifying the importance of CYP2C19 and PON1 in the mechanism of clopidogrel bioactivation and in vivo antiplatelet response. *Eur. Heart J.* **2012**, *33*, 2856–2864. [CrossRef] [PubMed]
7. Danielak, D.; Karaźniewicz-Łada, M.; Komosa, A.; Burchardt, P.; Lesiak, M.; Kruszyna, Ł.; Graczyk-Szuster, A.; Główka, F. Influence of genetic co-factors on the population pharmacokinetic model for clopidogrel and its active thiol metabolite. *Eur. J. Clin. Pharm.* **2017**, *73*, 1623–1632. [CrossRef] [PubMed]
8. Arinç, E. The role of polymorphic cytochrome P450 enzymes in drug design, development and drug interactions with a special emphasis on phenotyping. *J. Mol. Catal. B Enzym.* **2010**, *64*, 120–122. [CrossRef]

9. Reynald, R.L.; Sansen, S.; Stout, C.D.; Johnson, E.F. Structural characterization of human cytochrome P450 2C19: Active site differences between P450s 2C8, 2C9, and 2C19. *J. Biol. Chem.* **2012**, *287*, 44581–44591. [CrossRef]
10. Williams, P.A.; Cosme, J.; Ward, A.; Angove, H.C.; Matak Vinković, D.; Jhoti, H. Crystal structure of human cytochrome P450 2C9 with bound warfarin. *Nature* **2003**, *424*, 464–468. [CrossRef]
11. Wester, M.R.; Yano, J.K.; Schoch, G.A.; Yang, C.; Griffin, K.J.; Stout, C.D.; Johnson, E.F. The Structure of Human Cytochrome P450 2C9 Complexed with Flurbiprofen at 2.0-Å Resolution. *J. Biol. Chem.* **2004**, *279*, 35630–35637. [CrossRef] [PubMed]
12. Schoch, G.A.; Yano, J.K.; Sansen, S.; Dansette, P.M.; Stout, C.D.; Johnson, E.F. Determinants of cytochrome P450 2C8 substrate binding: Structures of complexes with montelukast, troglitazone, felodipine, and 9-cis-retinoic acid. *J. Biol. Chem.* **2008**, *283*, 17227–17237. [CrossRef] [PubMed]
13. Brändén, G.; Sjögren, T.; Schnecke, V.; Xue, Y. Structure-based ligand design to overcome CYP inhibition in drug discovery projects. *Drug Discov. Today* **2014**, *19*, 905–911. [CrossRef] [PubMed]
14. Skerratt, S.E.; de Groot, M.J.; Phillips, C. Discovery of a novel binding pocket for CYP 2C9 inhibitors: Crystallography, pharmacophore modelling and inhibitor SAR. *MedChemComm* **2016**, *7*, 813–819. [CrossRef]
15. Maekawa, K.; Adachi, M.; Matsuzawa, Y.; Zhang, Q.; Kuroki, R.; Saito, Y.; Shah, M.B. Structural Basis of Single-Nucleotide Polymorphisms in Cytochrome P450 2C9. *Biochemistry* **2017**, *56*, 5476–5480. [CrossRef] [PubMed]
16. Swain, N.A.; Batchelor, D.; Beaudoin, S.; Bechle, B.M.; Bradley, P.A.; Brown, A.D.; Brown, B.; Butcher, K.J.; Butt, R.P.; Chapman, M.L.; et al. Discovery of Clinical Candidate 4-[2-(5-Amino-1 H-pyrazol-4-yl)-4-chlorophenoxy]-5-chloro-2-fluoro-N-1,3-thiazol-4-ylbenzenesulfonamide (PF-05089771): Design and Optimization of Diaryl Ether Aryl Sulfonamides as Selective Inhibitors of Na V 1.7. *J. Med. Chem.* **2017**, *60*, 7029–7042. [CrossRef] [PubMed]
17. Liu, R.; Lyu, X.; Batt, S.M.; Hsu, M.-H.; Harbut, M.B.; Vilchèze, C.; Cheng, B.; Ajayi, K.; Yang, B.; Yang, Y.; et al. Determinants of the Inhibition of DprE1 and CYP2C9 by Antitubercular Thiophenes. *Angew. Chem. Int. Ed.* **2017**, *56*, 13011–13015. [CrossRef]
18. Ibeanu, G.C.; Ghanayem, B.I.; Linko, P.; Li, L.; Pedersen, L.G.; Goldstein, J.A. Identification of residues 99, 220, and 221 of human cytochrome P450 2C19 as key determinants of omeprazole hydroxylase activity. *J. Biol. Chem.* **1996**, *271*, 12496–12501. [CrossRef] [PubMed]
19. Jung, F.; Griffin, K.J.; Song, W.; Richardson, T.H.; Yang, M.; Johnson, E.F. Identification of Amino Acid Substitutions that Confer a High Affinity for Sulfaphenazole Binding and a High Catalytic Efficiency for Warfarin Metabolism To P450 2C19. *Biochemistry* **1998**, *37*, 16270–16279. [CrossRef]
20. Tsao, C.C.; Wester, M.R.; Ghanayem, B.; Coulter, S.J.; Chanas, B.; Johnson, E.F.; Goldstein, J.A. Identification of human CYP2C19 residues that confer S-mephenytoin 4′-hydroxylation activity to CYP2C9. *Biochemistry* **2001**, *40*, 1937–1944. [CrossRef]
21. Attia, T.Z.; Yamashita, T.; Hammad, M.A.; Hayasaki, A.; Sato, T.; Miyamoto, M.; Yasuhara, Y.; Nakamura, T.; Kagawa, Y.; Tsujino, H.; et al. Effect of Cytochrome P450 2C19 and 2C9 Amino Acid Residues 72 and 241 on Metabolism of Tricyclic Antidepressant Drugs. *Chem. Pharm. Bull.* **2014**, *62*, 176–181. [CrossRef] [PubMed]
22. Cojocaru, V.; Balali-Mood, K.; Sansom, M.S.P.; Wade, R.C. Structure and Dynamics of the Membrane-Bound Cytochrome P450 2C9. *PLoS Comput. Biol.* **2011**, *7*, e1002152. [CrossRef] [PubMed]
23. Cojocaru, V.; Winn, P.J.; Wade, R.C. The ins and outs of cytochrome P450s. *Biochim. Biophys. Acta Gen. Subj.* **2007**, *1770*, 390–401. [CrossRef] [PubMed]
24. Zawaira, A.; Ching, L.Y.; Coulson, L.; Blackburn, J.; Wei, Y.C. An expanded, unified substrate recognition site map for mammalian cytochrome P450s: Analysis of molecular interactions between 15 mammalian CYP450 isoforms and 868 substrates. *Curr. Drug Metab.* **2011**, *12*, 684–700. [CrossRef] [PubMed]
25. Mustafa, G.; Nandekar, P.P.; Yu, X.; Wade, R.C. On the application of the MARTINI coarse-grained model to immersion of a protein in a phospholipid bilayer. *J. Chem. Phys.* **2015**, *143*, 243139. [CrossRef] [PubMed]
26. Yu, X.; Nandekar, P.; Mustafa, G.; Cojocaru, V.; Lepesheva, G.I.; Wade, R.C. Ligand tunnels in T. brucei and human CYP51: Insights for parasite-specific drug design. *Biochim. Biophys. Acta Gen. Subj.* **2016**, *1860*, 67–78. [CrossRef]
27. Mustafa, G.; Nandekar, P.P.; Camp, T.J.; Bruce, N.J.; Gregory, M.C.; Sligar, S.G.; Wade, R.C. Influence of Transmembrane Helix Mutations on Cytochrome P450-Membrane Interactions and Function. *Biophys. J.* **2019**, *116*, 419–432. [CrossRef] [PubMed]

28. Williams, P.A.; Cosme, J.; Sridhar, V.; Johnson, E.F.; McRee, D.E. Mammalian microsomal cytochrome P450 monooxygenase: Structural adaptations for membrane binding and functional diversity. *Mol. Cell* **2000**, *5*, 121–131. [CrossRef]
29. Cojocaru, V.; Winn, P.; Wade, R. Multiple, ligand-dependent routes from the active site of cytochrome P450 2C9. *Curr. Drug Metab.* **2012**, *13*, 143–154. [CrossRef]
30. Lee, A.G. How lipids affect the activities of integral membrane proteins. *Biochim. Biophys. Acta Biomembr.* **2004**, *1666*, 62–87. [CrossRef]
31. Spijker, P.; van Hoof, B.; Debertrand, M.; Markvoort, A.J.; Vaidehi, N.; Hilbers, P.A.J. Coarse Grained Molecular Dynamics Simulations of Transmembrane Protein-Lipid Systems. *Int. J. Mol. Sci.* **2010**, *11*, 2393–2420. [CrossRef] [PubMed]
32. Sansom, M.S.P.; Scott, K.A.; Bond, P.J. Coarse-grained simulation: A high-throughput computational approach to membrane proteins. *Biochem. Soc. Trans.* **2008**, *36*, 27–32. [CrossRef] [PubMed]
33. de Jesus, A.J.; Allen, T.W. The determinants of hydrophobic mismatch response for transmembrane helices. *Biochim. Biophys. Acta Biomembr.* **2013**, *1828*, 851–863. [CrossRef] [PubMed]
34. Nussio, M.R.; Voelcker, N.H.; Miners, J.O.; Lewis, B.C.; Sykes, M.J.; Shapter, J.G. AFM study of the interaction of cytochrome P450 2C9 with phospholipid bilayers. *Chem. Phys. Lipids* **2010**, *163*, 182–189. [CrossRef] [PubMed]
35. Bayburt, T.H.; Sligar, S.G. Single-molecule height measurements on microsomal cytochrome P450 in nanometer-scale phospholipid bilayer disks. *Proc. Natl. Acad. Sci.* **2002**, *99*, 6725–6730. [CrossRef] [PubMed]
36. De Lemos-Chiarandini, C.; Frey, A.B.; Sabatini, D.D.; Kreibich, G. Determination of the Membrane Topology of the Phenobarbital-inducible Rat Liver Cytochrome P-450 Isoenzyme 4 Using Site-specific Antibodies. *J. Cell Biol.* **1987**, *104*, 209–219. [CrossRef] [PubMed]
37. Ozalp, C.; Szczesna-Skorupa, E.; Kemper, B. Identification of Membrane-Contacting Loops of the Catalytic Domain of Cytochrome P450 2C2 by Tryptophan Fluorescence Scanning. *Biochemistry* **2006**, *45*, 4629–4637. [CrossRef]
38. MacCallum, J.L.; Bennett, W.F.D.; Tieleman, D.P. Distribution of Amino Acids in a Lipid Bilayer from Computer Simulations. *Biophys. J.* **2008**, *94*, 3393–3404. [CrossRef]
39. Morita, M.; Katta, A.M.; Ahmad, S.; Mori, T.; Sugita, Y.; Mizuguchi, K. Lipid recognition propensities of amino acids in membrane proteins from atomic resolution data. *BMC Biophys.* **2011**, *4*, 21. [CrossRef]
40. Berka, K.; Hendrychová, T.; Anzenbacher, P.; Otyepka, M. Membrane position of ibuprofen agrees with suggested access path entrance to cytochrome P450 2C9 active site. *J. Phys. Chem. A* **2011**, *115*, 11248–11255. [CrossRef]
41. Berka, K.; Paloncýová, M.; Anzenbacher, P.; Otyepka, M. Behavior of Human Cytochromes P450 on Lipid Membranes. *J. Phys. Chem. B* **2013**, *117*, 11556–11564. [CrossRef] [PubMed]
42. Kučerka, N.; Tristram-Nagle, S.; Nagle, J.F. Structure of Fully Hydrated Fluid Phase Lipid Bilayers with Monounsaturated Chains. *J. Membr. Biol.* **2006**, *208*, 193–202. [CrossRef] [PubMed]
43. Lomize, M.A.; Lomize, A.L.; Pogozheva, I.D.; Mosberg, H.I. OPM: Orientations of proteins in membranes database. *Bioinformatics* **2006**, *22*, 623–625. [CrossRef] [PubMed]
44. McDougle, D.R.; Baylon, J.L.; Meling, D.D.; Kambalyal, A.; Grinkova, Y.V.; Hammernik, J.; Tajkhorshid, E.; Das, A. Incorporation of charged residues in the CYP2J2 F-G loop disrupts CYP2J2–lipid bilayer interactions. *Biochim. Biophys. Acta Biomembr.* **2015**, *1848*, 2460–2470. [CrossRef] [PubMed]
45. Hackett, J.C. Membrane-embedded substrate recognition by cytochrome P450 3A4. *J. Biol. Chem.* **2018**, *293*, 4037–4046. [CrossRef] [PubMed]
46. Denisov, I.G.; Grinkova, Y.V.; Baylon, J.L.; Tajkhorshid, E.; Sligar, S.G. Mechanism of drug-drug interactions mediated by human cytochrome P450 CYP3A4 monomer. *Biochemistry* **2015**, *54*, 2227–2239. [CrossRef] [PubMed]
47. van Meer, G.; Voelker, D.R.; Feigenson, G.W. Membrane lipids: Where they are and how they behave. *Nat. Mol. Cell. Biol.* **2008**, *9*, 112–124. [CrossRef]
48. Navrátilová, V.; Paloncýová, M.; Kajšová, M.; Berka, K.; Otyepka, M. Effect of Cholesterol on the Structure of Membrane-Attached Cytochrome P450 3A4. *J. Chem. Inf. Model.* **2015**, *55*, 628–635. [CrossRef]
49. Park, J.; Czapla, L.; Amaro, R.E. Molecular Simulations of Aromatase Reveal New Insights Into the Mechanism of Ligand Binding. *J. Chem. Inf. Model.* **2013**, *53*, 2047–2056. [CrossRef]

50. Robert, X.; Gouet, P. Deciphering key features in protein structures with the new ENDscript server. *Nucleic Acids Res.* **2014**, *42*, W320–W324. [CrossRef]
51. Nugent, T.; Ward, S.; Jones, D.T. The MEMPACK alpha-helical transmembrane protein structure prediction server. *Bioinformatics* **2011**, *27*, 1438–1439. [CrossRef] [PubMed]
52. Yachdav, G.; Kloppmann, E.; Kajan, L.; Hecht, M.; Goldberg, T.; Hamp, T.; Hönigschmid, P.; Schafferhans, A.; Roos, M.; Bernhofer, M.; et al. PredictProtein—An open resource for online prediction of protein structural and functional features. *Nucleic Acids Res.* **2014**, *42*, W337–W343. [CrossRef] [PubMed]
53. De Jong, D.H.; Singh, G.; Bennett, W.F.D.; Arnarez, C.; Wassenaar, T.A.; Scha, L.V.; Periole, X.; Tieleman, D.P.; Marrink, S.J. Improved Parameters for the Martini Coarse-Grained Protein Force Field. *J. Chem. Theory Comput.* **2013**, *9*, 687–697. [CrossRef] [PubMed]
54. Abraham, M.J.; Murtola, T.; Schulz, R.; Páll, S.; Smith, J.C.; Hess, B.; Lindahl, E. GROMACS: High performance molecular simulations through multi-level parallelism from laptops to supercomputers. *SoftwareX* **2015**, *1–2*, 19–25. [CrossRef]
55. Yu, X.; Cojocaru, V.; Mustafa, G.; Salo-Ahen, O.M.H.; Lepesheva, G.I.; Wade, R.C. Dynamics of CYP51: Implications for function and inhibitor design. *J. Mol. Recognit.* **2015**, *28*, 59–73. [CrossRef]
56. Wassenaar, T.A.; Pluhackova, K.; Böckmann, R.A.; Marrink, S.J.; Tieleman, D.P. Going Backward: A Flexible Geometric Approach to Reverse Transformation from Coarse Grained to Atomistic Models. *J. Chem. Theory Comput.* **2014**, *10*, 676–690. [CrossRef]
57. Maier, J.A.; Martinez, C.; Kasavajhala, K.; Wickstrom, L.; Hauser, K.E.; Simmerling, C. ff14SB: Improving the Accuracy of Protein Side Chain and Backbone Parameters from ff99SB. *J. Chem. Theory Comput.* **2015**, *11*, 3696–3713. [CrossRef]
58. Dickson, C.J.; Madej, B.D.; Skjevik, Å.A.; Betz, R.M.; Teigen, K.; Gould, I.R.; Walker, R.C. Lipid14: The Amber lipid force field. *J. Chem. Theory Comput.* **2014**, *10*, 865–879. [CrossRef]
59. Harris, D.L.; Park, J.-Y.; Gruenke, L.; Waskell, L. Theoretical study of the ligand-CYP2B4 complexes: Effect of structure on binding free energies and heme spin state. *Proteins* **2004**, *55*, 895–914. [CrossRef]
60. Price, D.J.; Brooks, C.L. A modified TIP3P water potential for simulation with Ewald summation. *J. Chem. Phys.* 2004. [CrossRef]
61. Phillips, J.C.; Braun, R.; Wang, W.; Gumbart, J.; Tajkhorshid, E.; Villa, E.; Chipot, C.; Skeel, R.D.; Kalé, L.; Schulten, K. Scalable molecular dynamics with NAMD. *J. Comput. Chem.* **2005**, *26*, 1781–1802. [CrossRef] [PubMed]
62. Humphrey, W.; Dalke, A.; Schulten, K. VMD: Visual molecular dynamics. *J. Mol. Graph.* **1996**, *14*, 33–38. [CrossRef]
63. Pravda, L.; Sehnal, D.; Toušek, D.; Navrátilová, V.; Bazgier, V.; Berka, K.; Svobodová Vareková, R.; Koca, J.; Otyepka, M. MOLEonline: A web-based tool for analyzing channels, tunnels and pores (2018 update). *Nucleic Acids Res.* **2018**, *46*, W368–W373. [CrossRef] [PubMed]

© 2019 by the authors. Licensee MDPI, Basel, Switzerland. This article is an open access article distributed under the terms and conditions of the Creative Commons Attribution (CC BY) license (http://creativecommons.org/licenses/by/4.0/).

Article

Polymorphisms of CYP2C8 Alter First-Electron Transfer Kinetics and Increase Catalytic Uncoupling

William R. Arnold [1], Susan Zelasko [1], Daryl D. Meling [1], Kimberly Sam [1] and Aditi Das [1,2,3,4,*]

[1] Department of Biochemistry, University of Illinois Urbana-Champaign, 3813 Veterinary Medicine Basic Sciences Building, 2001 South Lincoln Avenue, Urbana, IL 61802, USA; william.arnold@ucsf.edu (W.R.A.); susan.e.zelasko@gmail.com (S.Z.); dmeling@illinois.edu (D.D.M.); kimberlytsam@gmail.com (K.S.)
[2] Department of Comparative Biosciences, University of Illinois Urbana-Champaign, 3813 Veterinary Medicine Basic Sciences Building, 2001 South Lincoln Avenue, Urbana, IL 61802, USA
[3] Department of Bioengineering, University of Illinois Urbana-Champaign, Beckman Institute for Advanced Science and Technology, 3813 Veterinary Medicine Basic Sciences Building, 2001 South Lincoln Avenue, Urbana, IL 61802, USA
[4] Division of Nutritional Sciences, University of Illinois Urbana-Champaign, 3813 Veterinary Medicine Basic Sciences Building, 2001 South Lincoln Avenue, Urbana, IL 61802, USA
* Correspondence: aditidas@illinois.edu; Tel.: +1217-244-0630

Received: 30 August 2019; Accepted: 13 September 2019; Published: 18 September 2019

Abstract: Cytochrome P450 2C8 (CYP2C8) epoxygenase is responsible for the metabolism of over 60 clinically relevant drugs, notably the anticancer drug Taxol (paclitaxel, PAC). Specifically, there are naturally occurring polymorphisms, CYP2C8*2 and CYP2C8*3, that display altered PAC hydroxylation rates despite these mutations not being located in the active site. Herein, we demonstrate that these polymorphisms result in a greater uncoupling of PAC metabolism by increasing the amount of hydrogen peroxide formed per PAC turnover. Anaerobic stopped-flow measurements determined that these polymorphisms have altered first electron transfer kinetics, compared to CYP2C8*1 (wildtype), that suggest electron transfer from cytochrome P450 reductase (CPR) is disfavored. Therefore, these data demonstrate that these polymorphisms affect the catalytic cycle of CYP2C8 and suggest that redox interactions with CPR are disrupted.

Keywords: CYP2C8; polymorphisms; reactive oxygen species; paclitaxel; cytochrome P450 reductase; electron transfer

1. Introduction

Cytochrome P450 2C8 (CYP2C8) is a member of the cytochrome P450 (CYP) epoxygenase family that metabolizes over 60 clinically relevant drugs on the market [1–3]. For example, CYP2C8 is the primary enzyme involved in the metabolism of paclitaxel (PAC), a common chemotherapeutic that works by interfering with microtubule function [4]. CYP2C8 is primarily expressed hepatically, though it is also present in the vasculature and kidneys [5–7], where it metabolizes lipids, such as arachidonic acid (AA), to form biologically active epoxyeicosatrienoic acids (EETs). EETs are known to be anti-inflammatory [8], angiogenic [9], and inhibit vascular smooth muscle cell migration, implicating CYP2C8 in regulating kidney and vascular function [10].

Polymorphic variations in CYPs have been of clinical interest due to individual differences in drug metabolism. For example, CYP2D6 is among the most highly polymorphic CYPs that greatly contributes to the poor, intermediate, extensive, and ultra-rapid metabolizer phenotypes [11,12]. Two common, naturally occurring polymorphic variants of CYP2C8—CYP2C8*2 and CYP2C8*3—display altered drug elimination rates and EET production compared to CYP2C8*1 (wildtype, WT) [13,14]. The CYP2C8*3 polymorphism (R139K/K399R) is present in 2% of African-American and 13% of Caucasian populations [13]. CYP2C8*3 is associated with an increase in peripheral neuropathy in patients treated with PAC, presumably a result of slower PAC elimination by CYP2C8*3 [15,16]. However, some studies have suggested PAC metabolism is increased with CYP2C8*3 [17]. In vitro studies corroborate that CYP2C8*3 has only 30% and 15% of the activity compared to WT for the turnover of AA and PAC, respectively [13]. However, others report that PAC metabolism is not significantly affected by these polymorphisms [14,18], and one study observed greater PAC turnover by *3 as compared to WT [19]. Therefore, from previous reports, the effects of CYP2C8*3 on PAC metabolism are convoluted. The CYP2C8*2 (I269F) polymorphism, present in 18% of African-American populations [13], shows lower, albeit not always statistically significant, in vitro PAC turnover compared to WT [13,14,20].

Importantly, the amino acid residues that are different in the CYP2C8*2 and CYP2C8*3 (R139, K399, and I269) are not located within the enzyme active site of CYP2C8 but on the surface of the protein (Figure 1) [2,13,18,19]. This suggests that these mutations disrupt mechanisms of substrate metabolism that are not directly related to substrate binding. Indeed, CYP2C8*3 showed WT activity for the deethylation of amiodarone, and so it appears that this polymorphism does not affect substrate binding as a whole [21]. The CYP catalytic cycle is a complex series of redox reactions that require assistance from redox partners such as cytochrome P450 reductase (CPR) (Figure 1). The details of this complex cycle can be found in many reviews [22–25]. Therefore, these mutations may be affecting other steps in the CYP catalytic cycle, such as electron transfer from redox partners. Earlier work demonstrated that CYP2C8*3 has different binding affinities to its redox partners compared to WT. Particularly, CPR interacts with PAC-bound CYP2C8*3 better than WT as determined indirectly by PAC metabolism assays using varying CPR concentrations (apparent K_m = 5.5 ± 1.5 nM for CYP2C8*3 versus 35 ± 10 nM for WT) [19]. The greater apparent interaction with CPR ought to lead to a faster transfer of electrons and a greater substrate turnover. In the same study, the authors determined that PAC metabolism was increased with CYP2C8*3 compared to WT [19]. However, the consensus through other studies is that PAC metabolism is either lower or similar to WT [13,14,18]. Taken together, these data suggest that the CYP-CPR interaction may be disrupted in CYP2C8*3 and affects PAC metabolism.

CPR transfers two electrons to CYPs during the CYP catalytic cycle, with the first electron reducing the ferric heme to a ferrous heme and the second reducing the dioxygen-heme to a peroxy-heme (Figure 1). Many reactive oxidized intermediates are involved in the CYP catalytic cycle en route to substrate oxidation. These intermediates can sometimes decompose to form reactive oxygen species (ROS) instead of progressing towards substrate oxidation, a process known as uncoupling. These ROS, namely hydrogen peroxide (HOOH) and superoxide, are formed in large quantities by CYP2C enzymes [26,27]. ROS can induce mitochondrial dysfunction in cardiomyocytes, induce coronary artery vasoconstriction, and promote carcinogenesis [26,28–30], and ROS specifically generated by endothelial CYP2C8 has been shown to impair functional recovery after ischemia/reperfusion injury [27]. Another potential effect of these polymorphisms may therefore be on the coupling efficiency of PAC metabolism.

Herein, we determined the effects of the CYP2C8*2 and CYP2C8*3 polymorphisms in regards to first-electron transfer (FET) kinetics and PAC metabolism uncoupling. We tested CYP2C8*1, CYP2C8*2, and CYP2C8*3, as well as the single mutations of the CYP2C8*3 polymorphism (R139K and K399R). To study these polymorphisms, we utilized the Nanodisc technology to solubilize these CYP2C8 variants in a membrane mimic [31,32]. We find that CYP2C8*2 has a lower PAC turnover rate compared to WT. We further determined that CYP2C8*3 produces significantly more HOOH compared to WT, indicating a greater uncoupling of the catalytic cycle. Using stopped-flow measurements of the FET, we determined that the CYP2C8*2 and CYP2C8*3 have profoundly distinct and slower FET kinetics compared to WT. We determined that there is no change in the reduction potential of the polymorphisms compared to WT, which supports that the altered FET kinetics is due to an altered redox interaction with CPR. None of the single-mutant variants of CYP2C8*3 (R139K and K399R) reproduce the effects of the CYP2C8*3, indicating that the effects of this polymorphism are a synergism of both mutations. Taken together, these data demonstrate that these polymorphisms have altered FET kinetics, leading to an increase in HOOH production and greater PAC metabolism uncoupling.

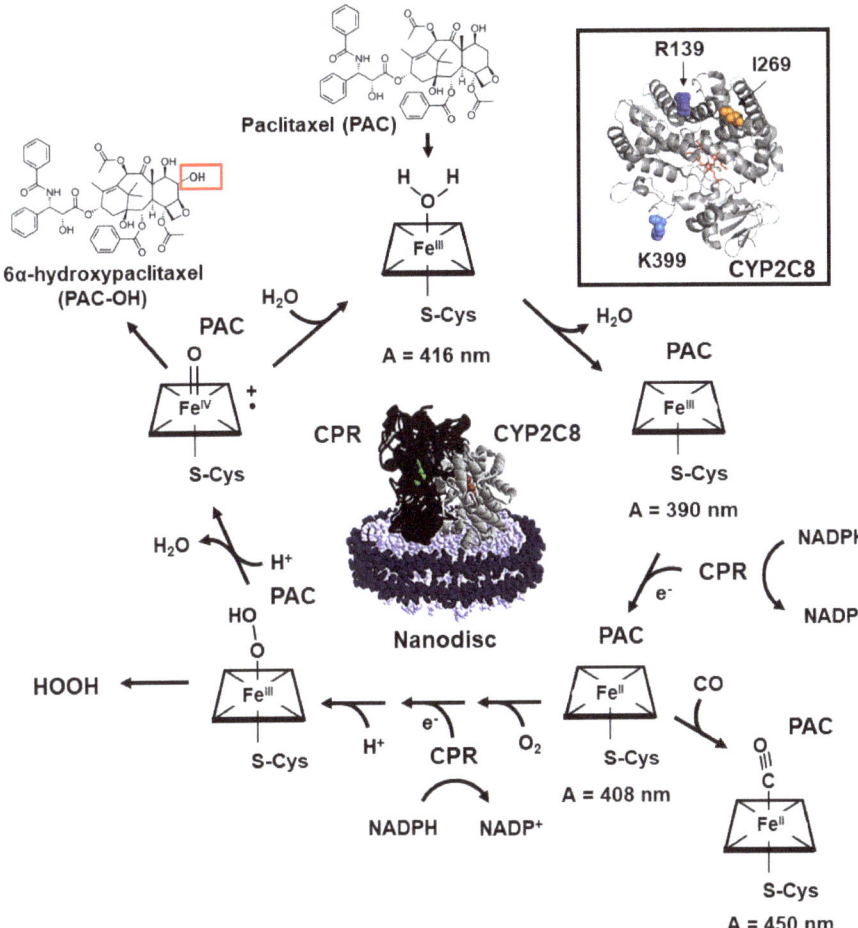

Figure 1. Schematic of the CYP catalytic cycle. Inset shows the structure of CYP2C8 with residues R139 (blue), I269 (orange), and K399 (light blue) highlighted. Structure was generated in PyMol v1.3r1 using the PDB entry 1PQ2. A schematic of CYP2C8 (grey) and CPR (black) incorporated into nanodiscs is shown in the center of the cycles. *Catalytic cycle.* Substrate (PAC) binds to the CYP active site, which perturbs the H_2O coordination to the iron heme. H_2O unbinds leaving a pentacoordinated high-spin iron heme. CPR reduces the iron heme using an electron obtained from NADPH. Under anaerobic conditions, CO ligates the heme to terminate the cycle. Under aerobic conditions, O_2 ligates the heme, followed by another one-electron reduction by CPR and the addition of a proton to produce a peroxy-heme. The peroxy-heme can decompose forming HOOH or eliminate an H_2O molecule to produce the catalytic ferryl iron heme radical (Compound I). Compound I can oxidize substrate (PAC) to product (PAC-OH, red box), followed by the coordination of an H_2O molecule to begin the cycle again. Spectroscopically visible species are indicated with their characteristic absorbance wavelength. More details of the cycle can be found in previous reviews [22–25].

2. Results and Discussion

2.1. P450 Characterization of CYP2C8*1/*2/*3/R139/K399-ND

In order to assess if these variants of CYP2C8 lead to unfolding of the protein and improper ligation of the heme group, we performed anaerobic CO-binding assays. All the variants showed a 90%–100% shift in the heme absorbance to 450 nm and resemble CYP2C8*1 characteristics (Figure 2 and Table S2). Therefore, these variants are well-folded.

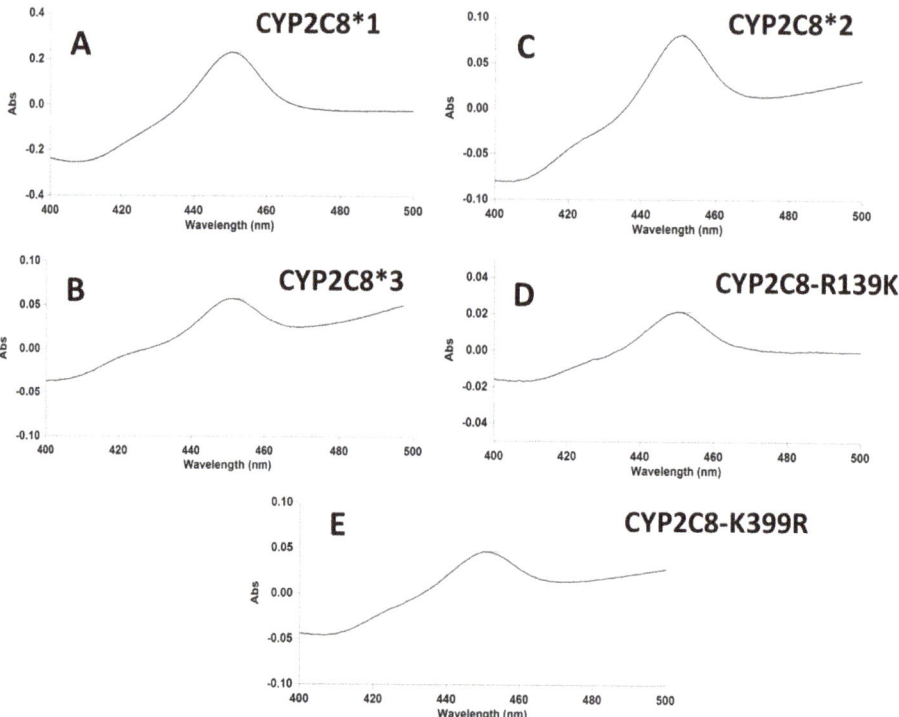

Figure 2. *CO-binding difference spectra.* (**A**) CYP2C8*1, (**B**) CYP2C8*2, (**C**) CYP2C8*3, (**D**) CYP2C8-R139K, and (**E**) CYP2C8-K399R.

2.2. Effect of Polymorphisms on PAC Metabolism

Next, we sought to examine the effect that these polymorphisms have on PAC turnover rates. Previous studies reported a poor solubility of PAC that precluded V_{max} determination for in vitro CYP2C8 [33], and so we analyzed our data based on the time linearity. Linearity was established for the hydroxylation of 70 µM PAC to 6α-hydroxypaclitaxel (PAC-OH) over a 20 min period. We can estimate the catalytic efficiency of the PAC metabolism among the variants by fitting the data to Equation (1)

$$[S] = [S_0]e^{-kt} \qquad (1)$$

where $[S_0]$ is the initial concentration of the substrate and $k = \frac{k_{cat}}{K_m}[E]$ (Figure S1) [34]. The catalytic efficiencies of these variants range from 0.207 to 1.02 $\text{min}^{-1}\,\text{nM}^{-1}$ (Figure S1, Table 1), which are lower than previously reported rates for CYP2C8-mediated in vitro PAC metabolism using a lipid-reconstituted system [33]. CYP2C8*2 showed a marked decrease in turnover rate (47.2% WT), consistent with earlier reports of its inefficient PAC metabolism [13,14]. Compared to WT, CYP2C8*3 had slightly, albeit not significantly, lower PAC turnover rates. These results for CYP2C8*3 agree with the findings of Yu et al. [14] and Soyama et al. [18]. However, they contradict the findings of Dai et al. [13], who were unable to measure PAC metabolism, and contradict the findings of Kaspera et al., who reported higher PAC metabolism [19]. The CYP2C8-K399R variant metabolized PAC with rates similar to WT, but interestingly the CYP2C8-R139K variant showed a remarkable increase in PAC turnover (265% compared to WT). Therefore, the effects of the CYP2C8*3 single mutations do not additively contribute to the CYP2C8*3 activity. Overall, we observed that the CYP2C8*3 polymorphism does not significantly affect PAC turnover, but CYP2C8*2 is half as efficient as WT. Since these polymorphisms do not occur in the active site of CYP2C8, it is unlikely that they directly affect PAC binding. Therefore, we further probed other steps of the CYP catalytic cycle to determine the mechanism through which these polymorphisms affect CYP2C8 activity.

Table 1. Paclitaxel (PAC) metabolism by CYP2C8 variants. Linear rates of 70 µM PAC metabolism by each CYP2C8 variant to PAC–OH was determined over a 20 min period. Estimates of the catalytic efficiencies were determined using Equation (1) as stated in the text. Error represents the SEM of three experiments.

Variant	Rate (pmol/min/nmol$_{prot}$)	%WT	k_{cat}/K_m (min^{-1} nM^{-1})	%WT
CYP2C8*1 (WT)	38.8 ± 0.2	100	0.381	100
CYP2C8*2	18.3 ± 3.5	47.2	0.207	54.4
CYP2C8*3	34.0 ± 3.0	88	0.321	84.4
CYP2C8-R139K	103 ± 2	265	1.02	269
CYP2C8-K399R	47.1 ± 4.3	121	0.441	116

2.3. Polymorphisms in CYP2C8 Lead to Greater HOOH Uncoupling

To assess the uncoupling efficiency of PAC metabolism, we next measured the rate of HOOH production over time by each variant. None of the variants showed a significant difference in the HOOH production rates in the presence of 70 µM PAC compared to without. This is likely due to the high amounts of HOOH produced compared to PAC turnover (Figure S2). The overall amount of HOOH linearly decreased for all CYP2C8 variants over time, indicating a burst of activity at the initiation of the reaction followed by decomposition of HOOH (Figure 3). For all time points, CYP2C8*3 displayed nearly 200% higher HOOH production compared to WT. Therefore, the CYP2C8*3 polymorphism leads to a greater ROS production and catalytic uncoupling. PAC and ROS have both been implicated in the pathogenesis of neuropathic pain [35,36], and therefore it would be interesting to see if the increase in HOOH production contributes to the neuropathy observed with CYP2C8*3 [15,16]. There was an increase in HOOH production with the CYP2C8-K399R variant at the start of the reaction and a minor decrease in HOOH production over time with the CYP2C8-R139K variant. Therefore, likewise to the PAC metabolism experiments, the individual mutations of the CYP2C8*3 do not additively contribute to the CYP2C8*3 phenotype. CYP2C8*2 did not have a significant difference in the HOOH production compared to WT, but this polymorphism also displayed half the PAC turnover as WT. By normalizing the amount of HOOH produced to the activity of the enzyme by looking at the HOOH produced per PAC turnover, we see that both CYP2C8*2 and CYP2C8*3 produce almost 200% more HOOH per PAC turnover than WT (Table 2). The CYP2C8*2 and CYP2C8*3 polymorphisms, therefore, are about twofold more uncoupled (i.e., produce twofold more ROS), with CYP2C8*3 producing significantly more HOOH than WT. ROS uncoupling can be caused by altered redox kinetics during the CYP2C8 catalytic cycle or electron transfer from CPR to CYP2C8 (Figure 1). Therefore, we next proceeded to determine if these polymorphisms alter the intrinsic redox potential of the CY2C8 heme.

Table 2. Hydrogen peroxide (HOOH) production per PAC turnover. The amount of HOOH formed at 20 min was divided by the amount of PAC–OH produced at 20 min. These values are compared to WT.

Variant	nmol$_{HOOH}$/pmol$_{PAC-OH}$ (at 20 min)	%WT
CYP2C8*1 (WT)	1.58	100
CYP2C8*2	2.96	187
CYP2C8*3	3.25	206
CYP2C8-R139K	0.46	29.6
CYP2C8-K399R	1.24	78.4

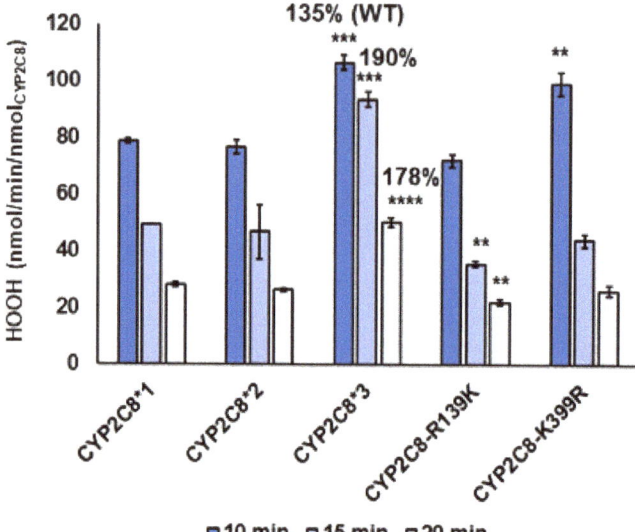

Figure 3. HOOH production rates. The rate of HOOH production by each CYP2C8 variant was measured using an Amplex Red peroxidase kit at 10, 15, and 20 min reaction times, in the presence of 70 μM PAC. Error represents the SEM of 3–4 experiments. Statistical significance was determined by comparing experiments to their respective WT controls. ** $p < 0.01$; *** $p = 0.0001$; **** $p < 0.0001$.

2.4. Spectral Characterization of Substrate Binding and Reduction Potentials of CYP2C8 Polymorphisms

We determined the reduction potential of the CYP2C8 variants using safranin T as a redox indicator as previously described [37]. Substrates binding to CYPs, such as CYP3A4, often perturbs water binding at the 6th coordinate position to produce a pentacoordinated high-spin heme (Figure 1). This results in an increase in the reduction potential of the heme and helps facilitate electron transfer from CPR [38]. The high-spin content can be observed as a shift in the heme absorbance from ~417 to ~390 nm. We did not observe a significant high-spin shift upon PAC binding to CYP2C8 in any of the variants, as was similarly observed for PUFAs binding to CYP2J2 [39]. We also determined that there is not a significant change in the reduction potential of CYP2C8*1 upon PAC binding (Figure 4A, Table 3), which correlates to the poor high-spin content of the PAC-bound protein. The poor high-spin content and minimal change to the reduction potential together support the slow metabolism of PAC by CYP2C8 compared to other CYP-mediated drug metabolisms. Compared to the polymorphisms, CYP2C8*2 had a slightly albeit not significantly lower reduction potential, and the reduction potential of CYP2C8*3 was similar to WT (Figure 4, Table 3). Altogether, there is not a significant change to the intrinsic redox properties of the heme in these polymorphisms. We next proceeded to measure the first-electron transfer (FET) kinetics between CPR and CYP2C8.

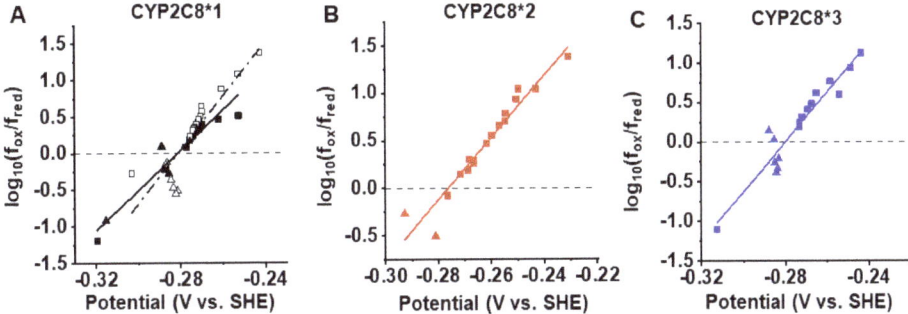

Figure 4. Reduction potential of CYP2C8 variants. Redox titration of CYP2C8 variants was conducted in 0.1 M phosphate buffer (pH 7.4 at 25 °C) with or without 70 µM PAC. The potential was measured spectroscopically by using a safranin T as the redox indicator. Reduction was achieved using light and dithionite and oxidation was achieved using $K_3[Fe(CN)_6)]$ as stated in the Methods section. Representative Nernst plots for (**A**) CYP2C8*1 without paclitaxel (open black) and with paclitaxel (solid black), (**B**) CYP2C8*2, and (**C**) CYP2C8*3 from two experiments are shown. Data obtained from reduction is given as squares and data from oxidation is given as triangles. The zero intercept gives $E°'$, the redox potential of the protein.

Table 3. Reduction potentials. Reduction potentials of the CYP2C8 variants was determined as described in the text. Error represents the SEM of two experiments.

Variant	Reduction Potential (V)
CYP2C8*1 No PAC	−0.283 ± 0.002
CYP2C8*1	−0.279 ± 0.002
CYP2C8*2	−0.297 ± 0.021
CYP2C8*3	−0.281 ± 0.001

2.5. Polymorphisms Show Altered First Electron Transfer (FET) Kinetics as Determined by CO Stopped-Flow

To further probe the effect of these polymorphisms on the CYP2C8 catalytic mechanism, we determined the kinetics of the FET from CPR to the CYP2C8 variants during the metabolism of 70 µM PAC. We firstly determined that there is not a significant change in the NADPH oxidation rates among the variants (Figure S3), which supports previous findings [19]. As NADPH is the initial step in the reaction, we then proceeded to conduct stopped-flow measurements in order to determine how CPR transfers electrons to the CYP2C8 variants.

CYPs display a characteristic shift in the heme absorbance to 450 nm upon CO binding to the heme. In order for CO to bind, the heme must be reduced to the ferrous state by CPR, which we determine by the appearance of the 450 nm absorbance band over time (Figure 5A). Therefore, the rate at which CO binds to the heme is directly related to the FET rate. The rate of CO binding was monophasic across all variants (Figure 5B–F). Compared to WT, CYP2C8*2 and CYP2C8*3 showed higher rates of CO binding, which may also explain the greater HOOH uncoupling and/or the higher levels of HOOH production by CYP2C8*3. Contrariwise, the CYP2C8-R139K and CYP2C8-K399R variants showed lower rates compared to WT (Table 4).

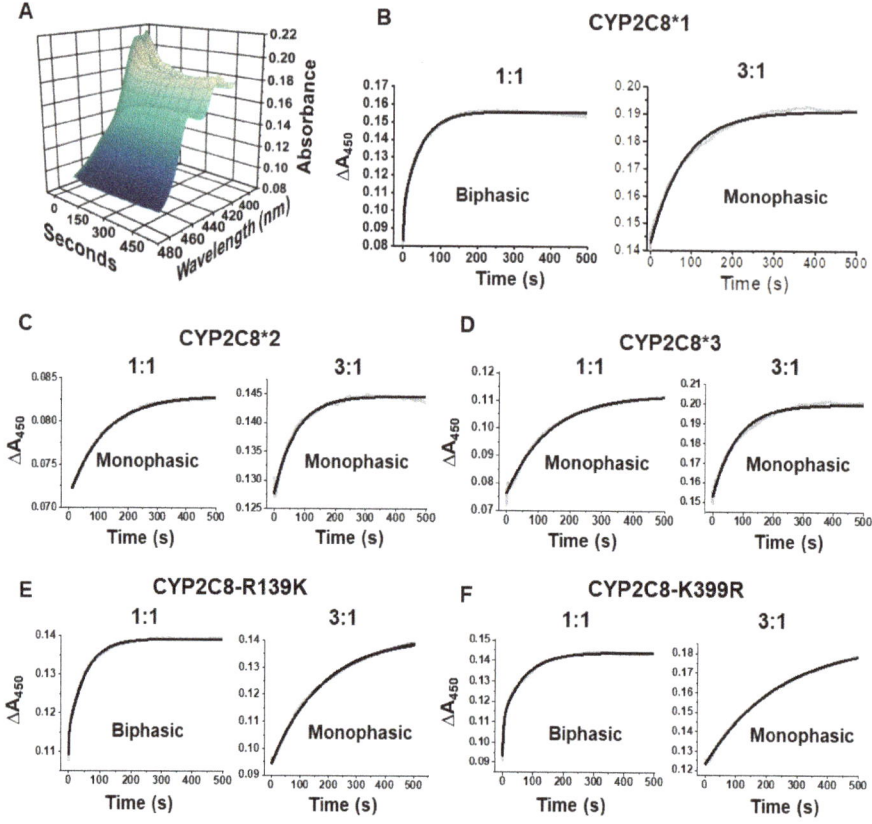

Figure 5. Stopped-flow electron transfer rate to CYP2C8 variants. (**A**) Three-dimensional plot of representative absorbance spectra from 0 to 500 s showing the increase in absorbance near 450 nm following the mixing of NADPH with a pre-incubated complex of paclitaxel-bound CYP2C8 and CPR. (**B–F**) Representative plots of the change in peak absorbance at 450 nm from 0 s to 500 s. Experiments were conducted at either a 1:1 CPR:CYP or a 3:1 CPR:CYP ratio. Data points (grey squares) are overlaid with the line of best fit (solid black line) derived from a fit of the data to either monophasic (single exponential) or biphasic (two exponential) kinetic equations in OriginLab as indicated.

Table 4. Stopped-flow CO binding kinetics. Fast (k_1) and slow (k_2) rates are in units of ms^{-1}. Experiments were conducted using a 1:1 and a 1:3 CYP:CPR ratios. Error represents the SEM of 3–6 experiments. Statistical significance compares to WT. ** $p < 0.01$, *** $p = 0.001$, **** $p < 0.0001$.

Variant	1:1 CPR:CYP		3:1 CPR:CYP
	k_1 (ms^{-1})	k_2 (ms^{-1})	k_2 (ms^{-1})
CYP2C8*1	468 ± 21	20.7 ± 0.4	9.70 ± 0.98
CYP2C8*2	—	8.67 ± 0.49 ****	17.4 ± 1.6 ***
CYP2C8*3	—	10.8 ± 1.84 ****	13.7 ± 0.7 ****
CYP2C8-R139K	406 ± 52	18.2 ± 0.65	6.02 ± 0.19 **
CYP2C8-K399R	445 ± 51	19.4 ± 0.99	3.90 ± 0.10 ***

CYP metabolism studies are typically performed using a saturating 3:1 CPR:CYP ratio in order to achieve a maximum pseudo-zero-order electron transport kinetics between CPR and the CYP. Hitherto, we have used a 3:1 CPR:CYP ratio in our experiments. To determine if the FET is dependent on the interactions of CYP2C8 with CPR, we repeated the experiments by lowering the CPR:CYP ratio from 3:1 to 1:1. CYP2C8*1, CYP2C8-R139K, and CYP2C8-K399R all showed biphasic CO binding at the 1:1 CPR:CYP ratio and had similar fast (k_1) and slow (k_2) kinetics compared (Figure 5B,E,F, Table 4). Altering the CPR:CYP ratio has previously been shown to change the number of phases of the FET in certain situations likely by forcing CPR to associate with CYPs in unproductive confirmations at higher CPR:CYP ratios [40]. Likewise, the 3:1 ratio with CYP2C8*1, CYP2C8-R139K, and CYP2C8-K399R may be promoting unproductive confirmations with CPR to produce these observations.

Interestingly, CYP2C8*2 and CYP2C8*3 only show the slow phase of the CO binding at the 1:1 ratio (Figure 5C,D, Table 4). The values of k_2 for these polymorphisms are about half that compared to WT. However, the FET rate of CYP2C8*2 and CYP2C8*3 rates in the 3:1 CPR:CYP experiments resemble those of the WT slow phase in the 1:1 ratio. Therefore, it appears that more CPR is required for the CYP2C8*2 and CYP2C8*3 polymorphisms to rescue the WT activity of this slow phase. Together, these data suggest that these polymorphisms reduce either the binding of CPR to CYP2C8 or the transfer of electrons from CPR to CYP2C8. We will refer to these two events as the CYP–CPR redox interaction. Kaspera, et al. had previously determined that the apparent affinity of CPR for CYP2C8*3 is greater than WT; however, the study also used a different recombinant system with cytochrome b5 as an auxiliary redox partner [19]. Homology modeling (Figure S4) reveals that R139, K399, and I269 all lie on the putative CYP-CPR binding interface, which means these mutations may be directly interfering with the CYP–CPR interaction. Further, K399 is located near the N-terminus at the protein-membrane interface (Figure 1A), and we showed previously that the N-terminus is essential for the CPR-mediated reduction of CYP2J2, which has a close homology to CYP2C8 [41].

3. Conclusions

CYP2C8*2 and CYP2C8*3 contain mutations that are not in the active site of CYP2C8. These mutations are instead located on the periphery of the protein. Therefore, these residues do not directly contribute to substrate binding and instead must affect CYP2C8 activity through their modulation of other steps in the CYP catalytic cycle. These mutations lie on the putative CYP–CPR interface and thus may be affecting the redox interactions between these proteins. We determined that CYP2C8*2 is 47% as active towards PAC turnover compared to WT and that CYP2C8*3 shows WT activity. The metabolism of PAC is 200% more uncoupled in the CYP2C8*2 and CYP2C8*3 polymorphisms, and CYP2C8*3 produces significantly more HOOH compared to WT. Stopped-flow kinetics of the FET suggest that the polymorphisms reduce the transfer of electrons by CPR to CYP2C8. In conclusion, these in vitro studies demonstrate that these polymorphisms of CYP2C8 do not directly affect PAC binding to CYP2C8 but may, in fact, be affecting the redox interaction between CYP2C8 and CPR. The CYP2C8*2 and CYP2C8*3 polymorphisms reduce the CYP–CPR redox interaction and promote greater uncoupling of PAC metabolism. Therefore, not only is the FET disrupted in these polymorphisms, the electrons are being utilized towards ROS formation in lieu of PAC turnover. However, we were unable to definitively determine how the redox interaction is being altered, i.e., whether the polymorphisms affect the ability of CPR to bind and dock to CYP2C8, how the electrons are shuttled through CYP2C8, or both.

Another important finding is that the effects of the CYP2C8*3 polymorphism on CYP2C8 activity cannot be explained by the additive contribution of the individual mutations themselves. In all experiments, CYP2C8*3 had distinct activities compared to the linear combination of the R139K and K399R individual data. In fact, the single mutations either showed WT activity (especially concerning FET kinetics) or were significantly different than either WT or CYP2C8*3 (e.g., the 265% increase in PAC metabolism by CYP2C8-R139K). Therefore, the mutation of these two residues do not additively contribute to the CYP2C8*3 phenotype.

There are many mechanisms by which these polymorphisms effect the CYP2C8 redox interaction. They may be directly affecting the binding of CPR to CYP2C8. The mutations may also be altering the architecture of CYP2C8 such that it disfavors redox interactions with CPR while also destabilizing the peroxy-heme intermediate to produce HOOH. CYP2C8*2 contains an I269F mutation, which is a significant change to the physical and chemical properties of the residue. It is possible that this mutation has a profound effect on the folding of CYP2C8 and the interaction with CPR. CYP2C8*3 contains a R139K/K399R double mutation, which interestingly swaps the Lys and Arg residues. While Lys and Arg are both positively charged residues, they differ significantly in their physical properties. For instance, Arg can form a greater number of electrostatic interactions and better maintains a positive charge compared to Lys. It has been shown that Arg substitution can increase the stability of GFP, which was shown in silico to be facilitated by a greater number of salt bridge interactions [42]. Arg also interacts differently to phospholipids and increases interfacial binding and perturbations to membranes [43]. Therefore, this polymorphism may be affecting how CYP2C8 associates with lipids as well as how it interacts with CPR.

4. Materials and Methods

4.1. Materials

The human CYP2C8 gene cloned into the Amp^R pAr5 (modified pCWOri+) plasmid was a gift from Dr. Eric Johnson. PCR reagents were purchased from New England Biolabs (Ipswich, MA, USA). Molecular biology enzymes and *E. coli* DH5α were purchased from Invitrogen (Waltham, MA, USA). Plasmid DNA was purified using a Qiagen Gel Extraction kit. Ampicillin (Amp), arabinose, chloramphenicol (Chlr), isopropyl β-D-1- thiogalactopyranoside (IPTG), and Ni-NTA resin were purchased from Gold Biotechnology (St. Louis, MO, USA). δ-Aminolevulinic acid (δ-ALA) was purchased from Frontier Scientific (Emeryville, CA, USA). 1-palmitoyl-2-oleoyl-sn-glycero-3-phosphocholine (POPC) and 1-palmitoyl-2-oleoyl-sn-glycero-3-phospho-L-serine (POPS) were purchased from Avanti Polar Lipids (Alabaster, AL, USA). Carbamazepine and paclitaxel were purchased from Cayman Chemicals (Ann Arbor, MI, USA). NADPH was purchased from P212121 Store.

4.2. Protein Engineering of CYP2C8*1/*2/*3/R139K/K399R

The CYP2C8 plasmid from Dr. Johnson was used directly for engineering the CYP2C8 variants. Plasmids were amplified and purified using a Qiagen Plasmid mini-prep kit (Valencia, CA, USA). The R139K and K399R single nucleotide substitutions were made using forward and reverse primers containing each mutation and an inserted BspQI restriction enzyme site (New England Biolabs) (Table S1). BspQI is a class II restriction enzyme that will create a unique sticky-end cut one nucleotide removed from the restriction enzyme site. The resulting gene of the R139K amplification and the K399R primers were then used to construct the CYP2C8*3 gene. A single *2 mutation was made similarly containing a single substitution at I269F. The PCR reaction consisted of 1 µM of forward and reverse primers in HF reaction buffer (New England Biolabs) containing 50 pg/µL CYP2C8-containg plasmid, 200 µM dNTPs, 5% DMSO, and Phusion DNA polymerase (10 U/mL). The PCR thermocycler was set to 95 °C for 3 min, 20 cycles (95 °C for 30 sec, 65 °C for 30 sec), and then 72 °C for 4 min. The mutated plasmids were then digested with BspQI and the resulting sticky ends were ligated to make complete plasmids. Chemically competent DH5α cells were transformed by heat shock at 42 °C for 45 min, and then set on ice. The addition of 1 mL of warm Super Optimal Broth (SOC) media was followed by shaking (250 rpm, 37 °C) for 1 hr. Cells were plated on an LB Amp plate to screen for the desired mutant plasmid. Mutant dsDNA was confirmed by DNA sequencing at the UIUC Core Sequencing Facility. Cells were co-transformed with pTGro7 plasmid containing the GroES-GroEL chaperonin system.

4.3. Protein Expression and Purification of CYP2C8*1/*2/*3/R139K/K399R

All CYP2C8 proteins were expressed according to the protocol used in CYP2J2 expression, as previously described [41,44]. The protein concentrations were determined using a UV–vis spectrophotometer (Agilent Technologies, Santa Clara, CA, USA) ($\varepsilon = 108$ mM^{-1}·cm^{-1}).

4.4. Expression of Cytochrome P450 Reductase

Expression of CPR from *Rattus norvegicus* was performed as previously described [41,44].

4.5. Assembly of CYP2C8-Nanodiscs

CYP2C8-ND were assembled as previously described [37,44,45] by mixing CYP2C8, membrane scaffold protein (MSP1E3D1), an 80:20 ratio of POPC:POPS lipids, and cholate, followed by detergent removal using Amberlite® beads, and purification by size exclusion chromatography [31,32].

4.6. Carbon Monoxide Binding Assay

The heme content of the purified CYP2C8 proteins was analyzed using UV–vis spectroscopy (Agilent Technologies) as previously described [37].

4.7. Paclitaxel Metabolism

Samples containing 0.1 µM of CYP2C8-ND (*1/*2/*3/R139K/ K399R) were incubated with CPR (0.3 µM), and PAC (70 µM) in 0.3 mL of 0.1 M potassium phosphate buffer (pH 7.4) for 5 min at 37 °C. NADPH (200 µM) was added and the mixture was incubated for 5, 10, and 20 min at 37 °C, then quenched with equivolume ethyl acetate. Samples were vortexed and thrice-extracted with ethyl acetate, dried under a stream of N_2 gas, and then resuspended in 180 proof ethanol for LC–MS/MS quantification.

4.8. Tandem LC–MS/MS for the Quantification of 6α-Hydroxypaclitaxel

Samples were analyzed with the 5500 QTRAP LC/MS/MS system (Sciex, Framingham, MA, USA) in Metabolomics Lab of Roy J. Carver Biotechnology Center, University of Illinois at Urbana-Champaign. Software Analyst 1.6.2 was used for data acquisition and analysis. The 1200 series HPLC system (Agilent Technologies) includes a degasser, an autosampler, and a binary pump. The LC separation was performed on an Agilent Eclipse XDB-C18 (4.6 × 150 mm, 5 µm) with mobile phase A (0.1% formic acid in water) and mobile phase B (0.1% formic acid in acetonitrile). The flow rate was 0.4 mL/min. The linear gradient was as follows: 0–2 min, 95%A; 8–15 min, 5%A; 15.5–22 min, 95%A. The autosampler was set at 15 °C. The injection volume was 5 µL. Mass spectra were acquired under positive electrospray ionization (ESI) with the ion spray voltage at +5000 V. The source temperature was 450 °C. The curtain gas, ion source gas 1, and ion source gas 2 were 32, 50, and 65, respectively. Multiple reaction monitoring (MRM) was used for quantitation: Paclitaxel m/z 854.4 → m/z 569.2; 6α-hydroxypaclitaxel m/z 870.4 → m/z 286.2. Internal standard carbamazepine was monitored at m/z 237.1 → m/z 194.0.

4.9. HOOH Measurements

Hydrogen peroxide measurements were made using an Amplex Red Hydrogen Peroxide/ Horseradish peroxidase (HRP) Kit (Life Technologies, Waltham, MA, USA) according to the published protocol. Amplex Red combined with HRP reacts with HOOH in a 1:1 stoichiometry producing the red-fluorescent oxidation product, resorufin (A_{560nm}). Samples containing 0.1 µM of CYP2C8-ND (*1/*2/*3/R139K/ K399R) were incubated with CPR (0.3 µM) in 0.3 mL of 0.1 M potassium phosphate buffer (pH 7.4), ± paclitaxel (70 µM), for 5 min at 37 °C. NADPH (200 µM) was added and the mixture was incubated for 10, 15, and 20 min at 37 °C, then quenched with equivolume ethyl acetate, vortexed thoroughly, and centrifuged at 3000 rpm at 4 °C for 5 min. The aqueous fraction containing HOOH was extracted and centrifuged at 10,000 rpm at 4 °C for 10 min to remove precipitated protein and lipids. Next, 50 µL of each sample was diluted eight-fold and sixteen-fold and combined with Amplex Red/HRP (10 mM Amplex Red, 10 U/mL HRP in 1× reaction buffer) in a clean, dry 96-well plate. Each sample was analyzed in triplicate. The reactions were incubated at room temperature for 30 min in the dark. The UV A_{560nm} was measured using a microplate reader. Baseline corrected absorbance values of each sample were compared to a standard curve ([HOOH] = 0 to 20 µM).

4.10. NADPH Assay

The rate of NADPH (ε = 6.2 108 mM^{-1}·cm^{-1} at A$_{340nm}$) consumption by each CYP2C8 variant (0.2 µM) incubated with CPR (0.6 µM) and PAC (70 µM) in 0.1 M potassium phosphate buffer and 200 µM NADPH was determined via UV–vis spectroscopy using a Cary 300 UV–vis spectrometer in kinetics mode (Agilent Technologies), as previously described [41,44].

4.11. Stopped-flow Kinetics of Electron Transfer

An Applied Photophysics SX-17 MV Spectrophotometer (Leatherhead UK) was used to monitor the reduction of CYP2C8*1/*2/*3/R139K/K399R, as previously described with the following modifications [41]. Reaction cell 1 containing CYP2C8 (2 µM) in 0.1% cholate, CPR (2 or 6 µM), paclitaxel (70 µM), glucose oxidase (1 U/mL), and glucose (10 mM) dissolved in 100 mM potassium phosphate buffer was kept anaerobic and CO-saturated. Reaction cell 2 containing excess NADPH (1 mM), paclitaxel (70 µM), glucose oxidase (1 U/mL), and glucose (10 mM) dissolved in 100 mM potassium phosphate buffer was also kept anaerobic. The reaction cells were kept at 4 °C until rapid mixing followed by absorbance readings at 37 °C.

4.12. Data Analysis of Stopped-Flow Experiments

The reduction of ferric CYP2C8 to a ferrous–CO complex was monitored near A$_{450nm}$ upon mixing the two separate reaction cells in logarithmic mode and analyzed as described previously, with the following changes [41]. All data indicate the average of 3–6 individual reactions fitted using either a monophasic or biphasic exponential equation using OriginPro 2017. The initial decrease in absorbance at 450 nm (A$_{450nm}$) corresponding to the rapid reduction of CPR were not included in these analyses due to spectroscopic noise. R^2 values for fits exceeded 0.99 in most cases. The errors reported are SEM.

4.13. Reduction Potential

Reduction potential of the CYP2C8 proteins was determined using safranin T as a redox probe as previously described [37]. Samples containing 5 µM of CYP2C8 variant, 20 nM paraquat (methyl viologen), 0.5 µM safarinin T, 10 mM EDTA, 50 µM of a 20% lipid reconstituted system [46], with or without 70 µM paclitaxel were prepared in 0.1 M potassium phosphate buffer, pH 7.4, in glass vials capped with septa. Samples were purged with N$_{2\,(g)}$ for 20 min and then loaded into a Coy anaerobic glove box. 0.5 mL of each samples was loaded into UV-invisible plastic cuvettes stopped with a septa. Reduction potential was determined spectroscopically using a Cary 300 UV–vis spectrometer (Agilent Technologies). Cuvettes were equilibrated at 25 °C for each reading. Safranin T was used as the redox indicator to measure the reduction potential of the solution. Oxidation of the protein was monitored at 417 nm (reduction at 408 nm) and compared to the oxidation of safranin T at 535 nm. Reduction of the samples was initially achieved by irradiating samples on ice with time points up to 5 min with a 250 W tungsten lamp. Samples were further reduced by titrating anaerobic dithionite from 8 and 80 mM stocks. Re-oxidation was achieved by titrating anaerobic K$_3$[Fe(CN)$_6$] from 10 mM stocks. Spectral data were then processed using a MATLAB (R2014a) subroutine and analyzed using the Nernst equation as previously described [37].

Supplementary Materials: Supplementary materials can be found at http://www.mdpi.com/1422-0067/20/18/4626/s1.

Author Contributions: W.R.A., S.Z., D.D.M., and A.D. all contributed to the design of the experiments. W.R.A., S.Z., and K.S. conducted experiments. Mutations were made by D.D.M. W.R.A. and S.Z. analyzed the data and made figures. W.R.A., S.Z., and A.D. contributed to writing the manuscript. All authors accept the final version of the manuscript.

Funding: This project was supported by the American Heart Association [15SDG25760064], and in part by National Institutes of Health Grants R01 GM1155884 and R03 DA 04236502.

Acknowledgments: We thank Eric F. Johnson for the CYP2C8 gene. We thank Sligar for providing the MSPE3D1 gene. We thank Lucas Li at the Roy J. Carver Metabolomics Center at UIUC for mass spectrometry analysis.

Conflicts of Interest: The authors declare no conflict of interest.

Abbreviations

Amp	Ampicillin
AA	Arachidonic acid
Chlr	Chloramphenicol
CYP	Cytochrome P450
CYP2C8	Cytochrome P450 2C8
CYP2C8*3	Cytochrome P450 2C8 R139K/K399R
CYP2C8*2	Cytochrome P450 2C8 I269F
CPR	Cytochrome P450 reductase
δ-ALA	δ-Aminolevulinic acid
EETs	Epoxyeicosatrienoic acids
FET	First electron transfer
HOOH	Hydrogen peroxide
IPTG	Isopropyl β-D-1-thiogalactopyranoside
ND	Nanodisc
POPC	1-palmitoyl-2-oleoyl-sn-glycero-3-phosphocholine
PAC	Paclitaxel
PAC-OH	6α-Hydroxypaclitaxel
POPS	1-Palmitoyl-2-oleoyl-sn-glycero-3-phospho-L-serine
ROS	Reactive oxygen species

References

1. VandenBrink, B.M.; Foti, R.S.; Rock, D.A.; Wienkers, L.C.; Wahlstrom, J.L. Evaluation of CYP2C8 inhibition in vitro: Utility of montelukast as a selective CYP2C8 probe substrate. *Drug Metab. Dispos.* **2011**, *39*, 1546–1554. [CrossRef] [PubMed]
2. Bahadur, N.; Leathart, J.B.S.; Mutch, E.; Steimel-Crespi, D.; Dunn, S.A.; Gilissen, R.; Houdt, J.V.; Hendrickx, J.; Mannens, G.; Bohets, H. Daly, CYP2C8 polymorphisms in caucasians and their relationship with paclitaxel 6α-hydroxylase activity in human liver microsomes. *Biochem. Pharmacol.* **2002**, *64*, 1579–1589. [CrossRef]
3. Lundblad, M.S.; Stark, K.; Eliasson, E.; Oliw, E.; Rane, A. Biosynthesis of epoxyeicosatrienoic acids varies between polymorphic CYP2C enzymes. *Biochem. Biophys. Res. Commun.* **2005**, *327*, 1052–1057. [CrossRef] [PubMed]
4. Marupudi, N.I.; Han, J.E.; Li, K.W.; Renard, V.M.; Tyler, B.M.; Brem, H. Paclitaxel: a review of adverse toxicities and novel delivery strategies. *Expert Opin. Drug Saf.* **2007**, *6*, 609–621. [CrossRef] [PubMed]
5. Klose, T.S.; Blaisdell, J.A.; Goldstein, J.A. Gene structure of CYP2C8 and extrahepatic distribution of the human CYP2Cs. *J. Biochem. Mol. Toxicol.* **1999**, *13*, 289–295. [CrossRef]
6. Enayetallah, A.E.; French, R.A.; Thibodeau, M.S.; Grant, D.F. Distribution of soluble epoxide hydrolase and of cytochrome P450 2C8, 2C9, and 2J2 in human tissues. *J. Histochem. Cytochem.* **2004**, *52*, 447–454. [CrossRef] [PubMed]
7. DeLozier, T.C.; Kissling, G.E.; Coulter, S.J.; Dai, D.; Foley, J.F.; Bradbury, J.A.; Murphy, E.; Steenbergen, C.; Zeldin, D.C.; Goldstein, J.A. Detection of human CYP2C8, CYP2C9, and CYP2J2 in cardiovascular tissues. *Drug Metab. Dispos.* **2007**, *35*, 682–688. [CrossRef] [PubMed]
8. Node, K.; Huo, Y.; Ruan, X.; Yang, B.; Spiecker, M.; Ley, K.; Zeldin, D.C.; Liao, J.K. Anti-inflammatory properties of cytochrome P450 epoxygenase-derived eicosanoids. *Science* **1999**, *285*, 1276–1279. [CrossRef]
9. Michaelis, U.R.; Fisslthaler, B.; Medhora, M.; Harder, D.; Fleming, I.; Busse, R. Cytochrome P450 2C9-derived epoxyeicosatrienoic acids induce angiogenesis via cross-talk with the epidermal growth factor receptor. *FASEB J.* **2003**. [CrossRef]
10. Sun, J.; Sui, X.; Bradbury, J.A.; Zeldin, D.C.; Conte, M.S.; Liao, J.K. Inhibition of vascular smooth muscle cell migration by cytochrome P450 epoxygenase-derived eicosanoids. *Circ. Res.* **2002**, *90*, 1020–1027. [CrossRef]
11. Zhou, S.F. Polymorphism of human cytochrome P450 2D6 and its clinical significance: Part I. *Clin. Pharmacokinet.* **2009**, *48*, 689–723. [CrossRef] [PubMed]

12. Zhou, S.F. Polymorphism of human cytochrome P450 2D6 and its clinical significance: part II. *Clin. Pharmacokinet.* **2009**, *48*, 761–804. [CrossRef] [PubMed]
13. Dai, D.; Zeldin, D.C.; Blaisdell, J.A.; Chanas, B.; Coulter, S.J.; Ghanayem, B.I.; Goldstein, J.A. Polymorphisms in human CYP2C8 decrease metabolism of the anticancer drug paclitaxel and arachidonic acid. *Pharmacogenet. Genomics* **2001**, *11*, 597–607. [CrossRef]
14. Yu, L.; Shi, D.; Ma, L.; Zhou, Q.; Zeng, S. Influence of CYP2C8 polymorphisms on the hydroxylation metabolism of paclitaxel, repaglinide and ibuprofen enantiomers in vitro. *Biopharm. Drug Dispos.* **2013**, *34*, 278–287. [CrossRef] [PubMed]
15. Hertz, D.; Roy, S.; Jack, J.; Motsinger-Reif, A.; Drobish, A.; Clark, L.S.; Carey, L.; Dees, E.C.; McLeod, H. Genetic heterogeneity beyond CYP2C8*3 does not explain differential sensitivity to paclitaxel-induced neuropathy. *Breast Cancer Res. Treat.* **2014**, *145*, 245–254. [CrossRef] [PubMed]
16. Hertz, D.L.; Roy, S.; Motsinger-Reif, A.A.; Drobish, A.; Clark, L.S.; McLeod, H.L.; Carey, L.A.; Dees, E.C. CYP2C8*3 increases risk of neuropathy in breast cancer patients treated with paclitaxel. *Ann. Oncol.* **2013**, *24*, 1472–1478. [CrossRef]
17. Marcath, L.A.; Kidwell, K.M.; Robinson, A.C.; Vangipuram, K.; Burness, M.L.; Griggs, J.J.; Poznak, C.V.; Schott, A.F.; Hayes, D.F.; Henry, N.L.; et al. Patients carrying CYP2C8*3 have shorter systemic paclitaxel exposure. *Pharmacogenomics* **2019**, *20*, 95–104. [CrossRef] [PubMed]
18. Soyama, A.; Saito, Y.; Hanioka, N.; Murayama, N.; Nakajima, O.; Katori, N.; Ishida, S.; Sai, K.; Ozawa, S.; Sawada, J.I. Non-synonymous single nucleotide alterations found in the CYP2C8 gene result in reduced in vitro paclitaxel metabolism. *Biol. Pharm. Bull.* **2001**, *24*, 1427–1430. [CrossRef]
19. Kaspera, R.; Naraharisetti, S.B.; Evangelista, E.A.; Marciante, K.D.; Psaty, B.M.; Totah, R.A. Drug metabolism by CYP2C8.3 is determined by substrate dependent interactions with cytochrome P450 reductase and cytochrome b5. *Biochem. Pharmacol.* **2011**, *82*, 681–691. [CrossRef]
20. Gao, Y.; Liu, D.; Wang, H.; Zhu, J.; Chen, C. Functional characterization of five CYP2C8 variants and prediction of CYP2C8 genotype-dependent effects on in vitro and in vivo drug–drug interactions. *Xenobiotica* **2010**, *40*, 467–475. [CrossRef]
21. Soyama, A.; Hanioka, N.; Saito, Y.; Murayama, N.; Ando, M.; Ozawa, S.; Sawada, J. Amiodarone N-deethylation by CYP2C8 and its variants, CYP2C8*3 and CYP2C8 P404A. *Pharmacol. Toxicol.* **2002**, *91*, 174–178. [CrossRef]
22. Meunier, B.; de Visser, S.P.; Shaik, S. Mechanism of oxidation reactions catalyzed by cytochrome p450 enzymes. *Chem. Rev.* **2004**, *104*, 3947–3980. [CrossRef] [PubMed]
23. Krest, C.M.; Onderko, E.L.; Yosca, T.H.; Calixto, J.C.; Karp, R.F.; Livada, J.; Rittle, J.; Green, M.T. Reactive intermediates in cytochrome p450 catalysis. *J. Biol. Chem.* **2013**, *288*, 17074–17081. [CrossRef] [PubMed]
24. Guengerich, F.P. Mechanisms of cytochrome P450-catalyzed oxidations. *ACS Catal.* **2018**, *8*, 10964–10976. [CrossRef] [PubMed]
25. Denisov, I.G.; Makris, T.M.; Sligar, S.G.; Schlichting, I. Structure and chemistry of cytochrome P450. *Chem. Rev.* **2005**, *105*. [CrossRef] [PubMed]
26. Fleming, I.; Michaelis, U.R.; Bredenkötter, D.; Fisslthaler, B.; Dehghani, F.; Brandes, R.P.; Busse, R. Endothelium-derived hyperpolarizing factor synthase (cytochrome P450 2C9) is a functionally significant source of reactive oxygen species in coronary arteries. *Circ. Res.* **2001**, *88*, 44–51. [CrossRef] [PubMed]
27. Edin, M.L.; Wang, Z.; Bradbury, J.A.; Graves, J.P.; Lih, F.B.; DeGraff, L.M.; Foley, J.F.; Torphy, R.; Ronnekleiv, O.K.; Tomer, K.B.; et al. Endothelial expression of human cytochrome P450 epoxygenase CYP2C8 increases susceptibility to ischemia-reperfusion injury in isolated mouse heart. *FASEB J.* **2011**, *25*, 3436–3447. [CrossRef] [PubMed]
28. Nesnow, S.; Grindstaff, R.D.; Lambert, G.; Padgett, W.T.; Bruno, M.; Ge, Y.; Chen, P.J.; Wood, C.E.; Murphy, L. Propiconazole increases reactive oxygen species levels in mouse hepatic cells in culture and in mouse liver by a cytochrome P450 enzyme mediated process. *Chem.-Biol. Interact.* **2011**, *194*, 79–89. [CrossRef]
29. Leung, T.; Rajendran, R.; Singh, S.; Garva, R.; Krstic-Demonacos, M.; Demonacos, C. Cytochrome P450 2E1 (CYP2E1) regulates the response to oxidative stress and migration of breast cancer cells. *Breast Cancer Res.* **2013**, *15*, R107. [CrossRef]
30. Hunter, A.L.; Bai, N.; Laher, I.; Granville, D.J. Cytochrome p450 2C inhibition reduces post-ischemic vascular dysfunction. *Vasc. Pharmacol.* **2005**, *43*, 213–219. [CrossRef]

31. Bayburt, T.H.; Grinkova, Y.V.; Sligar, S.G. Self-assembly of discoidal phospholipid bilayer nanoparticles with membrane scaffold proteins. *Nano Lett.* **2002**, *2*, 853–856. [CrossRef]
32. Orlando, B.J.; McDougle, D.R.; Lucido, M.J.; Eng, E.T.; Graham, L.A.; Schneider, C.; Stokes, D.L.; Das, A.; Malkowski, M.G. Cyclooxygenase-2 catalysis and inhibition in lipid bilayer nanodiscs. *Arch. Biochem. Biophys.* **2014**, *546*, 33–40. [CrossRef] [PubMed]
33. Schoch, G.A.; Yano, J.K.; Wester, M.R.; Griffin, K.J.; Stout, C.D.; Johnson, E.F. Structure of human microsomal cytochrome P450 2C8. Evidence for a peripheral fatty acid binding site. *J. Biol. Chem.* **2004**, *279*, 9497–9503. [CrossRef] [PubMed]
34. Copeland, R.A. *Enzymes: A Practical Introduction to Structure, Mechanism, and Data Analysis*; John Wiley & Sons: New York, NY, USA, 2004.
35. Kim, H.K.; Park, S.K.; Zhou, J.L.; Taglialatela, G.; Chung, K.; Coggeshall, R.E.; Chung, J.M. Reactive oxygen species (ROS) play an important role in a rat model of neuropathic pain. *Pain* **2004**, *111*, 116–124. [CrossRef] [PubMed]
36. Scripture, C.D.; Figg, W.D.; Sparreboom, A. Peripheral neuropathy induced by paclitaxel: recent insights and future perspectives. *Curr. Neuropharmacol.* **2006**, *4*, 165–172. [CrossRef] [PubMed]
37. Das, A.; Varma, S.S.; Mularczyk, C.; Meling, D.D. Functional investigations of thromboxane synthase (CYP5A1) in lipid bilayers of nanodiscs. *Chembiochem* **2014**, *15*, 892–899. [CrossRef] [PubMed]
38. Das, A.; Grinkova, Y.V.; Sligar, S.G. Redox potential control by drug binding to cytochrome P450 3A4. *J. Am. Chem. Soc.* **2007**, *129*, 13778–13779. [CrossRef] [PubMed]
39. Arnold, W.R.; Baylon, J.L.; Tajkhorshid, E.; Das, A. Asymmetric binding and metabolism of polyunsaturated fatty acids (PUFAs) by CYP2J2 epoxygenase. *Biochemistry* **2016**, *55*, 6969–6980. [CrossRef]
40. Reed, R.; Hollenberg, P.F. New perspectives on the conformational equilibrium regulating multi-phasic reduction of cytochrome P450 2B4 by cytochrome P450 reductase. *J. Inorg. Biochem.* **2003**, *97*, 276–286. [CrossRef]
41. Meling, D.D.; McDougle, D.R.; Das, A. CYP2J2 epoxygenase membrane anchor plays an important role in facilitating electron transfer from CPR. *J. Inorg. Biochem.* **2015**, *142*, 47–53. [CrossRef]
42. Sokalingam, S.; Raghunathan, G.; Soundrarajan, N.; Lee, S.G. A study on the effect of surface lysine to arginine mutagenesis on protein stability and structure using green fluorescent protein. *PLoS ONE* **2012**, *7*. [CrossRef] [PubMed]
43. Li, L.; Vorobyov, I.; Allen, T.W. The different interactions of lysine and arginine side chains with lipid membranes. *J. Phys. Chem.* **2013**, *117*, 11906–11920. [CrossRef]
44. McDougle, D.R.; Palaria, A.; Magnetta, E.; Meling, D.D.; Das, A. Functional studies of N-terminally modified CYP2J2 epoxygenase in model lipid bilayers. *Protein Sci.* **2013**. [CrossRef] [PubMed]
45. Zelasko, S.; Palaria, A.; Das, A. Optimizations to achieve high-level expression of cytochrome P450 proteins using Escherichia coli expression systems. *Protein Expr. Purif.* **2013**, *92*, 77–87. [CrossRef] [PubMed]
46. Arnold, W.R.; Baylon, J.L.; Tajkhorshid, E.; Das, A. Arachidonic acid metabolism by human cardiovascular CYP2J2 is modulated by doxorubicin. *Biochemistry* **2017**, *56*, 6700–6712. [CrossRef] [PubMed]

© 2019 by the authors. Licensee MDPI, Basel, Switzerland. This article is an open access article distributed under the terms and conditions of the Creative Commons Attribution (CC BY) license (http://creativecommons.org/licenses/by/4.0/).

Article

Probing the Role of the Hinge Segment of Cytochrome P450 Oxidoreductase in the Interaction with Cytochrome P450

Diana Campelo [1], Francisco Esteves [1], Bernardo Brito Palma [1], Bruno Costa Gomes [1], José Rueff [1], Thomas Lautier [2], Philippe Urban [2], Gilles Truan [2] and Michel Kranendonk [1,*]

1. Center for Toxicogenomics and Human Health, Genetics, Oncology and Human Toxicology, NOVA Medical School, Faculdade de Ciências Médicas, Universidade Nova de Lisboa, 1150-082 Lisbon, Portugal; diana.campelo@nms.unl.pt (D.C.); francisco.esteves@nms.unl.pt (F.E.); bernardo.palma@nms.unl.pt (B.B.P.); bruno.gomes@nms.unl.pt (B.C.G.); jose.rueff@nms.unl.pt (J.R.)
2. LISBP, Université de Toulouse, CNRS, INRA, INSA, 31077 Toulouse CEDEX 04, France; lautier@insa-toulouse.fr (T.L.); philippe.urban@insa-toulouse.fr (P.U.); gilles.truan@insa-toulouse.fr (G.T.)
* Correspondence: michel.kranendonk@nms.unl.pt; Tel.: +351-21-8803100

Received: 15 November 2018; Accepted: 5 December 2018; Published: 6 December 2018

Abstract: NADPH-cytochrome P450 reductase (CPR) is the unique redox partner of microsomal cytochrome P450s (CYPs). CPR exists in a conformational equilibrium between open and closed conformations throughout its electron transfer (ET) function. Previously, we have shown that electrostatic and flexibility properties of the hinge segment of CPR are critical for ET. Three mutants of human CPR were studied (S243P, I245P and R246A) and combined with representative human drug-metabolizing CYPs (isoforms 1A2, 2A6 and 3A4). To probe the effect of these hinge mutations different experimental approaches were employed: CYP bioactivation capacity of pre-carcinogens, enzyme kinetic analysis, and effect of the ionic strength and cytochrome b_5 (CYB5) on CYP activity. The hinge mutations influenced the bioactivation of pre-carcinogens, which seemed CYP isoform and substrate dependent. The deviations of Michaelis-Menten kinetic parameters uncovered tend to confirm this discrepancy, which was confirmed by CYP and hinge mutant specific salt/activity profiles. CPR/CYB5 competition experiments indicated a less important role of affinity in CPR/CYP interaction. Overall, our data suggest that the highly flexible hinge of CPR is responsible for the existence of a conformational aggregate of different open CPR conformers enabling ET-interaction with structural varied redox partners.

Keywords: NADPH-cytochrome P450 reductase (CPR); microsomal cytochrome P450 (CYP); Cytochrome b_5 (CYB5); protein dynamics; electron-transfer (ET); protein–protein interaction

---

1. Introduction

Microsomal cytochrome P450 (CYP) metabolism requires a coupled supply of electrons, which are donated by the auxiliary protein NADPH cytochrome P450 oxidoreductase (CPR). CPR, encoded by the *POR* gene, is a ~78-kDa electron-transferring diflavin enzyme anchored to the membrane of the endoplasmic reticulum [1]. CPR mediates a two-electron transfer (ET) per reaction cycle, originated from NADPH enabling CYP-mediated metabolism of many compounds. These include endobiotics, e.g., steroids, bile acids, vitamins and arachidonic acid metabolites, as well as many xenobiotics, including therapeutic drugs and environmental toxins [2,3]. Moreover, CPR is the unique electron supplier of heme oxygenase, squalene monooxygenase and fatty acid elongase [4], sustaining exclusively the activity of these enzymes. Cytochrome b_5 (CYB5) can donate the second electron to CYP, competing with CPR for the binding site on the proximal side of CYP [5]. CYB5's interaction may have either a stimulating, inhibiting or having no effect over CYP catalytic activity, which seems to be CYP isoform and even substrate dependent [6].

CPR comprises a number of structurally distinct domains namely an N-terminal hydrophobic membrane anchoring domain; two flavin binding domains for flavin adenine dinucleotide (FAD) and flavin mononucleotide (FMN); a linker domain joining the FMN and FAD domains, as the provider of structural flexibility; and an NADPH binding domain [7]. The hinge segment, a highly flexible stretch with no defined secondary structure links the FMN and the connecting/FAD domain [8–10]. Electrons are transferred from NADPH through FAD (reductase) and FMN (transporter) coenzymes of CPR to redox partners, such as to the heme group in the reactive center of CYP [11].

Initial structural studies of CPR identified compact conformations that allowed internal (inter-flavin) ET, but were unable to reduce external acceptors [8,12,13]. Subsequently, three separate studies identified different open structures of CPR that allowed ET to redox partners, indicative of domain motion of CPR [10,14,15]. It is now fairly established that CPR exists in a conformational equilibrium between open and closed states in its ET function, which is highly dependent on ionic strength conditions [10,16,17]. The transition between these states appears to occur through a rapid swinging and rotational movement [17,18]. Certain residues in the hinge region have been suggested to be of importance for these large conformational changes [19], and seem to form a conformational axis, involved in a partial rotational movement of the FMN domain relative to the remainder of the protein [10,18].

Analysis of CPR domain dynamics is pertinent to understand its role in the interactions with its natural redox partners and its gated ET function. The affinities between CPR and CYP have been indicated among the factors modulating the protein dynamics of CPR. Different CYP isoforms may be differently served by CPR gating its ET differentially [20,21]. Although advances obtained during the last decade, CPR's structural features controlling ET are not yet properly identified. CPR mutations may perturb specific structural requisites, necessary for the optimal transition between open and closed conformations, as well as disturb the interaction of CPR with its redox partners [10,20–22].

Previously, we have studied the effect of mutations in CPR on its redox partners [20,22,23] and the effect of alterations in the hinge segment in CPR-dependent cytochrome c reduction [24]. These hinge mutations showed differential effects on the conformational equilibrium of CPR and ET efficiency to cytochrome c, a non-physiological redox partner of CPR. Through modulation of the ionic strength conditions we demonstrated that electrostatic and flexibility properties of the hinge are critical for ET function, in which CPR's membrane anchoring was shown to play an important role [24]. Although frequently used as a surrogate, the soluble cytochrome c has been indicated to interact differently with CPR, when compared with interactions of natural membrane-bound partners, such as CYP [25]. The use of cytochrome c as redox partner may have obscured additional important clues on structural features of the hinge segment involved in CPR's open/closed dynamics and its gated ET function. To address this issue, three hinge mutants were selected from the initial eight mutants of our former study, based on their specific phenotypes in cytochrome c reduction. Human membrane bound CPR mutants S243P, I245P and R246A (numbering according to the human CPR consensus amino acid sequence NP_000932) were each combined with three different human CYPs, namely CYP1A2, 2A6 or 3A4, representatives of three major CYP families involved in drug metabolism [2,3]. The effect of the structural deviations of the three mutants was probed to obtain further insights on the role of the hinge segment of CPR in the interaction and ET with these physiological redox partners, using different experimental approaches.

2. Results

2.1. Bacterial Coexpression of CPR Mutants and CYP

Wild-type CPR and CPR hinge mutants S243P, I245P and R246A were separately introduced in the *E. coli* BTC strain and co-expressed with CYP1A2, 2A6 or 3A4, using methods described previously [26,27]. CYP expression levels were determined in bacterial whole-cells (Table 1). When co-expressed with CPR variants, CYP expression levels varied between 109–241 nM, 96–130 nM and 105–143 nM for CYP1A2, 2A6 and 3A4, respectively. Expression levels for these CYPs were comparable with those found previously with BTC strains [22,28,29]. More importantly, no large deviations were found in the CPR:CYP ratio between the four CPR variants, when expressed with each of the three CYPs (see Table 1). This enabled us to ascribe differences in activities of the CPR variants to the mutations, and not to variations in the stoichiometry between the two enzymes. These ratios were actually similar to those observed in our previous studies [22,28,29] and are in the range of those observed in human liver microsomes [30,31].

Table 1. Microsomal cytochrome P450 (CYP) and NADPH cytochrome P450 oxidoreductase (CPR) contents of BTC cultures and membrane fractions.

CYP Isoform	CPR Form	Whole-Cells		Membrane Fractions	
		CYP [1] (nM)	CYP [1]	CPR [1] (pmol/mg Protein)	CPR:CYP Ratios
CYP1A2	WT	109 ± 4	54 ± 1	4.1 ± 1.5	1:13
	S243P	241 ± 4	73 ± 4	7.7 ± 0.2	1:9
	I245P	206 ± 11	102 ± 1	6.1 ± 0.5	1:17
	R246A	176 ± 3	91 ± 2	5.4 ± 0.2	1:17
CYP2A6	WT	130 ± 2	139 ± 1	10.5 ± 1.3	1:13
	S243P	98 ± 1	106 ± 3	11.2 ± 0.5	1:9
	I245P	96 ± 7	102 ± 1	9.5 ± 0.9	1:11
	R246A	98 ± 2	146 ± 1	10.6 ± 1.5	1:14
CYP3A4	WT	105 ± 2	83 ± 3	19.8 ± 0.2	1:4
	S243P	122 ± 3	77 ± 2	22.5 ± 2.6	1:3
	I245P	128 ± 3	78 ± 2	18.3 ± 0.7	1:4
	R246A	143 ± 5	78 ± 1	21.4 ± 0.3	1:4

[1] CYP and CPR contents are mean ± sd.

2.2. CYP-Activities When Combined with the Three CPR Hinge Domain Mutant

2.2.1. Whole-Cell Bioactivation Assays

A whole cell/bioactivation assay was used for the first evaluation of the effect of the three hinge mutations on the activity of the three CYPs. This approach made use of the applicability of the BTC-CYP bacteria in mutagenicity testing [26,28]. The levels in CYP-dependent bioactivation of different pre-carcinogens, namely 2AA (2-aminoanthracene), IQ (2-amino-3-methylimidazo(4,5-*f*)quinolone), NNdEA (*N*-nitrosodiethylamine), NNK (4-(methylnitrosamino)-1-(3-pyridyl)-1-butanone) and AfB1 (aflatoxin B1)) were determined (Table 2; Figure 1). Interestingly, two of the three CPR hinge mutants lead to bioactivation capacities, which were either stimulated or equal, in comparison when CYPs were sustained by WT CPR, except for mutant I245P. This hinge mutant demonstrated a decrease for CYP1A2 and CYP3A4 mediated bioactivation of 2AA and AfB1, respectively. In contrast, CYP1A2 and CYP2A6 bioactivation capacities (for IQ and NNdEA, respectively) were increased when assayed with this mutant, with no significant differences in CYP2A6 bioactivation of NNK. Seemingly, the effect of CPR mutant I245P was CYP isoform and substrate dependent. The bioactivation capacity of all three CYPs was consistently augmented when assayed with CPR mutant R246A, i.e., all tested compounds demonstrated increased mutagenicity levels, in comparison with CYPs sustained by WT CPR. Mutant S243P demonstrated no significant differences in the bioactivation capacity of the three CYPs for the tested compounds, when compared with WT CPR.

Table 2. CYP-mediated bioactivation of pre-carcinogens.

CYP Isoform	Mutagen	CPR Form			
		WT	S243P	I245P	R246A
CYP1A2	2AA [1]	5643 ± 271	5666 ± 177	3339 ± 145	6274 ± 106
	IQ [1]	335 ± 8	323 ± 7	521 ± 104	1103 ± 253
CYP2A6	NNdEA [2]	537 ± 14	543 ± 21	885 ± 122	929 ± 118
	NNK [2]	770 ± 121	799 ± 110	788 ± 57	1014 ± 29
CYP3A4	AfB1 [1]	1129 ± 97	1109 ± 115	661 ± 185	1616 ± 163

Values are mean ± sd of three independent experiments, expressed as the number of revertant colonies per nmol [1] or per μmol [2] of pre-carcinogen.

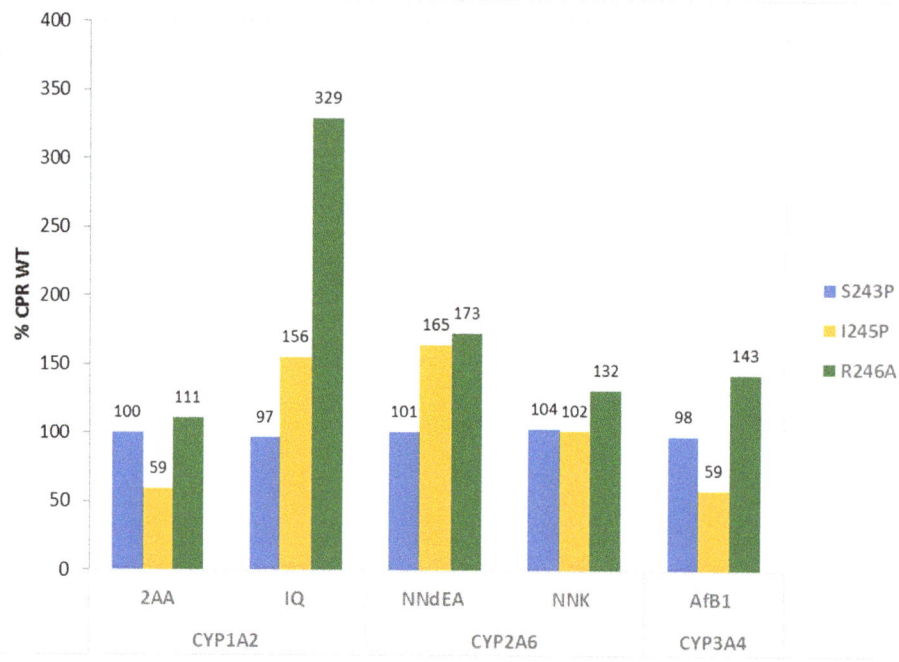

Figure 1. Bioactivation of pre-carcinogens mediated by CYP 1A2, 2A6 or 3A4 when combined with the three CPR hinge mutants. Bioactivation capacities were normalized with the one observed when CYPs were combined with WT CPR. (2AA: 2-aminoanthracene; IQ: 2-amino-3-methylimidazo(4,5-*f*)quinoline; NNdEA: N-nitrosodiethylamine; NNK: 4-(methylnitrosamino)-1-(3-pyridyl)-1-butanone; AfB1: aflatoxin B1).

2.2.2. Membrane Preparations

CYP-Enzyme Kinetic Analysis

Enzyme activities of CYP1A2, 2A6 and 3A4 were measured using specific fluorogenic probe substrates ethoxyresorufin (EthR), coumarin and dibenzylfluorescein (DBF), respectively). Reaction velocities could be plotted according to the Michaelis-Menten equation and kinetic parameters (k_{cat} and K_M) could be derived (Table 3). In general, CYP activities promoted by CPR mutants showed lower turn-over rates (k_{cat}) and affinities (K_M) when compared with WT that were also CYP form dependent. However, differences were minor in both constants and k_{cat}/K_M values were not significantly different from WT values for all tested CYPs (Table 3).

Table 3. Michaelis-Menten kinetic parameters of CYP activities.

CYP Isoform	CPR Form	k_{cat} [1] (Product Formed pmol·min^{-1}·pmol^{-1} CYP)	K_M [1] (μM)	Efficiency (k_{cat}/K_M) (% WT)
CYP1A2	WT	0.62 ± 0.02	1.94 ± 0.16	0.32 (1.00)
	S243P	0.63 ± 0.01	1.23 ± 0.05	0.51 (1.59)
	I245P	0.40 ± 0.01	0.89 ± 0.04	0.46 (1.44)
	R246A	0.43 ± 0.01	0.74 ± 0.06	0.59 (1.84)
CYP2A6	WT	1.37 ± 0.07	1.99 ± 0.34	0.69 (1.00)
	S243P	1.17 ± 0.08	2.03 ± 0.40	0.58 (0.84)
	I245P	1.20 ± 0.09	1.78 ± 0.39	0.67 (0.97)
	R246A	1.53 ± 0.08	1.98 ± 0.33	0.77 (1.12)
CYP3A4	WT	2.53 ± 0.16	3.75 ± 0.55	0.67 (1.00)
	S243P	1.83 ± 0.08	3.07 ± 0.34	0.60 (0.88)
	I245P	1.89 ± 0.13	2.85 ± 0.50	0.66 (0.98)
	R246A	2.18 ± 0.22	3.82 ± 0.87	0.57 (0.85)

[1] k_{cat} and K_M values are expressed as mean values of three independent experiments ± sd, determined at 0 M NaCl.

Ionic Strength Effect on CYP:CPR Interaction

We previously demonstrated that mutations in the hinge region of human CPR strongly influences ionic strength profiles of ET to cytochrome *c* [24]. Ionic strength dependency of CPR's ET was thus analyzed via CYP activities measurement in presence of increasing NaCl concentrations (0–1.25 M) (Figures 2 and 3). Control experiments showed no effect of the salt concentration on the fluorescence of the formed products (Figure S2) or on the pH of the reaction mixture (data not shown). Interestingly, the maximum velocities of the three CYPs seem to occur at lower ionic strength conditions for all CPR (WT or variants) when compared to their maximum velocities in cytochrome *c* reduction reported in our previous study [24].

The salt dependence of CYP1A2 activity demonstrated a bell-shaped EthR O-deethylase activity curve with all CPR tested (Figure 2A), analogous to the ones described in our former study when measuring cytochrome *c* reduction [24]. The maximum activity (maximum k_{obs}) was obtained, on average, at 100 mM NaCl. CYP1A2 activities dropped close to zero at the highest salt concentrations (1.25 M) for all CPR variants. CPR mutant S243P was more active than the WT, while other mutants showed lower activities when compared with WT.

For CYP2A6 (Figure 2B), the coumarin hydroxylase activity profiles showed less dependence on ionic strength as with CYP1A2 and showed no drop in activity at the highest salt concentrations. Maximum velocities were obtained approximately at 150 mM NaCl. All CPR mutants presented lower activities when compared with WT CPR.

CYP3A4 DBF O-debenzylase activity profiles (Figure 2C) were also bell-shaped, like CYP1A2, except for the CPR mutant R246A. The maximum activity was obtained, on average, at 100 mM NaCl. The activity dropped close to zero at the highest salt concentrations as was observed with CYP1A2. Interestingly, the CPR mutant R246A revealed a different salt profile with CYP3A4, leading to a continuous decrease of activity with increasing salt concentration, the maximal activity being even greater that the one obtained with WT CPR.

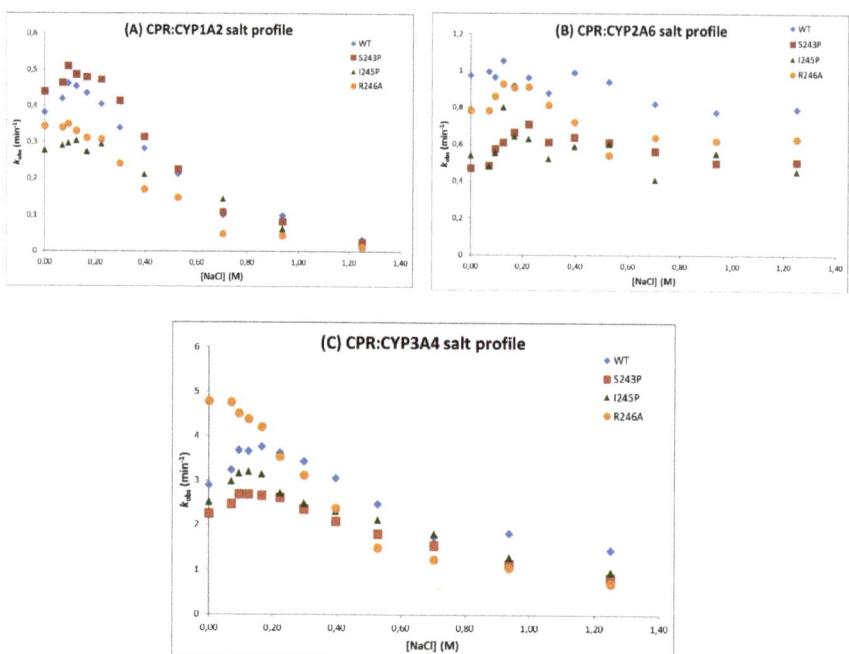

Figure 2. CYP reaction velocity (k_{obs}) in function of the NaCl concentration, for the WT and mutant forms of CPR with CYP1A2 (**A**), CYP2A6 (**B**) or CYP3A4 (**C**).

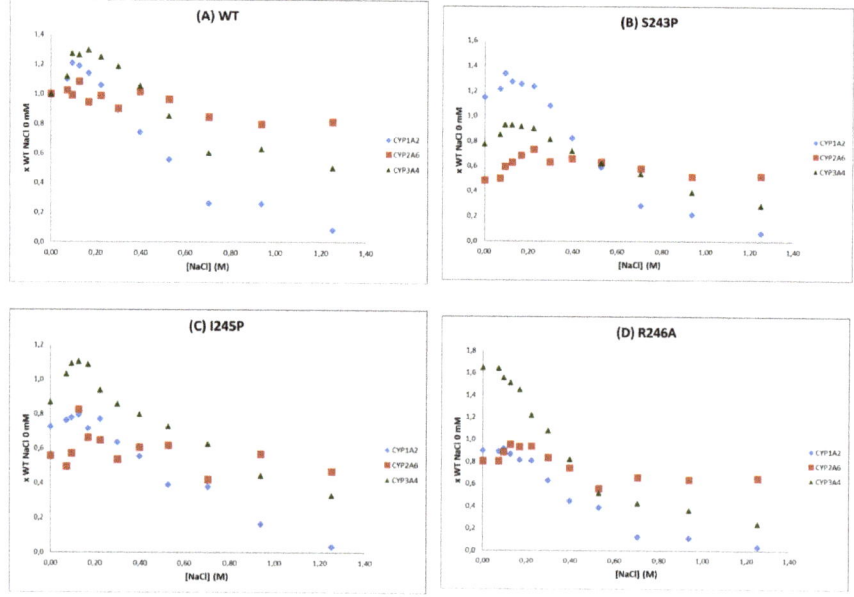

Figure 3. Relative CYP-reaction velocities (k_{obs}) of CYP1A2, 2A6 and 3A4 in function of the NaCl concentration, when combined with CPR WT (**A**) or mutants S234P (**B**), I245P (**C**) and R246A (**D**). Velocities were normalized with the one observed when combined with WT CPR at 0 M NaCl.

CYB5 Effect on Activity of CYP1A2, 2A6 and 3A4

CYB5 may act as an alternative donor of the second electron in the CYP catalytic cycle. We thus studied the effect of CPR hinge mutations when associated with CYB5, to determine if the presence of an alternative electron donor partner, capable of competing with CPR for CYPs is an important issue in the function of the hinge segment. The effects of CYB5 on CPR/CYP combinations were analyzed through enzyme activity assays in the presence of increasing concentrations of CYB5 and thus different CYB5:CPR ratios (Figure 4 and Table 4). Enzyme activities were normalized to the activity measured in the absence of CYB5. While the effect of CYB5 on CYP activities was not major, some interesting differences could be seen, notably in the CYB5 concentration giving the best stimulus. While for CYP1A2, the maximal effect was observed at 150 nM of CYB5, for CYP2A6 the concentration of CYB5 giving this maximal effect was dependent on the mutation, ranging from 50 nM with the WT CPR to 400 nM for the R246A mutant of CPR. For CYP3A4, the effect was relatively constant between all CPR mutants, but the concentration of CYB5 needed to obtain the maximal stimulation was much higher (400 nM). Overall, while CYP1A2 does not seem very sensitive to the presence of CYB5, a slight inhibition could be observed at high concentrations of CYB5. For the two other CYPs, the stimulation was quite pronounced, ranging from 2.6 to 3.4 or 1.8 to 2.6 for CYP2A6 and CYP3A4, respectively (Figure 4). No major differences were observed between CPR WT and mutants for each CPR/CYP combinations in term of the concentration of CYB5 to achieve the maximal effect, however, the intensity of the stimuli was different between CPR mutants.

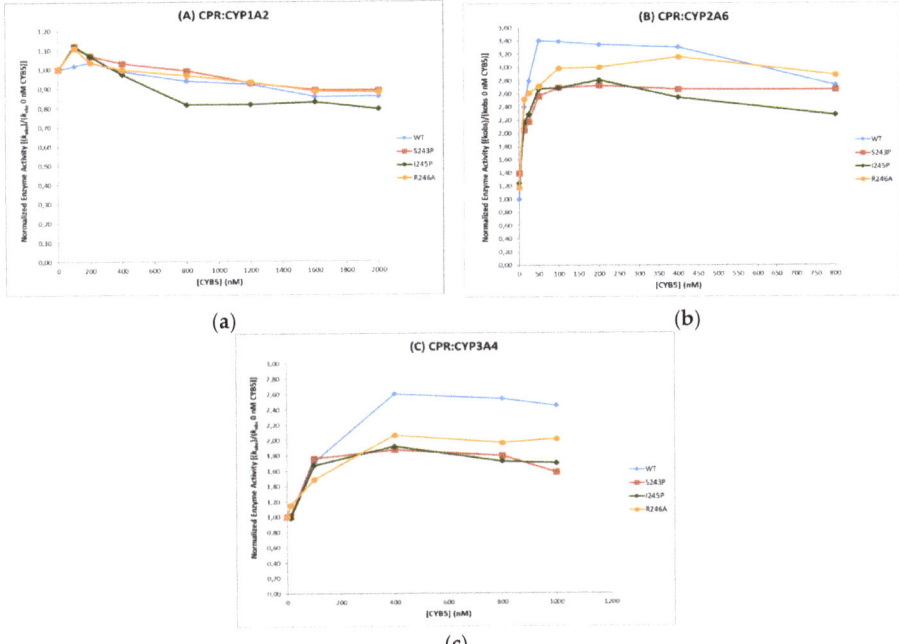

Figure 4. Effect of CYB5 concentration on maximum reaction velocities of CYP1A2 (A), 2A6 (B) and 3A4 (C) when sustained by WT CPR and the three hinge mutants. Activities were normalized to the one without CYB5 for each CPR mutant in comparison to WT.

Table 4. Effect of CYB5 on CPR/CYP combinations.

CYP Isoform	CPR Form	Maximal k_{obs} [1]	CYB5 Stimulus [2] (%)	CYB5:CPR Ratio [3]
CYP1A2	WT	1.04 ± 0.01	100	163
	S243P	1.11 ± 0.02	107	118
	I245P	1.12 ± 0.04	108	208
	R246A	1.12 ± 0.02	108	211
CYP2A6	WT	3.40 ± 0.04	100	26
	S243P	2.72 ± 0.05	80	106
	I245P	2.69 ± 0.04	79	43
	R246A	3.15 ± 0.04	93	211
CYP3A4	WT	2.60 ± 0.01	100	66
	S243P	1.87 ± 0.02	72	55
	I245P	1.92 ± 0.01	74	68
	R246A	2.06 ± 0.02	79	67

[1] Normalized maximum enzyme activity: $(k_{obs})/(k_{obs}$ at 0 nM CYB5). Values are expressed as the mean of three independent experiments ± sd. [2] Maximum CYB5 stimulus of enzyme activity when CYPs were combined with mutant CPRs, relative to the ones obtained with CPR WT. [3] The ratio at which maximum k_{obs} was reached. Grey shaded values: Maximum fold stimulation of enzyme activity by CYB5 when CYPs were combined with WT CPR.

3. Discussion

Microsomal CYP-mediated metabolism is dependent on ET through protein:protein interaction with its primary redox partner CPR. Traditionally, membrane anchoring and a negatively-charged surface patch of CPR have been considered to be major determinants of the proper alignment in this interaction, in which hydrophobic interactions have also been implicated [32,33]. Still, the view on CPR:CYP interaction for ET has become increasingly more complex with the recognition of CPR protein dynamics in ET, presenting an equilibrium between closed/locked and open/unlocked conformers in its function, in which the flexible hinge region seems to have a determinant role [10,16–18]. In a recent report we have shown the importance of this hinge segment and the effect of ionic strength in ET to the non-physiological redox partner cytochrome c, using eight different CPR mutants targeting four critical residues [24]. Data from this report lead us to hypothesize a hydrogen H-bond network around R246, important for the function of the open/closed dynamics.

Although the use of the soluble surrogate redox partner cytochrome c has been informative, the open/closed dynamics and thus ET may occur differently with CPR's natural membrane-bound electron acceptors, such as CYPs. Three hinge mutants were selected from our former set, namely mutant R246A, for R246's role in the suggested H-bond network, and mutants S243P and I245P demonstrating augmented cytochrome c reduction capabilities when compared with WT CPR. The CPR variants were combined with three different CYPs, representatives of three major CYP-families involved in drug metabolism. CYP families 1, 2 and 3 are responsible for 75% of all phase I metabolism of clinically used drugs: CYP3A4 is the major enzyme, and together these three CYPs are involved in almost 50% of metabolism of drugs [34]. Different experimental approaches were used to probe the effect of these hinge mutations.

The first general observation is that none of the three mutations caused a complete inactivation of CYP activity (i.e., obliterated CPR electron transfer to CYP), consistent with the data obtained using cytochrome c as electron acceptor [24]. We thus confirm that the three targeted positions in the hinge region are certainly part of a set of residues capable of controlling the CPR function and thus the effects of the mutations on internal and external ET are partially compensated by a larger network involved in the structural transitions (opening/closing).

The second interesting feature that these mutants demonstrated was their relative selectivity in inducing CYP-isoform dependent effects. This was first observed with CYP mediated bioactivation of pre-carcinogens, which seemed CYP isoform and in some cases substrate dependent. Still, the measurement of mutagenicity might be a quite indirect measurement of CYP activity, in generating

DNA damaging metabolites. However, the specific features of CPR hinge mutants were more clearly demonstrated by the ionic strength dependency of CYP activities. Salt activity profiles were deviated differently by the three CPR hinge mutations, in a CYP-isoform dependent manner (Figures 2 and 3). We previously demonstrated that the salt concentration at which the overall CPR to acceptor ET occurs depends on the relative difference in the rate of reduction of CPR and the rate of reduction of the electron acceptor [17]. The fact that all hinge mutations affect the salt concentration at which this maximal activity occur indicate that these CPR mutations modify the conformational equilibria. However, as noted above, all mutations have distinct salt-dependent signatures (Figure 3). We can thus hypothesize that the induced differences in conformational equilibria affect differently the three tested CYPs. Interestingly, all V_{max} and K_M values measured for the three CYPs were nearly identical between CPR mutants. This reinforces our current hypotheses addressing the conformational equilibria: The single hinge CPR mutants modify the salt-dependent conformational equilibria and not the intrinsic CYP properties. As such, they uncover the dependence of CYPs toward these conformational equilibria and provide a potential explanation of the mechanism by which CPR could, in certain conditions, favor one CYP over another.

An additional interesting feature of these salt profiles concerns the optimal salt concentration for CYP activity. Previously, when membrane-bound, WT CPR and the three mutants demonstrated very different salt concentrations for optimal cytochrome c reduction (ranging from 220–550 mM NaCl [24]). However, in the current study when combined with CYP1A2, 2A6 or 3A4, optimal activities were found at 100–200 mM NaCl, which together with the K/P reaction buffer approaches the ionic strength of physiological serum (154 mM KCl). This current data set exemplifies the difference in ET between the soluble surrogate electron acceptor cytochrome c and CPR's natural membrane-bound redox partners. Maximum ET rates of soluble WT CPR in cytochrome c reduction was found to occur when CPR was equally distributed between open and closed conformers at approximately 375 mM NaCl [17]. With membrane bound WT CPR, maximum cytochrome c reduction rates were shifted above 500 mM NaCl, namely at 527 mM [24]. As such, data from our previous and current study suggests that optimal ET to various acceptors, either natural or artificial, occurs at very different ionic strengths. However, under physiological conditions, membrane bound CPR is present in an equilibrium, which is mostly in the locked state, however maintaining a very fast rate in alternating between open and closed conformations. This reinforce the idea that additional factors, such as the presence of membrane bound redox-partners, with stoichiometry's favoring the electron acceptors (as is the case for CYPs, outnumbering CPR by a factor of 5–10), modulate the open/closed dynamics, as we put forward in our previous study [24].

As mentioned above, our data showed differences in the salt effect of the three different CYPs when combined with WT CPR, corroborating the study by Yun et al. [35] that ascribed the ionic strength effect observed mostly to CPR:CYP "interaction" while having relative minor effects on CYP protein conformation [36]. Still, results of the study of Voznesensky and Schenkman [37] indicated that charge pairing between CPR and CYP may not be the major determinant of the salt effect. Our current data,, as well as that of our previous study [24] confirms that the CPR:CYP interaction (i.e., electronic flow) dependency on ionic strength is mainly determined by its effect on the conformational equilibrium between locked and unlocked states of CPR and only in a minor manner by electrostatic interactions (affinity) between the FMN domain and the acceptor. Due to this major salt effect on CPR's open/closed dynamics, the identification of potential electrostatic interactions between CPR and CYP have therefore been obscured. In retrospect, the seminal studies of Voznesensky and Schenkman [37,38] and of others (reviewed in [32]) on the salt effect of CYP catalysis in the quest for electrostatic interactions between CPR and CYP have been hampered by the lack of knowledge, at that time, on the salt-dependent protein dynamics of CPR.

In vivo, under constant ionic strengths conditions, affinity parameters may become determining in the CPR:CYP interaction, modulating the open/closed dynamics as indicated above. CYB5, the optional electron donor, demonstrated also stimulation/inhibition profiles that were dependent on

CYPs, as well as on the various CPR used. This may also sign a relatively minor role for the affinity between CPR/CYP and CYB5/CYP interactions. Although CPR and CYB5 share similar, but not identical, binding-sites on the proximal side of CYP [39], the competition between the two electron donors does not seem to be a major factor in controlling CYP activities. Still, subtle differences in electrostatic interactions between the CPR/CYP and CYB5/CYP complexes have been shown [40,41], which seem even to be depending on substrate binding [42], indicating specific features in the sampling of the ensemble of open conformers of CPR.

Overall, our data point out that salt profiles are specific to CYP isoforms dependent interaction with CPR. The idea emerges for the existence of an ensemble of different unlocked CPR conformers that may be required for CYP-specific interactions. The highly flexible hinge region allows for a large ensemble of open conformations from which only a few or a subset may be required for ET to CYPs. CPR hinge mutants, by modifying the conformational equilibrium (as seen in salt profiles) may either promote or hinder specific conformations of the unlocked state, thus allowing or preventing interactions with (structurally) different redox partners. Such a selection of specific open conformations could explain the differential effect of the three hinge mutations on the activity of the different CYPs. Moreover, the soluble cytochrome c will sample these conformers quite differently when compared with the membrane-bound CYP redox partner, a plausible explanation for the difference in effects of the three hinge mutants when measuring ET to cytochrome c [24] or CYPs (this study).

From a structural perspective, it is clear that different CYP isoforms must share a common functional CPR binding surface. Although mitochondrial CYPs seem to have a signature of key basic amino acids on the proximal side for their interaction with the iron–sulfur protein adrenodoxin, such signature sequences do not exist for microsomal CYPs in their interaction with CPR [43]. This implies diversity in critical amino acids on their proximal side and suggests the possibility of affinity differences of microsomal CYPs for CPR, a plausible key element in the sampling of the open CPR conformers.

The conformational plasticity of CYPs could additionally play a role in this respect. Substrate binding has been shown to cause (subtle) conformational changes at the proximal site of CYPs, reviewed in Kandel and Lampe, 2014 [33], which may influence the affinity and thus sampling of open conformers of CPR for effective ET. In fact, this seems to be corroborated by our data of the hinge mutations, causing beside CYP-isoform dependent seemingly also substrate dependent effects, as described above.

It is tempting to speculate that CPR's protein dynamics, containing different ensembles of closed and open conformations, was Nature's way to enable CPR to be the "degenerated" electron supplier of so many (structural and functional) different redox-partners. In this respect it would be of interest to obtain insight on the evolutionary gain and thus the physiological relevance of such a universal electron donor for enzymes, involved in some many different and crucial metabolic pathways. In parallel to other central enzymes, the possibility may exist of a central hub- or central controller-function of CPR, for the fine tuning of multiple metabolic pathways and energy (NADPH) usage.

4. Materials and Methods

4.1. Reagents

L-Arginine, thiamine, chloramphenicol, ampicillin, kanamycin sulfate, isopropyl β-D-thiogalactoside (IPTG) (dioxane-free), δ-aminolevulinic acid, cytochrome c (horse heart), glucose 6-phosphate, glucose 6-phosphate dehydrogenase, nicotinamide adenine dinucleotide phosphate (NADP+ and NADPH), 2-aminoanthracene (2AA), N-nitrosodiethylamine (NNdEA), 4-(methylnitrosamino)-1-(3-pyridyl)-1-butanone (NNK), aflatoxin B1 (AfB1), resorufin, 7-hydroxy coumarine and fluorescein were obtained from Sigma-Aldrich (St. Louis, MO, USA). 2-amino-3-methylimidazo(4,5-f)quinoline (IQ) was obtained from Toronto Research Chemicals (North York, ON, Canada). LB Broth (Formedium, Norfolk, UK), bacto tryptone and bacto peptone were purchased from BD Biosciences (San Jose, CA, USA). Bacto yeast extract was obtained from Formedium (Norwich, England). EthR and coumarin were obtained from BD Biosciences (San Jose, CA, USA) and DBF from Santa Cruz Biotechnology (Santa Cruz, CA, USA). A polyclonal antibody from rabbit serum raised against recombinant human CPR obtained from Genetex (GTX101099) (Irvine, CA, USA) was used for immunodetection of the membrane-bound CPR.

4.2. Bacterial Expression of Human CYB5 and Purification

The cDNA of the open reading frame of human full length CYB5 was cloned in pET15b, as described in Nunez et al., 2010 [44], except that the full sequence was used instead of only the soluble part of it. The resulting plasmid was transformed into BL21-DE3 for expression. A single colony was grown in Terrific Browth medium containing 100 µg/mL ampicillin for 72 h with shaking at 22 °C. Cells were harvested by centrifugation at 4000× g, and the resulting pellet was resuspended and incubated for 30 min in 50 mM Tris-HCl, pH 7.4, containing 1 mM PMSF and 1 mg/mL lysozyme. Cells were lysed by sonication. Then 0.02 mg/mL RNase and 0.05 mg/mL DNase were added, and CYB5 was solubilized at 4 °C with 1% (w/v) sodium cholate, pH 7.4, for 1 h with moderate shaking. Supernatant was loaded onto a DEAE-cellulose anion-exchange column equilibrated with 0.2% (w/v) sodium cholate, 20 mM sodium/potassium phosphate buffer, pH 7.4. CYB5 was eluted with 0.5 M NaCl, 0.2% cholate, and 20 mM sodium/potassium phosphate buffer, pH 7.4. Fractions containing CYB5 were applied to a hydroxylapatite column equilibrated with 0.5 M NaCl and 20 mM sodium/potassium phosphate buffer, pH 7.4. Pure CYB5 was eluted with 0.1% (w/v) sodium cholate and 0.5 M sodium/potassium phosphate buffer and dialyzed against 0.1% (w/v) sodium cholate, 1 mM PMSF, and 20 mM sodium/potassium phosphate buffer. CYB5 content of samples was determined by spectrophotometric techniques as described previously [28,29]. CYB5 was concentrated and stored at -20 °C.

4.3. Bacterial Co-Expression of Human CPR Mutants and CYPs

CPR forms were expressed as a full-length membrane bound proteins using a dedicated E. coli host, the BTC strain, using the specialized bi-plasmid system adequate for co-expression of CPR with representative human CYP [26,27]. Plasmid pLCM_POR [20] was used for the expression of the membrane-bound, full-length WT form and mutants of human CPR [24]. The expression of human CYP-isoforms in the cell model was accomplished with plasmid pCWori containing human wildtype CYP cDNA (pCWh_1A2, pCWh_2A6 or pCWh_3A4) [45]. The pLCM_POR and the CYP plasmids pCWh were transfected through standard electroporation procedures [22]. Each strain was cultured in TB medium supplemented with peptone (2 g/L), thiamine (1 µg/mL), ampicillin (50 µg/mL), kanamycin (15 µg/mL), chloramphenicol (10 µg/mL), trace elements solution [46] (0.4 µL/mL), IPTG (0.2 mmol/L) and δ-Ala (100 µmol/L), final concentrations. Cultures were started with -80 °C glycerol stocks and cells were grown for 16 h at 28 °C with moderate agitation. CYP content of bacterial whole-cells preparations was determined using standard CO-difference spectrophotometry [47].

4.4. Whole-Cell Mutagenicity Assays

The mutagenicity assays were performed as described previously [20,23,45,47] using the liquid pre-incubation assay technique [48,49]). Briefly, BTC bacteria were grown for 18 h in TB medium supplemented with peptone (2 g/L), thiamine (1 µg/mL), ampicillin (50 µg/mL), kanamycin (15 µg/mL), a mixture of trace elements solution [46] (0.4 µL/mL) and IPTG (0.2 µmol/L), final concentrations. Pre-incubation was performed for 45 min in an orbital shaker at 37 °C before plating. Incubation buffer contained 10 mM glucose. Stock solutions of carcinogens were freshly made in dimethyl sulfoxide (DMSO) and working solutions were obtained by dilution in water. DMSO concentration in preincubations were ≤1.3%. Experiments were performed at least in triplicate. Revertant colonies on L-Arg selector plates were counted after 48 h incubation at 37 °C. Revertant colonies were determined by ProtoCOL 3 colony counter (Synbiosis, Cambridge, UK) using ProtoCOL

V0 1.0.6 Software. CYP-mediated bioactivation was expressed in terms of mutagenic activity [L-arginine prototrophic (revertant) colonies per nmole of test compound, or in revertant colonies per µmole of test compound], determined from the slope of the linear portion of the dose–response curve.

4.5. Membrane Preparation and Characterization

Membrane preparations were isolated as previously described [22,24]. Briefly, cultures were harvested at 2772× g for 20 min at 4 °C. The pellet was resuspended in Tris-sucrose buffer (50 mM Tris-HCl, 250 mM sucrose, pH 7.8). Lysozyme was added to a final concentration of 0.5 mg/mL, EDTA (0.5 mM), phenylmethanesulfonyl fluoride (0.5 mM) and benzonase (12.5 U/mL). Cells were incubated on a roller bench for 30 min at 4 °C. Cell lyses was performed by freezing (−80 °C) and thawing (1 cycle) and by subsequent several short rounds (30 s) of low-intensity sonication, interspersed with 60 s of ice-bath submersion. The suspension was centrifuged at 2772× g, for 20 min at 4 °C to eliminate unbroken cells. Membranes were pelleted by ultracentrifugation of the supernatant at 100,000× g at 4 °C for 60 min. Membranes were resuspended in TGE buffer (75 mM Tris-HCl, 10% (v/v) glycerol, 25 mM EDTA, pH 7.5) using a Potter homogenizer and stored at −80 °C. Protein concentrations were determined using the method described by Bradford, following the manufacturer's protocol from Bio-Rad (San Francisco, CA, USA), using bovine serum albumin as the standard. CYP contents of membrane preparations were determined using CO-difference spectrophotometry. Membrane proteins were separated by SDS–PAGE gel electrophoresis (10% polyacrylamide gel) and either stained with Coomassie blue or electro-transferred to PVDF membranes and further processed. CPR content of membrane fractions was quantified by immunodetection against a standard curve of purified human, full-length WT CPR, using polyclonal rabbit anti-CPR primary antibody and biotin-goat anti-rabbit antibody in combination with the fluorescent streptavidin conjugate (WesternDot 625 Western Blot Kit; Invitrogen, Eugene, OR, USA) (see Figure S1). Densitometry of CPR signals was performed using LabWorks 4.6 software (UVP, Cambridge, UK).

4.6. CYP-Enzyme Assays

Using membrane preparations, CYP-activities were assessed through determination of product formation by EthR O-deethylation (EROD; CYP1A2) (excitation 530 nm; emission 580 nm), coumarin 7-hydroxylation (CYP2A6) (excitation 330 nm; emission 460 nm) or O-debenzylation of DBF (CYP3A4) (excitation 485 nm; emission 535 nm) [20,22,26]. Assays were performed in 96-well format with a multi-mode microtiter plate reader (SpectraMax®i3x, Molecular Devices, USA) using SoftMax Pro 2.0 Software. Experiments were conducted with 8 nM CYP1A2, 100 nM CYP2A6, and 25 nM CYP3A4 (final well concentrations). Reactions were performed in 100 mM potassium phosphate buffer (without NaCl) (pH 7.6) supplemented with 3 mM MgCl$_2$ and an NADPH regenerating system (NADPH 200 µM, glucose 6-phosphate 500 µM and glucose 6-phosphate dehydrogenase 40 U·L^{-1}, final concentrations). Stock solutions of EthR were prepared in DMSO, while coumarin and DBF were prepared in acetonitrile (ACN). Final solvents concentrations were maintained constant throughout the experiment (0.2% (v/v) DMSO or 0.1% (v/v) ACN). Product formation was followed for 10 min at 37 °C and rates were calculated by using a standard curve of the products. Reactions were performed in triplicate with substrate concentrations ranging up to 5 µM EthR (CYP1A2), 20 µM coumarin (CYP2A6) or 10 µM DBF (CYP3A4). Velocity data were plotted according to the Michaelis–Menten equation with high confidence ($r^2 > 0.95$) using GraphPad Prism 5.01 Software (La Jolla, CA, USA) and kinetic parameters (k_{cat} and K_M) were derived [20,22,47]. Variance in data was analyzed using one-way ANOVA with Bonferroni's multiple comparison tests, with 95% confidence interval—GraphPad Prism 5.01 Software (La Jolla, CA, USA).

4.7. Ionic Strength Effect

Catalytic activity of CYP1A2, 2A6, and 3A4, sustained by WT CPR and CPR mutants, was assessed at various NaCl concentrations (0–1.25 M), using 5 µM EthR, 20 µM coumarin and 10 µM DBF, respectively, in 100 mM potassium phosphate buffer (pH 7.6), and NADPH regenerating system (NADPH 200 µM, glucose 6-phosphate 500 µM and glucose 6-phosphate dehydrogenase 40 U L^{-1}, final concentrations). Velocities were measured in triplicate in 96-well format using multi-mode microtiter plate reader (SpectraMax®i3x, Molecular Devices, San José, CA, USA; SoftMax Pro 2.0). Initial rates (picomoles of fluorescent product formed per picomoles of CYP per minute) were derived from the linear part of the kinetic traces using a standard curve of the respective products. Control experiments were conducted to assess the effect of ionic strength on the pH of the reaction matrix and the fluorescence of products.

4.8. CYP Activity Titration with CYB5

The CYB5 titration assay was performed in microplate format (96 wells) using the same enzyme assay conditions as described above (without NaCl), except substrate concentrations were hold constant (5 µM EthR, 20 µM coumarin or 7.5 µM DBF), applying a gradient of CYB5 (0–2000, 0–800 and 0–1000 nM for CYP1A2, 2A6 and 3A4, respectively).

Supplementary Materials: Supplementary materials can be found at http://www.mdpi.com/1422-0067/19/12/3914/s1.

Author Contributions: Conceptualization, M.K. and G.T.; methodology, D.C., B.C.G., B.B.P. and F.E.; formal analysis, M.K., G.T., T.L., P.U. and J.R.; writing—original draft preparation, D.C. and M.K.; writing—review and editing, all authors; project administration, M.K. and G.T.; funding acquisition, M.K. and G.T.

Funding: This work was in part financed by a joint project, funded by the Portuguese *Fundação para a Ciência e a Tecnologia* [Grant FCT-ANR/ BEXBCM/0002/2013, as well as Grant UID/BIM/0009/2016 of the Research Center for Toxicogenomics and Human Health (ToxOmics)] and the *L'Agence Nationale de la Recherche* [Grant ANR-13-ISV5-0001]. F.E. was supported with a post-doctoral fellowship grant of the Portuguese *Fundação para a Ciência e a Tecnologia* [Grant SFRH/BPD/110633/2015].

Conflicts of Interest: The authors declare no conflict of interest.

Abbreviations

2AA	2-Aminoanthracene
ACN	Acetonitrile
AfB1	Aflatoxin B1
CPR	NADPH-cytochrome P450 reductase
CYB5	Cytochrome b_5
CYP	Microsomal cytochrome P450
DBF	Dibenzylfluorescein
DMSO	Dimethyl sulfoxide
EROD	Ethoxyresorufin *O*-deethylation
ET	Electron transfer
EthR	Ethoxyresorufin
FAD	Flavin adenine dinucleotide
FMN	Flavin adenine dinucleotide
IPTG	Isopropyl β-D-thiogalactoside
IQ	2-Amino-3-methylimidazo(4,5-f)quinoline
NADP+/NADPH	Nicotinamide adenine dinucleotide phosphate
NNdEA	*N*-nitrosodiethylamine
NNK	4-(Methylnitrosamino)-1-(3-pyridyl)-1-butanone

References

1. Yasukochi, Y.; Masters, B.S. Some properties of a detergent-solubilized NADPH-cytochrome *c* (cytochrome P-450) reductase purified by biospecific affinity chromatography. *J. Biol. Chem.* **1976**, *251*, 5337–5344. [PubMed]
2. Rendic, S.; Guengerich, F.P. Survey of Human Oxidoreductases and Cytochrome P450 Enzymes Involved in the Metabolism of Xenobiotic and Natural Chemicals. *Chem. Res. Toxicol.* **2015**, *28*, 38–42. [CrossRef] [PubMed]
3. Guengerich, F.P. Human Cytochrome P450 Enzymes. In *Cytochrome P450: Structure, Mechanism, and Biochemistry*, 4th ed.; De Montellano, P.R.O., Ed.; Springer International Publishing: Basel, Switzerland, 2015; pp. 523–785.
4. Pandey, A.V.; Flück, C.E. NADPH P450 oxidoreductase: Structure, function, and pathology of diseases. *Pharmacol. Ther.* **2013**, *138*, 229–254. [CrossRef] [PubMed]
5. Vergéres, G.; Waskell, L. Cytochrome b5, its functions, structure and membrane topology. *Biochimie* **1995**, *77*, 604–620. [CrossRef]
6. Bart, A.G.; Scott, E.E. Structural and functional effects of cytochrome b5 interactions with human cytochrome P450 enzymes. *J. Biol. Chem.* **2017**, *292*, 20818–20833. [CrossRef] [PubMed]
7. Murataliev, M.B.; Feyereisen, R.; Walker, F.A. Electron transfer by diflavin reductases. *Biochim. Biophys. Acta* **2004**, *1698*, 1–26. [CrossRef] [PubMed]
8. Wang, M.; Roberts, D.L.; Paschke, R.; Shea, T.M.; Masters, B.S.; Kim, J.J. Three-dimensional structure of NADPH-cytochrome P450reductase: Prototype for FMN- and FAD-containing enzymes. *Proc. Natl. Acad. Sci. USA* **1997**, *94*, 8411–8416. [CrossRef] [PubMed]
9. Grunau, A.; Geraki, K.; Grossmann, J.G.; Gutierrez, A. Conformational dynamics and the energetics of protein ligand interactions: role of interdomain loop in human cytochrome P450 reductase. *Biochemistry* **2007**, *46*, 8244–8255. [CrossRef]
10. Ellis, J.; Gutierrez, A.; Barsukov, I.L.; Huang, W.C.; Grossmann, J.G.; Roberts, G.C. Domain motion in cytochrome P450 reductase: Conformational equilibria revealed by NMR and small angle xray scattering. *J. Biol. Chem.* **2009**, *284*, 36628–36637. [CrossRef]
11. Paine, M.J.I.; Scrutton, N.S.; Munro, A.W.; Gutierrez, A.; Roberts, G.C.K.; Wolf, C.R. Electron transfer partners of cytochrome P450. In *Cytochrome P450: Structure, Mechanism, and Biochemistry*, 3th ed.; De Montellano, P.R.O., Ed.; Springer International Publishing: Basel, Switzerland, 2005; pp. 115–148.
12. Xia, C.; Hamdane, D.; Shen, A.L.; Choi, V.; Kasper, C.B.; Pearl, N.M.; Zhang, H.; Im, S.C.; Waskell, L.; Kim, J.J. Conformational changes of NADPH-cytochrome P450 oxidoreductase are essential for catalysis and cofactor binding. *J. Biol. Chem.* **2011**, *286*, 16246–16260. [CrossRef]
13. Vincent, B.; Morellet, N.; Fatemi, F.; Aigrain, L.; Truan, G.; Guittet, E.; Lescop, E. The closed and compact domain organization of the 70-kDa human cytochrome P450 reductase in its oxidized state as revealed by NMR. *J. Mol. Biol.* **2012**, *420*, 296–309. [CrossRef] [PubMed]
14. Hamdane, D.; Xia, C.; Im, S.C.; Zhang, H.; Kim, J.; Waskell, L. Structure and function of an NADPH cytochrome P450 oxidoreductase in an open conformation capable of reducing cytochrome P450. *J. Biol. Chem.* **2009**, *284*, 11374–11384. [CrossRef] [PubMed]
15. Aigrain, L.; Pompon, D.; Truan, G.; Morera, S. Cloning, purification, crystallization and preliminary Xray analysis of a chimeric NADPH cytochrome P450 reductase. *Acta Crystallogr. Sect. F Struct. Biol. Cryst. Commun.* **2009**, *65 Pt 3*, 210–212. [CrossRef]
16. Huang, W.C.; Ellis, J.; Moody, P.C.; Raven, E.L.; Roberts, G.C. Redox linked domain movements in the catalytic cycle of cytochrome p450 reductase. *Structure* **2013**, *21*, 1581–1589. [CrossRef] [PubMed]
17. Frances, O.; Fatemi, F.; Pompon, D.; Guittet, E.; Sizun, C.; Pérez, J.; Lescop, E.; Truan, G. A Well-Balanced Preexisting Equilibrium Governs Electron Flux Efficiency of a Multidomain Diflavin Reductase. *Biophys. J.* **2015**, *108*, 1527–1536. [CrossRef] [PubMed]
18. Sündermann, A.; Oostenbrink, C. Molecular dynamics simulations give insight into the conformational change, complex formation, and electron transfer pathway for cytochrome p450 reductase. *Protein Sci.* **2013**, *22*, 1183–1195. [CrossRef]

19. Aigrain, L.; Pompon, D.; Truan, G. Role of the interface between the FMN and FAD domains in the control of redox potential and electronic transfer of NADPH-cytochrome P450 reductase. *Biochem. J.* **2011**, *435*, 197–206. [CrossRef]
20. Kranendonk, M.; Marohnic, C.C.; Panda, S.P.; Duarte, M.P.; Oliveira, J.S.; Masters, B.S.; Rueff, J. Impairment of human CYP1A2mediated xenobiotic metabolism by Antley Bixler syndrome variants of cytochrome P450 oxidoreductase. *Arch. Biochem. Biophys.* **2008**, *475*, 93–99. [CrossRef]
21. Miller, W.L.; Agrawal, V.; Sandee, D.; Tee, M.K.; Huang, N.; Choi, J.H.; Morrissey, K.; Giacomini, K.M. Consequences of POR mutations and polymorphisms. *Mol. Cell. Endocrinol.* **2011**, *336*, 174–179. [CrossRef]
22. Moutinho, D.; Marohnic, C.C.; Panda, S.P.; Rueff, J.; Masters, B.S.; Kranendonk, M. Altered human CYP3A4 activity caused by Antley Bixler syndrome related variants of NADPH cytochrome P450 oxidoreductase measured in a robust in vitro system. *Drug Metab. Dispos.* **2012**, *40*, 754–760. [CrossRef]
23. Marohnic, C.C.; Panda, S.P.; McCammon, K.; Rueff, J.; Masters, B.S.; Kranendonk, M. Human cytochrome P450 oxidoreductase deficiency caused by the Y181D mutation: Molecular consequences and rescue of defect. *Drug Metab. Dispos.* **2010**, *38*, 332–340. [CrossRef] [PubMed]
24. Campelo, D.; Lautier, T.; Urban, P.; Esteves, F.; Bozonnet, S.; Truan, G.; Kranendonk, M. The Hinge Segment of Human NADPH-Cytochrome P450 Reductase in Conformational Switching: The Critical Role of Ionic Strength. *Front. Pharmacol.* **2017**, *8*, 755. [CrossRef] [PubMed]
25. Shen, A.L.; Kasper, C.B. Role of acidic residues in the interaction of NADPH-cytochrome P450 oxidoreductase with cytochrome P450 and cytochrome c. *J. Biol. Chem.* **1995**, *270*, 27475–27480. [CrossRef] [PubMed]
26. Duarte, M.P.; Palma, B.B.; Gilep, A.A.; Laires, A.; Oliveira, J.S.; Usanov, S.A.; Rueff, J.; Kranendonk, M. The stimulatory role of human cytochrome b5 in the bioactivation activities of human CYP1A2, 2A6 and 2E1: A new cell expression system to study cytochrome P450 mediated biotransformation. *Mutagenesis* **2005**, *20*, 93–100. [CrossRef] [PubMed]
27. McCammon, K.M.; Panda, S.; Xia, C.; Kim, J.J.; Moutinho, D.; Kranendonk, M.; Auchus, R.J.; Lafer, E.M.; Ghosh, D.; Martasek, P.; et al. Instability of the Human Cytochrome P450 Reductase A287P Variant Is the Major Contributor to Its Antley-Bixler Syndrome-like Phenotype. *J. Biol. Chem.* **2016**, *291*, 20487–20502. [CrossRef] [PubMed]
28. Duarte, M.P.; Palma, B.B.; Laires, A.; Oliveira, J.S.; Rueff, J.; Kranendonk, M. Escherichia coli BTC, a human cytochrome P450 competent tester strain with a high sensitivity towards alkylating agents: Involvement of alkyltransferases in the repair of DNA damage induced by aromatic amines. *Mutagenesis* **2005**, *20*, 199–208. [CrossRef] [PubMed]
29. Palma, B.B.; Silva, E.S.M.; Urban, P.; Rueff, J.; Kranendonk, M. Functional characterization of eight human CYP1A2 variants: The role of cytochrome b5. *Pharmacogenet. Genomics* **2013**, *23*, 41–52. [CrossRef]
30. Venkatakrishnan, K.; von Moltke, L.L.; Court, M.H.; Harmatz, J.S.; Crespi, C.L.; Greenblatt, D.J. Comparison between cytochrome P450 (CYP) content and relative activity approaches to scaling from cDNA-expressed CYPs to human liver microsomes: ratios of accessory proteins as sources of discrepancies between the approaches. *Drug Metab. Dispos.* **2000**, *28*, 1493–1504.
31. Paine, M.F.; Khalighi, M.; Fisher, J.M.; Shen, D.D.; Kunze, K.L.; Marsh, C.L.; Perkins, J.D.; Thummel, K.E. Characterization of interintestinal and intraintestinal variations in human CYP3A-dependent metabolism. *J. Pharmacol. Exp. Ther.* **1997**, *283*, 1552–1562.
32. Hlavica, P.; Schulze, J.; Lewis, D.F. Functional interaction of cytochrome P450 with its redox partners: A critical assessment and update of the topology of predicted contact regions. *J. Inorg. Biochem.* **2003**, *96*, 279–297. [CrossRef]
33. Kandel, S.E.; Lampe, J.N. Role of protein-protein interactions in cytochrome P450-mediated drug metabolism and toxicity. *Chem. Res. Toxicol.* **2014**, *27*, 1474–1486. [CrossRef] [PubMed]
34. Zanger, U.M.; Schwab, M. Cytochrome P450 enzymes in drug metabolism: Regulation of gene expression, enzyme activities, and impact of genetic variation. *Pharmacol. Ther.* **2013**, *138*, 103–141. [CrossRef] [PubMed]
35. Yun, C.H.; Ahn, T.; Guengerich, F.P. Conformational change and activation of cytochrome P450 2B1 induced by salt and phospholipid. *Arch. Biochem. Biophys.* **1998**, *356*, 229–238. [CrossRef] [PubMed]
36. Yun, C.H.; Song, M.; Ahn, T.; Kim, H. Conformational change of cytochrome P450 1A2 induced by sodium chloride. *J. Biol. Chem.* **1996**, *271*, 31312–31316. [CrossRef] [PubMed]
37. Voznesensky, A.I.; Schenkman, J.B. The cytochrome P450 2B4-NADPH cytochrome P450 reductase electron transfer complex is not formed by charge-pairing. *J. Biol. Chem.* **1992**, *267*, 14669–14676. [PubMed]

38. Voznesensky, A.I.; Schenkman, J.B. Quantitative analyses of electrostatic interactions between NADPH-cytochrome P450 reductase and cytochrome P450 enzymes. *J. Biol. Chem.* **1994**, *269*, 15724–15731. [PubMed]
39. Waskell, L.; Kim, J.J.P. Electron Transfer Partners of Cytochrome P450. In *Cytochrome P450: Structure, Mechanism, and Biochemistry*, 4th ed.; De Montellano, P.R.O., Ed.; Springer International Publishing: Basel, Switzerland, 2015; pp. 33–68.
40. Im, S.C.; Waskell, L. The interaction of microsomal cytochrome P450 2B4 with its redox partners, cytochrome P450 reductase and cytochrome b(5). *Arch. Biochem. Biophys.* **2011**, *507*, 144–153. [CrossRef] [PubMed]
41. Zhang, M.; Huang, R.; Im, S.C.; Waskell, L.; Ramamoorthy, A. Effects of membrane mimetics on cytochrome P450-cytochrome b5 interactions characterized by NMR spectroscopy. *J. Biol. Chem.* **2015**, *290*, 12705–12718. [CrossRef] [PubMed]
42. Zhang, M.; Le Clair, S.V.; Huang, R.; Ahuja, S.; Im, S.C.; Waskell, L.; Ramamoorthy, A. Insights into the role of substrates on the interaction between cytochrome b5 and cytochrome P450 2B4 by NMR. *Sci. Rep.* **2015**, *5*, 8392. [CrossRef]
43. Pikuleva, I.A.; Cao, C.; Waterman, M.R. An additional electrostatic interaction between adrenodoxin and P450c27 (CYP27A1) results in tighter binding than between adrenodoxin and p450scc (CYP11A1). *J. Biol. Chem.* **1999**, *274*, 2045–2052. [CrossRef]
44. Nunez, M.; Guittet, E.; Pompon, D.; van Heijenoort, C.; Truan, G. NMR structure note: Oxidized microsomal human cytochrome b5. *J. Biomol. NMR* **2010**, *47*, 289–295. [CrossRef] [PubMed]
45. Kranendonk, M.; Carreira, F.; Theisen, P.; Laires, A.; Fisher, C.W.; Rueff, J.; Estabrook, R.W.; Vermeulen, N.P. Escherichia coli MTC, a human NADPH P450 reductase competent mutagenicity tester strain for the expression of human cytochrome P450 isoforms 1A1, 1A2, 2A6, 3A4, or 3A5: Catalytic activities and mutagenicity studies. *Mutat. Res.* **1999**, *441*, 73–83. [CrossRef]
46. Bauer, S.; Shiloach, J. Maximal exponential growth rate and yield of E. coli obtainable in a bench-scale fermentor. *Biotechnol. Bioeng.* **1974**, *16*, 933–941. [CrossRef] [PubMed]
47. Palma, B.B.; Silva, E.S.M.; Vosmeer, C.R.; Lastdrager, J.; Rueff, J.; Vermeulen, N.P.; Kranendonk, M. Functional characterization of eight human cytochrome P450 1A2 gene variants by recombinant protein expression. *Pharmacogenom. J.* **2010**, *10*, 478–488. [CrossRef] [PubMed]
48. Kranendonk, M.; Mesquita, P.; Laires, A.; Vermeulen, N.P.; Rueff, J. Expression of human cytochrome P450 1A2 in Escherichia coli: A system for biotransformation and genotoxicity studies of chemical carcinogens. *Mutagenesis* **1998**, *13*, 263–269. [CrossRef] [PubMed]
49. Maron, D.M.; Ames, B.N. Revised methods for the Salmonella mutagenicity test. *Mutat. Res.* **1983**, *113*, 173–215. [CrossRef]

© 2018 by the authors. Licensee MDPI, Basel, Switzerland. This article is an open access article distributed under the terms and conditions of the Creative Commons Attribution (CC BY) license (http://creativecommons.org/licenses/by/4.0/).

Review

The Role of CYP450 Drug Metabolism in Precision Cardio-Oncology

Olubadewa A. Fatunde [1] and Sherry-Ann Brown [2,*]

[1] Department of Medicine, University of Texas Health Science Center at Tyler–CHRISTUS Good Shepherd Medical Center, Longview, TX 75601, USA; olubadewa.fatunde@christushealth.org
[2] Department of Cardiovascular Diseases, Mayo Clinic, Rochester, MN 55905, USA
* Correspondence: brown.sherryann@mayo.edu; Tel.: +1-507-255-5123; Fax: +1-507-284-4362

Received: 2 December 2019; Accepted: 13 January 2020; Published: 17 January 2020

Abstract: As many novel cancer therapies continue to emerge, the field of Cardio-Oncology (or onco-cardiology) has become crucial to prevent, monitor and treat cancer therapy-related cardiovascular toxicity. Furthermore, given the narrow therapeutic window of most cancer therapies, drug-drug interactions are prevalent in the cancer population. Consequently, there is an increased risk of affecting drug efficacy or predisposing individual patients to adverse side effects. Here we review the role of cytochrome P450 (CYP450) enzymes in the field of Cardio-Oncology. We highlight the importance of cardiac medications in preventive Cardio-Oncology for high-risk patients or in the management of cardiotoxicities during or following cancer treatment. Common interactions between Oncology and Cardiology drugs are catalogued, emphasizing the impact of differential metabolism of each substrate drug on unpredictable drug bioavailability and consequent inter-individual variability in treatment response or development of cardiovascular toxicity. This inter-individual variability in bioavailability and subsequent response can be further enhanced by genomic variants in CYP450, or by modifications of CYP450 gene, RNA or protein expression or function in various 'omics' related to precision medicine. Thus, we advocate for an individualized approach to each patient by a multidisciplinary team with clinical pharmacists evaluating a treatment plan tailored to a practice of precision Cardio-Oncology. This review may increase awareness of these key concepts in the rapidly evolving field of Cardio-Oncology.

Keywords: CYP450; drug metabolism; precision Cardio-Oncology; precision medicine; systems medicine

---

1. Introduction

Cardio-Oncology is an emerging field that sits at the interface of Cardiology and Oncology and has close relationships with primary care specialties. A variety of oncology drugs can injure the cardiovascular system, causing various forms of cardiovascular toxicities. Further, cardiology drugs are widely used by the general population and by individuals with cancer. Many of these drugs are also frequently used for preventive cardioprotection or for the management of cardiotoxicity that has already occurred. In this review, we highlight several cytochrome P450 (CYP450) enzymes relevant to Cardio-Oncology (Figure 1). We classify drugs as CYP450 substrates, inducers or inhibitors, with an explanation of the three types of drug-enzyme interaction. Drug-drug interactions between Oncology and Cardiology drugs mediated by CYP450 enzymes are also surveyed. In addition, we discuss the fact that differential metabolism of each substrate drug in each specific individual can determine bioavailability. Examples from precision Cardio-Oncology are integrated to illustrate that inter-individual bioavailability can be further enhanced by genomic variation in CYP450 enzymes. Some variants enhance enzyme activity, while others do just the opposite. This helps to determine the level of drug available in the body. Not only genomic variants but also other modifications of the enzyme gene, RNA or protein, including those due to gene-environment interactions, can alter drug levels and individual response in precision medicine. We hope that this review can assist multidisciplinary teams in Cardio-Oncology with difficult drug-related decisions relevant to metabolism and bioavailability.

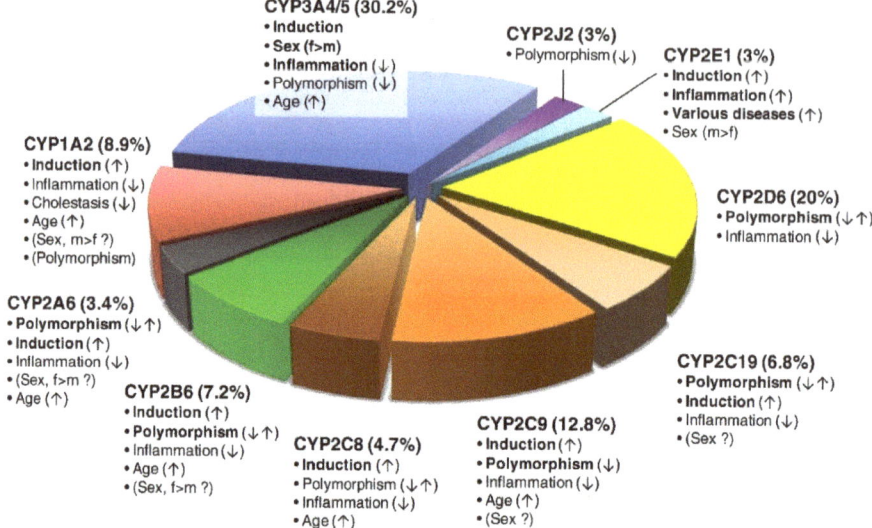

Figure 1. The pie chart depicts the various P450 isoforms, the percentage of clinically used drugs metabolized by the isoform and factors inducing or inhibiting the respective P450 enzyme, thereby influencing variability. The most important factors influencing variability are in bold, with a vertical arrow indicating increased activity (↑), decreased activity (↓) or both (↑↓). Biologic sex (female (f) or male (m)) and rarely polymorphism (CYP1A2) can be of controversial significance. In total, 248 CYP-related drug metabolism pathways were analyzed (excluding chemicals and endogenous substrates). Used with permission [1].

2. CYP450 Class of Enzymes

The CYP450 monooxygenase system consists of a family of enzymes that metabolize a variety of medications relevant to Cardiology and Oncology. The CYP450 enzymes are primarily located in the liver but can also be found in the small intestines, lungs, kidneys and even the heart [1–4]. These enzymes are responsible for the first pass metabolism and largely explain the higher pharmacokinetic variability of oral drugs compared to intravenous medications [5,6]. Their etymology derives from their intracellular, membrane-bound localization (i.e., cyto-), with a heme pigment forming part of the protein (i.e., chrome). The heme portion of the enzymes absorbs light at a maximum wavelength of 450 nm when complexed with carbon monoxide in the reduced state. In humans, more than 100 collective genes and pseudogenes encode over 50 CYP450 enzymes. CYP1A2, CYP2C9, CYP2C19, CYP2D6 and CYP3A4/5 metabolize over 90% of the substrate drugs and are the most extensively studied CYP450 enzymes [1–3,7] (Table 1).

Drug metabolism in the liver occurs in three major steps: hepatic (transporter-mediated) uptake, phase I reactions and phase II reactions. Hepatic uptake is responsible for a trivial amount of pharmacokinetic variability. In phase I reactions, the CYP450 enzymes oxidize, reduce or hydrolyze their substrates, resulting in loss of pharmacological activity or activation of prodrugs [3]. In phase II reactions, non-CYP450 enzymes conjugate phase I products by adding glucuronide, acetyl, methyl or sulphate groups to form usually inactive polar derivatives for renal or biliary elimination [3,8].

Table 1. Most common cytochrome P450 (CYP450) enzymes in humans.

Enzyme	Upper Limit of Normal Percentage of Total Hepatic CYP450 (%)
CYP3A4	37
CYP3A5	1
CYP2C9	29
CYP1A2	16.3
CYP2A6	14
CYP2B6	5.3
CYP2D6	4.3
CYP2C19	3.8

Note: As this is a range, values do not completely sum to 100%. Adapted from [1] with permission.

2.1. Phase I Enzymes

The most predominant CYP450 enzyme class involved in phase I reactions is the CYP3A family. CYP3A drugs metabolize 45–60% of all drugs currently in use [9,10], with CYP3A4 representing the most common allele. CYP3A4 is predominantly found in the liver [11,12] and intestines [13,14] and can also be found in the stomach, brain, breast and prostate [15]. A second phase I enzyme CYP3A5 is present in the liver and small intestine of 25–30% of individuals [1,3]. A third enzyme CYP3A7 is present predominantly in fetuses (50% of total expression), with expression typically shifting to CYP3A4 in adulthood [1]. CYP1A2 is constitutively expressed primarily in the liver, in significant quantities, measuring up to 16% of the total hepatic P450 pool in some individuals [1]. CYP2C9 is the second most common CYP450 enzyme found in the liver and extrahepatic tissues such as the intestines and endothelial cells; many CYP2C9 substrates (e.g., warfarin) have narrow therapeutic indices, requiring careful monitoring in patients taking these drugs. CYP2C19 is expressed in the liver and kidneys and is responsible for the metabolism clopidogrel, a drug commonly used in cardiology and with relevance for Cardio-Oncology; genetic variation is often associated with adverse drug effects. The CYP2D6 enzyme is primarily expressed outside of the liver and metabolizes approximately 15–25% of drugs from all therapeutic areas, including Cardiology (e.g., beta-blockers, antiarrhythmics) and Oncology (e.g., tamoxifen). There is substantial inter-individual variability with hepatic CYP450 enzyme activity, ranging from 30- to 40-fold variation for collective CYP3A enzymes [16–19] to 100-fold variation for CYP2D6 [16,19].

2.2. Phase II Enzymes

Phase II enzymes are non-CYP450 proteins that can indirectly exert influence on CYP450 enzyme activity. The most commonly occurring phase II reactions are glucoronidation and sulfonation (or sulfurylation), which are catalyzed by the enzymes uridine diphosphate glucoronysltransferase (UGTs) and sulfotransferases (SULTs), respectively. These enzymes are located throughout the gastrointestinal and genitourinary tracts.

The UGTs catalyze transfer of glucuronic acid to onto oxygen, nitrogen or sulfur on substrate drugs. Substrates range from endogenous substances (e.g., bilirubin, estradiol, serotonin) to exogenous substances (e.g., propofol, morphine). Interindividual variability in levels of UGT, stemming from a patient's age, sex, presence of enzyme inhibitors/inducers can contribute to drug-induced toxicity (slow metabolism) or ineffectual drug levels (rapid metabolism) [20].

Toxic drug metabolites of UGTs levels can lead to additive toxicity with P450 (phase I) drug-drug interactions. For example, the glucoronated products of gemfibrozil can inactivate CYP2C8, causing toxic levels of statins and significant rhabdomyolysis [21]. Likewise, clopidogrel can also inactivate CYP2C8, resulting in toxic levels of gemfibrozil [22]. Finally, UGT1 levels have been shown to inversely associate with development of a number of cancers (i.e., colon cancer, breast, bladder and biliary) in conditional UGT1 knockout mice [23].

Sulfonation reactions result in increased hydrophilicity and (usually) decreased pharmacological activity or inactivation of certain endogenous substances, such as thyroid hormones, steroids and monoamine transmitters. Inhibition of sulfonation by some compounds or metabolites can increase the toxicity of these substances [24,25]. Conversely, sulfonation can also bioactivate some substrates. This can result in a benign, more metabolically active form (e.g., minoxidil, morphine) [25] or can produce certain toxic metabolites, thereby increasing drug toxicity (e.g., tamoxifen) [24,26]. There are three main SULTs supergene families in humans—SULT1, SULT2, SULT4 [24–26]. SULT1A is most concentrated in the liver and has also been found in the kidney, lung, brain and gastrointestinal and genitourinary systems. The extensive expression of SULT1A1 and SULT1A3 in the intestines and lungs suggest they may play a role in extrahepatic drug detoxification and metabolism. SULT1B functions in thyroid hormone metabolism. SULT2A and SULT2B family are active in the metabolism of steroids and bile acids and are present throughout the body. Notably, products of sulfonation reactions catalyzed by various SULT enzymes (e.g., 1A1, 1A2, 1A3, 1C2, 1C4 and 2A1) can result in chemically reactive intermediate compounds that bind DNA, eliciting mutagenicity and carcinogenicity [25]. Interindividual variation in human sulfotransferase activity varies from 5- to 36-fold, largely explained by single nucleotide polymorphisms (SNPs) in the coding regions of SULT genes. This variation can play a complementary role to phase I reactions (largely catalyzed by P450 enzymes) in determining an individual's response to therapeutics.

2.3. Substrates, Inducers and Inhibitors

Drugs that interact with the CYP450 enzymes can be divided into three categories: substrates, inhibitors and inducers. Substrates are drugs upon which specific CYP450 enzyme acts. Inducers are drugs that increase enzyme activity. Inhibitors are drugs that decrease enzyme activity. Inhibitors compete with other drugs (typically substrates) for enzyme active sites, therefore altering the optimal level of a given substrate drug in the plasma. This alters the intended drug pharmacokinetics, rendering many prodrugs ineffective or conversely, potentially raising other drugs' plasma concentrations to toxic levels. A strong inhibitor is defined as one that increases plasma AUC substrate values greater than 5-fold or decreases substrate clearance to more than 80% of normal levels. A moderate inhibitor causes a greater than 2-fold increase in the plasma AUC values or a 50–80% decrease in drug clearance. A weak inhibitor causes a greater than 1.25-fold increase in plasma AUC values or a 20–50% decrease in drug clearance. Drugs commonly used in cardiology fall into all three categories (substrates, inducers and inhibitors) (Table 2) and can potentially interact with oncology drug substrates; the converse is also true. Being mindful of drug-drug interactions due to CYP450 activity related to substrates, inducers and inhibitors may help protect the hearts of patients undergoing cancer therapies.

Table 2. A list of CYP450 enzymes and many of their cardiac substrates, inducers and inhibitors.

Enzyme	Substrate Characteristics	Examples of Drugs Relevant to Cardiovascular Care	Inhibitors	Inducers
CYP3A4 and CYP3A5	Large & lipophilic molecules, with very diverse structures; includes over 50% of all clinically used drugs CCBs Statins Taxanes (paclitaxel and docetaxel) Sorafenib Dasatinib Cyclophosphamide Proteosome inhibitors (e.g., Bortezomib) Everolimus Cytarabine Dabrafenib Vemurafenib Irinotecan Imatinib/Ibrutinib	Antiarrhythmics: quinidine-3-OH (not 3A5) Calcium Channel Blockers: amlodipine, diltiazem, felodipine, lercanidipine, nifedipine2, nisoldipine, nitrendipine, verapamil HMG CoA Reductase Inhibitors: atorvastatin, cerivastatin, lovastatin, NOT pravastatin, NOT rosuvastatin, simvastatin Propranolol Cilostazol Eplerenone Fentanyl Lidocaine Others: Ondansetron Caffeine-trimethyluric acid Sorafenib	Strong: (Protease Inhibitors) indinavir, nelfinavir, ritonavir, saquinavir, (Antibacterials) clarithromycin, erythromycin, telithromycin, chloramphenicol (Antifungal): itraconazole, ketoconazole, fluconazole, voriconazole (Antidepressant): nefazodone (Vasopressin antagonist): conivaptan Moderate: (Antiemetic): aprepitant (Antibacterials): erythromycin, (Antifungal): fluconazole, (nDP-CCB): verapamil, diltiazem. Mibefradil (Immune modulating agents): Cyclosporine, Tacrolimus (Tyrosine kinase inhibitors -TKIs): Nilotinib, lapatinib (Hormonal agents): Enzalutamide, Bicalutamide (Chemotherapy): Sorafenib (Misc.): grapefruit juice, starfruit Weak: (H2 blockers): cimetidine (Topoisomerase inhibitors): Etoposide (Anthracyclines): Idarubicine (Alkylating agents): Cyclophosphamide (Antibacterials): ciprofloxacin, norfloxacin (Antifungal): voriconazole, ketoconazole, itraconazole, posaconazole, fluconazole (Antiarrhythmics): amiodarone (NNRTI): delavirdine (SSRIs): fluvoxamine, norfluoxetine (Protease Inhibitors): boceprevir, telaprevir (OCP):* gestodene, mifepristone (Chemotherapy): imatinib (Misc.): starfruit	(NNRTI): efavirenz, nevirapine, efavirenz/emtricitabine/tenofovir (GABA-Agonists): barbiturates, phenobarbital, (Anti-epileptics): carbamazepine, oxcarbazepine, phenytoin (Non-steroidal Anti-androgen): enzalutamide (Antibiotics): rifabutin, rifampin (Misc.): glucocorticoids, modafinil, St. John's Wort (Thiazolidinedione): pioglitazone troglitazone (Antimitotic agents): Paclitaxel (TKIs): Vemurafenib, Dabrafenib (Hormonal agents): enzalutamide (Angiogenesis inhibitor): Thalidomide (BRAF inhibitor): Vemurafenib
CYP2C9	Relatively large and weakly acidic molecules; includes antimalarials and oral antidiabetics Fluvastatin Nateglinide phenytoin-4-OH2 rosiglitazone	Angiotensin II Blockers: losartan irbesartan, valsartan Torsemide S-Warfarin Fluvastatin Rosiglitazone Others: NSAIDs, Sulfonylureas	Strong: fluconazole2 Moderate: amiodarone (NNRTI): efavirenz (Fibrate): fenofibrate (Antifungal): fluconazole, voriconazole (Statin): fluvastatin, lovastatin (SSRI): fluvoxamine2, paroxetine, sertraline (Antibiotic): isoniazid, metronidazole * phenylbutazone, sulfamethoxazole * sulfaphenazole, (Chemotherapeutic): teniposide, 5-flourouracil (Leukotriene receptor antag. LTRA): zafirlukast	(Non-steroidal Anti-androgen): enzalutamide (−3A4/5/7, 2C19) (NNRTI) Nevirapine (Antibiotics): Rifampin (Antiepileptics): phenobarbital, * secobarbital, carbamazepine (Misc.): St. John's Wort (−3A4,5,7)

Table 2. *Cont.*

Enzyme	Substrate Characteristics	Examples of Drugs Relevant to Cardiovascular Care	Inhibitors	Inducers
CYP2C8	Relatively large and weakly acidic molecules; includes antimalarials and oral antidiabetics Docetaxel Imatinib/Ibrutinib	Torsemide Cerivastatin Amiodarone (n) Thiazolidinedione (Pioglitazone, Rosiglitazone) Others: Repaglinide	Strong: gemfibrozil Moderate: trimethoprim (Thiazolidinediones): glitazones, (LTRA): montelukast (Plant flavonoid): quercetin (found in fruits, vegetables, leaves and grains; red onions and kale)	Rifampin
CYP2E1	Small, generally neutral and hydrophilic, planar molecules; includes aliphatic alcohols and halogenated alkanes Cisplatin	Ethanol	Disulfiram	Ethanol Isoniazid
CYP1A2	Planar, aromatic, polyaromatic and heterocyclic amides and amines	Caffeine Naproxen Ondansetron	Strong: fluvoxamine, ciprofloxacin Moderate: Vemurafenib Weak: cimetidine amiodarone, efavirenz, fluoroquinolones, fluvoxamine, furafylline1, interferon, * methoxsalen, * mibefradil, ticlopidine	(Food): broccoli, brussels sprouts, char-grilled meat (AEDs): carbamazepine (Diabetic meds) insulin (Misc): Modafinil (Antibiotic): Nafcillin, Rifampin (PPI): Omeprazole (Toxins): tobacco
CYP2A6	Nonplanar low molecular weight molecules usually with 2 hydrogen bond acceptors; includes ketones and nitrosamines	Ondansetron		
CYP2D6	Basic molecules with protonatable nitrogen atom(4-7) Å from the metabolism site; includes many plant alkaloids and antidepressants Proteosome inhibitors		Strong: bupropion, cinacalcet, fluoxetine, paroxetine, quinidine Moderate: duloxetine, sertraline, terbinafine, sorafenib Weak: amiodarone, cimetidine (NSAID): celecoxib (Antihistamine): chlorpheniramine, * clemastine, diphenhydramine, doxepin, histamine H1 receptor antagonists, hydroxyzine, promethazine, tripelennamine (Antipsychotic): chlorpromazine (SSRI): citalopram, escitalopram (TCA): clomipramine (ChemoRx): doxorubicin, imatinib (Antimalarial): halofantrine (Antipsychotic): haloperidol, levomepromazine, perphenazine (Opioids): methadone (DA agonist, Prokinetic): metoclopramide * mibefradil (Vasopressor): midodrine * moclobemide (H2 blocker): ranitidine (protease inhibitor): ritonavir (Antiplatelet): ticlopidine (Misc): cocaine	Dexamethasone Rifampin

Table 2. *Cont.*

Enzyme	Substrate Characteristics	Examples of Drugs Relevant to Cardiovascular Care	Inhibitors	Inducers
CYP2B6	Neutral or weakly basic, mostly lipophilic non-planar molecules with 1 to 2 hydrogen bond acceptors; includes anesthetics, insecticides and herbicides cyclophosphamide	N/A	Antiplatelets: clopidogrel, ticlopidine2, Antifungal: voriconazole, Chemotherapeutic: thiotepa	Artemisinin (AED): Carbamazepine, Phenobarbital, Phenytoin (NNRTI): Efavirenz, Nevirapine Rifampin (induces every listed CYP enzyme except 2EI)
CYP2C19	Neutral or weakly basic molecules or amides with 2 or 3 hydrogen bond acceptors; includes most proton pump inhibitors Proteosome inhibitors cyclophosphamide	Clopidogrel Labetalol Propranolol R-warfarin→8-OH Others: PPIs: Esomeprazole Lansoprazole Omeprazole2 Pantoprazole	(PPIs): esomeprazole, lansoprazole, omeprazole2, pantoprazole (Antibiotic): chloramphenicol, isoniazid (Antifungal): ketoconazole, voriconazole (H2 blocker): cimetidine (SSRI): fluoxetine, fluvoxamine (NSAID): indomethacin (Dopaminergic): modafinil oral contraceptives (Antiepileptics): oxcarbazepine topiramate (Antiplatelet): ticlopidine (Chemotherapy): sorafenib (Misc.): probenecid	(AED): carbamazepine (NNRTI): efavirenz (Protease Inhibitor): ritonavir (Non-steroidal Anti-androgen): enzalutamide (−3A4/5/7, 2C9) (NNRTI) (OCP): norethindrone (Misc.): prednisone, St. John's Wort (−3A4/5/7, 2C9) (Antibiotics): Rifampicin

Note: for medications not categorized as strong, moderate or weak inducers/inhibitors, there is insufficient evidence to further categorize them. Medications denoted with an asterisk (*) are not available in the US. Enzymes in bold denote the most commonly occurring CYP450 enzymes. Adapted from various sources [1,2,27–41]; used with permisson of the three primary sources [1,2,27].

3. Drug-Drug Interactions

Drug-drug interactions are fairly common in the oncologic patient with cardiac disease (Tables 2 and 3). In one study [42], 16% of patients receiving oral antineoplastic agents developed at least one major drug-drug interaction. This is of particular concern, given the narrow therapeutic window of many antineoplastic agents and some cardiology medications. In another study, a range of drug-drug interactions involving chemotherapeutic and common cardiac medications resulting from either pharmacokinetic (PK) interactions, pharmacodynamic (PD) interactions or a combination of the two was described [3,43]. The most common PK interactions in oncology involve the CYP450 enzymes and the efflux pump P-glycoprotein located in the intestine [43,44]. In essence, PK interactions describe the body's effect on a drug or substance, especially its absorption, distribution, metabolism or elimination [44,45]. On the other hand, PD interactions describe a drug's effect on the body. Drug-drug interactions in this arena are due to often unintentional additive effects of two agents with similar molecular targets, resulting in toxicity.

An example of a PD interaction in Cardio-Oncology occurs with concurrent use of beta-blockers (in Cardiology) and ceritinib/crizotinib (in Oncology); the latter is a combination chemotherapeutic drug used to treat metastatic (ALK-/ROS1-positive) non-small cell lung cancer. Co-administration of these medications can lead to symptomatic bradycardia, which can potentially be life-threatening [43]. Additionally, ceritinib/crizotinib can prolong the QT interval. Therefore, administration with other QT-prolonging medications that are often administered with chemotherapy, such as antiemetics, antibiotics and antidepressants, can potentially lead to malignant arrhythmias, including polymorphic ventricular tachycardia or 'torsades de pointes.' [43,46]. Consequently, beta-blockers or QT-prolonging medications should be used judiciously with ceritinib/crizotinib if co-administered with any drugs that inhibit CYP3A, as both ceritinib and crizotinib are extensively metabolized by CYP3A in the liver [47,48].

An example of a PK interaction in Cardio-Oncology involves the moderate inhibition of CYP3A4 by diltiazem/verapamil. When co-administered with chemotherapeutic agents metabolized by the same pathway, such as doxorubicin, imatinib or ibrutinib, this could lead to increased chemotherapy drug concentration. This can be accompanied by several adverse effects, including QT prolongation, gastrointestinal symptoms, shortness of breath, edema, chest pain, hepatotoxicity or bone marrow suppression [43]. This can be managed by using alternative medications for chronotropy or blood pressure control during the expected course of chemotherapy or appropriately decreasing the dose of administered chemotherapy if absolutely necessary [43,46].

Antiplatelet agents are a mainstay of atherosclerotic cardiovascular disease treatment and account for 40.4% of drug sales in cardiovascular disease [49]. PK interactions between antiplatelet agents and chemotherapeutics can alter the level of functioning of one or both drugs. For example, the chemotherapeutic combination agent enzalutamide/dasatinib can decrease the level of antiplatelet medication in the blood, causing disastrous/catastrophic consequences following cardiac catheterization [43]. When doxorubicin is administered with ticagrelor, CYP3A4 inhibition by ticagrelor can lead to an increase in doxorubicin exposure, placing patients at increased risk for known toxicities of the drug [43]. Based on the indicated chemotherapy regimen, the antiplatelet agent can usually be adjusted, taking into account patient characteristics.

Anticoagulants are indicated for patients with malignancies who develop deep venous thromboses (DVT) or pulmonary emboli (PE), which complicate the clinical course of approximately 5–10% of all cancer patients [50]. In patients with malignancies who also have cardiovascular disease requiring anticoagulation (e.g., atrial fibrillation, mechanical valves, mechanical support devices), the number of concurrent medications can increase the risk of drug-drug interaction. Historically, warfarin has been the most commonly used anticoagulant and remains in frequent use due to familiarity, cost and patient preference. Most drug-drug interactions involving chemotherapeutic agents and warfarin are due to a reduction in warfarin metabolism, often from CYP450 inhibition, leading to increased risk of bleeding (Table 2) [43]. As warfarin operates via vitamin K inhibition, oncologic patients with numerous reasons for vitamin K deficiency, such as diarrhea from chemotherapy or radiation or antibiotics for infections due to immunosuppression, face an additionally increased risk of bleeding [43]. The current standard of care for management of cancer-associated venous thromboembolism (VTE) is low molecular weight heparin [51–55]. Direct oral anticoagulants (DOACs) are emerging as potentially equally efficacious alternatives to low molecular weight heparin, with ideal bioavailability and mode of administration (orally) [29]. However, DOACs should be used cautiously in certain populations at increased risk for bleeding complications (e.g., gastrointestinal malignancies, advanced age and frailty) [43,50]. It is of note that DOACs are variably metabolized by CYP450 enzymes (dabigatran 0%, edoxaban < 4%, apixaban 15%, rivaroxaban 66%) [29]. Therefore, in cancer patients concurrently treated with strong CYP3A4 inducers or inhibitors (Table 2), dabigatran (or possibly edoxaban) may become the DOAC of choice; such decision-making may benefit close collaboration with a clinical pharmacist [29,56]. Nevertheless, both LMWH and DOACs have fewer drug-drug interactions than warfarin [43,56–58].

4. Precision Cardio-Oncology

4.1. Variability in Concentration and Activity

There can be wide variation in the concentration levels and activity of CYP450 between and within populations, as illustrated in the following examples. In the general population, there is up to 50-fold variation in the levels of CYP3A4 among individuals [15,59]. In a small study investigating the metabolism of cyclophosphamide in patients with lung, breast and gastrointestinal malignancies, the level of CYP2C19 was lower in patients with cancer compared to the general population [60]. Conversely, in a study investigating the role of miRNA in the regulation of CYP1B1, higher levels of CYP1B1 were noted particularly among individuals with estrogen-sensitive cancers [60,61]. Indeed, some CYP450 enzymes are preferentially upregulated by cancerous cells. For example, CYP1B1, mainly expressed in the ovary, uterus and breast tissue [61,62], is upregulated in malignant cells [63] to catalyze the metabolic activation of pro-carcinogens such as polycyclic aromatic hydrocarbons, aryl nitrate and 4-hydroxyestradiol [61]; 4-hydroxyestradiol is a catechol metabolite of 17β-estradiol, which generates free radicals, resulting in DNA damage [64,65]. Additionally, among patients with cancer, there can be wide variation in CYP450 enzyme activity.

4.2. Interindividual and Genetic Variability

Genetic variability, for example, of promoter or coding regions, may in part explain why different individuals have varying responses to the same drugs. Single nucleotide polymorphisms among CYP450 enzymes affect metabolism and therefore bioavailability of substrate drugs. Polymorphisms in CYP1A2 and CYP2B6 can result in decreased nicotine metabolism in smokers and have been associated with increased susceptibility to cancers and possibly atherosclerotic cardiovascular disease [1]. In particular, different combinations of alleles of CYP450 enzymes resulting in absent, low or high levels of enzymatic activity can lead to differential responses (e.g., toxicity, underdosing) to drug regimens, due to variable drug clearance [1,42,43].

4.3. Genomic Profiling

Genomic profiling in general characterizes an individual's complement of genes [66]. Next generation sequencing and genome-wide association studies (GWAS), which correlate SNPs with disease phenotypes, have revolutionized the speed at which such information can be analyzed, investigated and translated into a component of clinical care [67]. Genomic profiling is most commonly utilized in oncology—particularly for breast, ovarian, colon and lung cancers [67]. The applicability in cardiology is increasing, from medication metabolism to treatment of cardiomyopathy and inherited arrhythmias [67]. Genomic profiling has a number of potential applications, including the study of genetic variations that influence individual response to drugs (i.e., pharmacogenomics), precision medicine and new modalities to diagnose and treat disease.

4.3.1. Genomic Variation in CYP450

CYP450 genomics examine how modifications in the genes encoding the CYP450 complex affect enzyme function in metabolism and downstream drug therapeutic response. This has direct clinical applicability. This can be illustrated by examining two CYP450 enzymes relevant to Cardio-Oncology—CYP2C19 and CYP2D6 [7,16,68].

4.3.2. Genomic Variation in CYP2C19

Over 50 CYP2C19 genomic variants have been identified [7]. The CYP2C19 gene is located on chromosome 10q24.1q24.3, is composed of nine exons and produces a medium-sized protein (55.93 kDa) from 490 amino acids [7]. Homozygosity for loss-of-function alleles confers poor metabolism, heterozygosity for loss-of-function (LOF) alleles confers intermediate metabolism, wild type alleles (*1/*1) confer what may be considered 'normal' metabolism and homozygosity for a gain-of-function (GOF) allele confers ultra-rapid metabolism [7]. The four major phenotypes listed above correspond to selected permutations of the most common CYP2C19 genetic variants (*1, *2, *3, *17) in the dose-response relationship (Table 3). The frequency of these phenotypes seems to differ with ethnicity. Approximately 2% of Europeans (the most widely studied population) are poor CYP2C19 metabolizers, while up to 20% of Asians are poor metabolizers [1,2,69], underscoring the need for further study in all populations. Poor or intermediate metabolism of clopidogrel may lead to persistently elevated platelet function in spite of treatment (i.e., high on-treatment platelet reactivity or HTPR) in individuals treated for acute coronary syndromes (ACS) [7]. These individuals remain at high-risk for ischemia, limited post-PCI myocardial flow and adverse cardiovascular outcomes (e.g., stent thrombosis, myocardial infarction, stroke and death) [7]. The most common SNPs responsible for the poor metabolizer phenotype result from premature stop codons due to the presence of Adenine in lieu of Guanine on nucleotide 681 of exon 5 (CYP2C19*2) and on nucleotide 636 of exon 3 (CYP2C19*2). While there are other alleles associated with CYP2C19 LOF (CYP2C19*4, *5, *6, *7, *8), they comprise less than one percent of the known CYP2C19 alleles [7]. Only one GOF variant (CYP2C19*17) has been identified.

Table 3. Anticipated CYP2C19 phenotypes corresponding to genotypes. Used with permission of Creative Commons [7], copyright 2019; and adapted from Reference [70], used with permission of John Wiley and Sons; copyright 2013.

Phenotype	Example Genotypes	Enzyme Activity
Ultra-rapid metabolizer (UM)	*1/*17 *17/*17	Normal or increased
Extensive metabolizer (EM)	*1/*1 (wild type)	Normal
Intermediate metabolizer (IM)	*1/*2 *1/*3 *2/*17 *3/17	Intermediate Likely intermediate Likely intermediate
Poor metabolizer (PM)	*2/*2 *3/*3 *2/*3	Low or absent

4.3.3. Genomic Variation in CYP2D6

CYP2D6 phenotypes can also be categorized into 4 different groups based on enzyme activity: poor metabolizer, intermediate metabolizer, extensive metabolizer and ultra-rapid metabolizer [16,71–75]. The poor metabolizer phenotype is in part explained by genetic alterations, namely frame-shift mutations or splicing defects, yielding minimal expression of the CYP2D6 protein or a nonfunctional CYP2D6 protein [16]. There are over 100 CYP2D6 genomic variants, of which nearly half—through either decreased (CYP2D6*9, *17, *41, etc.) or no functional enzyme (CYP2D6 *4, *6, *15, etc.)—are phenotypically poor metabolizers [68]. The remaining known variants are phenotypically normal or as of yet undetermined. Less than 10% of the population possesses the 2D6 poor metabolizer phenotype and CYP2D6 metabolizes up to 20% of medications currently in use in the general population [16]. More research is needed to similarly characterize mechanisms behind the extensive and ultra-rapid CYP2D6 metabolizer phenotypes [16]. Further, translational and post-translational modifications may complement the contribution of genomics to interindividual variability. Among individuals with confirmed wild type ('normal') gene copies of CYP2D6, a remarkable extent of interindividual variability of phenotypic CYP2D6 activity was noted [16,76].

4.4. Systems Approach

While genetic variation plays a distinct role, this does not completely account for interindividual differences in CYP450 enzyme concentration level and activity [7,16]. The contribution of variation in transcriptional regulators and (more recently) posttranscriptional protein modifications (especially alterations in microRNA) affecting CYP450 expression are gaining increasing recognition [15]. Further, ontogenic changes to CYP450 activity, due to xenobiotic exposure (e.g., phenobarbital) during early development and the early postnatal period, have been shown to have long-lasting effects in mouse models. Importantly, efforts to achieve both early diagnosis and optimal treatment of disease as well as prevent and mitigate cardiac adverse effects in oncology patients has led to study and potential application of multi-omic disciplines to the field of Cardio-Oncology [7,77]. Genomics, epigenomics, transcriptomics, proteomics, miRNAomics, metabolomics and microbiomics have the potential to more precisely guide the clinical management of these patients [7,78]. The integration of multi-omics with systems biology and incorporating disciplines such as mobile health (mHealth), pharmacogenomics, mathematical and computational modeling is depicted in Figures 2 and 3 and has been well described [7,77,78].

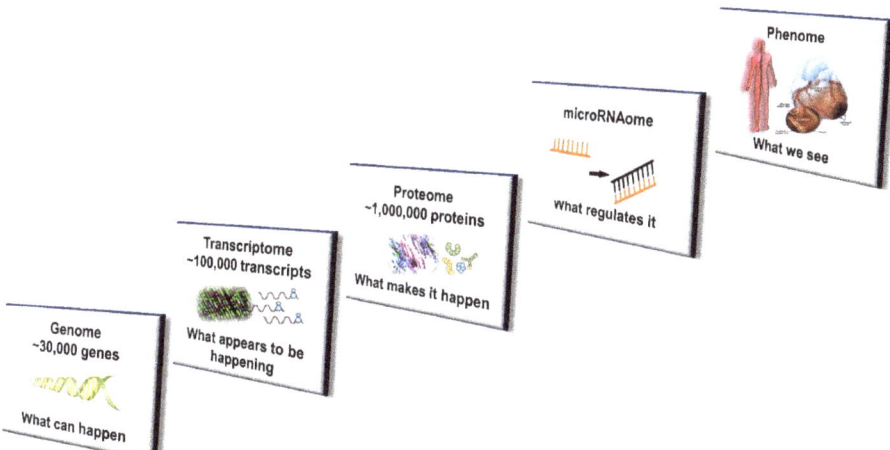

Figure 2. Multi-level systems-based approach to CYP450 expression, activity and regulation in Precision Cardio-Oncology. These multi-omics and additionally epigenomics, metabolomics, microbiomics and other personalization tools will likely be integrated in the future with mobile health, informatics and other emerging technologies for precision patient care relevant to Cardiology, Oncology and Cardio-Oncology.

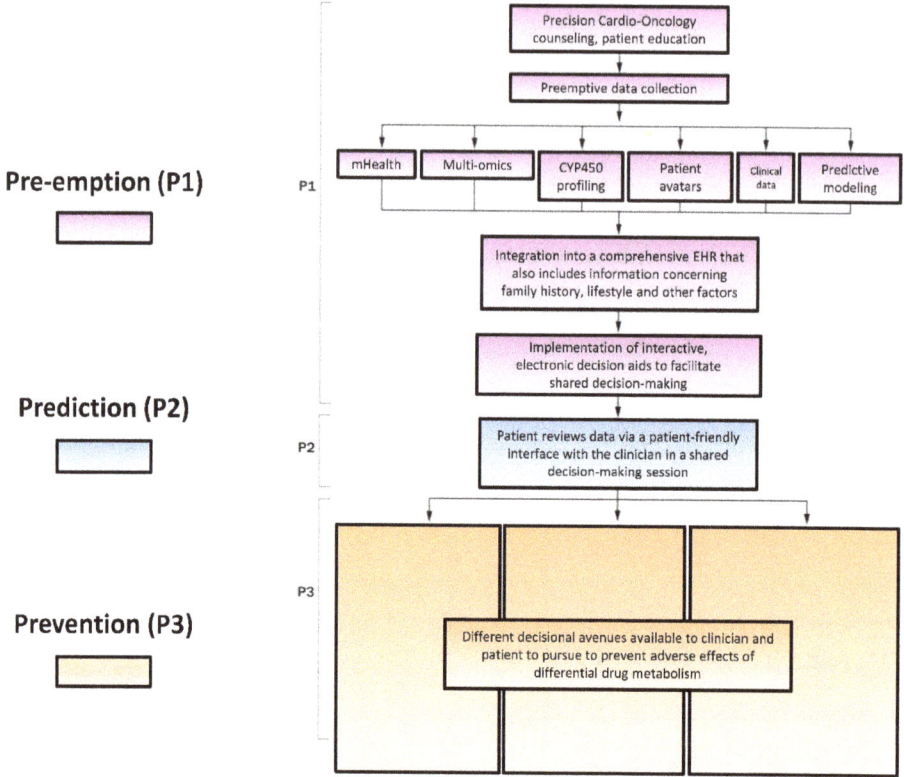

Figure 3. The P*3 Precision Medicine approach to individualizing therapy involving P450 enzymes. P1, Pre-empt: Pre-emption encompasses incorporation of elements of systems-based medicine, such as patients' at-risk genetic polymorphisms, into the EHR seamlessly alongside clinical data. P2, Predict: Prediction describes the process of utilizing systems medicine data to estimate patients' risk of developing cardiotoxicity. P3, Prevent: Prevention is proactively adjusting patients' treatment plans based on their cardiovascular toxicity risk to prevent de novo damage or mitigate further damage from antineoplastic therapy. Adapted with permission [78]. CYP450 = cytochrome P450; EHR = electronic health record.

4.4.1. Transcriptomics

Transcriptomics or gene expression profiling, is the study of the complement of genes expressed, that is, mRNA [79]. Transcribed gene expression profiles can exhibit wide variability, depending on the physiologic stimuli. Genetic variation in promotor regions and transcriptional regulators of P450 enzymes can result in altered rates of gene expression and in turn, drug metabolism [16,80]. Approximately 10% of variation in target P450 enzyme expression is explained by shared hepatic P450s transcriptional regulators (e.g., hepatocyte nuclear factor (HNF) 4β, forkhead box proteins (FOX) A2 and A3 and pregnane X receptor (PXR) polymorphisms) [16,80].

4.4.2. Epigenomics

Epigenomics is the study of physical modifications made to DNA molecules without alterations in DNA sequence. These modifications are most commonly but not limited to, acetylation and histone modification, methylation and transcription factor binding of DNA [7]. Both intrinsic and extrinsic factors can lead to epigenomic modifications of genes encoding CYP450 enzymes, thereby contributing to interindividual variability of therapeutic response [16,81]. Intrinsic factors include old age, congestive heart failure, cancer or pregnancy [7,82–84], while extrinsic factors include xenobiotics, tobacco smoke and air pollution [79]. Drugs can act as ligands to activate transcription factors, consequently inducing the expression of P450 genes. Phenobarbital and rifampin, for example, induce CYP2B6, CYP2C9 or CYP3A4 via activation of the nuclear receptors PXR or constitutive androstane receptor (CAR) [16]. It is hypothesized that epigenetic modifications in the promotor or coding region, in response to certain physiologic or pathophysiologic states, can either activate or silence certain CYP450 enzymes. For example, methylation of CpG motifs (CpG11, CpG12, CpG13) in the promoter region for CYP2C19 was associated with HTPR and increased risk of ischemic events on clopidogrel [7,85,86]. If an allele is preferentially silenced via epigenetic modifications in the coding region, the complementary allele will be predominantly expressed. Notably, DNA methylation and histone acetylation modification patterns have been found to be heritable in certain cases [79].

4.4.3. Proteomics

Proteomics is the study of the complement of proteins and can also sometimes refer to chemical modifications (e.g., phosphorylation, acetylation, nitrosylation, etc.) undergone by proteins during and after translation, due to specific (patho)physiological states and their effects on protein structure and function [7,79]. These changes can occur hours following an exposure (e.g., drug administration). In one study, a decrease in peroxiredoxin-4 (a molecule associated with HTPR) was observed just 24 h following the administration of a loading dose of clopidogrel [7,87].

4.4.4. Metabolomics

Proteins with enzymatic activity produce metabolites. Metabolomics is the study of all measurable metabolites resulting from biotransformation of compounds, often by CYP450 enzymes, corresponding to specific physiologic or pathophysiologic states [7,88]. Homeostasis of an individual's metabolome is impacted by both intrinsic (e.g., resting metabolic rate, age, genotype, etc.) and extrinsic (e.g., xenobiotics, microbiome, surrounding environment, etc.) factors [7,84,87,88]. In cardioncology patients, metabolic analysis can identify a specific metabolic pattern—a metabotype—corresponding to specific (patho)physiologic states (e.g., heart failure, myocardial infarction, myocarditis, etc.) and may serve as a 'yellow-brick road,' of sorts, offering insights into the present state and potentially, the future arc of disease [7,77]. Indeed, changes in a patient's metabolic profile over time may help identify predictive biomarkers associated with different clinical trajectories, while helping to identify an individual's response to therapy (or lack thereof), preceding phenotypic changes, helping to guide therapy [7,77].

4.4.5. Microbiomics

Microbiomics studies the impact of the human microbiome on disease states. As 70% of the millions of bacteria inhabiting the human body reside in the gut, the metabolic products of these bacteria play a role in modulating the body's inflammatory response and impacting chronic disease [7]. For example, trimethylamine N-oxide (TMAO) is a product of dietary phosphatidylcholine and gut microbe metabolism of carnitine and betaine [7,88]. High levels of TMAO are independently associated with platelet hyperactivity and thrombosis of atherosclerotic plaques [7,89–91], increasing risk of cardiovascular events [7,89] and predicting all-cause mortality [7,91]. Additionally, aspirin has been observed to alter the composition of the gut microbiota. This likely accounts for its ability to abate the effect of TMAO on platelets [7,91,92].

4.4.6. MicroRNAomics

The field of miRNAomics is quickly emerging and encompasses a diverse array of applications ranging from regulation of translation by noncoding miRNAs to small RNA therapies for challenging drug targets. Micro-RNAs (miRNAs) are a class of noncoding RNAs that play an important role in selective gene silencing, through which miRNAs regulate the diverse functions of target genes ranging from proliferation and apoptosis to differentiation. Specifically, mature miRNAs function by binding to complementary regions of transcripts (via the RNA induced silencing complex or RISC) to either (mechanically) prevent translation by the ribosome or cut target mRNAs into fragments, thereby precipitating their degradation inside the cell [15,61]. The former regulatory method of translational prevention is the predominant one in mammals (i.e., humans), while the latter method resulting in mRNA degradation is used in plants [15]. Over 1900 miRNAs have been identified in humans [15,93]. As a class, miRNAs are predicted to regulate up to 60% of the human genome [15,94,95], with miRNAs acting on one or multiple genes and single genes often targeted by one or more miRNAs [15,96].

The role of miRNAs in regulating CYP450 enzymes was first demonstrated nearly 15 years ago [61]. Since that time, the number of known miRNAs has nearly doubled and innovative techniques such as computer modeling (i.e., *in silico*) studies are being used to discover more miRNAs and determine their associated function at an exponential rate [15,80]. Data from such computer modeling, integrated with in vitro and in vivo techniques, revealed that translation of CYP3A4 mRNA is repressed by multiple miRNAs, including hsa-miR-577, hsa-miR-1, hsa-miR-532-3p and hsa-miR-627 [15]. Notably, another study found that specific miRNA profiles affected response to chemotherapy, with hsa-miR-577 and hsa-miR-1 significantly improving chemosensitivity following administration of chemotherapy for gastric cancer [15,97]. More studies are needed to confirm whether this increase in chemosensitivity is related to an inhibitory effect of hsa-miR-577 and hsa-miR-1 on CYP3A4 transcripts.

Various miRNAs are generally downregulated in various cancers, often to limit translation and promulgate the action of CYP450 enzymes promoting metabolic production of pro-carcinogens, facilitating cancer growth [61]. For example, expression of CYP1B1 is post-transcriptionally regulated by miR-27b. Studies have found that the downregulation [61] or actual deletion [98] of the 9q22.1 gene locus coding for miR-27b allows for the high expression of CYP1B1, which can facilitate tumor growth. While this inverse relationship of miR-27b downregulation and CYP1B1 upregulation expression mainly occurs in estrogen-sensitive malignancies, it is also found in urothelial or bladder cancers [61,98,99].

In addition, miR-34a has an inverse relationship with expression of CYP3A4 and CYP2C19 proteins, as well as transcription factors and other proteins that indirectly affect CYP450 levels or activity (e.g., AKR1D1 and SLC10A1) and is generally found at higher levels in females than males [80]. Females are also known to have higher CYP3A4 levels than males, suggesting miR-34a as a possible etiology for intersex differences in CYP3A4 activity [80]. The miR-34a is upregulated in several cancers, including hepatocellular cancer [80,100]. Given the inhibition of CYP3A4 and CYP2C19 by miR-34a, patients with high levels of miR-34a are potentially at an increased risk of drug toxicity. Conversely, miR148a increases levels of CYP3A4 and CYP2C19 and is downregulated in a number of cancers, including most of the lower gastrointestinal (GI) tumors, along with head & neck cancers, breast cancer, lung cancer and melanoma [80].

A myriad of miRNAs are being considered for clinical application as biomarkers or therapeutics for a variety of diseases [101]. Perhaps miRNAs that repress CYP450 enzymes to enhance tumor growth could be used as biomarkers if homeostatically their levels vary with concentration or activity of CYP450 induced by exogenous methods. For example, hsa-miR-577, hsa-miR-1, hsa-miR-532-3p and hsa-miR-627, could be used as a biomarker to gauge response to any therapeutics involving CYP3A4. Similarly miR-27b could be used as a biomarker for therapeutics related to CYP1B1 and miR-34a for CYP3A4 and CYP2C19 in addition to transcription factors and other proteins that indirectly affect CYP450 levels or activity (e.g., AKR1D1 and SLC10A1). Conversely, perhaps miRNAs that are endogenously downregulated to increase levels of CYP450 enzymes (thereby enhancing tumor growth) could be used as therapeutic options in various cancers. For instance, exogenous miR148a could be administered to boost its levels and effect on CYP3A4 and CYP2C19 in GI tumors, head & neck cancers, breast cancer, lung cancer and melanoma [80].

4.4.7. Small RNA Therapeutics

In addition to miRNA, the entities antisense oligonucleotides (ASOs), aptamers, siRNAs and synthetic mRNAs are collectively termed small RNAs and are used in the development of RNA interference therapeutics [102]. ASOs are single-stranded deoxyribonucleotides, which bind complementary mRNA targets [103]. This leads to cleavage of the mRNA-DNA heteroduplex by RNase H endonuclease, which prevents translation of the target mRNA and thereby downregulates expression of the corresponding target protein [103]. Aptamers are single-stranded oligonucleotides that bind target mRNA with high affinity and specificity, through physiochemical mechanisms such as hydrophobic, electrostatic, hydrogen bonding, van der Waals forces, base stacking and shape complementarity interactions [104,105]. Due to their desirable tissue penetrability and high affinity and specificity target-binding, aptamers are expected to become a widely used platform for delivery of therapeutic small RNAs [105]. Synthetic small interfering RNAs (siRNAs) are short double-stranded RNAs that contain a guiding strand that binds target mRNA more completely than miRNA and similarly limit translation of the target mRNA [106].

Small RNAs are currently being investigated for RNA-targeting therapeutics to treat (or prevent) illness by limiting translation of mRNA destined to become disease-relevant proteins [102]. Downregulation of disease-causing genes, that is, gene silencing, occurs via binding of the RISC complex to endogenous mRNA. Small RNA can be introduced into the cell via viral vectors or by directly insertion into cytoplasm, where the antisense siRNA-RISC complex then forms and blocks translation of target complementary mRNA. The direct interaction of these small RNA (in the RISC complex) with endogenous mRNA vastly expands the repertoire of possible therapeutics for previously 'undruggable' targets. Traditionally, drugs have been developed based on their interactions as ligands for proteins, particularly those with enzyme binding sites [107]. Expanding drug development to small RNAs opens up a new frontier in therapeutics targeting nucleic acids instead of proteins, with potential for overcoming barriers to treating previously intractable diseases [102]. The promise of small RNAs therefore may address close to 85% of the human proteome lacking ligand-binding domains or enzyme binding sites [108,109]. Development of small RNA therapeutics is not without challenges. For decades, development of ASOs and siRNAs has been tempered by immunogenicity, limited potency and poor focused delivery to the cytoplasm of the right cells and tissue [102]. Other setbacks for small RNA therapeutics have included unintended off-target effects on homologous RNA sequences and premature metabolism and excretion of the small RNA. Dozens of trials have been designed to addresses these limitations. While progress is being made to reduce these barriers, more research is needed for application of siRNA therapies to become more widespread [110]. The Food and Drug Administration (FDA) has begun to approve small RNAs as biomarkers to gauge response to therapy or as therapeutics specifically targeting previously undruggable targets, such as the RAS oncogene, with ramification for patients with lung and pancreatic cancer. This is only the beginning of a new era.

4.4.8. Integration of 'Omics'

In summary, multiple 'omics' systems within the individual operate independently and synergistically to modulate various states of health and disease. While it has been known for some time that variations in genomics incompletely account for interindividual variability and genotype-phenotype discordance in P450 enzyme expression, the underlying mechanisms have been unclear. The effect of multiple 'omics' on various states of health and disease is slowly being elucidated with continued investigation. The application of systematic multi-omics approaches to precision medicine and systems biology has great potential to improve the care of patients in Cardio-Oncology.

5. Clinical Implementation

5.1. Pharmacogenomics

Interindividual variability is largely heritable [1,2,7,69,111], thus stimulating interest in utilizing personalization tools such as pharmacogenomics (i.e., the impact of genome variations on individual response to therapeutics) in patient care [2,7,112–116].

5.1.1. Master Regulators

Studies identifying the polymorphisms of "master regulator" genes (affecting multiple CYP enzymes) have revealed key insights, as follows [16]. The 3'-untranslated region (UTR) of the aldo-keto reductase 1D1 (AKR1D1) gene is significantly associated with mRNA expression and enzyme activity of CYP2B6, CYP2C19, CYP2C8 and CYP3A4; the AKR1D1 SNP (rs1872930) yields higher constitutive mRNA expression [16]. Identifying such master regulator genes and categorizing notable SNPs may help clinicians and pharmacists predict levels and activity of CYP450 enzymes and corresponding therapeutic consequences. As CYP450 enzymes metabolize a significant majority of medications, there is a potential role for pharmacogenomics to optimize each individual's therapeutic response and prevent adverse effects.

5.1.2. Warfarin

To illustrate, a notable example in precision cardiovascular medicine is CYP2C9 genotype-guided dosing of the commonly used anticoagulant warfarin [112,117]. The Clinical Pharmacogenetic Implementation Consortium (CPIC) is an initiative focused on integrating pharmacogenomics into routine clinical care. The consortium has provided guidelines recommending the use of pharmacogenetic dosing algorithms to assist with warfarin dose based on CYP2C9 and VKORC1 genotypes, with consideration of clinical factors. Specifically, CYP2C9 alleles *2 and *3 in addition to *5, *6, *8 and *11 are associated with lower warfarin dosing requirements, due to decreased clearance of the S-enantiomer of warfarin [117,118]. While more is known about alleles *2 and *3, alleles *5, *6, *8 and *11 occur more frequently in the African American populations [117,119]. In spite of these guidelines, at present, patient dosing across much of clinical practice is guided by a dose—response—adjust cycle.

Typically, an empiric dose of warfarin is administered and patient response is tracked by measuring the international normalized ratio (INR). The dose is subsequently adjusted to fit a target threshold depending on the indication for anticoagulation. This may be, at least in part, due to the seemingly inconsistent results of various studies examining the potential impact of genetic testing to guide warfarin dosing. For example, the EU-PACT, GIFT and COAG studies are studies all investigating the utility, efficacy, cost and benefit of genetic testing to guide warfarin dosing decisions [112,117,118,120,121]. The two former studies were conducted largely in homogenous populations (>90% European ancestry), whereas the COAG study was conducted in a more diverse population in the US (28% of trial participants were African American) and did not show a great benefit to warfarin pharmacogenomics [117,120]. However, while EU-PACT administered a loading dose (according to American College of Chest Physicians guidelines) upon initiation of warfarin, the COAG study did not [121–123]. Warfarin-related variants have been less studied among African and Hispanic populations, which is notable as there is higher dose variability in these ethnic populations [112,117]. To date, no study has accounted for CYP2C9 variants more common in African Americans (*5, *6, *8, *11). It is unclear whether genetic samples currently in use for warfarin pharmacogenomics accurately represent a diverse US population. The generalizability of these trials may therefore be limited. While guidelines of governing clinical bodies at best recommend weak support of pharmacogenomic testing of warfarin dosing (Class IIB recommendation for primary stroke prevention by the American Heart Association) [124], a small study has shown promise (efficient achievement of therapeutic anticoagulation, fewer supratherapeutic INR values and a shorter duration of low molecular weight heparin) using genotype-guided dosing [117,125]. Further investigation is required.

5.1.3. Clopidogrel

To further illustrate this the potential impact of pharmacogenomics in precision medicine, some individuals with specific CYP2C19 gene mutations are poor metabolizers of the clopidogrel prodrug (the most commonly used antiplatelet agent), which can limit safety of coronary stents after percutaneous coronary intervention [7]. Some studies have demonstrated the possible cost-effectiveness of genotype-guided antiplatelet strategy [126–129], underscoring the potential utility of precision medicine to tailor medication regimens and achieve optimal patient care. In the genotype-guided antiplatelet strategy, antiplatelet utilization following acute coronary syndrome (ACS) and percutaneous coronary intervention (PCI) would be determined based on CYP2C19 genotype obtained prior to the procedure.

The desire to incorporate genotyping in routine clinical care arose from reports of clopidogrel resistance due to polymorphisms in CYP2C19 [7,117]. The CYP2C19 allele primarily responsible for decreased clopidogrel metabolism is CYP2C19*2. Additionally, multiple alleles encoding deficient enzyme activity can manifest as a true loss-of-function and an inability to metabolize the clopidogrel prodrug. Recently, results of the POPular Genetics trial—a multicenter randomized, open label study of almost 2500 patients based in the Netherlands—demonstrated noninferiority between performing genetic testing for clopidogrel resistance prior to clinical use and using another P2Y12 inhibitor (e.g., ticagrelor or prasugrel) among patients undergoing PCI [130]. As clopidogrel (available as a generic medication) is the least expensive P2Y12 inhibitor, with the lowest bleeding risk [131,132], these results represent a promising option to reduce rates of clopidogrel treatment failure, a boon particularly to those with limited financial resources. TAILOR-PCI is a currently ongoing trial, investigating the effect of the knowledge of genotype to help guide choice of P2Y12 inhibitor In one arm of the trial prospective genotyping is pursued, with individuals possessing loss-of-function (LOF) mutations receiving an alternative P2Y12 inhibitor (e.g., ticagrelor). In the control or conventional care arm, all patients received clopidogrel after PCI and their genotype is only obtained 12 months later after completing therapy. The results of this trial could shed more light on the effect of precision medicine on patient outcomes.

On a population level, CYP450 enzyme polymorphisms can account for part of the variations in drug response common among different ethnic groups. For example, 20% of Asians are poor metabolizers of CYP2C19 (responsible for metabolizing clopidogrel, as well as phenytoin/phenobarbital, omeprazole and so on), while 7% of whites are poor metabolizers of CYP2D6 (responsible for metabolizing beta-blockers, antidepressants, opioids and so on) [1,2,69]. Discussion of CYP450 variations between ethnic groups also underscores the disparities in current literature. For example, while CYP2C9 variants are more frequent in populations of African ancestry, this population is infrequently included in clinical trials [112].

It is important to note that genetic polymorphisms in CYP2C19 do not cause clopidogrel resistance in a vacuum. Comprehensive interactions among pharmacogenomics, patient characteristics and other factors affecting the activity of P450 enzymes largely determine the level of platelet response to inhibition [7]. In addition to the presence of two or more CYP2C19 LOF alleles, type 2 diabetes and increased body mass index (BMI) are likely the most important risk factors, as they independently predict and synergistically contribute to clopidogrel resistance [7,133]. Other important modifiers include chronic kidney disease, hyperlipidemia and age > 65. Lifestyle factors can either potentiate (e.g., diet, caffeine and smoking) or interfere with (e.g., grapefruit) platelet response to clopidogrel, due to interactions with CYP450 enzyme activity [7,133]. Interestingly, a large study in patients with advanced solid tumors showed a 14-fold variation in CYP3A activity, not entirely explained by genetic polymorphisms alone [134]. Potential etiologies include increased chronic inflammatory response ((IL)-1β, TNF-α and IL-6 and cytokine activity), as well as decreased liver function [3,135,136]. This differential CYP450 enzyme activity contributes to interindividual variation, medication bioavailability and therapeutic, null or toxic effects of medications.

5.2. Modified P*3 Pathway

An illustration of the optimal interactions of these myriad patient characteristics with the P450 enzyme system is the P*3 pathway [78] (Figure 3). The individual elements of the P*3 pathway—P1, Pre-empt; P2, Predict; P3, Prevent—represent a systems-based approach to patient care. In the pathway, a patient first receives precision counseling, which is akin to genetic counseling but involves digestible information about a suite of tests in precision medicine that can coalesce to form a comprehensive risk assessment. For example, after receiving precision counseling, a patient with breast cancer may in the future receive test results that include a high-risk genetic profile—mutations in TOP2A/B (mediate response to chemotherapy), RAC2 (associates with acute cardiotoxicity phenotype) and NCF4 (associates with chronic cardiotoxicity phenotype) [137]. Systems-based techniques could integrate these separate pieces of information via mathematical modeling into a single risk factor profile accessible by clinicians to help guide the patient towards optimal therapy while minimizing cardiovascular risk [78]. This information would be made available to the patient's oncologist, cardiologist and primary care provider, all of whom ideally use the same electronic medical record (EMR). With shared decision-making, prior to chemotherapy, her cardiologist may recommend cardioprotective measures with potential clinical utility, such as prophylactic angiotensin converting enzyme inhibitor, statin, beta-blocker or dexrazoxane administered prior to each course of doxorubicin. Her oncologist can also take additional precautions, including using liposomal doxorubicin preparations, avoiding concurrent trastuzumab and anthracycline use or considering alternative therapies [78]. Patient data can then be fed back into the predictive computational models to ensure continuous learning to identify actionable items for implementation in precision Cardio-Oncology. Further, patients at high versus low risk for toxicity from various anti-neoplastic agents (e.g., tyrosine kinase inhibitors, anthracyclines, monoclonal antibodies, immunotherapies) and responders versus non-responders to cardioprotective therapy, could be stratified and identified [137].

In the case of differential metabolism, genomic and other variation leading to altered P450 enzyme activity potentially resulting in drug toxicity may serve as a single hit along the path to cardiotoxicity. The presence of patient characteristics and baseline cardiovascular risk factors or existing cardiovascular disease can increase the risk for cardiovascular toxicity. When a patient's genotype, drug exposure and other factors accumulate, such multiple hits can further increase the risk of drug toxicity and consequent cardiovascular toxicity [137,138]. Using precision medicine to, for example, uncover known causes of cardiomyopathy (e.g., Titin-truncation mutations) or other cardiovascular diseases in phenotypically normal individuals, may alert patients' cardiologists and oncologists to take measures to avoid incurring additional hits. In addition to consideration of prophylactic use of ACE inhibitors, statins, beta-blockers or dexrazoxane prior to each course of doxorubicin, other cardioprotective measures may include liposomal doxorubicin preparations, avoiding multiple cardiotoxic regimens or discussion of alternative therapies, thus potentially preventing cardiotoxicity [78,139].

Data from the entire genome can potentially eventually be combined with information about a patient's transcriptome, proteome, methylome, microbiome, metabolome, environmentome, mutanome, interactome and so on in the P*3 pathway to potentially facilitate delivery of the right therapy to the right patient or group of patients at the right time [112]. Advances in the fields of Cardio-Oncology, precision medicine and Information Technology are therefore coalescing to create new possibilities for prevention and management of cardiac dysfunction from cancer therapy-related adverse effects. Cardioprotection in the oncologic patient involves initiation of cardiac medications in order to minimize or treat cardiotoxicity from cancer therapies, while maximizing administration of indicated cancer treatment. These medications—beta-blockers, angiotensin converting enzyme inhibitors or angiotensin receptor blockers and statins—comprise the cornerstone of cardiovascular disease risk management and treatment in the general population as well as in Cardio-Oncology [140,141]. Perhaps using systems-based approaches in precision medicine to guide the use of these cardioprotective therapies is the panacea of prevention in Cardio-Oncology—in Preventive Cardio-Oncology.

Precision Cardio-Oncology is thus a burgeoning field that seeks to further personalize the cardiovascular care of patients in oncology for decisions related to both management and prevention of cardiovascular toxicities [142]. The goal is to achieve a maximal amount of indicated chemotherapy administered with minimal interruption, while avoiding toxicity. Other notable goals include delivering more effective and efficient care, reducing patient harm and limiting healthcare costs from inappropriate treatment [43]. Indeed, precision medicine has great potential in the care of patients taking medications metabolized by CYP450 enzymes, not only in Cardio-Oncology but in all fields of medicine.

6. Conclusions

Metabolism by CYP450 enzymes can determine the bioavailability and thereby efficacy of several drugs in Cardiology and Oncology and in the emergent field of Cardio-Oncology. These enzymes can also affect drug-drug interactions between Oncology and Cardiology drugs. This can compound the use of cardiology drugs for protection from or treatment of cardiovascular toxicity. Differential metabolism of each drug can determine to a certain degree unpredictable bioavailability of the drugs in a specific individual. This can be further impacted by variations in the genome, in the context of the broader epigenome, transcriptome, proteome, microRNA regulome, microbiome, metabolome, environmentome, populome and other components of the individual as a whole organism or system, with multiple parts that can be perturbed by various cardiology or oncology drugs. Increasing knowledge and implementation of the multidimensional impact of endogenous regulatory systems on CYP450-mediated drug metabolism may help preempt drug-drug interactions, predict variations in CYP450 enzymes and prevent complications from subtherapeutic or supratherapeutic drug levels. Such a systems-based view should be considered as we move towards clinical and research practice of Precision Cardiovascular oncology, with particular attention to the role of CYP450 enzymes.

Author Contributions: Conceptualization, S.-A.B.; Writing—Original Draft Preparation, O.A.F. and S.-A.B.; Writing—Review & Editing, O.A.F. and S.-A.B. All authors have read and agreed to the published version of the manuscript.

Acknowledgments: The authors are grateful to Giselle A. Suero-Abreu at Rutgers New Jersey Medical School in Newark, N.J. for helpful discussions, suggestions and assistance with the manuscript. We are also thankful to Diana Mechelay, PharmD at CHRISTUS Good Shepherd Medical Center in Longview, T.X. for helpful discussions, suggestions and reading the manuscript.

Conflicts of Interest: The authors declare no conflict of interest.

References

1. Zanger, U.M.; Schwab, M. Cytochrome P450 enzymes in drug metabolism: Regulation of gene expression, enzyme activities, and impact of genetic variation. *Pharmacol. Ther.* **2013**, *138*, 103–141. [CrossRef]
2. Lynch, T.; Price, A. The effect of cytochrome P450 metabolism on drug response, interactions, and adverse effects. *Am. Fam. Physician* **2007**, *76*, 391–396.
3. Undevia, S.; Gomez-Abuin, G.; Ratain, M. Pharmacokinetic Variability of Anticancer Agents. Available online: www.medscape.com/viewarticle/506712_1 (accessed on 12 September 2019).
4. Chaudhary, K.R.; Batchu, S.N.; Seubert, J.M. Cytochrome P450 enzymes and the heart. *IUBMB Life* **2009**, *61*, 954–960. [CrossRef]
5. Carreca, I.; Balducci, L. Oral chemotherapy of cancer in the elderly. *Am. J. Cancer* **2002**, *1*, 101–108. [CrossRef]
6. DeMario, M.D.; Ratain, M.J. Oral chemotherapy: Rationale and future directions. *J. Clin. Oncol.* **1998**, *16*, 2557–2567. [CrossRef]
7. Brown, S.-A.; Pereira, N. Pharmacogenomic Impact of CYP2C19 Variation on Clopidogrel Therapy in Precision Cardiovascular Medicine. *J. Pers. Med.* **2018**, *8*, 8. [CrossRef]
8. Kutsuno, Y.; Itoh, T.; Tukey, R.H.; Fujiwara, R. Glucuronidation of drugs and drug-induced toxicity in humanized UDP-glucuronosyltransferase 1 mice. *Drug Metab. Dispos.* **2014**, *42*, 1146–1152. [CrossRef]

9. Li, A.P.; Kaminski, D.L.; Rasmussen, A. Substrates of human hepatic cytochrome P450 3A4. *Toxicology* **1995**, *104*, 1–8. [CrossRef]
10. Evans, W.E.; Relling, M.V. Pharmacogenomics: Translating functional genomics into rational therapeutics. *Science* **1999**, *286*, 487–491. [CrossRef]
11. Wrighton, S.A.; Brian, W.R.; Sari, M.A.; Iwasaki, M.; Guengerich, F.P.; Raucy, J.L.; Molowa, D.T.; Vandenbranden, M. Studies on the expression and metabolic capabilities of human liver cytochrome P450IIIA5 (HLp3). *Mol. Pharmacol.* **1990**, *38*, 207–213.
12. Shimada, T.; Yamazaki, H.; Mimura, M.; Inui, Y.; Guengerich, F.P. Interindividual variations in human liver cytochrome P-450 enzymes involved in the oxidation of drugs, carcinogens and toxic chemicals: Studies with liver microsomes of 30 Japanese and 30 Caucasians. *J. Pharmacol. Exp. Ther.* **1994**, *270*, 414–423.
13. Watkins, P.B.; Wrighton, S.A.; Schuetz, E.G.; Molowa, D.T.; Guzelian, P.S. Identification of glucocorticoid-inducible cytochromes P-450 in the intestinal mucosa of rats and man. *J. Clin. Investig.* **1987**, *80*, 1029–1036. [CrossRef]
14. Kolars, J.C.; Schmiedlin-Ren, P.; Schuetz, J.D.; Fang, C.; Watkins, P.B. Identification of rifampin-inducible P450IIIA4 (CYP3A4) in human small bowel enterocytes. *J. Clin. Investig.* **1992**, *90*, 1871–1878. [CrossRef]
15. Wei, Z.; Jiang, S.; Zhang, Y.; Wang, X.; Peng, X.; Meng, C.; Liu, Y.; Wang, H.; Guo, L.; Qin, S.; et al. The effect of microRNAs in the regulation of human CYP3A4: A systematic study using a mathematical model. *Sci. Rep.* **2014**, *4*, 4283. [CrossRef]
16. Tracy, T.S.; Chaudhry, A.S.; Prasad, B.; Thummel, K.E.; Schuetz, E.G.; Zhong, X.B.; Tien, Y.C.; Jeong, H.; Pan, X.; Shireman, L.M.; et al. Interindividual Variability in Cytochrome P450-Mediated Drug Metabolism. *Drug Metab. Dispos.* **2016**, *44*, 343–351. [CrossRef]
17. Westlind, A.; Löfberg, L.; Tindberg, N.; Andersson, T.B.; Ingelman-Sundberg, M. Interindividual differences in hepatic expression of CYP3A4: Relationship to genetic polymorphism in the 5′-upstream regulatory region. *Biochem. Biophys. Res. Commun.* **1999**, *259*, 201–205. [CrossRef]
18. Lamba, J.K.; Lin, Y.S.; Schuetz, E.G.; Thummel, K.E. Genetic contribution to variable human CYP3A-mediated metabolism. *Adv. Drug Deliv. Rev.* **2002**, *54*, 1271–1294. [CrossRef]
19. Hart, S.N.; Wang, S.; Nakamoto, K.; Wesselman, C.; Li, Y.; Zhong, X.B. Genetic polymorphisms in cytochrome P450 oxidoreductase influence microsomal P450-catalyzed drug metabolism. *Pharm. Genom.* **2008**, *18*, 11–24. [CrossRef]
20. Court, M.H. Interindividual variability in hepatic drug glucuronidation: Studies into the role of age, sex, enzyme inducers, and genetic polymorphism using the human liver bank as a model system. *Drug Metab. Rev.* **2010**, *42*, 209–224. [CrossRef]
21. Foti, R.S.; Dalvie, D.K. Cytochrome P450 and Non-Cytochrome P450 Oxidative Metabolism: Contributions to the Pharmacokinetics, Safety, and Efficacy of Xenobiotics. *Drug Metab. Dispos.* **2016**, *44*, 1229–1245. [CrossRef]
22. Tornio, A.; Filppula, A.M.; Kailari, O.; Neuvonen, M.; Nyrönen, T.H.; Tapaninen, T.; Neuvonen, P.J.; Niemi, M.; Backman, J.T. Glucuronidation converts clopidogrel to a strong time-dependent inhibitor of CYP2C8: A phase II metabolite as a perpetrator of drug-drug interactions. *Clin. Pharmacol. Ther.* **2014**, *96*, 498–507. [CrossRef] [PubMed]
23. Fujiwara, R.; Yoda, E.; Tukey, R.H. Species differences in drug glucuronidation: Humanized UDP-glucuronosyltransferase 1 mice and their application for predicting drug glucuronidation and drug-induced toxicity in humans. *Drug Metab. Pharm.* **2018**, *33*, 9–16. [CrossRef] [PubMed]
24. Diao, X.; Pang, X.; Xie, C.; Guo, Z.; Zhong, D.; Chen, X. Bioactivation of 3-n-butylphthalide via sulfation of its major metabolite 3-hydroxy-NBP: Mediated mainly by sulfotransferase 1A1. *Drug Metab. Dispos.* **2014**, *42*, 774–781. [CrossRef] [PubMed]
25. Wang, L.-Q.; James, M. Inhibition of Sulfotransferases by Xenobiotics. *Curr. Drug Metab.* **2006**, *7*, 83–104. [CrossRef]
26. Gamage, N.; Barnett, A.; Hempel, N.; Duggleby, R.G.; Windmill, K.F.; Martin, J.L.; McManus, M.E. Human sulfotransferases and their role in chemical metabolism. *Toxicol. Sci.* **2006**, *90*, 5–22. [CrossRef]
27. Ipe, J.; Lu, J.; Nguyen, A. P450 Drug Interactions—Flockhart Table™. Available online: https://drug-interactions.medicine.iu.edu/MainTable.aspx (accessed on 5 August 2019).

28. Chang, H.M.; Okwuosa, T.M.; Scarabelli, T.; Moudgil, R.; Yeh, E.T.H. Cardiovascular Complications of Cancer Therapy: Best Practices in Diagnosis, Prevention, and Management: Part 2. *J. Am. Coll. Cardiol.* **2017**, *70*, 2552–2565. [CrossRef]
29. Mosarla, R.C.; Vaduganathan, M.; Qamar, A.; Moslehi, J.; Piazza, G.; Giugliano, R.P. Anticoagulation Strategies in Patients With Cancer: JACC Review Topic of the Week. *J. Am. Coll. Cardiol.* **2019**, *73*, 1336–1349. [CrossRef]
30. van Eijk, M.; Boosman, R.J.; Schinkel, A.H.; Huitema, A.D.R.; Beijnen, J.H. Cytochrome P450 3A4, 3A5, and 2C8 expression in breast, prostate, lung, endometrial, and ovarian tumors: Relevance for resistance to taxanes. *Cancer Chemother. Pharmacol.* **2019**, *84*, 487–499. [CrossRef]
31. Gunes, A.; Coskun, U.; Boruban, C.; Gunel, N.; Babaoglu, M.O.; Sencan, O.; Bozkurt, A.; Rane, A.; Hassan, M.; Zengil, H.; et al. Inhibitory effect of 5-fluorouracil on cytochrome P450 2C9 activity in cancer patients. *Basic Clin. Pharmacol. Toxicol.* **2006**, *98*, 197–200. [CrossRef]
32. Flaherty, K.T.; Lathia, C.; Frye, R.F.; Schuchter, L.; Redlinger, M.; Rosen, M.; O'Dwyer, P.J. Interaction of sorafenib and cytochrome P450 isoenzymes in patients with advanced melanoma: A phase I/II pharmacokinetic interaction study. *Cancer Chemother. Pharmacol.* **2011**, *68*, 1111–1118. [CrossRef]
33. Chang, H.M.; Moudgil, R.; Scarabelli, T.; Okwuosa, T.M.; Yeh, E.T.H. Cardiovascular Complications of Cancer Therapy: Best Practices in Diagnosis, Prevention, and Management: Part 1. *J. Am. Coll. Cardiol.* **2017**, *70*, 2536–2551. [CrossRef] [PubMed]
34. Bullock, K.E.; Petros, W.P.; Younis, I.; Uronis, H.E.; Morse, M.A.; Blobe, G.C.; Zafar, S.Y.; Gockerman, J.P.; Lager, J.J.; Truax, R.; et al. A Phase I Study of Bevacizumab (B) in Combination with Everolimus (E) and Erlotinib (E) in Advanced Cancer (BEE). *Cancer Chemother. Pharmacol.* **2014**, *67*, 465–474. [CrossRef] [PubMed]
35. Wang, L.; Christopher, L.J.; Cui, D.; Li, W.; Iyer, R.; Humphreys, W.G.; Zhang, D. Identification of the human enzymes involved in the oxidative metabolism of dasatinib: An effective approach for determining metabolite formation kinetics. *Drug Metab. Dispos. Biol. Fate Chem.* **2008**. [CrossRef] [PubMed]
36. Colburn, D.E.; Giles, F.J.; Oladovich, D.; Smith, J.A. In vitro evaluation of cytochrome P450-mediated drug interactions between cytarabine, idarubicin, itraconazole and caspofungin. *Hematology* **2004**, *9*, 217–221. [CrossRef] [PubMed]
37. Quintanilha, J.C.F.; de Sousa, V.M.; Visacri, M.B.; Amaral, L.S.; Santos, R.M.M.; Zambrano, T.; Salazar, L.A.; Moriel, P. Involvement of cytochrome P450 in cisplatin treatment: Implications for toxicity. *Cancer Chemother. Pharmacol.* **2017**, *80*, 223–233. [CrossRef] [PubMed]
38. Murayama, N.; van Beuningen, R.; Suemizu, H.; Guillouzo, C.G.; Shibata, N.; Yajima, K.; Utoh, M.; Shimizu, M.; Chesné, C.; Nakamura, M.; et al. Thalidomide increases human hepatic cytochrome P450 3A enzymes by direct activation of the pregnane X receptor. *Chem. Res. Toxicol.* **2014**, *27*, 304–308. [CrossRef]
39. Lawrence, S.; Nguyen, D.; Bowen, C.; Richards-Peterson, L.; Sordos, K. The metabolic drug-drug interaction profile of Dabrafenib: In vitro investigations and quantitative extrapolation of the P450-mediated DDI risk. *Drug Metab. Dispos.* **2014**, *42*, 1180–1190. [CrossRef]
40. Zhang, W.; Heinzmann, D.; Grippo, J.F. Clinical Pharmacokinetics of Vemurafenib. *Clin. Pharm.* **2017**, *56*, 1033–1043. [CrossRef]
41. Whirl-Carrillo, M.; McDonagh, E.M.; Hebert, J.M.; Gong, L.; Sangkuhl, K.; Thorn, C.F.; Altman, R.B.; Klein, T.E. Pharmacogenomics knowledge for personalized medicine. *Clin. Pharmacol. Ther.* **2012**, *92*, 414–417. [CrossRef]
42. van Leeuwen, R.W.; Brundel, D.H.; Neef, C.; van Gelder, T.; Mathijssen, R.H.; Burger, D.M.; Jansman, F.G. Prevalence of potential drug-drug interactions in cancer patients treated with oral anticancer drugs. *Br. J. Cancer* **2013**, *108*, 1071–1078. [CrossRef]
43. Zukkoor, S.; Thohan, V. Drug-Drug Interactions of Common Cardiac Medications and Chemotherapeutic Agents. Available online: www.acc.org/latest-in-cardiology/articles/2018/12/21/09/52/drug-drug-interactions-of-common-cardiac-medications-and-chemotherapeutic-agents (accessed on 25 August 2019).
44. Sasu-Tenkoramaa, J.; Fudin, J. Drug Interactions in Cancer Patients Requiring Concomitant Chemotherapy and Analgesics. *Prac. Pain Manag.* **2013**, *13*, 50–64.
45. Kennedy, C.; Brewer, L.; Williams, D. Drug interactions. *Medicine* **2016**, *44*, 422–426. [CrossRef]
46. Lexicomp Medication Database. *Lexicomp®Online™*, 2013.

47. Yamazaki, S.; Johnson, T.R.; Smith, B.J. Prediction of Drug-Drug Interactions with Crizotinib as the CYP3A Substrate Using a Physiologically Based Pharmacokinetic Model. *Drug Metab. Dispos.* **2015**, *43*, 1417–1429. [CrossRef] [PubMed]
48. Khozin, S.; Blumenthal, G.M.; Zhang, L.; Tang, S.; Brower, M.; Fox, E.; Helms, W.; Leong, R.; Song, P.; Pan, Y.; et al. FDA approval: Ceritinib for the treatment of metastatic anaplastic lymphoma kinase-positive non-small cell lung cancer. *Clin. Cancer Res.* **2015**, *21*, 2436–2439. [CrossRef] [PubMed]
49. Fan, P.; Gao, Y.; Zheng, M.; Xu, T.; Schoenhagen, P.; Zhaohui, J. Recent Progress and Market Analysis of Anticoagulant Drugs. *J. Thorac. Dis.* **2018**, *10*, 2011–2025. [CrossRef] [PubMed]
50. Al-Samkari, H.; Connors, J.M. The Role of Direct Oral Anticoagulants in Treatment of Cancer-Associated Thrombosis. *Cancers* **2018**, *10*, 271. [CrossRef] [PubMed]
51. Engman, C.A.; Zacharski, L.R. Low molecular weight heparins as extended prophylaxis against recurrent thrombosis in cancer patients. *J. Natl. Compr. Cancer Netw.* **2008**, *6*, 637–645. [CrossRef]
52. Lyman, G.H.; Khorana, A.A.; Kuderer, N.M.; Lee, A.Y.; Arcelus, J.I.; Balaban, E.P.; Clarke, J.M.; Flowers, C.R.; Francis, C.W.; Gates, L.E.; et al. Venous thromboembolism prophylaxis and treatment in patients with cancer: American Society of Clinical Oncology clinical practice guideline update. *J. Clin. Oncol.* **2013**, *31*, 2189–2204. [CrossRef]
53. Farge, D.; Debourdeau, P.; Beckers, M.; Baglin, C.; Bauersachs, R.M.; Brenner, B.; Brilhante, D.; Falanga, A.; Gerotzafias, G.T.; Haim, N.; et al. International clinical practice guidelines for the treatment and prophylaxis of venous thromboembolism in patients with cancer. *J. Thromb. Haemost.* **2013**, *11*, 56–70. [CrossRef]
54. Kearon, C.; Akl, E.; Ornelas, J.; Blaivas, A.; Jimenez, D.; Bounameaux, H.; Huisman, M.; King, C.; Morris, T.; Sood, N.; et al. Correction to Grade in: Antithrombotic Therapy for VTE Disease: CHEST Guideline and Expert Panel Report. *Chest* **2016**, *150*, 315–352. [CrossRef]
55. Samuelson Bannow, B.T.; Lee, A.; Khorana, A.A.; Zwicker, J.I.; Noble, S.; Ay, C.; Carrier, M. Management of Cancer-Associated Thrombosis in Patients with Thrombocytopenia: Guidance from the SSC of the ISTH. *J. Thromb. Haemost.* **2018**, *16*, 1246–1249. [CrossRef] [PubMed]
56. Short, N.J.; Connors, J.M. New oral anticoagulants and the cancer patient. *Oncologist* **2014**, *19*, 82–93. [CrossRef] [PubMed]
57. Lakkireddy, D.; Karst, E.; Mahapatra, S.; Winterfield, J.; Mansour, M. Lower Adherence Direct Oral Anticoagulants Use Is Associated With Increased Risk Of Thromboembolic Events Than Warfarin—Understanding The Real-World Performance Of Systemic Anticoagulation In Atrial Fibrillation. In Proceedings of the Heart Rhythm Society's 39th Annual Scientific Sessions, Boston, MA, USA, 9–12 May 2018.
58. Burn, J.; Pirmohamed, M. Correction: Direct Oral Anticoagulants versus Warfarin: Is New Always Better than the Old? *Open Heart* **2018**, *5*, 1–5. [CrossRef] [PubMed]
59. Ozdemir, V.; Kalow, W.; Tang, B.K.; Paterson, A.D.; Walker, S.E.; Endrenyi, L.; Kashuba, A.D. Evaluation of the genetic component of variability in CYP3A4 activity: A repeated drug administration method. *Pharmacogenetics* **2000**, *10*, 373–388. [CrossRef] [PubMed]
60. Helsby, N.A.; Lo, W.Y.; Sharples, K.; Riley, G.; Murray, M.; Spells, K.; Dzhelai, M.; Simpson, A.; Findlay, M. CYP2C19 pharmacogenetics in advanced cancer: Compromised function independent of genotype. *Br. J. Cancer* **2008**, *99*, 1251–1255. [CrossRef] [PubMed]
61. Tsuchiya, Y.; Nakajima, M.; Takagi, S.; Taniya, T.; Yokoi, T. MicroRNA regulates the expression of human cytochrome P450 1B1. *Cancer Res.* **2006**, *66*, 9090–9098. [CrossRef]
62. Shimada, T.; Hayes, C.L.; Yamazaki, H.; Amin, S.; Hecht, S.S.; Guengerich, F.P.; Sutter, T.R. Activation of chemically diverse procarcinogens by human cytochrome P-450 1B1. *Cancer Res.* **1996**, *56*, 2979–2984.
63. Murray, G.I.; Taylor, M.C.; McFadyen, M.C.; McKay, J.A.; Greenlee, W.F.; Burke, M.D.; Melvin, W.T. Tumor-specific expression of cytochrome P450 CYP1B1. *Cancer Res.* **1997**, *57*, 3026–3031.
64. Newbold, R.R.; Liehr, J.G. Induction of uterine adenocarcinoma in CD-1 mice by catechol estrogens. *Cancer Res.* **2000**, *60*, 235–237.
65. Han, X.; Liehr, J.G. DNA single-strand breaks in kidneys of Syrian hamsters treated with steroidal estrogens: Hormone-induced free radical damage preceding renal malignancy. *Carcinogenesis* **1994**, *15*, 997–1000. [CrossRef]

66. Genomic profiling. In *NCI Dictionary of Cancer Terms*; National Cancer Institute at the National Institutes of Health. Available online: https://www.cancer.gov/publications/dictionaries/cancer-terms/def/genomic-profiling (accessed on 20 September 2019).
67. Puckelwartz, M.J.; McNally, E.M. Genetic profiling for risk reduction in human cardiovascular disease. *Genes* **2014**, *5*, 214–234. [CrossRef] [PubMed]
68. Pharmacogene Variation Consortium (PharmVar). Available online: https://www.pharmvar.org/ (accessed on 20 September 2019).
69. Weinshilboum, R. Inheritance and drug response. *N. Engl. J. Med.* **2003**, *348*, 529–537. [CrossRef] [PubMed]
70. Scott, S.A.; Sangkuhl, K.; Stein, C.M.; Hulot, J.S.; Mega, J.L.; Roden, D.M.; Klein, T.E.; Sabatine, M.S.; Johnson, J.A.; Shuldiner, A.R.; et al. Clinical Pharmacogenetics Implementation Consortium guidelines for CYP2C19 genotype and clopidogrel therapy: 2013 update. *Clin. Pharmacol. Ther.* **2013**, *94*, 317–323. [CrossRef] [PubMed]
71. Hou, Z.Y.; Pickle, L.W.; Meyer, P.S.; Woosley, R.L. Salivary analysis for determination of dextromethorphan metabolic phenotype. *Clin. Pharmacol. Ther.* **1991**, *49*, 410–419. [CrossRef] [PubMed]
72. Bertilsson, L.; Dahl, M.; Dalén, P.; Al-Shurbaji, A. Molecular genetics of CYP2D6: Clinical relevance with focus on psychotropic drugs. *Br. J. Clin. Pharmacol.* **2002**, *53*, 111–122. [CrossRef] [PubMed]
73. Dahl, M.-L.; Johansson, I.; Palmertz, M.P.; Ingelman-Sundberg, M.; Sjöqvist, F. Analysis of the CYP2D6 gene in relation to debrisoquin and desipramine hydroxylation in a Swedish population. *Clin. Pharmacol. Ther.* **1992**, *51*, 12–17. [CrossRef] [PubMed]
74. Sachse, C.; Brockmöller, J.; Bauer, S.; Roots, I. Cytochrome P450 2D6 variants in a Caucasian population: Allele frequencies and phenotypic consequences. *Am. J. Hum. Genet.* **1997**, *60*, 284–295.
75. Zanger, U.M.; Fischer, J.; Raimundo, S.; Stüven, T.; Evert, B.O.; Schwab, M.; Eichelbaum, M. Comprehensive analysis of the genetic factors determining expression and function of hepatic CYP2D6. *Pharmacogenetics* **2001**, *11*, 573–585. [CrossRef]
76. Gaedigk, A.; Simon, S.; Pearce, R.; Bradford, L.; Kennedy, M.; Leeder, J. The CYP2D6 Activity Score: Translating Genotype Information into a Qualitative Measure of Phenotype. *Clin. Pharmacol. Ther.* **2007**, *83*, 234–242. [CrossRef]
77. Dreyfuss, A.D.; Bravo, P.E.; Koumenis, C.; Ky, B. Precision Cardio-Oncology. *J. Nucl. Med.* **2019**, *60*, 443–450. [CrossRef]
78. Brown, S.A.; Sandhu, N.; Herrmann, J. Systems biology approaches to adverse drug effects: The example of cardio-oncology. *Nat. Rev. Clin. Oncol.* **2015**, *12*, 718–731. [CrossRef] [PubMed]
79. Cappola, T.; Margulies, K. Functional Genomics Applied to Cardiovascular Medicine. *Circulation* **2011**, *124*, 87–94. [CrossRef] [PubMed]
80. Lamba, V.; Ghodke, Y.; Guan, W.; Tracy, T.S. microRNA-34a is associated with expression of key hepatic transcription factors and cytochromes P450. *Biochem. Biophys. Res. Commun.* **2014**, *445*, 404–411. [CrossRef]
81. Lin, J.H.; Lu, A.Y. Interindividual variability in inhibition and induction of cytochrome P450 enzymes. *Annu. Rev. Pharmacol. Toxicol.* **2001**, *41*, 535–567. [CrossRef] [PubMed]
82. Williams, M.L.; Bhargava, P.; Cherrouk, I.; Marshall, J.L.; A Flockhart, D.; Wainer, I.W. A discordance of the cytochrome P450 2C19 genotype and phenotype in patients with advanced cancer. *Br. J. Clin. Pharmacol.* **2000**, *49*, 485–488. [CrossRef] [PubMed]
83. Frye, R.F.; Schneider, V.M.; Frye, C.S.; Feldman, A.M. Plasma levels of TNF-alpha and IL-6 are inversely related to cytochrome P450-dependent drug metabolism in patients with congestive heart failure. *J. Card. Fail.* **2002**, *8*, 315–319. [CrossRef] [PubMed]
84. Ishizawa, Y.; Yasui-Furukori, N.; Takahata, T.; Sasaki, M.; Tateishi, T. The effect of aging on the relationship between the cytochrome P450 2C19 genotype and omeprazole pharmacokinetics. *Clin. Pharm.* **2005**, *44*, 1179–1189. [CrossRef] [PubMed]
85. Su, J.; Li, X.; Yu, Q.; Liu, Y.; Wang, Y.; Song, H.; Cui, H.; Du, W.; Fei, X.; Liu, J.; et al. Association of P2Y12 Gene Promoter DNA Methylation with the Risk of Clopidogrel Resistance in Coronary Artery Disease Patients. *BioMed Res. Int.* **2014**, *2014*, 1–8.
86. Li, X.G.; Ma, N.; Wang, B.; Li, X.Q.; Mei, S.H.; Zhao, K.; Wang, Y.J.; Li, W.; Zhao, Z.G.; Sun, S.S.; et al. The impact of P2Y12 promoter DNA methylation on the recurrence of ischemic events in Chinese patients with ischemic cerebrovascular disease. *Sci. Rep.* **2016**, *6*, 34570. [CrossRef]

87. Caruso, R.; Rocchiccioli, S.; Gori, A.M.; Cecchettini, A.; Giusti, B.; Parodi, G.; Cozzi, L.; Marcucci, R.; Parolini, M.; Romagnuolo, I.; et al. Inflammatory and antioxidant pattern unbalance in "clopidogrel-resistant" patients during acute coronary syndrome. *Mediat. Inflamm.* **2015**, *2015*, 710123. [CrossRef]
88. Goodacre, R. Metabolomics of a superorganism. *J. Nutr.* **2007**, *137*, 259S–266S. [CrossRef] [PubMed]
89. Senthong, V.; Wang, Z.; Li, X.S.; Fan, Y.; Wu, Y.; Tang, W.H.W.; Hazen, S.L. Intestinal Microbiota-Generated Metabolite Trimethylamine- N- Oxide and 5-Year Mortality Risk in Stable Coronary Artery Disease: The Contributory Role of Intestinal Microbiota in a COURAGE-Like Patient Cohort. *J. Am. Hear. Assoc.* **2016**, *5*, e002816. [CrossRef] [PubMed]
90. Zhu, W.; Gregory, J.C.; Org, E.; Buffa, J.A.; Gupta, N.; Wang, Z.; Li, L.; Fu, X.; Wu, Y.; Mehrabian, M.; et al. Gut Microbial Metabolite TMAO Enhances Platelet Hyperreactivity and Thrombosis Risk. *Cell* **2016**, *165*, 111–124. [CrossRef] [PubMed]
91. Zhu, W.; Wang, Z.; Tang, W.H.W.; Hazen, S.L. Gut Microbe-Generated Trimethylamine N-Oxide From Dietary Choline is Prothrombotic in Subects. *Circulation* **2017**, *135*, 1671–1673. [CrossRef] [PubMed]
92. Rogers, M.A.M.; Aronoff, D.M. The influence of non-steroidal anti-inflammatory drugs on the gut microbiome. *Clin. Microbiol. Infect.* **2016**, *22*, e171–e178. [CrossRef] [PubMed]
93. Bentwich, I.; Avniel, A.; Karov, Y.; Aharonov, R.; Gilad, S.; Barad, O.; Barzilai, A.; Einat, P.; Einav, U.; Meiri, E.; et al. Identification of hundreds of conserved and nonconserved human microRNAs. *Nat. Genet.* **2005**, *37*, 766–770. [CrossRef]
94. Lewis, B.P.; Burge, C.B.; Bartel, D.P. Conserved seed pairing, often flanked by adenosines, indicates that thousands of human genes are microRNA targets. *Cell* **2005**, *120*, 15–20. [CrossRef]
95. Friedman, R.C.; Farh, K.K.; Burge, C.B.; Bartel, D.P. Most mammalian mRNAs are conserved targets of microRNAs. *Genome Res.* **2009**, *19*, 92–105. [CrossRef]
96. Wu, S.; Huang, S.; Ding, J.; Zhao, Y.; Liang, L.; Liu, T.; Zhan, R.; He, X. Multiple microRNAs modulate p21Cip1/Waf1 expression by directly targeting its 3' untranslated region. *Oncogene* **2010**, *29*, 2302–2308. [CrossRef]
97. Kim, C.H.; Kim, H.K.; Rettig, R.L.; Kim, J.; Lee, E.T.; Aprelikova, O.; Choi, I.J.; Munroe, D.J.; Green, J.E. miRNA signature associated with outcome of gastric cancer patients following chemotherapy. *BMC Med. Genom.* **2011**, *4*, 79. [CrossRef]
98. Calin, G.A.; Sevignani, C.; Dumitru, C.D.; Hyslop, T.; Noch, E.; Yendamuri, S.; Shimizu, M.; Rattan, S.; Bullrich, F.; Negrini, M.; et al. Human microRNA genes are frequently located at fragile sites and genomic regions involved in cancers. *Proc. Natl. Acad. Sci. USA* **2004**, *101*, 2999–3004. [CrossRef] [PubMed]
99. Carnell, D.M.; Smith, R.E.; Daley, F.M.; Barber, P.R.; Hoskin, P.J.; Wilson, G.D.; Murray, G.I.; Everett, S.A. Target validation of cytochrome P450 CYP1B1 in prostate carcinoma with protein expression in associated hyperplastic and premalignant tissue. *Int. J. Radiat Oncol. Biol. Phys.* **2004**, *58*, 500–509. [CrossRef] [PubMed]
100. Dang, Y.; Luo, D.; Rong, M.; Chen, G. Underexpression of miR-34a in hepatocellular carcinoma and its contribution towards enhancement of proliferating inhibitory effects of agents targeting c-MET. *PLoS ONE* **2013**, *8*, e61054. [CrossRef] [PubMed]
101. Hanna, J.; Hossain, G.S.; Kocerha, J. The Potential for microRNA Therapeutics and Clinical Research. *Front. Genet.* **2019**, *10*, 478. [CrossRef]
102. Delivering the promise of RNA therapeutics. *Nat. Med.* **2019**, *25*, 1321. [CrossRef]
103. Di Fusco, D.; Dinallo, V.; Marafini, I.; Figliuzzi, M.M.; Romano, B.; Monteleone, G. Antisense Oligonucleotide: Basic Concepts and Therapeutic Application in Inflammatory Bowel Disease. *Front Pharmacol.* **2019**, *10*, 305. [CrossRef]
104. Kalra, P.; Dhiman, A.; Cho, W.C.; Bruno, J.G.; Sharma, T.K. Simple Methods and Rational Design for Enhancing Aptamer Sensitivity and Specificity. *Front Mol Biosci.* **2018**, *5*, 41. [CrossRef]
105. Soldevilla, M.M.; Meraviglia-Crivelli de Caso, D.; Menon, A.P.; Pastor, F. Aptamer-iRNAs as Therapeutics for Cancer Treatment. *Pharmaceuticals (Basel)* **2018**, *11*, 108. [CrossRef]
106. Li, J.; Wu, C.; Wang, W.; He, Y.; Elkayam, E.; Joshua-Tor, L. Structurally modulated codelivery of siRNA and Argonaute 2 for enhanced RNA interference. *Proc. Natl. Acad. Sci. USA* **2018**, *115*, E2696–E2705. [CrossRef]
107. Lieberman, J. Tapping the RNA world for therapeutics. *Nat. Struct. Mol. Biol.* **2018**, *25*, 357–364. [CrossRef]
108. Owens, J. Determining druggability. *Nat. Rev. Drug Discov.* **2007**, *6*, 187. [CrossRef]

109. Dance, A. Drug Discovery Techniques Open the Door to RNA-targeted Drugs. The Scientist: Exploring Life, Inspiring Innovation. 1 June 2019. Available online: https://www.the-scientist.com/lab-tools/drug-discovery-techniques-open-the-door-to-rna-targeted-drugs-65903 (accessed on 20 September 2019).
110. Peer, D.; Lieberman, J. Special delivery: Targeted therapy with small RNAs. *Gene Ther.* **2011**, *18*, 1127–1133. [CrossRef] [PubMed]
111. Wilkinson, G.R. Drug metabolism and variability among patients in drug response. *N. Engl. J. Med.* **2005**, *352*, 2211–2221. [CrossRef] [PubMed]
112. Dainis, A.M.; Ashley, E.A. Cardiovascular Precision Medicine in the Genomics Era. *JACC Basic Transl. Sci* **2018**, *3*, 313–326. [CrossRef]
113. Phillips, K.A.; Veenstra, D.L.; Oren, E.; Lee, J.K.; Sadee, W. Potential role of pharmacogenomics in reducing adverse drug reactions: A systematic review. *JAMA* **2001**, *286*, 2270–2279. [CrossRef]
114. Bradford, L.D. CYP2D6 allele frequency in European Caucasians, Asians, Africans and their descendants. *Pharmacogenomics* **2002**, *3*, 229–243. [CrossRef]
115. Bernard, S.; Neville, K.; Nguyen, A.; Flockhart, D. Inter-ethnic differences in genetic polymorphisms of CYP2D6 in the U.S. population: Clinical implications. *Oncologist* **2006**, *11*, 126–135. [CrossRef]
116. Abraham, B.K.; Adithan, C.; Mohanasundaram, J.; Shashindran, C.H.; Koumaravelou, K.; Asad, M. Genetic polymorphism of CYP2D6 in Tamil population. *Eur. J. Clin. Pharmacol.* **2001**, *56*, 849–850. [CrossRef]
117. Cavallari, L.H.; Mason, D.L. Cardiovascular Pharmacogenomics—Implications for Patients With CKD. *Adv. Chronic Kidney Dis.* **2016**, *23*, 82–90. [CrossRef]
118. Johnson, J.A.; Gong, L.; Whirl-Carrillo, M.; Gage, B.F.; Scott, S.A.; Stein, C.M.; Anderson, J.L.; Kimmel, S.E.; Lee, M.T.; Pirmohamed, M.; et al. Clinical Pharmacogenetics Implementation Consortium Guidelines for CYP2C9 and VKORC1 genotypes and warfarin dosing. *Clin. Pharmacol. Ther.* **2011**, *90*, 625–629. [CrossRef]
119. Cavallari, L.H.; Perera, M.A. The future of warfarin pharmacogenetics in under-represented minority groups. *Future Cardiol.* **2012**, *8*, 563–576. [CrossRef] [PubMed]
120. Kimmel, S.E.; French, B.; Kasner, S.E.; Johnson, J.A.; Anderson, J.L.; Gage, B.F.; Rosenberg, Y.D.; Eby, C.S.; Madigan, R.A.; McBane, R.B.; et al. A pharmacogenetic versus a clinical algorithm for warfarin dosing. *N. Engl. J. Med.* **2013**, *369*, 2283–2293. [CrossRef] [PubMed]
121. Pirmohamed, M.; Burnside, G.; Eriksson, N.; Jorgensen, A.L.; Toh, C.H.; Nicholson, T.; Kesteven, P.; Christersson, C.; Wahlström, B.; Stafberg, C.; et al. A randomized trial of genotype-guided dosing of warfarin. *N. Engl. J. Med.* **2013**, *369*, 2294–2303. [CrossRef] [PubMed]
122. Holbrook, A.; Schulman, S.; Witt, D.M.; Vandvik, P.O.; Fish, J.; Kovacs, M.J.; Svensson, P.J.; Veenstra, D.L.; Crowther, M.; Guyatt, G.H. Evidence-based management of anticoagulant therapy: Antithrombotic Therapy and Prevention of Thrombosis, 9th ed: American College of Chest Physicians Evidence-Based Clinical Practice Guidelines. *Chest* **2012**, *141*, e152S–e184S. [CrossRef] [PubMed]
123. Gage, B.F.; Bass, A.R.; Lin, H.; Woller, S.C.; Stevens, S.M.; Al-Hammadi, N.; Li, J.; Rodríguez, T.; Miller, J.P.; McMillin, G.A.; et al. Effect of Genotype-Guided Warfarin Dosing on Clinical Events and Anticoagulation Control Among Patients Undergoing Hip or Knee Arthroplasty: The GIFT Randomized Clinical Trial. *JAMA* **2017**, *318*, 1115–1124. [CrossRef] [PubMed]
124. Meschia, J.F.; Bushnell, C.; Boden-Albala, B.; Braun, L.T.; Bravata, D.M.; Chaturvedi, S.; Creager, M.A.; Eckel, R.H.; Elkind, M.S.; Fornage, M.; et al. Guidelines for the primary prevention of stroke: A statement for healthcare professionals from the American Heart Association/American Stroke Association. *Stroke* **2014**, *45*, 3754–3832. [CrossRef] [PubMed]
125. Nutescu, E.; Duarte, J.; Cheng, W.; Sarangpur, S.; Gor, D.; Drozda, K.; Galanter, W.; Stamos, T.; Peace, D.; Garofalo, J.; et al. Abstract 16119: Novel Genotype Guided Personalized Warfarin Service Improves Outcomes in an Ethnically Diverse Population. CORE 2. EPIDEMIOLOGY AND PREVENTION OF CV DISEASE: PHYSIOLOGY, PHARMACOLOGY AND LIFESTYLE. *Circulation* **2014**, *130*, A16119.
126. Kim, K.; Lee, T.A.; Touchette, D.R.; DiDomenico, R.J.; Ardati, A.K.; Walton, S.M. Contemporary Trends in Oral Antiplatelet Agent Use in Patients Treated with Percutaneous Coronary Intervention for Acute Coronary Syndrome. *J. Manag. Care Spec. Pharm.* **2017**, *23*, 57–63. [CrossRef]
127. Mauskopf, J.A.; Graham, J.B.; Bae, J.P.; Ramaswamy, K.; Zagar, A.J.; Magnuson, E.A.; Cohen, D.J.; Meadows, E.S. Cost-effectiveness of prasugrel in a US managed care population. *J. Med. Econ.* **2012**, *15*, 166–174. [CrossRef]

128. Coleman, C.I.; Limone, B.L. Cost-effectiveness of universal and platelet reactivity assay-driven antiplatelet therapy in acute coronary syndrome. *Am. J. Cardiol.* **2013**, *112*, 355–362. [CrossRef]
129. Patel, V.; Lin, F.J.; Ojo, O.; Rao, S.; Yu, S.; Zhan, L.; Touchette, D.R. Cost-utility analysis of genotype-guided antiplatelet therapy in patients with moderate-to-high risk acute coronary syndrome and planned percutaneous coronary intervention. *Pharm. Pract.* **2014**, *12*, 438. [CrossRef] [PubMed]
130. Claassens, D.M.F.; Vos, G.J.A.; Bergmeijer, T.O.; Hermanides, R.S.; van 't Hof, A.W.J.; van der Harst, P.; Barbato, E.; Morisco, C.; Tjon Joe Gin, R.M.; Asselbergs, F.W.; et al. A Genotype-Guided Strategy for Oral P2Y. *N. Engl. J. Med.* **2019**, *381*, 1621–1631. [CrossRef] [PubMed]
131. Koski, R.; Kennedy, B. Comparative Review of Oral P2Y. *Pharm Ther.* **2018**, *43*, 352–357.
132. Roden, D.M. Clopidogrel Pharmacogenetics—Why the Wait? *N. Engl. J. Med.* **2019**, *381*, 1677–1678. [CrossRef] [PubMed]
133. Liu, T.; Yin, T.; Li, Y.; Song, L.-Q.; Yu, J.; Si, R.; Zhang, Y.-M.; He, Y.; Guo, W.-Y.; Wang, H.-C. CYP2C19 polymorphisms and coronary heart disease risk factors synergistically impact clopidogrel response variety after percutaneous coronary intervention. *Coron. Artery Dis.* **2014**, *25*, 412–420. [CrossRef] [PubMed]
134. Baker, S.D.; van Schaik, R.H.; Rivory, L.P.; Ten Tije, A.J.; Dinh, K.; Graveland, W.J.; Schenk, P.W.; Charles, K.A.; Clarke, S.J.; Carducci, M.A.; et al. Factors affecting cytochrome P-450 3A activity in cancer patients. *Clin. Cancer Res.* **2004**, *10*, 8341–8350. [CrossRef] [PubMed]
135. Slaviero, K.A.; Clarke, S.J.; Rivory, L.P. Inflammatory response: An unrecognised source of variability in the pharmacokinetics and pharmacodynamics of cancer chemotherapy. *Lancet. Oncol.* **2003**, *4*, 224–232. [CrossRef]
136. Shedlofsky, S.I.; Israel, B.C.; McClain, C.J.; Hill, D.B.; Blouin, R.A. Endotoxin administration to humans inhibits hepatic cytochrome P450-mediated drug metabolism. *J. Clin. Investig.* **1994**, *94*, 2209–2214. [CrossRef]
137. Brown, S.-A.; Nhola, L.; Herrmann, J. Cardiovascular Toxicities of Small Molecule Tyrosine Kinase Inhibitors: An Opportunity for Systems-Based Approaches. *Clin. Pharmacol. Ther.* **2017**, *101*, 65–80. [CrossRef]
138. Jones, L.W.; Haykowsky, M.J.; Swartz, J.J.; Douglas, P.S.; Mackey, J.R. Early breast cancer therapy and cardiovascular injury. *J. Am. Coll. Cardiol.* **2007**, *50*, 1435–1441. [CrossRef]
139. Garcia-Pavia, P.; Kim, Y.; Restrepo-Cordoba, M.A.; Lunde, I.G.; Wakimoto, H.; Smith, A.M.; Toepfer, C.N.; Getz, K.; Gorham, J.; Patel, P.; et al. Genetic Variants Associated With Cancer Therapy-Induced Cardiomyopathy. *Circulation* **2019**, *140*, 31–41. [CrossRef] [PubMed]
140. McGowan, J.V.; Chung, R.; Maulik, A.; Piotrowska, I.; Walker, J.M.; Yellon, D.M. Anthracycline Chemotherapy and Cardiotoxicity. *Cardiovasc. Drugs Ther.* **2017**, *31*, 63–75. [CrossRef] [PubMed]
141. Saidi, A.; Alharethi, R. Management of chemotherapy induced cardiomyopathy. *Curr. Cardiol. Rev.* **2011**, *7*, 245–249. [CrossRef] [PubMed]
142. Pasipoularides, A. Implementing genome-driven personalized cardiology in clinical practice. *J. Mol. Cell Cardiol.* **2018**, *115*, 142–157. [CrossRef]

© 2020 by the authors. Licensee MDPI, Basel, Switzerland. This article is an open access article distributed under the terms and conditions of the Creative Commons Attribution (CC BY) license (http://creativecommons.org/licenses/by/4.0/).

Review

Molecular Functionality of Cytochrome P450 4 (CYP4) Genetic Polymorphisms and Their Clinical Implications

Yazun Bashir Jarrar [1] and Su-Jun Lee [2,*]

[1] Department of Pharmacy, College of Pharmacy, Alzaytoonah University of Jordan, 11734 Amman, Jordan
[2] Department of Pharmacology and Pharmacogenomics Research Center, Inje University College of Medicine, Inje University, Busan 47392, Korea
* Correspondence: 2sujun@inje.ac.kr; Tel.: +82-(051)-890-8665; Fax: +82-0504-290-5739

Received: 18 July 2019; Accepted: 28 August 2019; Published: 31 August 2019

Abstract: Enzymes in the cytochrome P450 4 (CYP4) family are involved in the metabolism of fatty acids, xenobiotics, therapeutic drugs, and signaling molecules, including eicosanoids, leukotrienes, and prostanoids. As CYP4 enzymes play a role in the maintenance of fatty acids and fatty-acid-derived bioactive molecules within a normal range, they have been implicated in various biological functions, including inflammation, skin barrier, eye function, cardiovascular health, and cancer. Numerous studies have indicated that genetic variants of *CYP4* genes cause inter-individual variations in metabolism and disease susceptibility. Genetic variants of *CYP4A11, 4F2* genes are associated with cardiovascular diseases. Mutations of *CYP4B1, CYP4Z1*, and other *CYP4* genes that generate 20-HETE are a potential risk for cancer. *CYP4V2* gene variants are associated with ocular disease, while those of *CYP4F22* are linked to skin disease and *CYP4F3B* is associated with the inflammatory response. The present study comprehensively collected research to provide an updated view of the molecular functionality of *CYP4* genes and their associations with human diseases. Functional analysis of *CYP4* genes with clinical implications is necessary to understand inter-individual variations in disease susceptibility and for the development of alternative treatment strategies.

Keywords: *CYP4* genes; genetic polymorphisms; 20-HETE; fatty acid; arachidonic acid; SNPs; molecular functionality; metabolism; lamellar ichthyosis; Bietti's crystalline dystrophy

1. Introduction

Cytochrome P450s (CYPs) are a superfamily of enzymes located either in the inner membrane of mitochondria or in the endoplasmic reticulum membrane of eukaryotic cells. There are 57 CYP proteins encoded in the human genome, which are responsible for the metabolism of numerous endogenous and exogenous compounds [1–3]. CYPs mainly oxidize these compounds to generate more hydrophilic metabolites, enhancing their excretion outside the body and thus playing a major role in the detoxification of toxic chemicals [1]. Generally, CYP families 1, 2, and 3 include major xenobiotic-metabolizing enzymes responsible for their major roles in pharmacogenomics risk, while CYP4 enzymes are involved in the metabolism of fatty acids, with their close links to genetic disease risk. Fatty acid metabolism by CYP4 enzymes is responsible for the elimination of excess free fatty acids from the body, as well as for the synthesis of proper levels of bioactive fatty acid molecules [4]. The present review focused on the CYP4 family of enzymes in terms of their functional roles, genetic variations, and influences on human diseases.

2. Classification and Tissue Distribution of the CYP4 Family

Although there are more than 11 subfamilies of CYP4 in different species, only 6 subfamilies of *CYP4* genes have been reported in humans. The human CYP4 subfamilies are CYP4A, B, F, V, X, and Z [2]. Seven *CYP4F* isoforms, *CYP4F2, CYP4F3A, CYP4F3B, CYP4F8, CYP4F11, CYP4F12,* and *CYP4F22*, are clustered on chromosome 19 and are encoded by six genes. *CYP4A* isoforms include *CYP4A11* and *CYP4A22* on chromosome 1 [3], and the remaining *CYP4* subfamily genes are *CYP4B1, CYP4V2, CYP4X1,* and *CYP4Z1* [4]. The major sites of CYP4A11 expression are the liver and kidney [5]. However, Jarrar et al. found that CYP4A11 protein was also highly expressed in human platelets to a similar level as in the human liver [6]. CYP4A22 expression has been reported in the human liver at very low levels, with poor enzyme activity compared to that of CYP4A11 [7]. Among the seven *CYP4F* genes, *CYP4F2, CYP4F3B, CYP4F11,* and *CYP4F12* are mainly expressed in the liver and kidney [8–12]. However, their relative contributions to the total amount of CYP4 enzymes in tissues are difficult to determine, as the high structural homology of these four enzymes has hampered the production of specific antibodies for the detection of each enzyme. In addition to the lack of specific antibodies, genetic polymorphisms and differing profiles of up- and downregulation among CYP4 enzymes have further complicated determination of the intrinsic amount of each enzyme in tissues. Currently, mass spectrometry is used to detect target proteins through measurement of specific peptides of the target protein [13,14]. The total amount of CYP4F protein in human liver was estimated as 18–128 pmol/mg liver microsomal protein [15]. One of the most abundant P450s, CYP3A4, was estimated at 64 pmol/mg liver microsomal protein [13], indicating that the contribution of CYP4F to the total P450 level is large. CYP4F3A is expressed in neutrophils and plays a major role in inflammation [16]. CYP4F8 is expressed in the prostate and seminal vesicles [17]. CYP4F22 is expressed in human skin and plays a major role in formation of the skin lipid barrier [18]. CYP4V2 is widely expressed in the liver and ophthalmic tissues and CYP4V2 defect has been linked to ophthalmic diseases, such as Bietti's crystalline dystrophy [19]. CYP4B1 is expressed mainly in the lung and bladder tissues, and in smaller amounts in the liver [20]. CYP4X1 is expressed in the brain and bronchial airways [21], while CYP4Z1 is expressed in mammary tissue; these proteins are also overexpressed in cancer compared to normal cells [22]. Expression levels of CYP4 proteins are summarized in Table 1.

Table 1. Substrates and major expression tissues of cytochrome P450 4 (CYP4) enzymes in humans.

CYP4 Enzyme	Expression Tissues	Substrates	Reference
CYP4A11	Platelets, liver, kidney	Lauric acid, myristic acid, arachidonic acid	[6,23,24]
CYP4A22	Low level in liver	Arachidonic acid	[25,26]
CYP4B1	Lung, bladder, fat tissues	2-aminofluorene, 2-naphthylamine, benzidine, arachidonic acid	[27,28]
CYP4F2	Liver, platelet, kidney	Arachidonic acid, lauric acid, vitamin K, leukotriene	[6,29]
CYP4F3A	Neutrophils, monocytes, eosinophils	Leukotriene B4	[12,30,31]
CYP4F3B	Liver, kidney, trachea, gastro intestinal tract	Eicosapentaenoic acid, arachidonic acid	[12,32]
CYP4F8	Prostate, seminal vesicles, epidermis, hair follicles, sweat glands, corneal epithelium, proximal renal tubules, epithelial linings of gut and urinary tract	Arachidonic acid, prostaglandin H, prostaglandin E2	[33–35]
CYP4F11	Liver, kidney, heart, skeletal muscle, gall bladder, keratinocytes	Vitamin K, erythromycin, arachidonic acid	[11,29,36–38]
CYP4F12	Liver, kidney, colon, small intestine, heart, eosinophils, neutrophils	Arachidonic acid, leukotriene B4, ebastine	[8,31,33,39]
CYP4F22	Skin	Ultra-long-chain fatty acid (acylceramide production)	[18]
CYP4X1	Skin, breast, brain, heart, liver, prostate, trachea, aorta	Anandamide, arachidonic acid	[21,40]
CYP4V2	Macrophages, retina cells, cornea cells	Arachidonic acid, lauric acid, eicosapentanoic acid, docosahexanoic acid	[19,41]
CYP4Z1	Mammary tissues, cancer cells	Lauric acid, myristic acid	[21,42]

3. Role of the CYP4 Family in the Metabolism of Endogenous Compounds

The CYP4 family plays a major role in the metabolism of fatty acids, in most cases through oxidation of fatty acids and subsequent catalysis in the mitochondria to produce cellular energy. CYP4B metabolizes short-chain fatty acids (approximately 7 to 10 carbon fatty acids) [20], while CYP4A and CYP4V metabolize intermediate-chain fatty acids (C10 to 16) [19] and CYP4F catalyzes long-chain fatty acids (C16 to 26), such as prostanoids [43]. Decreasing the expression levels of the CYP4 family was associated with accumulation of fats in tissues such as the liver [44]. Therefore, decreased levels of CYP4 family proteins reduce the capacity for fat removal from tissues. Jarrar et al. [44] found that non-steroidal anti-inflammatory drugs caused fatty livers in treated mice, which were associated with significant downregulation of mouse *cyp4a12* gene expression in liver tissues. CYP4F2, 4F3B, 4A11, and 4V2 were found to oxidize arachidonic acid through ω-hydroxylation to 20-hydroxyeicosatetraenoic acid (20-HETE) [6,9,45,46], which is a vasoconstrictor and activator of platelet aggregation [47]. Several studies have reported that CYP4F and CYP4A are overexpressed in cardiovascular diseases, wherein they are correlated with 20-HETE production [48–50]. In addition, doxorubicin-induced cardiotoxicity was associated with increased 20-HETE production due to increased mRNA expression of rat CYP4A and CYP4F enzymes [51]. CYP4A11 and 4V2 oxidize saturated fatty acids such as lauric acid [23,41,52]. In addition to the metabolism of arachidonic acid and omega-3 polyunsaturated fatty acids, CYP4F2 has been reported to ω-hydroxylate leukotriene (LTA) 4 [53]. CYP4F3A in white blood cells catalyzes the ω-hydroxylation of leukotriene B4 to 20-hydroxy leukotriene B4, which is an important regulatory step of the inflammatory response [54]. Instead of ω-hydroxylation, CYP4F8 has been reported to hydroxylate prostaglandin (PG) E2 at position 19 [17]. Although epoxyeicosatrienoic acids (EETs) are synthesized by the CYP2C subfamily [55], they can be further ω-hydroxylated by CYP4 enzymes to 20-hydroxyepoxyeicosatrienoic acids (HEETs) [56].

4. Role of the CYP4 Family in the Metabolism of Drugs

The roles of most CYP4 family proteins in the metabolism of drugs and xenobiotic compounds appear to be minor compared to those of CYP1, 2, and 3. However, CYP4F2 metabolizes the ester prodrug of gemcitabine and the antiparasitic pafuramidine [57]. In addition, CYP4A11 exhibited metabolism of the immune suppressant tacrolimus to an inactive form [58]. Although the turnover rates were low compared to those of CYP3A4, CYP4F11 exhibited catalytic activity towards commonly used drugs such as erythromycin, benzphetamine, and chlorpromazine [36,37,59]. CYP4F12 has been reported to slowly metabolize the antihistamine ebastine [60] and the antifungal terfenadine [61]. CYP4 enzymes are indirectly involved in drug metabolism and drug responses. For example, CYP4F2 and CYP4F11 are involved in the metabolism of vitamin K, facilitating vitamin K inactivation and elimination [29,62]. The amount of active vitamin K is important for maintenance of warfarin dosing, as it is metabolized strongly by CYP2C9 [63,64], indicating that CYP4 enzymes are indirectly involved in warfarin dose maintenance. CYP4 enzymes show catalytic activity toward various fatty acids and their metabolites have the potential to act as ligands or activators of nuclear receptors, such as peroxisome proliferator-activated receptors (PPARs) [65,66]. Therefore, drugs targeting the activation or inactivation of PPARs may show altered pharmacokinetics or toxic responses [67,68]. Such indirect involvement may affect the drug response to conditions such as fatty liver diseases, diabetic diseases, and inflammatory diseases.

5. The CYP4 Family and Inflammation

CYP4 enzymes are involved in inflammation through the metabolism of inflammatory molecules. They metabolize inflammatory mediators such as leukotrienes (LTs) and also produce 20-HETE [53]. While CYP4F11 possesses lower affinity toward leukotriene B4 (LTB4), neutrophilic CYP4F3A has the highest affinity for LTB4 ω-hydroxylation [36]. CYP4F3A metabolizes LTB4 into the inactive form 20-hydroxy leukotriene B4, mediating a critical step in regulation of the inflammatory response.

However, CYP4A11 has shown low activity toward LTB4 using in vitro methods [69]. CYP4F3B ω-hydroxylates omega-3 eicosapentaenoic acid (EPA) and docosahexaenoic acid (DHA) to their 20-hydroxy and 22-hydroxy metabolites, respectively [32], which are lipid mediators that can activate inflammatory PPARs [70]. Studies have shown that hepatic and renal rat *CYP4F* genes were upregulated under inflammatory conditions following treatment with barium sulfate [71]. On the other hand, rat hepatic CYP4A mRNAs were downregulated in response to lipopolysaccharides used as a model of inflammation [72]. Human CYP4V2 was first identified in inflammatory cell macrophages, and its gene expression was reduced following selective treatment with a PPARγ agonist [41]. Depending on the clinical situation, ω-hydroxylase activity associated with the CYP4 family could be considered as a potential drug target for reducing the inflammatory response, providing a novel mechanism for future anti-inflammatory drugs.

6. The CYP4 Family and Cancers

Induction of CYP4 family members, including CYP4F2, CYP4F3, CYP4A11, and CYP4Z1, has been reported in various types of cancer [73,74]. Upregulation of CYPF2 and CYP4A11 was confirmed through Western blot assays in human thyroid, ovarian, breast, pancreatic, and colon cancer tissues [75]. CYP4Z1 is expressed in mammary tissue and upregulated in breast cancer tissue [74]. These findings suggest that ω-hydroxylase activity may be a biomarker of cancer prognosis. Evaluation of the CYP4 expression profile in hepatocellular carcinoma (HCC) showed that CYP4F2, CYP4F12, and CYP4V2 mRNA levels were negatively correlated with cell-cycle-associated genes, suggesting that these *CYP4* genes are favorable prognostic factors in HCC [76]. In addition, expression of CYP4 has been reported to be associated with angiogenesis through production of 20-HETE, which activates vascular endothelial receptors in arteries and thus increases blood supply to cancer cells [77]. Among CYP4 enzymes, CYP4F3B, CYP4A11, and CYP4F2 are major enzymes involved in the generation of 20-HETE, which plays an important role in tumor progression and angiogenesis. Therefore, their tissue expression and omega-hydroxylase activity levels play roles in cancer progression. CYP4B1 metabolizes several protoxic xenobiotics, including 2-aminofluorine, 2-naphthylamine, 4-ipomeanol, and benzidine [78–81]. Therefore, CYP4B1 involvement in cancers has been suggested based on its expression levels and metabolism of pro-carcinogens in the bladder and lung [78,82]. CYP4B1 may play a role in detoxification or activation in tissues. Sasaki et al. reported that the individuals carrying the *CYP4B1*2* allele have an increased risk of bladder cancer [27]. However, it has also been reported that there is no association between the *CYP4B1* genotype and the risk of lung cancer in the Japanese population [83]. Downregulation of CYP4B1 proteins represented an unfavorable indicator in patients with urothelial carcinomas of the upper urinary tract and bladder, indicating a protective role of CYP4B1 in patients with urotherial carcinomas [84]. Involvement of CYP4Z1 in breast cancer has been suggested, as it was identified in breast tissue and upregulated in breast carcinoma [74,76]. Therefore, CYP4Z1 was proposed as a biomarker for malignancy and/or progression of ovarian and prostate cancer [85]. It was reported that breast cancer cells exhibited the abnormal translocation of CYP4Z1 protein to the plasma membrane instead of targeting to the intracellular membrane of the endoplasmic reticulum, which caused the CYP4Z1 autoantibody production that might serve as a biomarker for the diagnosis [86]. Expression of CYP4Z1 has been reported to promote angiogenesis and tumor growth by increasing 20-HETE synthesis [74]. However, a recent functional study of CYP4Z1 in a recombinant enzyme system indicated that 20-HETE was not detected in the CYP4Z1 reaction with arachidonic acid, and suggested that CYP4Z1 may modulate breast cancer without direct 20-HETE synthesis [87]. Further studies are needed to clarify the roles of CYP4Z1 in carcinogenesis in various tissues.

7. The CYP4 Family and Cardiovascular Diseases

Several studies have shown that *CYP4* family genes are associated with cardiovascular diseases, including hypertension and myocardial infarction, through the production of 20-HETE or perturbation of fatty acid metabolism [88,89]. Multiple aspects of the mechanism underlying the effect of 20-HETE on the cardiovascular complex have been reported. In a metabolomics study in mice, increased 20-HETE levels in the blood (>120-fold) with chronic rofecoxib treatment were associated with reduced bleeding time and increased platelet aggregation [47]. Additionally, 20-HETE has been suggested to mediate androgen-induced hypertension through increasing the level of Cyp4a12 in a mouse study [90], wherein the increased level of Cyp4a12 produced more eicosanoids, which were predicted to mediate androgen-induced hypertension. In the kidney, however, 20-HETE exerts anti-hypertensive effects through inhibition of sodium reabsorption in the proximal tubule and thick ascending limb of Henle [91]. Furthermore, 20-HETE was found to act as a vasoconstrictor of vascular smooth muscle cells by allowing increased calcium entry into cells and enhanced phosphorylation of contractile elements [92–94]. Several studies have suggested interplay between 20-HETE and the renin–angiotensin aldosterone system (RAAS) in hypertension. Briefly, angiotensinogen II has been reported to increase renal production of 20-HETE [95], and 20-HETE can activate the RAAS by inducing angiotensin-converting enzyme [96,97]. Further investigations are needed to fully elucidate the mechanistic link between 20-HETE and the RAAS in humans. Rat CYP4A was downregulated in the kidney of hypertensive rats, which was associated with reduced formation of 20-HETE in the kidney and reduction of the diuretic effect [98]. CYP4A was upregulated in studies of doxorubicin-induced cardiotoxicity, where it was associated with myocardial infarction and increased 20-HETE synthesis [51]. Furthermore, Jarrar et al. found that heart cyp4a12 was highly upregulated in mice after cardiac toxicity induced by non-steroidal anti-inflammatory drugs [44]. Thus, targeting of 20-HETE synthesis or modulation of eicosanoid levels through manipulation of CYP4 enzymes can decrease the cardiotoxicity of such drugs. This application should be considered in future development of the drug for cardiovascular health care.

8. Role of the CYP4 Family in Other Diseases

Bietti's crystalline dystrophy (BCD) is an autosomal recessive disease characterized by the presence of numerous small, yellow or white crystal-like deposits of fatty compounds in the light-sensitive retina tissue [52,99,100]. These deposits damage the retina, resulting in progressive atrophy of the retinal pigment epithelium and progressive vision loss at approximately 40 or 50 years of age [101,102]. The occurrence of BCD is more common in East Asian populations than other ethnic groups [103,104]. BCD is caused by mutations in the *CYP4V2* gene, which is comprised of 11 exons encoding a 525 amino acid protein on chromosome 4 [99,105,106]. CYP4V2 is known to metabolize fatty acids, and thus CYP4V2 in the retina is most likely involved in the breakdown and elimination of fatty acids from the retina [52]. Impaired CYP4V2 function due to genetic mutations may affect lipid metabolism and elimination from the retina. The severity and progression of BCD symptoms varies widely among patients. These variations may be influenced by differing levels of defectiveness in CYP4V2 function caused by mutations of different severities. Various mutations in *CYP4V2* have been found, including stop codon creation, an amino acid change in an important region, destruction of a splice site, and a frameshift in the CYP4V2 protein-coding cDNA. More than 60 mutations of the *CYP4V2* gene have been reported in BCD patients [99,103,105,107–112]. A number of mutations of *CYP4V2* have significant impacts on CYP4V2 activity. The most common mutation in BCD is an insertion–deletion mutation at the end of intron 6 and the beginning of exon 7 (IVS6-8del17insGC, c.802-8del17/insGC) [103,105,106,108,109,111–125]. This mutation causes the deletion of exon 7 in the CYP4V2 protein, resulting in a major structural change and the complete loss of CYP4V2 activity.

Type 3 lamellar ichthyosis, a skin keratinization disease, was found to be caused by genetic mutation of *CYP4F22* [126]. Since the discovery that *CYP4F22* is one of the causative genes for ichthyosis, the molecular mechanisms underlying the role of CYP4F22 in the etiology of ichthyosis have remained largely unknown until recently. Acylceramide is an important lipid of the skin permeability barrier, and patients with ichthyosis show strongly repressed acylceramide production [127–130]. Ohno et al. (2015) reported that CYP4F22 is responsible for the generation of acylceramide through ω-hydroxylation of long-chain fatty acids [18]. Recently, a *CYP4F22* genetic variant associated with lamellar ichthyosis was reported in a Tunisian family [131]. A missense mutation in exon 8, CYP4F22

Arg243Leu, was suggested to be linked to lamellar ichthyosis and predicted to be a functionally defective variant based on in silico analysis. Genetic screening for *CYP4F22* mutations associated with lamellar ichthyosis should be extended in future works.

9. Genetic Variants of the *CYP4* Family

9.1. Genetic Variants of *CYP4B1*

The first screening study for genetic polymorphism of *CYP4B1* was performed in French Caucasians and identified the new *CYP4B1* alleles *CYP4B1*2*, *3*, *4*, and *5* based on the P450 Nomenclature Committee [132]. Among them, *CYP4B1*2* caused a frameshift and premature stop codon, resulting in complete loss of CYP4B1 function. Two more alleles with frequencies <1%, *CYP4B1*6* and *CYP4B1*7*, were identified using a denaturing high-performance liquid chromatography method for 192 Japanese individuals [133]. Since CYP4B1 is involved in the metabolism of pro-carcinogens, its association with bladder cancer was investigated in a Japanese population, and subjects carrying the *CYP4B1*1/*2* or *CYP4B1*2/*2* genotypes exhibited a 1.75-fold increased risk of bladder cancer [27]. This finding might be explained as the loss of function allele *CYP4B1*2* providing lower capacity for activation of carcinogenic compounds. However, a lung cancer risk study of *CYP4B1*1–*7* showed no association with lung cancer in a Japanese population [83]. Further studies are needed to determine its association with lung cancer using a large cohort. Study of structure–function relationships has been essential to understanding the efficiency of catalytic activity as well as to explaining the varying degrees of molecular defectiveness of the protein mutants. Investigation of local peptide structures on the CYP4B1 protein and their roles in heme stability with catalytic function has been reported [134–136], and these data will be important to understand inter-individual variations in the activity of CYP4B1 coding variants.

9.2. Genetic Variants of *CYP4A11*, *CYP4F2*, *4F11*, and *CYP4F22*

Among *CYP4* family genes, *CYP4A11* and *CYP4F2* have been extensively studied in association with warfarin dosage and the cardiovascular complex. Genetic variants of *CYP4F2* and *CYP4A11* genes are reportedly associated with cardiovascular diseases such as hypertension [137–139]. More than 3400 single nucleotide polymorphisms (SNPs) of human *CYP4A11* and 5900 SNPs of the *CYP4F2* gene have been reported in the NCBI database to date. However, only a small number of the SNPs have been shown to have clinical associations with functional changes. One of the most extensively studied SNPs of *CYP4A11* is a variant of rs1126742 that causes an amino acid change of Phe434 to Ser, leading to reduced 20-HETE synthesis from arachidonic acid [140,141]. Since the discovery of the functional role of CYP4A11 in the synthesis of 20-HETE, the association of *CYP4A11* polymorphisms with cardiovascular risk has been studied extensively in humans [142–148]. The US Food and Drug Administration recommends genotyping of *CYP4F2* variants for determination of warfarin doses [149,150]. The *CYP4F2* genetic variant rs2108622 is a non-synonymous variant that causes a change in the amino acid sequence of valine to methionine and exhibits reduced enzymatic activity toward the metabolism of vitamin K [62]. Since individuals with reduced activity of CYP4F2 for vitamin K inactivation may have higher levels of warfarin than individuals with *CYP4F2*1/*1*, higher maintenance dosages of warfarin have been recommended for individuals with reduced *CYP4F2* alleles [149]. Many studies have attempted to develop an accurate warfarin dosing algorithm using multiple genes, such as *CYP2C9*, *VKORC1*, and *CYP4F2* [151–155]. Studies regarding *CYP4A22* genetic polymorphisms have been limited to certain populations, such as Japanese and French populations [25,156]. The association of *CYP4A22* variants with human diseases has still not been investigated, which might be due to low expression levels of the *CYP4A22* gene. The *CYP4F3* gene undergoes alternative splicing to form the CYP4F3A and CYP4F3B enzymes, depending on the cell type [157]. Genome-wide investigation showed that the functional SNP *CYP4F3* rs4646904 was associated with lung cancer, especially in smokers [30]. However, the functionality of this SNP in lung cancer pathology remains unidentified. In addition, a high intake of polyunsaturated fatty acids was associated with reduced risk of ulcerative colitis in patients with *CYP4F3* rs4646904 GG/AG, but not those with the AA genotype [158]. Regarding the *CYP4F11* gene, Yi et al. found through in vitro methods that CYP4F11 D315N protein showed approximately 50% and 32% decreases in intrinsic clearance of erythromycin and arachidonic acid, respectively, compared to the wild type [37]. The *CYP4F11* variant (rs1060463) was associated with small bowel bleeding risk induced by aspirin [159]. Seven variants with amino acid changes in the *CYP4F12* gene were identified and functional changes were investigated using ebastine as a substrate [160]. In their report, two coding variants, Val90Ile and Arg188Cys, exhibited significantly decreased activity toward ebastine hydroxylation. The intronic variant *CYP4F12* rs11085971, which contains a nucleotide substitution of guanine to thymine, was identified as a candidate oxidative-stress-related genetic marker for the development of type 1 lesions in cerebral cavernous malformation, and could serve as an early objective predictor of disease outcome [161]. Since the discovery of *CYP4F22* was linked to its association with lamellar ichthyosis [18], genetic studies of *CYP4F22* polymorphisms have been undertaken. A *CYP4F22* variant, CYP4F22 Arg243Leu, was associated with lamellar ichthyosis in a Tunisian family [131], and further genetic studies should be conducted in clinical settings.

9.3. Genetic Variants of Other CYP4 Genes

Genetic polymorphism studies of *CYP4V2* with respect to BCD are described above. In addition to BCD, genome-wide analysis found that a *CYP4V2* genetic variant was strongly associated with deep vein thrombosis [162], which was confirmed later in multiple studies [163,164]. Association of the genetic variant 7234C>A (rs13146272) on exon 6 of the *CYP4V2* gene with the risk of deep venous thrombosis and tamoxifen-induced venous thrombosis has been reported [165]. The exact mechanism through which the *CYP4V2* gene defect increases the risk of deep vein thrombosis remains poorly understood. This genetic variant substitutes polar glutamine with basic lysine at position 259 of the CYP4V2 amino acid sequence, which may influence its activity. Genetic studies of *4Z1* and *4X1* are scarce, as these genes were recently identified and their physiological roles remain unclear. CYP4X1 was found to convert the endocannabinoid anandamide, an important signaling molecule in the neurovascular cascade, into a single monooxygenated product (14,15-epoxyeicosatrienoic ethanolamide), suggesting a potential role in brain signaling [40]. High levels of mRNA expression of CYP4X1 were found in the skin, brain, heart, liver, prostate, and breast [40] and CYP4Z1 mRNA was preferentially expressed in mammary tissue [21]. Functional analysis of CYP4Z1 indicated that CYP4Z1 has catalytic activity toward lauric and arachidonic acids, but 20-HETE was not detected in arachidonic acid metabolism [87]. Major genetic polymorphisms in *CYP4* genes with clinical impact were summarized in Table 2.

Table 2. Representatives of genetic polymorphisms in CYP4 genes with clinical impact and their frequencies in different ethnic groups.

Gene	SNP	Location	Mutation	Effect	Frequency [a] European	Frequency [a] African	Frequency [a] Asian	Functional Effect
CYP4A11	rs1126742	Exon	A > G	Phe330Ser	0.15	0.36	0.25	It was associated with hypertension in white individuals, most probably through decreased production of 20-HETE in the kidney [137].
CYP4F2	rs2108622	Exon	C > T	Val433Met	0.27	0.06	0.26	It reduced the metabolism of vitamin K. Therefore, patients carrying this genetic variant needed a higher dose of warfarin, in order to keep the targeted anticoagulant effect [149–151].
CYP4F3	rs4646904	Exon	A > G	Val358Val	0.65	0.35	0.34	It was associated with lung cancer, especially in smokers [30] and ulcerative colitis [158].
CYP4F11	rs200033002	Exon	C > T	Asp315Asn	0	0	0.01	It decreased the metabolism of erythromycin and arachidonic acid compared to the wild type in vitro [37].
CYP4B1	rs3215983	Frameshift variant	AT881–882del	Produces premature stop codon	0.15	ND	0.33	It was reported to increase the risk of bladder cancer, because it has lower capacity to metabolize the carcinogenic compounds [27].
CYP4V2	rs13146272	Exon	C > A	Gln259Lys	0.36	0.4	0.6	It was associated with the risk of deep venous thrombosis and tamoxifen-induced venous thrombosis [162–165].
CYP4V2	rs199476197	Exon	A > C	His331Pro	0	0	0.0004	It decreased CYP4V2 protein expression and activity toward fatty acid metabolism. Therefore, this genetic variant may cause accumulation of fatty acids in the retina [19,166].
CYP4V2	IVS6-8del17insGC	Intron 6, exon 7	Insertion/deletion	Exon7 del	ND	ND	ND	It causes deletion of exon 7 in the CYP4V2 protein, resulting in a complete loss of CYP4V2 activity. It is the most common mutations in BCD patients [52,99].

[a] Data regarding the frequency of genetic variants among different ethnic groups were obtained from Ensemble database. ND, not determined. BCD, Bietti's crystalline dystrophy.

10. Linkage Disequilibrium among *CYP4* Genes

Five *CYP4* genes, *CYP4A22, CYP4A11, CYP4B1, CYP4X1,* and *CYP4Z1,* are located on chromosome 1 [4]. A number of studies based on next generation sequencing tools and a 1000-genome project have identified SNPs in these genes. However, their functional roles, clinical relationships, and linkage disequilibrium (LD) are poorly characterized. From the 1000-genome database, a total of 14 coding SNPs with > 5% global minimum allele frequency were identified for *CYP4A22, CYP4A11,* and *CYP4B1,* and this LD block was analyzed (Figure 1A). Ethnically distinct populations exhibited differing LD blocks and haplotype structures. No strong LD was found among these three *CYP4* genes that are clustered on chromosome 1. Six *CYP4F* genes, including *CYP4F2, CYP4F3, CYP4F8, CYP5F11, CYP4F12,* and *CYP4F22,* are located on the same chromatid of chromosome 19 [4]. Using the same method, coding variants with > 5% global frequency were selected from a 1000-genome database and their haplotypes and LD were analyzed (Figure 1B). As illustrated in Figure 1A, ethnically distinct groups showed differing frequencies and LD structures. An LD block covering more than one *CYP4* gene was not observed for *CYP4F* genes in coding variant analysis. Instead, a strong linkage was found between *CYP4F2* (rs2074900) and *CYP4F11* (rs8104361) in a Western European population. Since *CYP4* genes on the same chromosome with highly similar DNA structures can act as a linkage unit or as independent genes, further linkage analysis using more validated SNPs over all regions of *CYP4* genes is needed to improve the current knowledge of *CYP4* genetics.

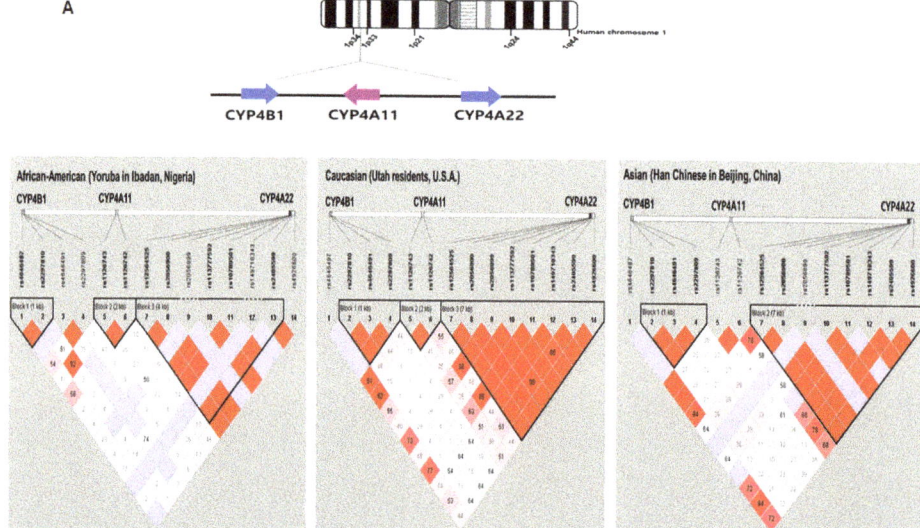

Figure 1. *Cont.*

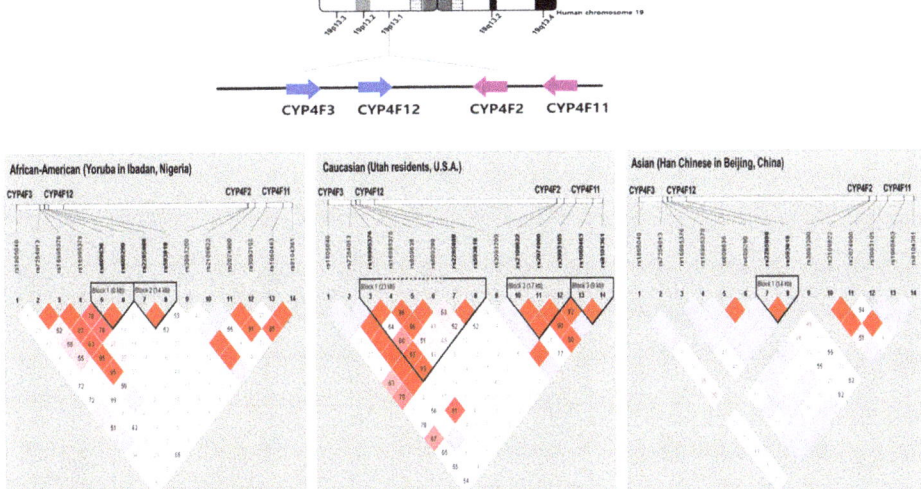

Figure 1. Linkage disequilibrium (LD) plots of *CYP4* genetic variants in African, Caucasian, and Asian populations. Populations in Yoruba, Utah, and Beijing represent African, Caucasian, and Asian populations, respectively. The coding single nucleotide polymorphisms (SNPs) with a minor allele frequency (MAF) of 0.05 or greater in the 1000 genome data base were selected to avoid estimation errors in linkage analysis. (**A**) LD structures of *CYP4A11*, *CYP4A22*, and *CYP4B1* with common coding SNPs. *CYP4A11*, *CYP4A22*, and *CYP4B1* are clustered on chromosome 1. The SNPs, shown from left to right within the figure, are as follows: rs4646487, rs2297810, rs4646491, rs2297809, rs1126743, rs1126742, rs12564525, rs2056900, rs2056899, rs113777592, rs10789501, rs149718343, rs2405599, and rs4926600. (**B**) LD structures of *CYP4F2*, *CYP4F3*, *CYP4F11*, and *CYP4F12* using common coding SNPs. *CYP4F2*, *CYP4F3*, *CYP4F11*, and *CYP4F12* are clustered on chromosome 19. The SNPs, shown from left to right within the figure, are as follows: rs1805040, rs7254013, rs16995376, rs16995378, rs609636, rs609290, rs2285888, rs593818, rs3093200, rs2108622, rs2074900, rs3093105, rs1060463, and rs8104361. The numbers in squares refer to pairwise LD values, measured as D′ (coefficient of linkage disequilibrium). Red depicts a significant linkage between a pair of SNPs. Numbers inside squares indicate the D′ value multiplied by 100.

11. Conclusions and Future Prospects

CYP4 enzymes are responsible for the metabolism of fatty acids and play important roles in the homeostasis of fatty acids and fatty-acid-derived biomolecules such as leukotriene, prostanoid, and 20-HETE. Thus, CYP4 enzymes make important contributions to human health, including cardiovascular health, skin barrier maintenance, eye function, and cancer protection. However, the lack of research into certain aspects of the CYP4 family must be overcome. First, a specific antibody for the detection of each CYP4 protein and a specific substrate for each enzyme function must be developed to clearly determine the expression levels of these enzymes in different tissues under various induction, inhibition, and genetic conditions. High similarity of protein structures, overlapping substrates, co-expression in the same tissues, and genetic differences among individuals have interfered with the identification and characterization of *CYP4* genes. For targeted therapy and targeted delivery of drugs into cells or specific tissues, accurate measurement of CYP4 activity in tissues is essential. Second, further functional studies of *CYP4* genetic variants are needed. A growing number of genetic mutations of *CYP4* genes have been identified using high-throughput sequencing techniques. However, most of their functional changes compared to the wild type remain unknown. Only a small number of high-frequency genetic variants with known functional information have been investigated in multiple populations, likely due to their high statistical power, which enables publication. Although in silico tools are useful for the prediction of functional changes, in silico prediction does not yet perfectly reflect in vivo conditions. Therefore, various commercial software programs often provide inconsistent predictions for the same genetic mutations. Development of high-throughput techniques for in vitro functional study and improvement of in silico methods are needed to elucidate the functional changes caused by mutations. Third, globally standardized values for CYP4 activity must be developed for application in artificial intelligence technology and algorithms used for the prediction of CYP4-related human diseases or the progression of disease states. As shown in Figure 2, large variations in CYP4-mediated metabolism, genetic variants of *CYP4* and other genes, and differing environmental conditions have been observed among individuals. Data integration to support correct diagnosis in humans is currently not possible, but is the ultimate goal of such research. To achieve this goal, accurate molecular tools for characterization of each CYP4 enzyme, functional information about *CYP4* genetic variants, and a standardized system for the application of CYP4 functional values in artificial intelligence or machine-learning tools are needed for personalized health care.

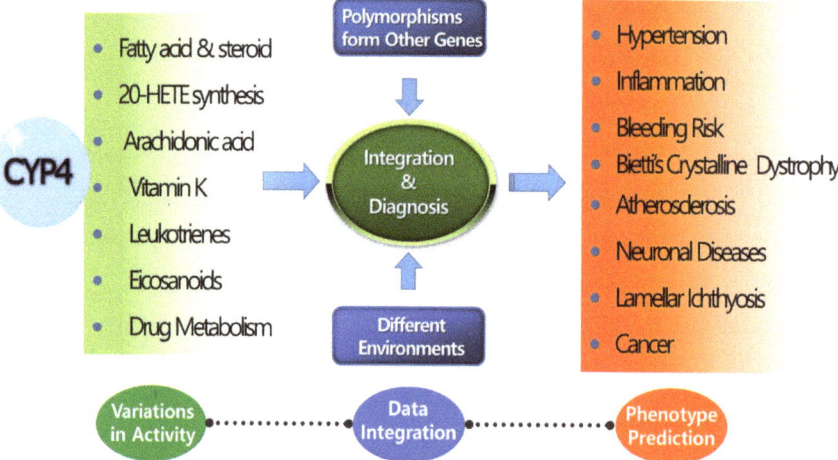

Figure 2. Correlation of *CYP4* genes with phenotypic outcomes. Most *CYP4* genes share similar structures and overlapping metabolic substrates. Phenotypic outcome prediction is difficult with a single or few *CYP4* genetic studies. Phenotypic outcomes are affected by genetic polymorphisms of various genes and dynamic environmental factors. Fundamental research into *CYP4* genes is essential to provide the data integration necessary for more accurate phenotype prediction than can be obtained using conventional methods.

Funding: This work was supported by the National Research Foundation of Korea (NRF) grant funded by the Korea government (MSIT) (No.2018R1A5A2021242) and by the National Research Foundation of Korea grant funded by the Korea government (NRF-2017R1D1A3B03031007).

Conflicts of Interest: The authors declare no conflict of interest.

References

1. Gonzalez, F.J.; Nebert, D.W. Evolution of the P450 gene superfamily: animal-plant 'warfare', molecular drive and human genetic differences in drug oxidation. *Trends Genet.* **1990**, *6*, 182–186. [CrossRef]
2. Edson, K.Z.; Rettie, A.E. CYP4 enzymes as potential drug targets: focus on enzyme multiplicity, inducers and inhibitors, and therapeutic modulation of 20-hydroxyeicosatetraenoic acid (20-HETE) synthase and fatty acid omega-hydroxylase activities. *Curr. Top. Med. Chem.* **2013**, *13*, 1429–1440. [CrossRef] [PubMed]

3. Drolet, B.; Pilote, S.; Gélinas, C. Altered Protein Expression of Cardiac CYP2J and Hepatic CYP2C, CYP4A and CYP4F in a Mouse Model of Type II Diabetes-A Link in the Onset and Development of Cardiovascular Disease? *Pharmaceutics* **2017**, *9*, 44. [CrossRef] [PubMed]
4. Hsu, M.H.; Savas, U.; Griffin, K.J.; Johnson, E.F. Human cytochrome p450 family 4 enzymes: function, genetic variation and regulation. *Drug Metab. Rev.* **2007**, *39*, 515–538. [CrossRef] [PubMed]
5. Savas, U.; Hsu, M.H.; Johnson, E.F. Differential regulation of human CYP4A genes by peroxisome proliferators and dexamethasone. *Arch. Biochem. Biophys.* **2003**, *409*, 212–220. [CrossRef]
6. Jarrar, Y.B.; Cho, S.A.; Oh, K.S.; Kim, D.H.; Shin, J.G.; Lee, S.J. Identification of cytochrome P450s involved in the metabolism of arachidonic acid in human platelets. *Prostaglandins Leukot Essent. Fatty Acids* **2013**, *89*, 227–234. [CrossRef] [PubMed]
7. Hsu, M.H.; Savas, U.; Griffin, K.J.; Johnson, E.F. Regulation of human cytochrome P450 4F2 expression by sterol regulatory element-binding protein and lovastatin. *J. Biol. Chem.* **2007**, *282*, 5225–5236. [CrossRef] [PubMed]
8. Bylund, J.; Bylund, M.; Oliw, E.H. cDNA cloning and expression of CYP4F12, a novel human cytochrome. *Biochem. Biophys. Res. Commun.* **2001**, *280*, 892–897. [CrossRef] [PubMed]
9. Powell, P.K.; Wolf, I.; Jin, R.; Lasker, J.M. Metabolism of arachidonic acid to 20-hydroxy-5,8,11, 14-eicosatetraenoic acid by P450 enzymes in human liver: involvement of CYP4F2 and CYP4A11. *J. Pharmacol. Exp. Ther.* **1998**, *285*, 1327–1336.
10. Lasker, J.M.; Chen, W.B.; Wolf, I.; Bloswick, B.P.; Wilson, P.D.; Powell, P.K. Formation of 20-hydroxyeicosatetraenoic acid, a vasoactive and natriuretic eicosanoid, in human kidney. Role of Cyp4F2 and Cyp4A11. *J. Biol. Chem.* **2000**, *275*, 4118–4126. [CrossRef]
11. Cui, X.; Nelson, D.R.; Strobel, H.W. A novel human cytochrome P450 4F isoform (CYP4F11): cDNA cloning, expression, and genomic structural characterization. *Genomics* **2000**, *68*, 161–166. [CrossRef] [PubMed]
12. Christmas, P.; Jones, J.P.; Patten, C.J.; Rock, D.A.; Zheng, Y.; Cheng, S.M.; Weber, B.M.; Carlesso, N.; Scadden, D.T. Rettie AE Alternative splicing determines the function of CYP4F3 by switching substrate specificity. *J Biol Chem.* **2001**, *276*, 38166–38172. [PubMed]
13. Kawakami, H.; Ohtsuki, S.; Kamiie, J.; Suzuki, T.; Abe, T.; Terasaki, T. Simultaneous absolute quantification of 11 cytochrome P450 isoforms in human liver microsomes by liquid chromatography tandem mass spectrometry with in silico target peptide selection. *J. Pharm. Sci.* **2011**, *100*, 341–352. [CrossRef] [PubMed]
14. Groer, C.; Busch, D.; Patrzyk, M.; Beyer, K.; Busemann, A.; Heidecke, C.D.; Drozdzik, M.; Siegmund, W.; Oswald, S. Absolute protein quantification of clinically relevant cytochrome P450 enzymes and UDP-glucuronosyltransferases by mass spectrometry-based targeted proteomics. *J. Pharm. Biomed. Anal.* **2014**, *100*, 393–401. [CrossRef] [PubMed]
15. Jin, Y.; Zollinger, M.; Borell, H.; Zimmerlin, A. Patten CJ CYP4F enzymes are responsible for the elimination of fingolimod (FTY720), a novel treatment of relapsing multiple sclerosis. *Drug Metab. Dispos.* **2011**, *39*, 191–198. [CrossRef] [PubMed]
16. Christmas, P.; Carlesso, N.; Shang, H.; Cheng, S.M.; Weber, B.M.; Preffer, F.I.; Scadden, D.T.; Soberman, R.J. Myeloid expression of cytochrome P450 4F3 is determined by a lineage-specific alternative promoter. *J. Biol. Chem.* **2003**, *278*, 25133–25142. [CrossRef] [PubMed]
17. Bylund, J.; Hidestrand, M.; Ingelman-Sundberg, M.; Oliw, E.H. Identification of CYP4F8 in human seminal vesicles as a prominent 19-hydroxylase of prostaglandin endoperoxides. *J. Biol. Chem.* **2000**, *275*, 21844–21849. [CrossRef] [PubMed]
18. Ohno, Y.; Nakamichi, S.; Ohkuni, A.; Kamiyama, N.; Naoe, A.; Tsujimura, H.; Yokose, U.; Sugiura, K.; Ishikawa, J.; Akiyama, M.; et al. Essential role of the cytochrome P450 CYP4F22 in the production of acylceramide, the key lipid for skin permeability barrier formation. *Proc. Natl. Acad. Sci. USA* **2015**, *112*, 77. [CrossRef] [PubMed]
19. Nakano, M.; Kelly, E.J.; Wiek, C.; Hanenberg, H.; Rettie, A.E. cyp4v2 in Bietti's crystalline dystrophy: ocular localization, metabolism of omega-3-polyunsaturated fatty acids and functional deficit of the p.H331P variant. *Mol. Pharmacol.* **2012**, *82*, 679–686. [CrossRef]
20. Wiek, C.; Schmidt, E.M.; Roellecke, K.; Freund, M.; Nakano, M.; Kelly, E.J.; Kaisers, W.; Yarov-Yarovoy, V.; Kramm, C.M.; Rettie, A.E.; et al. Identification of amino acid determinants in CYP4B1 for optimal catalytic processing of 4-ipomeanol. *Biochem. J.* **2015**, *465*, 103–114. [CrossRef]

21. Savas, U.; Hsu, M.H.; Griffin, K.J.; Bell, D.R.; Johnson, E.F. Conditional regulation of the human CYP4X1 and CYP4Z1 genes. *Arch. Biochem. Biophys.* **2005**, *436*, 377–385. [CrossRef] [PubMed]
22. Rieger, M.A.; Ebner, R.; Bell, D.R.; Kiessling, A.; Rohayem, J.; Schmitz, M.; Temme, A.; Rieber, E.P.; Weigle, B. Identification of a novel mammary-restricted cytochrome P450, CYP4Z1, with overexpression in breast carcinoma. *Cancer Res.* **2004**, *64*, 2357–2364. [CrossRef] [PubMed]
23. Powell, P.K.; Wolf, I.; Lasker, J.M. Identification of CYP4A11 as the major lauric acid omega-hydroxylase in human liver microsomes. *Arch. Biochem. Biophys.* **1996**, *335*, 219–226. [CrossRef] [PubMed]
24. Crespi, C.L.; Chang, T.K.; Waxman, D.J. Determination of CYP4A11-catalyzed lauric acid 12-hydroxylation by high-performance liquid chromatography with radiometric detection. *Methods Mol. Biol.* **2006**, *320*, 137–143.
25. Lino Cardenas, C.L.; Renault, N.; Farce, A.; Cauffiez, C.; Allorge, D.; Lo-Guidice, J.M.; Lhermitte, M.; Chavatte, P.; Broly, F.; Chevalier, D. Genetic polymorphism of CYP4A11 and CYP4A22 genes and in silico insights from comparative 3D modelling in a French population. *Gene* **2011**, *487*, 10–20. [CrossRef] [PubMed]
26. Gajendrarao, P.; Krishnamoorthy, N.; Sakkiah, S.; Lazar, P.; Lee, K.W. Molecular modeling study on orphan human protein CYP4A22 for identification of potential ligand binding site. *J. Mol. Graph. Model.* **2010**, *28*, 524–532. [CrossRef] [PubMed]
27. Sasaki, T.; Horikawa, M.; Orikasa, K.; Sato, M.; Arai, Y.; Mitachi, Y.; Mizugaki, M.; Ishikawa, M.; Hiratsuka, M. Possible relationship between the risk of Japanese bladder cancer cases and the CYP4B1 genotype. *Jpn. J. Clin. Oncol.* **2008**, *3*, 634–640. [CrossRef]
28. Ashkar, S.; Mesentsev, A.; Zhang, W.X.; Mastyugin, V.; Dunn, M.W. Laniado-Schwartzman M. Retinoic acid induces corneal epithelial CYP4B1 gene expression and stimulates the synthesis of inflammatory 12-hydroxyeicosanoids. *J. Ocul. Pharmacol. Ther.* **2004**, *20*, 65–74. [CrossRef]
29. Edson, K.Z.; Prasad, B.; Unadkat, J.D.; Suhara, Y.; Okano, T.; Guengerich, F.P.; Rettie, A.E. Cytochrome P450-dependent catabolism of vitamin K: omega-hydroxylation catalyzed by human CYP4F2 and CYP4F11. *Biochemistry* **2013**, *52*, 8276–8285. [CrossRef]
30. Yin, J.; Liu, H.; Liu, Z.; Owzar, K.; Han, Y.; Su, L.; Wei, Y.; Hung, R.J.; Brhane, Y.; McLaughlin, J. Pathway-analysis of published genome-wide association studies of lung cancer: A potential role for the CYP4F3 locus. *Mol. Carcinog.* **2017**, *56*, 1663–1672. [CrossRef]
31. Kikuta, Y.; Mizomoto, J.; Strobel, H.W.; Ohkawa, H. Expression and physiological function of CYP4F subfamily in human eosinophils. *Biochim. Biophys Acta* **2007**, *1771*, 1439–1445. [CrossRef] [PubMed]
32. Harmon, S.D.; Fang, X.; Kaduce, T.L.; Hu, S.; Raj Gopal, V.; Falck, J.R.; Spector, A.A. Oxygenation of omega-3 fatty acids by human cytochrome P450 4F3B: effect on 20-hydroxyeicosatetraenoic acid production. *Prostaglandins Leukot. Essent. Fatty Acids* **2006**, *75*, 169–177. [CrossRef] [PubMed]
33. Stark, K.; Wongsud, B.; Burman, R.; Oliw, E.H. Oxygenation of polyunsaturated long chain fatty acids by recombinant CYP4F8 and CYP4F12 and catalytic importance of Tyr-125 and Gly-328 of CYP4F8. *Arch. Biochem. Biophys.* **2005**, *441*, 174–181. [CrossRef] [PubMed]
34. Bylund, J.; Finnstrom, N.; Oliw, E.H. Gene expression of a novel cytochrome P450 of the CYP4F subfamily in human seminal vesicles. *Biochem. Biophys. Re.s Commun.* **1999**, *261*, 169–174. [CrossRef] [PubMed]
35. Stark, K.; Törmä, H.; Cristea, M.; Oliw, E.H. Expression of CYP4F8 (prostaglandin H 19-hydroxylase) in human epithelia and prominent induction in epidermis of psoriatic lesions. *Arch. Biochem. Biophys.* **2003**, *409*, 188–196. [CrossRef]
36. Kalsotra, A.; Turman, C.M.; Kikuta, Y.; Strobel, H.W. Expression and characterization of human cytochrome P450 4F11: Putative role in the metabolism of therapeutic drugs and eicosanoids. *Toxicol. Appl. Pharmacol.* **2004**, *199*, 295–304. [CrossRef] [PubMed]
37. Yi, M.; Cho, S.A.; Min, J.; Kim, D.H.; Shin, J.G.; Lee, S.J. Functional characterization of a common CYP4F11 genetic variant and identification of functionally defective CYP4F11 variants in erythromycin metabolism and 20-HETE synthesis. *Arch. Biochem. Biophys.* **2017**, *620*, 43–51. [CrossRef] [PubMed]
38. Wang, Y. Gene regulation of CYP4F11 in human keratinocyte HaCaT cells. *Drug Metab. Dispos.* **2010**, *38*, 100–107. [CrossRef] [PubMed]
39. Hashizume, T.; Imaoka, S.; Hiroi, T.; Terauchi, Y.; Fujii, T.; Miyazaki, H.; Kamataki, T.; Funae, Y. cDNA cloning and expression of a novel cytochrome p450 (cyp4f12) from human small intestine. *Biochem. Biophys. Res. Commun.* **2001**, *280*, 1135–1141. [CrossRef]
40. Stark, K.; Dostalek, M.; Guengerich, F.P. Expression and purification of orphan cytochrome P450 4X1 and oxidation of anandamide. *Febs. J.* **2008**, *275*, 3706–3717. [CrossRef]

41. Yi, M.; Shin, J.G.; Lee, S.J. Expression of CYP4V2 in human THP1 macrophages and its transcriptional regulation by peroxisome proliferator-activated receptor gamma. *Toxicol. Appl. Pharmacol.* **2017**, *330*, 100–106. [CrossRef] [PubMed]
42. Zollner, A.; Dragan, C.A.; Pistorius, D.; Müller, R.; Bode, H.B.; Peters, F.T.; Maurer, H.H.; Bureik, M. Human CYP4Z1 catalyzes the in-chain hydroxylation of lauric acid and myristic acid. *Biol. Chem.* **2009**, *390*, 313–317. [CrossRef] [PubMed]
43. Kim, W.Y.; Lee, S.J.; Min, J.; Oh, K.S.; Kim, D.H.; Kim, H.S.; Shin, J.G. Identification of novel CYP4F2 genetic variants exhibiting decreased catalytic activity in the conversion of arachidonic acid to 20-hydroxyeicosatetraenoic acid (20-HETE). *Prostaglandins Leukot Essent. Fatty Acids* **2018**, *131*, 6–13. [CrossRef] [PubMed]
44. Jarrar, Y.B.; Jarrar, Q.; Abed, A.; Abu-Shalhoob, M. Effects of nonsteroidal anti-inflammatory drugs on the expression of arachidonic acid-metabolizing Cyp450 genes in mouse hearts, kidneys and livers. *Prostaglandins Other Lipid Mediat.* **2019**, *141*, 14–21. [CrossRef] [PubMed]
45. Nakano, M.; Kelly, E.J.; Rettie, A.E. Expression and characterization of CYP4V2 as a fatty acid omega-hydroxylase. *Drug Metab. Dispos.* **2009**, *37*, 2119–2122. [CrossRef] [PubMed]
46. Antoun, J.; Goulitquer, S.; Amet, Y.; Dreano, Y.; Salaun, J.P.; Corcos, L.; Plée-Gautier, E. CYP4F3B is induced by PGA1 in human liver cells: a regulation of the 20-HETE synthesis. *J. Lipid Res.* **2008**, *49*, 2135–21341. [CrossRef] [PubMed]
47. Liu, J.Y.; Li, N.; Yang, J.; Li, N.; Qiu, H.; Ai, D.; Chiamvimonvat, N.; Zhu, Y.; Hammock, B.D. Metabolic profiling of murine plasma reveals an unexpected biomarker in rofecoxib-mediated cardiovascular events. *Proc. Natl. Acad. Sci. USA* **2010**, *107*, 17017–17022. [CrossRef] [PubMed]
48. Costa, T.J.; Ceravolo, G.S.; Echem, C.; Hashimoto, C.M.; Costa, B.P.; Santos-Eichler, R.A.; Oliveira, M.A.; Jiménez-Altayó, F.; Akamine, E.H.; Dantas, A.P.; et al. Detrimental Effects of Testosterone Addition to Estrogen Therapy Involve Cytochrome P-450-Induced 20-HETE Synthesis in Aorta of Ovariectomized Spontaneously Hypertensive Rat (SHR), a Model of Postmenopausal Hypertension. *Front. Physiol.* **2018**, *9*, 490. [CrossRef]
49. Pavek, P.; Dvorak, Z. Xenobiotic-induced transcriptional regulation of xenobiotic metabolizing enzymes of the cytochrome P450 superfamily in human extrahepatic tissues. *Curr. Drug Metab.* **2008**, *9*, 129–143. [CrossRef]
50. Joseph, G. Elevated 20-HETE impairs coronary collateral growth in metabolic syndrome via endothelial dysfunction. *Am. J. Physiol. Heart Circ. Physiol.* **2017**, *312*, H528–H540. [CrossRef]
51. Zordoky, B.N.; Anwar-Mohamed, A.; Aboutabl, M.E.; El-Kadi, A.O. Acute doxorubicin cardiotoxicity alters cardiac cytochrome P450 expression and arachidonic acid metabolism in rats. *Toxicol. Appl. Pharmacol.* **2010**, *242*, 38–46. [CrossRef] [PubMed]
52. Kelly, E.J.; Nakano, M.; Rohatgi, P.; Yarov-Yarovoy, V.; Rettie, A.E. Finding homes for orphan cytochrome P450s: CYP4V2 and CYP4F22 in disease states. *Mol. Interv.* **2011**, *11*, 124–132. [CrossRef] [PubMed]
53. Kikuta, Y.; Kusunose, E.; Kusunose, M. Characterization of human liver leukotriene B(4) omega-hydroxylase P450 (CYP4F2). *J. Biochem.* **2000**, *127*, 1047–1052. [CrossRef] [PubMed]
54. Kikuta, Y.; Kusunose, E.; Endo, K.; Yamamoto, S.; Sogawa, K.; Fujii-Kuriyama, Y.; Kusunose, M. A novel form of cytochrome P-450 family 4 in human polymorphonuclear leukocytes. cDNA cloning and expression of leukotriene B4 omega-hydroxylase. *J. Biol. Chem.* **1993**, *268*, 9376–9380. [PubMed]
55. Spector, A.A. Arachidonic acid cytochrome P450 epoxygenase pathway. *J. Lipid Res.* **2009**, *50*, S52–S56. [CrossRef] [PubMed]
56. Le Quere, V.; Plée-Gautier, E.; Potin, P.; Madec, S.; Salaün, J.P. Human CYP4F3s are the main catalysts in the oxidation of fatty acid epoxides. *J. Lipid Res.* **2004**, *45*, 1446–1458. [CrossRef]
57. Wang, Y.; Li, Y.; Lu, J.; Qi, H.; Cheng, I.; Zhang, H. Involvement of CYP4F2 in the Metabolism of a Novel Monophosphate Ester Prodrug of Gemcitabine and Its Interaction Potential In Vitro. *Molecules* **2018**, *23*, 1195. [CrossRef] [PubMed]
58. Wang, J.; Li, K.; Zhang, X.; Teng, D.; Ju, M.; Jing, Y.; Zhao, Y. The correlation between the expression of genes involved in drug metabolism and the blood level of tacrolimus in liver transplant receipts. *Sci. Rep.* **2017**, *7*, 3429. [CrossRef] [PubMed]

59. Tang, Z.; Salamanca-Pinzón, S.G.; Wu, Z.L.; Xiao, Y.; Guengerich, F.P. Human cytochrome P450 4F11: heterologous expression in bacteria, purification and characterization of catalytic function. *Arch. Biochem. Biophys.* **2010**, *494*, 86–93. [CrossRef]
60. Hashizume, T.; Imaoka, S.; Mise, M.; Terauchi, Y.; Fujii, T.; Miyazaki, H.; Kamataki, T.; Funae, Y. Involvement of CYP2J2 and CYP4F12 in the metabolism of ebastine in human intestinal microsomes. *J. Pharmacol. Exp. Ther.* **2002**, *300*, 298–304. [CrossRef]
61. Evangelista, E.A.; Kaspera, R.; Mokadam, N.A.; Jones, J.P.; Totah, R.A. Activity, inhibition, and induction of cytochrome P450 2J2 in adult human primary cardiomyocytes. *Drug Metab. Dispos.* **2013**, *41*, 2087–2094. [CrossRef] [PubMed]
62. McDonald, M.G.; Rieder, M.J.; Nakano, M.; Hsia, C.K.; Rettie, A.E. CYP4F2 is a vitamin K1 oxidase: An explanation for altered warfarin dose in carriers of the V433M variant. *Mol. Pharmacol.* **2009**, *75*, 1337–1346. [CrossRef] [PubMed]
63. Kaminsky, L.S.; Zhang, Z.Y. Human P450 metabolism of warfarin. *Pharmacol. Ther.* **1997**, *73*, 67–74. [CrossRef]
64. Hermans, J.J.; Thijssen, H.H. Human liver microsomal metabolism of the enantiomers of warfarin and acenocoumarol: P450 isozyme diversity determines the differences in their pharmacokinetics. *Br. J. Pharmacol.* **1993**, *110*, 482–490. [CrossRef] [PubMed]
65. Souza-Mello, V. Peroxisome proliferator-activated receptors as targets to treat non-alcoholic fatty liver disease. *World J. Hepatol.* **2015**, *7*, 1012–1019. [CrossRef] [PubMed]
66. Hong, F.; Xu, P.; Zhai, Y. The Opportunities and Challenges of Peroxisome Proliferator-Activated Receptors Ligands in Clinical Drug Discovery and Development. *Int. J. Mol. Sci.* **2018**, *19*, 2189. [CrossRef] [PubMed]
67. Hardwick, J.P.; Osei-Hyiaman, D.; Wiland, H.; Abdelmegeed, M.A.; Song, B.J. PPAR/RXR Regulation of Fatty Acid Metabolism and Fatty Acid omega-Hydroxylase (CYP4) Isozymes: Implications for Prevention of Lipotoxicity in Fatty Liver Disease. *PPAR Res.* **2009**, *2009*, 952734. [CrossRef]
68. Peraza, M.A.; Burdick, A.D.; Marin, H.E.; Gonzalez, F.J.; Peters, J.M. The toxicology of ligands for peroxisome proliferator-activated receptors (PPAR). *Toxicol. Sci.* **2006**, *90*, 269–295. [CrossRef]
69. Kikuta, Y.; Kusunose, E.; Kondo, T.; Yamamoto, S.; Kinoshita, H.; Kusunose, M. Cloning and expression of a novel form of leukotriene B4 omega-hydroxylase from human liver. *FEBS Lett.* **1994**, *348*, 70–74. [CrossRef]
70. Edwards, I.J.; O'Flaherty, J.T. Omega-3 Fatty Acids and PPARgamma in Cancer. *PPAR Res.* **2008**, *2008*, 358052. [CrossRef]
71. Kalsotra, A.; Cui, X.; Antonovic, L.; Robida, A.M.; Morgan, E.T.; Strobel, H.W. Inflammatory prompts produce isoform-specific changes in the expression of leukotriene B(4) omega-hydroxylases in rat liver and kidney. *FEBS Lett.* **2003**, *555*, 236–242. [CrossRef]
72. Muntane, J. Effect of carrageenan-induced granuloma on hepatic cytochrome P-450 isozymes in rats. *Inflammation* **1995**, *19*, 143–156. [CrossRef] [PubMed]
73. Alexanian, A.; Sorokin, A. Targeting 20-HETE producing enzymes in cancer - rationale, pharmacology, and clinical potential. *Onco. Targets Ther.* **2013**, *6*, 243–255. [PubMed]
74. Yu, W.; Chai, H.; Li, Y.; Zhao, H.; Xie, X.; Zheng, H.; Wang, C.; Wang, X.; Yang, G.; Cai, X.; et al. Increased expression of CYP4Z1 promotes tumor angiogenesis and growth in human breast cancer. *Toxicol. Appl. Pharmacol.* **2012**, *264*, 73–83. [CrossRef] [PubMed]
75. Alexanian, A.; Miller, B.; Roman, R.J.; Sorokin, A. 20-HETE-producing enzymes are up-regulated in human cancers. *Cancer Genom. Proteom.* **2012**, *9*, 163–169.
76. Eun, H.S.; Cho, S.Y.; Lee, B.S.; Seong, I.O.; Kim, K.H. Profiling cytochrome P450 family 4 gene expression in human hepatocellular carcinoma. *Mol. Med. Rep.* **2018**, *18*, 4865–4876. [CrossRef] [PubMed]
77. Shibuya, M. Vascular endothelial growth factor and its receptor system: physiological functions in angiogenesis and pathological roles in various diseases. *J. Biochem.* **2013**, *153*, 13–19. [CrossRef]
78. Imaoka, S.; Yoneda, Y.; Sugimoto, T.; Hiroi, T.; Yamamoto, K.; Nakatani, T.; Funae, Y. CYP4B1 is a possible risk factor for bladder cancer in humans. *Biochem. Biophys. Res. Commun.* **2000**, *277*, 776–780. [CrossRef]
79. Vanderslice, R.R.; Boyd, J.A.; Eling, T.E.; Philpot, R.M. The cytochrome P-450 monooxygenase system of rabbit bladder mucosa: enzyme components and isozyme 5-dependent metabolism of 2-aminofluorene. *Cancer Res.* **1985**, *45*, 5851–5858.
80. Imaoka, S.; Yoneda, Y.; Matsuda, T.; Degawa, M.; Fukushima, S.; Funae, Y. Mutagenic activation of urinary bladder carcinogens by CYP4B1 and the presence of CYP4B1 in bladder mucosa. *Biochem. Pharmacol.* **1997**, *54*, 677–683. [CrossRef]

81. Hsu, H.; Rainov, N.G.; Quinones, A.; Eling, D.J.; Sakamoto, K.M.; Spear, M.A. Combined radiation and cytochrome CYP4B1/4-ipomeanol gene therapy using the EGR1 promoter. *Anticancer Res.* **2003**, *23*, 2723–2728. [PubMed]
82. Choudhary, D.; Jansson, I.; Stoilov, I.; Sarfarazi, M.; Schenkman, J.B. Expression patterns of mouse and human CYP orthologs (families 1-4) during development and in different adult tissues. *Arch. Biochem. Biophys.* **2005**, *436*, 50–61. [CrossRef] [PubMed]
83. Tamaki, Y.; Arai, T.; Sugimura, H.; Sasaki, T.; Honda, M.; Muroi, Y.; Matsubara, Y.; Kanno, S.; Ishikawa, M.; Hirasawa, N.; et al. Association between cancer risk and drug-metabolizing enzyme gene (CYP2A6, CYP2A13, CYP4B1, SULT1A1, GSTM1 and GSTT1) polymorphisms in cases of lung cancer in Japan. *Drug Metab. Pharmacokinet.* **2011**, *26*, 516–522. [CrossRef] [PubMed]
84. Lin, J.T. Downregulation of the cytochrome P450 4B1 protein confers a poor prognostic factor in patients with urothelial carcinomas of upper urinary tracts and urinary bladder. *Apmis* **2019**, *127*, 170–180. [CrossRef] [PubMed]
85. Downie, D.; McFadyen, M.C.; Rooney, P.H.; Cruickshank, M.E.; Parkin, D.E.; Miller, I.D.; Telfer, C.; Melvin, W.T.; Murray, G.I. Profiling cytochrome P450 expression in ovarian cancer: identification of prognostic markers. *Clin. Cancer Res.* **2005**, *11*, 7369–7375. [CrossRef] [PubMed]
86. Khayeka-Wandabwa, C.; Ma, X.; Cao, X.; Nunna, V.; Pathak, J.L.; Bernhardt, R.; Cai, P.; Bureik, M. Plasma membrane localization of CYP4Z1 and CYP19A1 and the detection of anti-CYP19A1 autoantibodies in humans. *Int. Immunopharmacol.* **2019**, *73*, 64–71. [CrossRef] [PubMed]
87. McDonald, M.G.; Ray, S.; Amorosi, C.J.; Sitko, K.A.; Kowalski, J.P.; Paco, L.; Nath, A.; Gallis, B.; Totah, R.A.; Dunham, M.J.; et al. Expression and Functional Characterization of Breast Cancer-Associated Cytochrome P450 4Z1 in Saccharomyces cerevisiae. *Drug Metab. Dispos.* **2017**, *45*, 1364–1371. [CrossRef]
88. Wu, C.C.; Schwartzman, M.L. The role of 20-HETE in androgen-mediated hypertension. *Prostaglandins Other Lipid Mediat.* **2011**, *96*, 45–53. [CrossRef]
89. Rocic, P.; Schwartzman, M.L. 20-HETE in the regulation of vascular and cardiac function. *Pharmacol. Ther.* **2018**, *192*, 74–87. [CrossRef]
90. Wu, C.C.; Ei, S.; Cheng, J.; Ding, Y.; Weidenhammer, A.; Garcia, V.; Zhang, F.; Gotlinger, K.; Manthati, V.L.; Falck, J.R.; et al. Androgen-sensitive hypertension associates with upregulated vascular CYP4A12-20-HETE synthase. *J. Am. Soc. Nephrol.* **2013**, *24*, 1288–1296. [CrossRef]
91. Zhang, C.; Booz, G.W.; Yu, Q.; He, X.; Wang, S.; Fan, F. Conflicting roles of 20-HETE in hypertension and renal end organ damage. *Eur. J. Pharmacol.* **2018**, *833*, 190–200. [CrossRef] [PubMed]
92. Fan, F.; Sun, C.W.; Maier, K.G.; Williams, J.M.; Pabbidi, M.R.; Didion, S.P.; Falck, J.R.; Zhuo, J.; Roman, R.J. 20-Hydroxyeicosatetraenoic acid contributes to the inhibition of K+ channel activity and vasoconstrictor response to angiotensin II in rat renal microvessels. *PLoS ONE* **2013**, *8*, e82482. [CrossRef] [PubMed]
93. Gebremedhin, D.; Lange, A.R.; Narayanan, J.; Aebly, M.R.; Jacobs, E.R.; Harder, D.R. Cat cerebral arterial smooth muscle cells express cytochrome P450 4A2 enzyme and produce the vasoconstrictor 20-HETE which enhances L-type Ca2+ current. *J. Physiol.* **1998**, *507*, 771–781. [CrossRef] [PubMed]
94. Roman, R.J. P-450 metabolites of arachidonic acid in the control of cardiovascular function. *Physiol. Rev.* **2002**, *82*, 131–185. [CrossRef] [PubMed]
95. Alonso-Galicia, M.; Maier, K.G.; Greene, A.S.; Cowley, A.W., Jr.; Roman, R.J. Role of 20-hydroxyeicosatetraenoic acid in the renal and vasoconstrictor actions of angiotensin II. *Am. J. Physiol. Regul. Integr. Comp. Physiol.* **2002**, *283*, R60–R68. [CrossRef] [PubMed]
96. Garcia, V.; Shkolnik, B.; Milhau, L.; Falck, J.R.; Schwartzman, M.L. 20-HETE Activates the Transcription of Angiotensin-Converting Enzyme via Nuclear Factor-kappaB Translocation and Promoter Binding. *J. Pharmacol. Exp. Ther.* **2013**, *56*, 525–533.
97. Sodhi, K.; Wu, C.C.; Cheng, J.; Gotlinger, K.; Inoue, K.; Goli, M.; Falck, J.R.; Abraham, N.G. Schwartzman ML CYP4A2-induced hypertension is 20-hydroxyeicosatetraenoic acid- and angiotensin II-dependent. *Hypertension* **2010**, *56*, 871–878. [CrossRef] [PubMed]
98. Gerhold, D.; Bagchi, A.; Lu, M.; Figueroa, D.; Keenan, K.; Holder, D.; Wang, Y.; Jin, H.; Connolly, B.; Austin, C.; et al. Androgens drive divergent responses to salt stress in male versus female rat kidneys. *Genomics* **2007**, *89*, 731–744. [CrossRef] [PubMed]
99. Ng, D.S.; Lai, T.Y.; Ng, T.K.; Pang, C.P. Genetics of Bietti Crystalline Dystrophy. *Asia Pac. J. Ophthalmol. (Phila.)* **2016**, *5*, 245–252. [CrossRef]

100. Fong, A.M.; Koh, A.; Lee, K.; Ang, C.L. Bietti's crystalline dystrophy in Asians: clinical, angiographic and electrophysiological characteristics. *Int. Ophthalmol.* **2009**, *29*, 459–470. [CrossRef]
101. Mansour, A.M.; Uwaydat, S.H.; Chan, C.C. Long-term follow-up in Bietti crystalline dystrophy. *Eur. J. Ophthalmol.* **2007**, *17*, 680–682. [CrossRef] [PubMed]
102. Vargas, M.; Mitchell, A.; Yang, P.; Weleber, R. *Bietti Crystalline Dystrophy*; Adam, M.P., Ardinger, H.H., Pagon, R.A., Wallace, S.E., Bean, L.J.H., Stephens, K., Amemiya, A., Eds.; GeneReviews® [Internet]; University of Washington: Seattle, WA, USA, 2019.
103. Lin, J.; Nishiguchi, K.M.; Nakamura, M.; Dryja, T.P.; Berson, E.L.; Miyake, Y. Recessive mutations in the CYP4V2 gene in East Asian and Middle Eastern patients with Bietti crystalline corneoretinal dystrophy. *J. Med. Genet.* **2005**, *42*, e38. [CrossRef] [PubMed]
104. Mataftsi, A.; Zografos, L.; Millá, E.; Secrétan, M.; Munier, F.L. Bietti's crystalline corneoretinal dystrophy: a cross-sectional study. *Retina* **2004**, *24*, 416–426. [CrossRef] [PubMed]
105. Shan, M.; Dong, B.; Zhao, X.; Wang, J.; Li, G.; Yang, Y.; Li, Y. Novel mutations in the CYP4V2 gene associated with Bietti crystalline corneoretinal dystrophy. *Mol. Vis.* **2005**, *11*, 738–743. [PubMed]
106. Li, A.; Jiao, X.; Munier, F.L.; Schorderet, D.F.; Yao, W.; Iwata, F.; Hayakawa, M.; Kanai, A.; Shy Chen, M.; Alan Lewis, R.; et al. Bietti crystalline corneoretinal dystrophy is caused by mutations in the novel gene CYP4V2. *Am. J. Hum. Genet.* **2004**, *74*, 817–826. [CrossRef] [PubMed]
107. Zenteno, J.C.; Ayala-Ramirez, R.; Graue-Wiechers, F. Novel CYP4V2 gene mutation in a Mexican patient with Bietti's crystalline corneoretinal dystrophy. *Curr. Eye Res.* **2008**, *33*, 313–318. [CrossRef] [PubMed]
108. Lee, K.Y.; Koh, A.H.; Aung, T.; Yong, V.H.; Yeung, K.; Ang, C.L.; Vithana, E.N. Characterization of Bietti crystalline dystrophy patients with CYP4V2 mutations. *Invest. Ophthalmol. Vis. Sci.* **2005**, *46*, 3812–3816. [CrossRef]
109. Gekka, T.; Hayashi, T.; Takeuchi, T.; Goto-Omoto, S.; Kitahara, K. CYP4V2 mutations in two Japanese patients with Bietti's crystalline dystrophy. *Ophthalmic. Res.* **2005**, *37*, 262–269. [CrossRef]
110. Wada, Y.; Itabashi, T.; Sato, H.; Kawamura, M.; Tada, A.; Tamai, M. Screening for mutations in CYP4V2 gene in Japanese patients with Bietti's crystalline corneoretinal dystrophy. *Am. J. Ophthalmol.* **2005**, *139*, 894–899. [CrossRef]
111. Jin, Z.B.; Ito, S.; Saito, Y.; Inoue, Y.; Yanagi, Y.; Nao, I.N. Clinical and molecular findings in three Japanese patients with crystalline retinopathy. *Jpn. J. Ophthalmol.* **2006**, *50*, 426–431. [CrossRef]
112. Lai, T.Y.; Ng, T.K.; Tam, P.O.; Yam, G.H.; Ngai, J.W.; Chan, W.M.; Liu, D.T.; Lam, D.S.; Pang, C.P. Genotype phenotype analysis of Bietti's crystalline dystrophy in patients with CYP4V2 mutations. *Invest. Ophthalmol. Vis. Sci.* **2007**, *48*, 5212–5220. [CrossRef] [PubMed]
113. Li, Q.; Li, Y.; Zhang, X.; Xu, Z.; Zhu, X.; Ma, K.; She, H.; Peng, X. Utilization of fundus autofluorescence, spectral domain optical coherence tomography and enhanced depth imaging in the characterization of Bietti crystalline dystrophy in different stages. *Retina* **2015**, *35*, 2074–2084. [CrossRef] [PubMed]
114. Gocho, K. High-Resolution Imaging of Patients with Bietti Crystalline Dystrophy with CYP4V2 Mutation. *J. Ophthalmol.* **2014**, *2014*, 283603. [PubMed]
115. Halford, S.; Liew, G.; Mackay, D.S.; Sergouniotis, P.I.; Holt, R.; Broadgate, S.; Volpi, E.V.; Ocaka, L.; Robson, A.G.; Holder, G.E.; et al. Detailed phenotypic and genotypic characterization of bietti crystalline dystrophy. *Ophthalmology* **2014**, *121*, 1174–1184. [CrossRef] [PubMed]
116. Yin, H.; Jin, C.; Fang, X.; Miao, Q.; Zhao, Y.; Chen, Z.; Su, Z.; Ye, P.; Wang, Y.; Yin, J. Molecular analysis and phenotypic study in 14 Chinese families with Bietti crystalline dystrophy. *PLoS ONE* **2014**, *9*, 94960. [CrossRef] [PubMed]
117. Xiao, X.; Mai, G.; Li, S.; Guo, X.; Zhang, Q. Identification of CYP4V2 mutation in 21 families and overview of mutation spectrum in Bietti crystalline corneoretinal dystrophy. *Biochem. Biophys. Res. Commun.* **2011**, *409*, 18–186. [CrossRef] [PubMed]
118. Meng, X.H. Identification of novel CYP4V2 gene mutations in 92 Chinese families with Bietti's crystalline corneoretinal dystrophy. *Mol. Vis.* **2014**, *20*, 1806–1814. [PubMed]
119. Astuti, G.D. Novel insights into the molecular pathogenesis of CYP4V2-associated Bietti's retinal dystrophy. *Mol. Genet. Genomic. Med.* **2015**, *3*, 14–29. [CrossRef] [PubMed]
120. Nakamura, M.; Lin, J.; Nishiguchi, K.; Kondo, M.; Sugita, J.; Miyake, Y. Bietti crystalline corneoretinal dystrophy associated with CYP4V2 gene mutations. *Adv. Exp. Med. Biol.* **2006**, *572*, 49–53. [PubMed]

121. Chung, J.K.; Shin, J.H.; Jeon, B.R.; Ki, C.S.; Park, T.K. Optical coherence tomographic findings of crystal deposits in the lens and cornea in Bietti crystalline corneoretinopathy associated with mutation in the CYP4V2 gene. *Jpn. J. Ophthalmol.* **2013**, *57*, 447–450. [CrossRef] [PubMed]
122. Liu, D.N.; Liu, Y.; Meng, X.H.; Yin, Z.Q. The characterization of functional disturbances in Chinese patients with Bietti's crystalline dystrophy at different fundus stages. *Graefes Arch. Clin. Exp. Ophthalmol.* **2012**, *250*, 191–200. [CrossRef] [PubMed]
123. Tian, R.; Wang, S.R.; Wang, J.; Chen, Y.X. Novel CYP4V2 mutations associated with Bietti crystalline corneoretinal dystrophy in Chinese patients. *Int. J. Ophthalmol.* **2015**, *8*, 465–469. [PubMed]
124. Wada, Y. Specular microscopic findings of corneal deposits in patients with Bietti's crystalline corneal retinal dystrophy. *Br. J. Ophthalmol.* **1999**, *83*, 1095. [CrossRef] [PubMed]
125. Yokoi, Y.; Nakazawa, M.; Mizukoshi, S.; Sato, K.; Usui, T.; Takeuchi, K. Crystal deposits on the lens capsules in Bietti crystalline corneoretinal dystrophy associated with a mutation in the CYP4V2 gene. *Acta Ophthalmol.* **2010**, *88*, 607–609. [CrossRef] [PubMed]
126. Lefevre, C.; Bouadjar, B.; Ferrand, V.; Tadini, G.; Mégarbané, A.; Lathrop, M.; Prud'homme, J.F.; Fischer, J. Mutations in a new cytochrome P450 gene in lamellar ichthyosis type 3. *Hum. Mol. Genet.* **2006**, *15*, 767–776. [CrossRef] [PubMed]
127. Wertz, P.W.; Cho, E.S.; Downing, D.T. Effect of essential fatty acid deficiency on the epidermal sphingolipids of the rat. *Biochim. Biophys. Acta* **1983**, *753*, 350–355. [CrossRef]
128. Imokawa, G. Decreased level of ceramides in stratum corneum of atopic dermatitis: an etiologic factor in atopic dry skin? *J. Invest. Dermatol.* **1991**, *96*, 523–526. [CrossRef] [PubMed]
129. Ishikawa, J.; Narita, H.; Kondo, N.; Hotta, M.; Takagi, Y.; Masukawa, Y.; Kitahara, T.; Takema, Y.; Koyano, S.; Yamazaki, S. Changes in the ceramide profile of atopic dermatitis patients. *J. Invest. Dermatol.* **2010**, *130*, 2511–2514. [CrossRef]
130. Janssens, M.; van Smeden, J.; Gooris, G.S.; Bras, W.; Portale, G.; Caspers, P.J.; Vreeken, R.J.; Hankemeier, T.; Kezic, S.; Wolterbeek, R.; et al. Increase in short-chain ceramides correlates with an altered lipid organization and decreased barrier function in atopic eczema patients. *J. Lipid Res.* **2012**, *53*, 2755–2766. [CrossRef]
131. Sayeb, M.; Riahi, Z.; Laroussi, N.; Bonnet, C.; Romdhane, L.; Mkaouar, R.; Zaouak, A.; Marrakchi, J.; Abdessalem, G.; Messaoud, O. A Tunisian family with a novel mutation in the gene CYP4F22 for lamellar ichthyosis and co-occurrence of hearing loss in a child due to mutation in the SLC26A4 gene. *Int. J. Dermatol.* **2019**. [CrossRef]
132. Lo-Guidice, J.M.; Allorge, D.; Cauffiez, C.; Chevalier, D.; Lafitte, J.J.; Lhermitte, M.; Broly, F. Genetic polymorphism of the human cytochrome P450 CYP4B1: evidence for a non-functional allelic variant. *Pharmacogenetics* **2002**, *12*, 367–374. [CrossRef] [PubMed]
133. Hiratsuka, M.; Nozawa, H.; Konno, Y.; Saito, T.; Konno, S.; Mizugaki, M. Human CYP4B1 gene in the japanese population analyzed by denaturing HPLC. *Drug Metab. Pharmacokinet.* **2004**, *19*, 114–119. [CrossRef] [PubMed]
134. Jennings, G.K.; Hsu, M.H.; Shock, L.S.; Johnson, E.F.; Hackett, J.C. Noncovalent interactions dominate dynamic heme distortion in cytochrome P450 4B1. *J. Biol. Chem.* **2018**, *293*, 11433–11446. [CrossRef] [PubMed]
135. Scott, E.E. Omega-versus (omega-1)-hydroxylation: Cytochrome P450 4B1 sterics make the call. *J. Biol. Chem.* **2017**, *292*, 5622–5623. [CrossRef] [PubMed]
136. Hsu, M.H. The Crystal Structure of Cytochrome P450 4B1 (CYP4B1) Monooxygenase Complexed with Octane Discloses Several Structural Adaptations for omega-Hydroxylation. *J. Biol. Chem.* **2017**, *292*, 5610–5621. [CrossRef] [PubMed]
137. Sirotina, S.; Ponomarenko, I.; Kharchenko, A.; Bykanova, M.; Bocharova, A.; Vagaytseva, K.; Stepanov, V.; Churnosov, M.; Solodilova, M.; Polonikov, A. A Novel Polymorphism in the Promoter of the CYP4A11 Gene Is Associated with Susceptibility to Coronary Artery Disease. *Dis. Markers* **2018**, *2018*, 5812802. [CrossRef] [PubMed]
138. Yan, H.Q.; Yuan, Y.; Zhang, P.; Huang, Z.; Chang, L.; Gui, Y.K. CYP4F2 gene single nucleotide polymorphism is associated with ischemic stroke. *Genet. Mol. Res.* **2015**, *14*, 659–664. [CrossRef] [PubMed]
139. Fava, C. The V433M variant of the CYP4F2 is associated with ischemic stroke in male Swedes beyond its effect on blood pressure. *Hypertension* **2008**, *52*, 373–380. [CrossRef] [PubMed]

140. Gainer, J.V.; Bellamine, A.; Dawson, E.P.; Womble, K.E.; Grant, S.W.; Wang, Y.; Cupples, L.A.; Guo, C.Y.; Demissie, S.; O'Donnell, C.J. Functional variant of CYP4A11 20-hydroxyeicosatetraenoic acid synthase is associated with essential hypertension. *Circulation* **2005**, *111*, 63–69. [CrossRef]
141. Mayer, B.; Lieb, W.; Götz, A.; König, I.R.; Aherrahrou, Z.; Thiemig, A.; Holmer, S.; Hengstenberg, C.; Doering, A.; Loewel, H.; et al. Association of the T8590C polymorphism of CYP4A11 with hypertension in the MONICA Augsburg echocardiographic substudy. *Hypertension* **2005**, *46*, 766–771. [CrossRef]
142. Mayer, B.; Lieb, W.; Götz, A.; König, I.R.; Kauschen, L.F.; Linsel-Nitschke, P.; Pomarino, A.; Holmer, S.; Hengstenberg, C.; Doering, A. Association of a functional polymorphism in the CYP4A11 gene with systolic blood pressure in survivors of myocardial infarction. *J. Hypertens* **2006**, *24*, 1965–1970. [CrossRef] [PubMed]
143. Yu, K.; Zhang, T.; Li, X. Genetic role of CYP4A11 polymorphisms in the risk of developing cardiovascular and cerebrovascular diseases. *Ann. Hum. Genet.* **2018**, *82*, 371–380. [CrossRef] [PubMed]
144. Liang, J.Q.; Yan, M.R.; Yang, L.; Suyila, Q.; Cui, H.W.; Su, X.L. Association of a CYP4A11 polymorphism and hypertension in the Mongolian and Han populations of China. *Genet. Mol. Res.* **2014**, *13*, 508–517. [CrossRef] [PubMed]
145. Yan, H.C.; Liu, J.H.; Li, J.; He, B.X.; Yang, L.; Qiu, J.; Li, L.; Ding, D.P.; Shi, L.; Zhao, S.J. Association between the CYP4A11 T8590C variant and essential hypertension: new data from Han Chinese and a meta-analysis. *PLoS ONE* **2013**, *8*, e80072. [CrossRef] [PubMed]
146. Williams, J.S.; Hopkins, P.N.; Jeunemaitre, X.; Brown, N.J. CYP4A11 T8590C polymorphism, salt-sensitive hypertension, and renal blood flow. *J. Hypertens* **2011**, *29*, 1913–1918. [CrossRef] [PubMed]
147. Ding, H.; Cui, G.; Zhang, L.; Xu, Y.; Bao, X.; Tu, Y.; Wu, B.; Wang, Q.; Hui, R.; Wang, W. Association of common variants of CYP4A11 and CYP4F2 with stroke in the Han Chinese population. *Pharmacogenet Genomics* **2010**, *20*, 187–194. [CrossRef]
148. Laffer, C.L. The T8590C polymorphism of CYP4A11 and 20-hydroxyeicosatetraenoic acid in essential hypertension. *Hypertension* **2008**, *51*, 767–772. [CrossRef]
149. Caldwell, M.D.; Awad, T.; Johnson, J.A.; Gage, B.F.; Falkowski, M.; Gardina, P.; Hubbard, J.; Turpaz, Y.; Langaee, T.Y.; Eby, C. CYP4F2 genetic variant alters required warfarin dose. *Blood* **2008**, *111*, 4106–4112. [CrossRef]
150. Dean, L. Warfarin Therapy and VKORC1 and CYP Genotype. In *Medical Genetics Summaries*; Pratt, V., Ed.; Bethesda: Rockville, MD, USA, 2012.
151. Borgiani, P.; Ciccacci, C.; Forte, V.; Sirianni, E.; Novelli, L.; Bramanti, P.; Novelli, G. CYP4F2 genetic variant (rs2108622) significantly contributes to warfarin dosing variability in the Italian population. *Pharmacogenomics* **2009**, *10*, 261–266. [CrossRef]
152. Takeuchi, F.; McGinnis, R.; Bourgeois, S.; Barnes, C.; Eriksson, N.; Soranzo, N.; Whittaker, P.; Ranganath, V.; Kumanduri, V.; McLaren, W. A genome-wide association study confirms VKORC1, CYP2C9, and CYP4F2 as principal genetic determinants of warfarin dose. *PLoS Genet.* **2009**, *5*, e1000433. [CrossRef]
153. Cohen, J.L.; Thompson, E.; Sinvani, L.; Kozikowski, A.; Qiu, G.; Pekmezaris, R.; Spyropoulos, A.C.; Wang, J.J. Assessment of warfarin algorithms for hospitalized adults: searching for a safe dosing strategy. *J. Thromb. Thrombolysis* **2019**. [CrossRef] [PubMed]
154. Sasano, M.; Ohno, M.; Fukuda, Y.; Nonen, S.; Hirobe, S.; Maeda, S.; Miwa, Y.; Yokoyama, J.; Nakayama, H.; Miyagawa, S.; et al. Verification of pharmacogenomics-based algorithms to predict warfarin maintenance dose using registered data of Japanese patients. *Eur. J. Clin. Pharmacol.* **2019**, *75*, 901–911. [CrossRef] [PubMed]
155. Horne, B.D.; Lenzini, P.A.; Wadelius, M.; Jorgensen, A.L.; Kimmel, S.E.; Ridker, P.M.; Eriksson, N.; Anderson, J.L.; Pirmohamed, M.; Limdi, N.A.; et al. Pharmacogenetic warfarin dose refinements remain significantly influenced by genetic factors after one week of therapy. *Thromb. Haemost.* **2012**, *107*, 232–240. [PubMed]
156. Hiratsuka, M.; Nozawa, H.; Katsumoto, Y.; Moteki, T.; Sasaki, T.; Konno, Y.; Mizugaki, M. Genetic polymorphisms and haplotype structures of the CYP4A22 gene in a Japanese population. *Mutat. Res.* **2006**, *599*, 98–104. [CrossRef] [PubMed]
157. Nelson, D.R.; Zeldin, D.C.; Hoffman, S.M.; Maltais, L.J.; Wain, H.M.; Nebert, D.W. Comparison of cytochrome P450 (CYP) genes from the mouse and human genomes, including nomenclature recommendations for genes, pseudogenes and alternative-splice variants. *Pharmacogenetics* **2004**, *14*, 1–18. [CrossRef] [PubMed]

158. Ananthakrishnan, A.N.; Khalili, H.; Song, M.; Higuchi, L.M.; Lochhead, P.; Richter, J.M.; Chan, A.T. Genetic Polymorphisms in Fatty Acid Metabolism Modify the Association Between Dietary n3: n6 Intake and Risk of Ulcerative Colitis: A Prospective Cohort Study. *Inflamm. Bowel. Dis.* **2017**, *23*, 1898–1904. [CrossRef] [PubMed]
159. Shiotani, A.; Murao, T.; Fujita, Y.; Fujimura, Y.; Sakakibara, T.; Nishio, K.; Haruma, K. Novel single nucleotide polymorphism markers for low dose aspirin-associated small bowel bleeding. *PLoS ONE* **2013**, *8*, e84244. [CrossRef] [PubMed]
160. Cauffiez, C.; Klinzig, F.; Rat, E.; Tournel, G.; Allorge, D.; Chevalier, D.; Pottier, N.; Lovecchio, T.; Colombel, J.F.; Lhermitte, M.; et al. P4F12 genetic polymorphism: identification and functional characterization of seven variant allozymes. *Biochem. Pharmacol.* **2004**, *68*, 2417–2425. [CrossRef]
161. Choquet, H.; Trapani, E.; Goitre, L.; Trabalzini, L.; Akers, A.; Fontanella, M.; Hart, B.L.; Morrison, L.A.; Pawlikowska, L.; Kim, H. Cytochrome P450 and matrix metalloproteinase genetic modifiers of disease severity in Cerebral Cavernous Malformation type 1. *Free Radic. Biol. Med.* **2016**, *92*, 100–109. [CrossRef]
162. Austin, H.; De Staercke, C.; Lally, C.; Bezemer, I.D.; Rosendaal, F.R.; Hooper, W.C. New gene variants associated with venous thrombosis: A replication study in White and Black Americans. *J. Thromb. Haemost.* **2011**, *9*, 489–495. [CrossRef]
163. Fiatal, S. Genetic profiling revealed an increased risk of venous thrombosis in the Hungarian Roma population. *Thromb. Res.* **2019**, *179*, 37–44. [CrossRef] [PubMed]
164. Yue, Y.; Sun, Q.; Man, C.; Fu, Y. Association of the CYP4V2 polymorphism rs13146272 with venous thromboembolism in a Chinese population. *Clin. Exp. Med.* **2019**, *19*, 159–166. [CrossRef] [PubMed]
165. Glurich, I.; Chyou, P.H.; Engel, J.M.; Cross, D.S.; Onitilo, A.A. Tamoxifen-induced venothromboembolic events: exploring validation of putative genetic association. *Clin. Med. Res.* **2013**, *11*, 16–25. [CrossRef] [PubMed]
166. Song, Y.; Mo, G.; Yin, G. A novel mutation in the CYP4V2 gene in a Chinese patient with Bietti's crystalline dystrophy. *Int. Ophthalmol.* **2013**, *33*, 269–276. [CrossRef] [PubMed]

© 2019 by the authors. Licensee MDPI, Basel, Switzerland. This article is an open access article distributed under the terms and conditions of the Creative Commons Attribution (CC BY) license (http://creativecommons.org/licenses/by/4.0/).

Review

Circulating Extracellular Vesicles Containing Xenobiotic Metabolizing CYP Enzymes and Their Potential Roles in Extrahepatic Cells Via Cell–Cell Interactions

Kelli Gerth, Sunitha Kodidela, Madeline Mahon, Sanjana Haque, Neha Verma and Santosh Kumar *

Department of Pharmaceutical sciences, University of Tennessee Health Science Center, 881 Madison Avenue, Memphis, TN 38163, USA; kgerth1@uthsc.edu (K.G.); skodidel@uthsc.edu (S.K.); mmahon1@uthsc.edu (M.M.); shanque8@uthsc.edu (S.H.); neha.verma@smail.astate.edu (N.V.)
* Correspondence: ksantosh@uthsc.edu.com

Received: 1 October 2019; Accepted: 4 December 2019; Published: 7 December 2019

Abstract: The cytochrome P450 (CYP) family of enzymes is known to metabolize the majority of xenobiotics. Hepatocytes, powerhouses of CYP enzymes, are where most drugs are metabolized into non-toxic metabolites. Additional tissues/cells such as gut, kidneys, lungs, blood, and brain cells express selective CYP enzymes. Extrahepatic CYP enzymes, especially in kidneys, also metabolize drugs into excretable forms. However, extrahepatic cells express a much lower level of CYPs than hepatocytes. It is possible that the liver secretes CYP enzymes, which circulate via plasma and are eventually delivered to extrahepatic cells (e.g., brain cells). CYP circulation likely occurs via extracellular vesicles (EVs), which carry important biomolecules for delivery to distant cells. Recent studies have revealed an abundance of several CYPs in plasma EVs and other cell-derived EVs, and have demonstrated the role of CYP-containing EVs in xenobiotic-induced toxicity via cell–cell interactions. Thus, it is important to study the mechanism for packaging CYP into EVs, their circulation via plasma, and their role in extrahepatic cells. Future studies could help to find novel EV biomarkers and help to utilize EVs in novel interventions via CYP-containing EV drug delivery. This review mainly covers the abundance of CYPs in plasma EVs and EVs derived from CYP-expressing cells, as well as the potential role of EV CYPs in cell–cell communication and their application with respect to novel biomarkers and therapeutic interventions.

Keywords: extracellular vesicles; exosomes; cytochrome P450; extrahepatic tissues; plasma; circulatory CYPs

1. Introduction

The cytochrome P450 (CYP) superfamily is a group of Phase I mono-oxidase enzymes with broad substrate specificity that is responsible for the majority of xenobiotic metabolism [1]. CYP enzymes are also involved in vital endogenous pathways, including prostaglandin metabolism and steroid hormone biosynthesis [2]. Gene names are determined according to a standardized nomenclature. Using CYP3A4 as an example, "CYP" refers to the superfamily of cytochrome P450 genes, "3" refers to the family designation (<40% amino acid identity with other CYPs), "A" refers to the subfamily designation (40–55% amino acid identity with other CYPs in that family), and "4" refers to the specific gene within the subfamily with >55% sequence identity [3]. This enzyme superfamily is ubiquitous in nature—it is present in bacteria, fungi, plants, and animals—with varying expression patterns [4]. In fact, many drug metabolizing isoforms including CYP 1A, 2C, and 3A show significant interspecies differences in enzymatic activity while CYP 2E1 does not [5].

Hepatocytes express an abundance of drug metabolizing CYP enzymes and demonstrate the greatest capacity for Phase I xenobiotic biotransformation, followed by the small intestine and kidneys [6–8]. Of the 57 known human CYP enzymes, five CYP isoenzymes from CYP 1–3 families metabolize the majority of clinically used drugs—CYP3A4/5, CYP2D6, CYP2C9, CYP2C19, and CYP1A2 [9]. Although CYP 1–3 families predominantly aid in xenobiotic detoxification, CYP enzymes are also involved in the bioactivation of xenobiotics, resulting in the formation of toxic intermediates. CYPs 1B1, 1A1, and 2A6 are involved in the bioactivation pathways of cigarette smoke constituents [10,11], while CYP2E1 is involved in the metabolism of alcohol and acetaminophen [12]. These are associated with the generation of hepatotoxic or carcinogenic metabolites and promote reactive oxygen species (ROS) production in vitro and in vivo [13–15], which eventually causes organ damage and cancers.

CYP enzymes are generally upregulated by their own substrates/drugs, resulting in enhanced metabolism and suboptimal plasma concentrations of concurrent drugs [16]. Conversely, CYP inhibition by various drugs contributes to supratherapeutic drug levels and drug-induced toxicity, thus preventing CYPs from performing their protective role in detoxification [16]. In the case of prodrugs that require CYP-mediated conversion to their active form, the reverse is true. Prototypical pharmacologic CYP enzyme inducers and inhibitors that are commonly implicated in drug–drug interactions include rifampin (inducer) and azole antifungals (inhibitors), which may interact with multiple CYP isoenzymes to varying degrees [9]. Further, pharmacogenetic variations in CYP activity may result in reduced, absent or increased metabolic capacity. Drug metabolizing isoforms with functionally relevant polymorphisms include CYPs 1A2, 2B6, 2C8, 2C9, 2C19, 2D6, and 3A4/5 [17,18]. Certain isoforms are more highly polymorphic than others [17] and are associated with clinically significant effects, such as toxicity or lack of therapeutic response [17]. Furthermore, while CYP enzymes are most abundantly expressed in the liver, they are also found in extrahepatic tissues throughout the body. Although the majority of extrahepatic CYPs are involved in endogenous pathways [2], drug metabolism also occurs outside the liver. The small intestine and kidneys are the primary sites of extrahepatic drug metabolizing CYP enzymes [6,7]; however, drug metabolizing CYPs are also expressed in the lungs, blood (monocytes, lymphocytes), brain, and heart [19–22]. Extrahepatic CYP enzymes contribute to cell-specific biotransformation, albeit to a lesser extent than hepatic CYPs. While extrahepatic CYP expression and metabolic capacity are not able to mediate total body clearance of xenobiotics, the enzymes may play a significant role in local tissue exposure and toxicity [19].

Recently, we provided the first evidence that functional CYP isoforms are packaged into extracellular vesicles (EVs) derived from human plasma of healthy volunteers, as well as in EVs derived from hepatic cell lines [23]. Extracellular vesicles (EVs) are nanosized, membrane-bound particles that are secreted from most cell types into biological fluids, namely plasma, and are taken up by other cells [24]. EV cargo includes a heterogeneous array of biomolecules, e.g., lipids, carbohydrates, cytokines, proteins, and nucleic acids—mRNAs, miRNAs, etc. [24–26]. Thus, EVs are thought to be critical in cell-to-cell signaling, protein transfer, and nucleic acid shuttling [24,27,28]. These characteristics suggest that EVs might be potential biomarkers, therapeutic targets, and drug-delivery systems [24].

It is important to note that exosomes are a subgroup of extracellular vesicles with a distinct biogenesis pathway [29]. Although much of the literature prior to 2018 refers to "exosomes", distinguishing exosomes from other EVs has proved challenging due to overlaps in size, composition, and marker proteins [30]. Therefore, in accordance with current ISEV guidelines [30], this review will exclusively refer to "EVs" even when published reports refer to "exosomes".

As drug metabolic capacity is limited in extrahepatic tissues, it is possible that CYP-containing plasma EVs are secreted from the liver, circulate via plasma, and are delivered to distant sites (e.g., brain cells), where they may aid in extrahepatic drug metabolism, detoxification, and may also influence toxicity at these sites (Figure 1). It is also possible that extrahepatic cells in the kidneys, lungs, blood, heart, and brain also secrete EVs that are pooled in the plasma and cerebral spinal fluid (CSF), making an "EV-depot". These EVs may then be delivered to other cells as needed and under specific conditions, to perform biological functions. This review will provide an overview of the contribution of CYPs to drug metabolism in extrahepatic tissues. Since our primary goal is to discuss relevant drug-metabolizing CYP enzymes and their xenobiotic substrates, discussion pertaining to endogenous pathways is largely outside the scope of this review. Importantly, we will summarize the most recent literature pertaining to CYPs and EVs, the relative abundance of CYPs in human plasma-derived EVs, and potential implications of CYP-containing EVs in xenobiotic biotransformation/bioactivation. Further, we will discuss the potential role of EV CYP enzymes as biomarkers in various pathological conditions and xenobiotic exposure/drug use, as well as suggest novel therapeutic interventions.

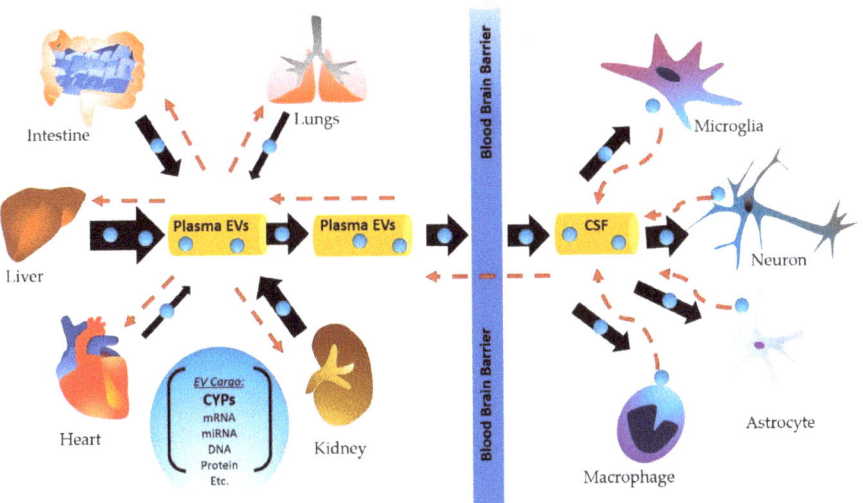

Figure 1. Cytochrome P450 (CYP)-containing plasma extracellular vesicles (EVs)/exosomes are secreted from the liver and other peripheral organs, circulate via plasma, and are delivered to distant sites (e.g., brain cells), where they may aid in extrahepatic drug metabolism, detoxification, and may also influence toxicity at these sites. Similarly, secretion of EVs from extrahepatic cells, including brain cells are also likely to contain CYPs in addition to other biomolecules, which would also be circulated via plasma and delivered to other distant cells.

2. Expression of CYP Enzymes in Extrahepatic Tissues

Although the metabolic capacity of extrahepatic CYPs is relatively low compared to the capacity of hepatic CYPs, extrahepatic CYPs may still influence local tissue function and drug exposure, as well as drug biotransformation and bioactivation at these sites. Further, extrahepatic CYP enzymes may alter overall systemic exposure to xenobiotics, with corresponding elevations in toxicity [19,31]. Interestingly, some CYPs are expressed preferentially in extrahepatic tissues, which may lead to unique extrahepatic metabolites and tissue-specific consequences in cellular toxicity and organ pathology. The CYPs expressed in extrahepatic tissues are shown in Table 1.

Table 1. Select xenobiotic metabolizing cytochrome P450 enzymes expressed in human extrahepatic tissues.

Tissues/Organs	CYPs Detected	References
Small intestine	3A4/5$^{+/++/+++}$, 1A1$^{+/++}$, 1B1^{+}, 2C9$^{+/++/+++}$, 2C19$^{+/++/+++}$, 2D6^{+}, 2E1^{+}	[32–34]
Kidney	2B6$^{+/++/+++}$, 3A5$^{+/++/+++}$	[6,35–37]
Lungs	1A1$^{+/++/+++}$, 1A2$^{+/++/+++}$, 1B1$^{+/++}$, 2A6$^{+/++/+++}$, 2B6^{+}, 2C$^{+/++}$, 2D6$^{+/++/+++}$, 2E1$^{+/++/+++}$ 3A4/5$^{+/++/+++}$	[38–44]
Heart	1A1^{+}, 1A2$^{+/++}$, 1B1^{+}, 2C8$^{+/++}$, 2C9$^{+/++}$, 2J2$^{+/++}$, 2B6/7^{+}, 2D6^{+}, 2E1$^{+/++}$, 3A4^{++}	[31,45–48]
Blood (monocytes and lymphocytes)	1A1$^{+/++}$, 1B1$^{+/++}$, 2A6$^{+/++/+++}$, 2B6$^{+/++}$, 2D6$^{+/++}$, 2E1$^{+/++}$, and 3A4/5$^{+/++/+++}$	[13,49–52]
Brain	1A1$^{+/++}$, 1A2$^{+/++}$, 1B1$^{+/++}$, 2A6$^{+/++/+++}$, 2B6^{++}, 2C8^{+}, 2D6$^{+/++/+++}$, 2E1$^{+/++/+++}$, 3A4/5$^{+/++}$	[13–15,53]

Key: + mRNA, ++ protein, +++ activity.

2.1. Small Intestine

CYP3A4 is the main CYP isoform in the small intestine, accounting for roughly 82% of CYP enzymes expressed in gut tissue [8]. Many drugs, which are 3A4 substrates, have low oral bioavailability due in part to CYP-mediated intestinal first pass metabolism [8]. Due to CYP3A4's broad substrate specificity and high expression in gut tissue, certain foods and dietary supplements can cause significant drug interactions. Grapefruit juice, a potent inhibitor of intestinal 3A4, is known to increase the plasma concentration of common 3A4 substrates, e.g., statins, calcium channel blockers, protease inhibitors, and many others [54]. Inhibition by grapefruit juice does not affect hepatic 3A4 but may decrease intestinal 3A4 function by 62% [32].

2.2. Kidneys

It has been estimated that the human kidney contains anywhere from 4–20% of hepatic CYP protein content [8]. Only CYPs 2B6 and 3A5 have been confirmed in the human kidney, and evidence for CYPs 3A4, 2C9, and 2C8 is equivocal [6]. Of these, CYP3A5 is the most prevalent isoform [19]. Renal CYP3A5 is highly polymorphic [19]. In fact, genetic variations in renal 3A5 expression may influence nephrotoxicity associated with the immunosuppressive agent, tacrolimus [55], as well as with the anticancer agent, ifosfamide [56]. Further, CYP3A forms are expressed consistently in renal cancer cells and may be involved in renal cancer development and multidrug resistance [57]. Nevertheless, renal CYP3A enzymes may also help suppress cancer via bioactivation of certain agents, forming metabolites that are cytotoxic to tumor cells and benign in noncancer cells [57].

2.3. Lungs

Lung tissue expresses CYPs 1A1, 1B1, 2A6, 2E1, 3A4, and 3A5 among others [8]. Many respiratory tract CYPs are linked to bioactivation of the constituents of cigarette smoke and enhanced toxicity and carcinogenicity. CYP1A1 is mostly expressed in smokers [8]. Both CYPs 1A1 and 1B1 isoforms are induced by compounds found in tobacco smoke, particularly Benzo(a)pyrene (Bap) [10,58]. Bap is a polycyclic aromatic hydrocarbon (PAH) carcinogen that is converted to DNA-reactive intermediates in a process dependent on CYP1A1 and CYP1B1 metabolic pathways [10]. CYP2A6 is mostly expressed in the trachea and is also thought to be involved in bioactivation of carcinogens from tobacco smoke [8]. CYP2A6 mainly metabolizes nicotine, the primary constituent in cigarette smoke, into cotinine and nicotine-derived nitrosamine ketone (NNK) [59]. Of note, CYP2A6 polymorphisms are involved in the development of lung cancer and nicotine dependence [8,60].

2.4. Heart

Multiple CYP enzymes relevant to drug metabolism or bioactivation are present in cardiac tissue in low or moderate amounts, including CYPs 1A1, 1B1, 2C8, 2C9, 2D6, 2E1, and 3A4 [31]. Interestingly, higher CYP mRNA expression (e.g., 2D6, 2C) in right ventricular tissue vs. left ventricular tissue, which is indicative of increased drug inactivation at this site, has led some researchers to suggest that differential expression may contribute to therapeutic failure in pharmacological treatment of right ventricular hypertrophy [45]. However, the most prevalent CYP isoform in the heart is CYP2J2, an enzyme involved in metabolizing the anticancer drug, doxorubicin [31]. One common adverse effect of doxorubicin is cardiotoxicity, an effect that may be mitigated in cases of CYP2J2 over-production in cardiomyocytes [61]. Further, cardiac CYPs 1A1 and 2J2 have been shown to be induced in mice following treatment with cocaine and Bap [62].

2.5. Blood

CYP mRNA and protein, including that of CYPs 3A4, 2A6, 2E1, 1A1, and 1B1, have been observed in human monocytic and lymphocytic cells [13,21,22,50,51], with 2A6 being the most abundantly expressed isoform in monocyte-derived macrophages [21]. Monocytes are part of the mononuclear phagocyte system, a family of myeloid lineage that also includes macrophages and dendritic cells [63]. These blood cells are critical in host defenses against pathogens, as well as in maintaining tissue homeostasis [64]. Blood monocytes may mature into macrophages under inflammatory conditions and migrate to tissues, where they can synthesize and secrete inflammatory mediators [63]. CYP2A6 induction has been observed in monocytes derived from the plasma of smokers [50], and 2A6-mediated metabolism of nicotine is associated with increased oxidative stress and DNA damage in monocytic cells [13,50]. Similarly, cigarette smoke condensate (CSC) induced CYPs 1A1 and 1B1 in in vitro monocyte studies [51]. In addition, alcohol-inducible CYP2E1, which is known to metabolize alcohol in the liver, was also found to be expressed and induced by alcohol in monocytes/macrophages, leading to oxidative stress [21].

2.6. Brain

Total CYP protein content in the brain is substantially lower compared to the liver and is estimated at 0.5%–2% of hepatic CYP content [65]. Although the contribution to systemic drug metabolism may be minimal, brain CYP activity and variation may have a significant impact on local metabolism and the therapeutic efficacy of centrally acting drugs, including antidepressants, antipsychotics, drugs of abuse, and carcinogens [66]. CYP enzymes are differentially expressed in some regions of the brain and in neurons and glial cells. The highest CYP content is found in the brain stem and cerebellum, and the lowest in the striatum and hippocampus [67]. The proposed functions of cerebral CYPs vary by cell type and location.

2.6.1. Neurons

Drug metabolizing CYPs 1A1, 2B6, 2E1, and 3A4 are found primarily in neurons [65], while CYP2D6 is expressed in pyramidal neurons in addition to glial cells [68]. CYP3A4 is involved in psychotropic drug metabolism, including antiepileptic metabolism [65,68]. Considering that alcohol dehydrogenase (ADH) is not expressed in the brain, while CYP2E1 is constitutively expressed in various brain regions, it is probable that CYP2E1 is the brain's major alcohol metabolizing enzyme [69].

2.6.2. Monocytes and Glial Cells

CYPs are abundant in astrocytes at the blood–brain barrier (BBB), aiding in the regulation of xenobiotic influx into the CNS, blood flow modulation, and signaling during inflammatory conditions [70]. Notably, CYP1B1 is expressed on cerebral micro-vessels and astrocytes at the BBB interface [71,72], and in conjunction with membrane transporters, may aid in regulating xenobiotic passage into and out of the brain. CYP2D6 is expressed in neurons and glial cells [68]. CYP2D6 is involved in opioid metabolism and that of many antidepressants, antipsychotics, and detoxification of pesticides [66,73]. Further, in vitro studies have shown that CYPs 3A4, 2A6, 1A1, and 2E1 are expressed in human monocyte-derived-macrophages as well as in astrocytic cell lines [13,49–52]. CYPs 1A1 and 2A6 account for the majority of CYP content in SVGA astrocytes, while CYP2A6 is the predominant isoform in U937 macrophages [13,21]. Monocytes are known to enter the brain from the periphery and differentiate into perivascular macrophages and microglia [74], making their presence in the CNS an important target for CYP activity.

Relatively low levels of brain CYP content may have neurotoxic ramifications. For example, low cerebral 3A4 may make brain tissue more sensitive to pharmaceutical inhibition of psychotropic drug metabolism, as illustrated by ritonavir-mediated inhibition of carbamazepine and consequent ataxia [75]. Moreover, a possible explanation for nicotine-mediated induction of 1A1 and 2A6 observed in astrocytes, but not hepatocytes, may be that hepatic 1A1 and 2A6 are already expressed at maximal levels [13]. Indeed, brain CYPs seem to be particularly sensitive to xenobiotic-mediated induction. Increased expression of CYPs 2E1, 2B6, and 2D6 has been observed in the brain tissue of smokers and alcoholics [14,69,76–78], but changes in hepatic 2B6 and 2D6 are undetectable [78,79]. Our in vitro study showed that CYP2E1 is induced by alcohol exposure in both astrocytes and monocytes via oxidative stress-mediated protein kinase C/c-Jun N-terminal kinase/specificity protein1 (PKC/JNK/SP1) pathways, which eventually causes cellular toxicity. Although evidence suggests CYP2E1-mediated mechanisms of cellular toxicity, including neurotoxicity and contribution to neurological diseases (HAND, PD), it is possible that that cellular CYP content in the CNS is insufficient to mediate these effects. Therefore, it is worth considering the possibility of additional sources of CYP enzymes, which may be transferred to brain cells, when evaluating xenobiotic-induced toxicities and therapeutic efficacy in brain cells.

3. Circulating CYP Enzymes and Their Role in Cell–Cell Communication

3.1. EVs and Their Origin

EVs are small, membrane-bound vehicles of intercellular communication that carry various types of cellular information throughout the body. The role of EVs in cell–cell interactions is rather complicated; thus, EVs may influence the pathophysiology of recipient cells in either positive or negative ways [80]. EVs comprise a heterogenous group and are therefore classified into three major categories based on their biogenesis pathway: exosomes, micro-vesicles, and apoptotic bodies [29]. Exosomes (<200 nm), which originate from endosomal compartments, are secreted from cells when micro-vesicular bodies (MVBs) fuse with the plasma membrane, whereas micro-vesicles (50–1000 nm) are routinely shed from cell membranes, and apoptotic bodies (50–5000 nm) are released during membrane blebbing upon programed cell death [29]. Due to overlaps in size, composition, and marker proteins, exosome characterization has proved to be challenging for researchers [30]. Therefore, we will exclusively refer to "EVs", even when published data refers to "exosomes".

3.2. Role of Circulatory CYPs in Drug Metabolism and in Cell–Cell Communication

Produced by and secreted from cells into extracellular plasma, EVs transmit genetic material, proteins, and other biological cargos that reflects the function of the organ from which they originate [24,81]. Once EVs exit the cell by exocytosis, they travel to distant cells via biological fluids such as plasma, cerebrospinal fluid, and urine, where they fuse with recipient cells [81]. The cargo is then released and is free to exert its effects on target cells [81,82]. This transmission of EVs throughout the body provides a means of communication between cells—offering a new source of biomarkers, as well as a potential tool in characterizing variability in drug exposure and therapeutic intervention [24,82].

Furthermore, studies have revealed that EVs carry a multitude of drug metabolizing enzymes, including members of the CYP enzyme group [23,83,84] (Table 2). However, the presence and amount of CYP enzymes is likely to vary greatly depending on the EV source, whether the EVs are isolated from plasma or a specific cell line, as well as the physiological condition of the cells from which they originate [85].

Table 2. Select xenobiotic metabolizing cytochrome P450 enzymes expressed in EVs.

Tissue/Organ/Fluid	Human/Animal	CYPs Detected	Isolation Method/References
Plasma	Human	$1A1^{++}$, $1A2^{+/++}$, $1B1^{+/++}$, $2B6^{+/++}$, $2A6^{+/++}$, $2C8^{+/++}$, $2C9^{+/++}$, $2C19^{+/++}$, $2D6^{+/++}$, $2E1^{+/++/+++}$, $2J2^{+/++}$, $3A4/5^{+/++/+++}$	Total Exosome Isolation Kit (from plasma) [23], exoEasy Kit (membrane affinity spin column) [86]
	Mouse	$2E1^{++}$, $3A4^{++}$	Total Exosome Isolation Kit (from plasma) [87] (unpublished data)
	Rat	$1A1^{++}$, $1A2^{++}$, $2E1^{++}$, $4A^{++}$, $4B^{++}$	Ultracentrifugation [88]
Hepatocytes	Human	$2E1^{+/++}$	Total Exosome Isolation Kit (from cell culture media) [23]
	Rat	$2A1^{++}$, $2A2^{++}$, $4A2^{++}$, $2B3^{++}$, $2C11^{++}$, $2D1^{++/+++}$, $2D3^{++}$, $2D18^{++}$, $2D10^{++}$, $2D26^{++}$	ExoQuick Kit [89] Ultracentrifugation [84,90,91]
Monocytes/Macrophages	Human	$2E1^{+/++/+++}$, $1A1^{+/++}$, $2A6^{+/++}$	Total Exosome Isolation Kit (from cell culture media [23,92]

Key: + mRNA, ++ protein, +++ activit.

Our group recently detected CYPs 1B1, 2A6, 2E1, and 3A4 mRNA in plasma-derived EVs from healthy subjects, with 2E1 displaying > 500-fold higher expression than the other CYPs identified. We also detected CYPs 1A1, 1B1, 2A6, 2E1, and 3A4 at the protein level [23]. In our studies, plasma EVs were isolated using 0.22 μm filtration, followed by different methods including single and double isolations with a commercial kit [23,26], in addition to the ultracentrifugation method [92]. Further, absolute spectra revealed a higher level of CYPs in plasma-derived EVs versus plasma alone, which indicates specific packaging of CYPs within circulating plasma EVs. Importantly, activity assays confirmed the enzymatic activity of EV CYP2E1 and 3A4. Interestingly, our finding indicated a higher level of CYP2E1 in plasma EVs than in liver cells/EVs, which is the powerhouse of CYP enzymes. The plasma EV CYP2E1 level was also higher than alcohol-induced CYP2E1 in monocytes. Together, these findings suggest that EV packaging is carefully regulated. A study performed by Rowland et al. further strengthened our findings, demonstrating the presence of peptides and mRNA of CYPs, 1A2, 2B6, 2C8, 2C9, 2C19, 2D6, 2E1, 2J2, 3A4 and 3A5, UGT 1A1, 1A3, 1A4, 1A6, 1A9, 2B4, 2B7, 2B10 and 2B15, and NADPH-cytochrome CYP reductase in plasma-derived exosomes [86]. As EVs act as intercellular messengers, this differential packaging has a crucial impact on the pathophysiology of the recipient cells, and an abundance of CYP enzymes in EVs suggests their necessity at points across the body.

3.2.1. Alcohol, Acetaminophen and EV CYP2E1

Circulating CYPs likely play a role in extrahepatic metabolism. Once secreted from the liver, EVs can deliver CYP enzymes to distant sites, where they then affect the target cells by influencing the metabolism of pharmaceutical drugs, drugs of abuse, and other xenobiotics [23]. CYP2E1 is mainly found in the liver, where it is known to be a major metabolizer of alcohol and acetaminophen (APAP) [12]. Importantly, CYP2E1-mediated metabolism produces reactive oxygen species (ROS) responsible for oxidative liver damage and cellular toxicity [12]. While alcohol and APAP misuse are significant contributors to liver damage, they also affect extrahepatic tissues [93]. Moreover, alcohol induces EV release from hepatocytes in association with ROS [94]. Our study has also shown that alcohol-exposure to mice induces the level of EV CYP2E1 in the plasma [87]. We have further demonstrated that when plasma EVs containing increased levels of CYP2E1 are exposed to hepatic and monocytic cells, they exacerbate alcohol- and APAP-induced toxicity [87]. We have also shown that the toxicity is mainly caused by CYP2E1, as a CYP2E1-selective inhibitor significantly reduced EV-exacerbated toxicity by both alcohol and APAP. Together, the results suggest that EVs containing CYP2E1 can cause both intra- and intercellular communication. Thus, CYP2E1's presence extrahepatically and its role in cell–cell communication suggests that alcohol and APAP are also being metabolized at different locations throughout the body. It is likely that plasma EVs transmit CYP2E1 from the liver to targets throughout the body, e.g., the brain, where 2E1 can then participate in xenobiotic metabolism and bioactivation of toxic metabolites. EVs carrying CYP2E1 to distant cells may help explain alcohol-induced cellular injury occurring outside the liver.

3.2.2. Tobacco Smoking and EV CYP2A6, 1B1, 1A1

CYP2A6, CYP1B1, and CYP1A1, all of which play a vital role in the metabolism and bioactivation of tobacco/cigarette smoke constituents, were also detected in EVs [23]. Although CYP2A6 and 1B1 are mainly expressed in the respiratory system, they can also be found in liver cells. Thus, EV CYP2A6 and 1B1 may originate from either organ. Both produce toxic metabolites in association with their roles in metabolizing nicotine and PAHs, respectively [10]. Our recent studies suggest that EVs may also play a defensive role, specifically in protecting against smoking-induced HIV-1 pathogenesis [92,95]. CYP-mediated elevations in oxidative stress that accompany tobacco smoking, also promote HIV-1 replication [13,14,50,51]. Additionally, cigarette smoking is associated with EV release in smokers and in various cell types in vitro [96]. Our study revealed that EVs from CSC-treated cells were found to alter their antioxidant capacity and packaging—showing a protective effect against toxicity and viral replication in the early stages of HIV-1 replication [92].

3.2.3. Drug Metabolism and EV CYP3A4

Importantly, we also detected metabolically active CYP3A4 enzyme in plasma exosomes [23]. Being the major drug-metabolizing CYP enzyme, the presence of CYP3A4 in EVs has clinical significance in terms of therapeutics. During drug development, the focus is traditionally on hepatic drug metabolism; however, failing to account for circulating CYPs may result in unintended drug–drug interactions or toxicity. Furthermore, EV CYP3A4 can be used as a biological marker, specifically in examining the metabolism of pharmacological or illicit drugs. Rowland et al. demonstrated a strong relationship between EV CYP3A4 and drug clearance in patients, which suggests that EVs can be a potential tool for identifying variability in drug exposure [86]. The study also found that EV CYP3A4 exhibits comparable kinetics to microsomes taken from liver samples [86]. The circulation of EVs in bodily fluids allows for greater accessibility in terms of isolating these biomarkers. Sometimes called "liquid biopsy," this form of sample collection does not require the use of invasive techniques such as tissue biopsy or liver resection. Rather, to assess the expression of CYP3A4 mRNAs, they can simply be isolated from a blood sample. These findings suggest that EV CYPs may provide a new and easier way to explore variability in pharmaceutical drug metabolism and exposure.

3.2.4. Biological and Clinical Significance of CYP Packaging/Circulation in Plasma EVs

Although it is recognized that EVs can envelope functionally active CYP enzymes, the specific mechanistic pathway of this differential packaging is still under investigation. Circulating EVs with metabolically active CYP enzymes may have a considerable impact on neighboring and distant cells and tissue systems. For example, CYP enzymes carried within EVs might influence the metabolism of endogenous and xenobiotic compounds. EVs could be way of removing unwanted CYP enzymes from the cells. Further investigations are warranted to fully appreciate the impact that these modified EVs may have in the body [83]. EVs derived from patient hepatocytes can be utilized as a non-invasive tool to characterize variability in drug response—one way in which EVs may be used as potential biomarkers [83]. Further, EVs are already under investigation to be used as drug delivery systems, designed to contain specific content for transport to different cell types [97]. Thus, EVs might be a useful tool in combating xenobiotic-induced toxicity by controlled alteration of their contents.

4. Potential Applications of EVs Containing CYP Enzymes

4.1. Circulating CYP Enzymes as Biological Markers of Drug-Induced Toxicity

The current gold standard biomarkers for hepatic injury are based on the measurements of hepatic enzymes levels, including ALT, AST, etc., in plasma or serum. However, the ALT levels do not always correlate with various stages of liver disease due to its relatively short half-life [98]. Therefore, specific components in circulating EVs may have great utility as non-invasive biomarkers for diagnosis and during treatment of hepatic injury. For example, alcohol use increases CYP2E1 expression in plasma EVs [88] and there is a correlation between increased CYP2E1 level and alcohol-induced liver injury [99]. Furthermore, alcohol exposure increases EV release into the circulation [88], making circulatory EVs a potential source of biomarkers in the setting of drug-induced liver injury [84,100].

Recent studies show that much like alcohol's effect on the liver, alcohol can also alter EV cargo [93]. Furthermore, EVs derived from alcohol-treated cells have been shown to exacerbate disease progression through the delivery of altered cellular material to target cells. We previously observed that EVs collected from mouse and human plasma aggravated alcohol and APAP-induced toxicity [87]. In a similar study, Cho et al. demonstrated that CYP2E1-rich EVs from alcohol-exposed rats and patients induced hepatic cell death [88]. These findings further highlight the potential value of CYP2E1-containing EVs as noninvasive, diagnostic biomarkers in alcoholism and microsomal stress [101].

Similarly, smoking induces CYP1A1 [58], CYP1B1 [102], and the activity of these enzymes can exacerbate smoking-related toxicity by providing additional oxidative stress [58]. Since these CYPs are present in EVs [23,86], EVs can serve as markers to diagnose smoking-induced tissue toxicity. Similarly, a strong relationship between EV-derived CYP3A4 and drug clearance in patients [86] suggests that EV CYP3A4 can be used as a biological marker, specifically in examining the metabolism of pharmacological or illicit drugs.

4.2. Use of EV CYPs in Synthetic Biology

CYPs can catalyze the specific addition of oxygen atoms to chemical scaffolds, which would be very challenging and expensive by traditional methods. Several CYPs and engineered variants are now used to synthesize and produce various compounds on a larger scale and for diverse purposes, including drug discovery and development [1]. For instance, artemisinic acid, a precursor for the *Artemisia annua*-derived antimalarial drug, artemisinin, has been synthesized using an engineered form of the plant's CYP71AV1 enzyme [103]. Furthermore, engineered CYPs have utility in statin synthesis. Compactin is a naturally-occurring HMG-CoA reductase inhibitor originally isolated from *Penicillium citrinium* by Endo et al., 1976 [104]. Using an engineered version of *Amycolatopsis orientalis*-derived CYP105AS1 in *Penicillium chrysogenum* fungi, researchers are now able to synthesize pravastatin from compactin [105]. Moreover, as the proteins packaged in EVs are stable and protected from degradation [106], engineered CYPs can be loaded in EVs (Figure 2), a step that would improve their stability and subsequent activity in the production of various therapeutic molecules.

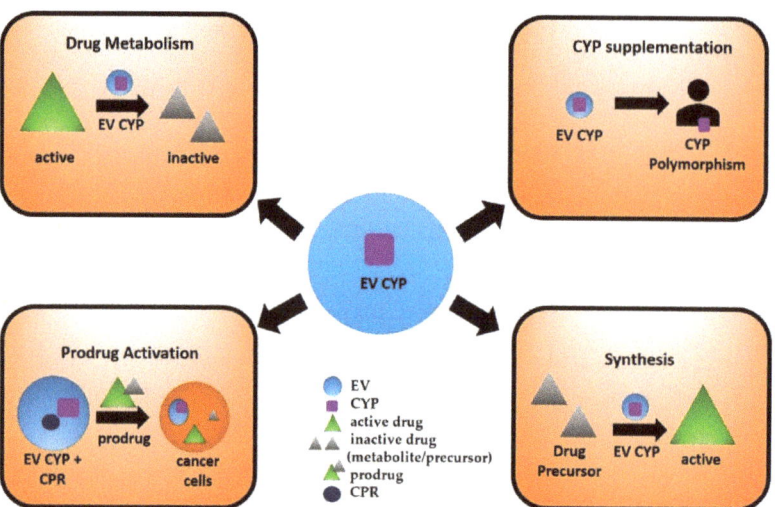

Figure 2. Potential applications of extracellular vesicles containing CYP enzymes include drug metabolism, prodrug activation, supplementation of CYP to subjects with genetic polymorphisms, and industrial synthesis of biomolecules.

4.3. Targeted Delivery of EV CYPs for Prodrug Activation

The ability of EVs to package and transport a variety of biological cargos has prompted investigators to examine the possibility of loading EVs with specific therapeutic content [107]. Several methods of EV loading have been developed, including electroporation, transfection, and incubation, among others [108]. Utilizing one of these methods, it is conceivable that EVs may be loaded with CYP enzyme, along with CPR. The EV-loaded CYP and prodrugs can be directly administered to the site of disease. For example, in the case of solid cancer, EV CYP can activate anticancer prodrugs at the disease site, reducing toxicity in healthy cells caused by anticancer drugs. Engineering such a delivery system could enhance the efficacy and bioavailability of certain prodrugs, cancer treatments (including brain cancer), and neurological disease therapies.

Previously, gene-directed enzyme prodrug therapy (GDEPT), which utilizes gene transfer of CYP enzyme and cytochrome CPR within a viral vector, has been proposed as a novel way to increase therapeutic efficacy and decrease systemic side effects of anticancer prodrugs, e.g., cyclophosphamide (CPA) and ifosfamide (IFA) [1]. The purpose of CYP-based GDEPT is to facilitate local CPA/IFA bioactivation by expressing CYP enzymes directly within tumor cells [109]. While initial trials of CYP-based GDEPT systems have demonstrated safety and enhanced chemosensitivity to tumors, no GDEPT products are currently on the market [110,111]. As EVs are already under investigation as potential delivery systems [97], it is possible that EVs loaded with CYP and CPR could replace the viral vector in CYP-based GDEPT systems (Figure 2). The approach of loading EVs with CYP and CPR would be safe and economical due to the biological origin of EVs. Moreover, several therapeutics fail to achieve optimal concentrations in the CNS due to their inability to cross the BBB. In such instances, EVs can be engineered to target CNS cells and deliver their contents. For example, in delivering a prodrug along with its activating CYP enzyme to microglial cells, EVs could be conjugated with anti-TEME119 antibody [112], which is specific to microglia, to target and deliver EV cargo to these cells. Further, bacterial CYP enzymes have been expressed and engineered to activate prodrugs [1]. Loading of these CYPs in EVs targeted to a particular tissue can increase their stability and further promote their prodrug-converting activity in target sites.

4.4. Delivery of EV CYPs to Supplement Naturally Inactive CYPs

EV-loaded CYPs can also be administered to subjects with loss of function polymorphisms for particular CYP enzymes (Figure 2). Genetic polymorphisms of drug-metabolizing enzymes can result in either decreased, increased, or complete lack of activity of an enzyme, leading to disease susceptibility [113,114] or variability in drug response [115–117]. CYP2D6, which is known to metabolize approximately 20% of drugs, is the most polymorphic CYP enzyme in many ethnic populations and varies from 1–50% [118]. Several CYP2D6 variants cause very low to no activity with several drugs [119]. Similarly, CYP3A5 contributes significantly to drug metabolism in humans and is not expressed in 90% of Caucasians [120]. Thus, certain drugs, e.g., tacrolimus and sirolimus, that are metabolized by CYP3A5, tend to accumulate and cause toxicity in most Caucasians [120]. Thus, administrating EVs loaded with CYP3A5 to Caucasians could be helpful in metabolizing 3A5 substrates and decreasing their respective toxicities.

4.5. Current Challenges Associated with Using EVs as Therapeutics

Although EVs have advantageous properties over synthetic delivery systems in terms of their biological source and ability to deliver functional cargo, clinical translation of EVs as diagnostic or prognostic markers of pathological states remains a challenge due to various reasons. One reason might be the lack of uniformity in isolation, characterization and analysis methods of EVs. This can lead to variations in EV counts and phenotypes between different laboratories, making data analysis and clinical translation difficult. Furthermore, the half-life of exosomes in athymic nude mice was reported to be 30 min, and clearance was estimated to be 6 h after intravenous injections [121]. However, due to compartmental changes as EVs travel throughout the human body, it is difficult to estimate the half-life of EVs in blood. Moreover, EVs from different cells and EVs with different sizes possess different biodistribution profiles. In addition to all these concerns, most studies regarding the physiological or pathological effects of EVs have been done in cell culture models. However, cells under in vivo conditions are under a constant steady-state exposure to EVs. Therefore, the extent to which controlled EV exposures under in vitro conditions corresponds to the in vivo environment remains unclear.

The purification and detection of EVs is improving with the help of technological advancements. Moreover, International Society for Extracellular Vesicles (ISEV) attempts to provide guidelines to isolate and characterize EVs in order to improve reproducibility and to avoid ambiguity in the identification of EVs [122,123]. EVs can be engineered in order to increase their circulation time and improve their delivery to target tissues. For example, EVs can be coated with polyethylene glycol, which is known to increase the half-life of nanoparticles [124]. Increased expression of CD47 on the EV surface can also improve the circulation time of EVs by opposing the actions of phosphatidylserine, which promotes the initiation of phagocytosis and subsequent removal from the circulation by macrophages [125,126]. Therefore, exploring EV circulation kinetics, targeting, internalization, and cell–cell trafficking routes will be useful in engineering EVs for therapeutic purposes.

5. Conclusions

Considering the profound contribution of CYP enzymes in mediating xenobiotic metabolism and bioactivation of toxicants, the presence of CYP enzymes in EVs and their biological significance cannot be ignored. As EVs circulate throughout the body via biological fluids and participate in cellular communication, they may be clinically useful as biomarkers for drug-induced toxicity, synthesis of drug/metabolite synthesis, and targeted prodrug activation. Thus, further investigating the roles of circulating CYPs in extrahepatic cells would help generate novel treatment options for neurological diseases, cancer, and more.

Funding: The authors acknowledge the funding from the National Institutes of Health Grant DA047178 to S.Ku.

Conflicts of Interest: The authors declare no conflict of interest.

Abbreviations

EVs	Extracellular vesicles
CYP	Cytochrome P450
CPR	CYP reductase
APAP	Acetaminophen
ROS	Reactive oxygen species
CSF	Cerebral spinal fluid
Bap	Benzo(a)pyrene
PAH	Polycyclic aromatic hydrocarbon
NNK	Nicotine-derived nitrosamine ketone
CSC	Cigarette smoke condensate
ADH	Alcohol dehydrogenase
BBB	Blood-brain barrier
CNS	Central nervous system
HAND	HIV-associated neurocognitive disorders
PD	Parkinson's disease
MVBs	Micro-vesicular bodies
ALT	Alanine aminotransferase
AST	Aspartate aminotransferase
GDEPT	Gene-directed enzyme prodrug therapy
CPA	Cyclophosphamide
IFA	Ifosfamide

References

1. Kumar, S. Engineering cytochrome P450 biocatalysts for biotechnology, medicine and bioremediation. *Expert Opin. Drug Metab. Toxicol.* **2010**, *6*, 115–131. [CrossRef] [PubMed]
2. Rendic, S.P.; Peter Guengerich, F. Human cytochrome P450 enzymes 5-51 as targets of drugs and natural and environmental compounds: mechanisms, induction, and inhibition - toxic effects and benefits. *Drug Metab. Rev.* **2018**, *50*, 256–342. [CrossRef] [PubMed]
3. Nelson, D.R. Cytochrome P450 nomenclature, 2004. *Methods Mol. Biol.* **2006**, *320*, 1–10. [PubMed]
4. Nelson, D.R. Cytochrome P450 diversity in the tree of life. *Biochim. Biophys. Acta Proteins Proteom.* **2018**, *1866*, 141–154. [CrossRef] [PubMed]
5. Martignoni, M.; Groothuis, G.M.M.; de Kanter, R. Species differences between mouse, rat, dog, monkey and human CYP-mediated drug metabolism, inhibition and induction. *Expert Opin. Drug Metab. Toxicol.* **2006**, *2*, 875–894. [CrossRef] [PubMed]
6. Knights, K.M.; Rowland, A.; Miners, J.O. Renal drug metabolism in humans: The potential for drug-endobiotic interactions involving cytochrome P450 (CYP) and UDP-glucuronosyltransferase (UGT). *Br. J. Clin. Pharmacol.* **2013**, *76*, 587–602. [CrossRef] [PubMed]
7. Thelen, K.; Dressman, J.B. Cytochrome P450-mediated metabolism in the human gut wall. *J. Pharm. Pharmacol.* **2009**, *61*, 541–558. [CrossRef]
8. Gundert-Remy, U.; Bernauer, U.; Blömeke, B.; Döring, B.; Fabian, E.; Goebel, C.; Hessel, S.; Jäckh, C.; Lampen, A.; Oesch, F.; et al. Extrahepatic metabolism at the body's internal-external interfaces. *Drug Metab. Rev.* **2014**, *46*, 291–324. [CrossRef]
9. Sychev, D.A.; Ashraf, G.M.; Svistunov, A.A.; Maksimov, M.L.; Tarasov, V.V.; Chubarev, V.N.; Otdelenov, V.A.; Denisenko, N.P.; Barreto, G.E.; Aliev, G. The cytochrome P450 isoenzyme and some new opportunities for the prediction of negative drug interaction in vivo. *Drug Des. Dev. Ther.* **2018**, *12*, 1147–1156. [CrossRef]
10. Henkler, F.; Stolpmann, K.; Luch, A. Exposure to polycyclic aromatic hydrocarbons: Bulky DNA adducts and cellular responses. *Exp. Suppl.* **2012**, *101*, 107–131.
11. Benowitz, N.L. Pharmacology of nicotine: Addiction, smoking-induced disease, and therapeutics. *Annu. Rev. Pharmacol. Toxicol.* **2009**, *49*, 57–71. [CrossRef] [PubMed]
12. Chen, J.; Jiang, S.; Wang, J.; Renukuntla, J.; Sirimulla, S.; Chen, J. A comprehensive review of cytochrome P450 2E1 for xenobiotic metabolism. *Drug Metab. Rev.* **2019**, *51*, 178–195. [CrossRef] [PubMed]

13. Jin, M.; Earla, R.; Shah, A.; Earla, R.L.; Gupte, R.; Mitra, A.K.; Kumar, A.; Kumar, S. A LC-MS/MS method for concurrent determination of nicotine metabolites and role of CYP2A6 in nicotine metabolism in U937 macrophages: Implications in oxidative stress in HIV + smokers. *J. Neuroimmune Pharmacol.* **2012**, *7*, 289–299. [CrossRef] [PubMed]
14. Ande, A.; Earla, R.; Jin, M.; Silverstein, P.S.; Mitra, A.K.; Kumar, A.; Kumar, S. An LC-MS/MS method for concurrent determination of nicotine metabolites and the role of CYP2A6 in nicotine metabolite-mediated oxidative stress in SVGA astrocytes. *Drug Alcohol. Depend.* **2012**, *125*, 49–59. [CrossRef] [PubMed]
15. Jin, M.; Ande, A.; Kumar, A.; Kumar, S. Regulation of cytochrome P450 2e1 expression by ethanol: Role of oxidative stress-mediated pkc/jnk/sp1 pathway. *Cell Death Dis.* **2013**, *4*, e554. [CrossRef] [PubMed]
16. Manikandan, P.; Nagini, S. Cytochrome P450 Structure, Function and Clinical Significance: A Review. *Curr. Drug Targets* **2018**, *19*, 38–54. [CrossRef]
17. Tracy, T.S.; Chaudhry, A.S.; Prasad, B.; Thummel, K.E.; Schuetz, E.G.; Zhong, X.-B.; Tien, Y.-C.; Jeong, H.; Pan, X.; Shireman, L.M.; et al. Interindividual Variability in Cytochrome P450-Mediated Drug Metabolism. *Drug Metab. Dispos.* **2016**, *44*, 343–351. [CrossRef]
18. Satyanarayana, C.R.U.; Devendran, A.; Sundaram, R.; Gopal, S.D.; Rajagopal, K.; Chandrasekaran, A. Genetic variations and haplotypes of the 5′ regulatory region of CYP2C19 in South Indian population. *Drug Metab. Pharmacokinet.* **2009**, *24*, 185–193. [CrossRef]
19. Pavek, P.; Dvorak, Z. Xenobiotic-induced transcriptional regulation of xenobiotic metabolizing enzymes of the cytochrome P450 superfamily in human extrahepatic tissues. *Curr. Drug Metab.* **2008**, *9*, 129–143. [CrossRef]
20. Rao, P.S.S.; Kumar, S. Chronic Effects of Ethanol and/or Darunavir/Ritonavir on U937 Monocytic Cells: Regulation of Cytochrome P450 and Antioxidant Enzymes, Oxidative Stress, and Cytotoxicity. *Alcohol. Clin. Exp. Res.* **2016**, *40*, 73–82. [CrossRef]
21. Jin, M.; Arya, P.; Patel, K.; Singh, B.; Silverstein, P.S.; Bhat, H.K.; Kumar, A.; Kumar, S. Effect of alcohol on drug efflux protein and drug metabolic enzymes in U937 macrophages. *Alcohol. Clin. Exp. Res.* **2011**, *35*, 132–139. [CrossRef] [PubMed]
22. Ande, A.; McArthur, C.; Kumar, A.; Kumar, S. Tobacco smoking effect on HIV-1 pathogenesis: Role of cytochrome P450 isozymes. *Expert Opin. Drug Metab. Toxicol.* **2013**, *9*, 1453–1464. [CrossRef] [PubMed]
23. Kumar, S.; Sinha, N.; Gerth, K.A.; Rahman, M.A.; Yallapu, M.M.; Midde, N.M. Specific packaging and circulation of cytochromes P450, especially 2E1 isozyme, in human plasma exosomes and their implications in cellular communications. *Biochem. Biophys. Res. Commun.* **2017**, *491*, 675–680. [CrossRef] [PubMed]
24. Barile, L.; Vassalli, G. Exosomes: Therapy delivery tools and biomarkers of diseases. *Pharmacol. Ther.* **2017**, *174*, 63–78. [CrossRef] [PubMed]
25. Kodidela, S.; Ranjit, S.; Sinha, N.; McArthur, C.; Kumar, A.; Kumar, S. Cytokine profiling of exosomes derived from the plasma of HIV-infected alcohol drinkers and cigarette smokers. *PLoS ONE* **2018**, *13*, e0201144. [CrossRef] [PubMed]
26. Kodidela, S.; Wang, Y.; Patters, B.J.; Gong, Y.; Sinha, N.; Ranjit, S.; Gerth, K.; Haque, S.; Cory, T.; McArthur, C.; et al. Proteomic Profiling of Exosomes Derived from Plasma of HIV-Infected Alcohol Drinkers and Cigarette Smokers. *J. Neuroimmune Pharmacol.* **2019**. [CrossRef]
27. Théry, C.; Zitvogel, L.; Amigorena, S. Exosomes: Composition, biogenesis and function. *Nat. Rev. Immunol.* **2002**, *2*, 569–579. [CrossRef]
28. Record, M.; Subra, C.; Silvente-Poirot, S.; Poirot, M. Exosomes as intercellular signalosomes and pharmacological effectors. *Biochem. Pharmacol.* **2011**, *81*, 1171–1182. [CrossRef]
29. Hartjes, T.A.; Mytnyk, S.; Jenster, G.W.; van Steijn, V.; van Royen, M.E. Extracellular Vesicle Quantification and Characterization: Common Methods and Emerging Approaches. *Bioengineering* **2019**, *6*, 7. [CrossRef]
30. Théry, C.; Witwer, K.W.; Aikawa, E.; Alcaraz, M.J.; Anderson, J.D.; Andriantsitohaina, R.; Antoniou, A.; Arab, T.; Archer, F.; Atkin-Smith, G.K.; et al. Minimal information for studies of extracellular vesicles 2018 (MISEV2018): A position statement of the International Society for Extracellular Vesicles and update of the MISEV2014 guidelines. *J. Extracell Vesicles* **2018**, *7*, 1535750. [CrossRef]
31. Chaudhary, K.R.; Batchu, S.N.; Seubert, J.M. Cytochrome P450 enzymes and the heart. *IUBMB Life* **2009**, *61*, 954–960. [CrossRef] [PubMed]

32. Ding, X.; Kaminsky, L.S. Human extrahepatic cytochromes P450: Function in xenobiotic metabolism and tissue-selective chemical toxicity in the respiratory and gastrointestinal tracts. *Annu. Rev. Pharmacol. Toxicol.* **2003**, *43*, 149–173. [CrossRef] [PubMed]
33. Zhang, Q.Y.; Dunbar, D.; Ostrowska, A.; Zeisloft, S.; Yang, J.; Kaminsky, L.S. Characterization of human small intestinal cytochromes P-450. *Drug Metab. Dispos.* **1999**, *27*, 804–809. [PubMed]
34. Obach, R.S.; Zhang, Q.Y.; Dunbar, D.; Kaminsky, L.S. Metabolic characterization of the major human small intestinal cytochrome p450s. *Drug Metab. Dispos.* **2001**, *29*, 347–352. [PubMed]
35. Aleksa, K.; Matsell, D.; Krausz, K.; Gelboin, H.; Ito, S.; Koren, G. Cytochrome P450 3A and 2B6 in the developing kidney: implications for ifosfamide nephrotoxicity. *Pediatr. Nephrol.* **2005**, *20*, 872–885. [CrossRef] [PubMed]
36. Gervot, L.; Rochat, B.; Gautier, J.C.; Bohnenstengel, F.; Kroemer, H.; de Berardinis, V.; Martin, H.; Beaune, P.; de Waziers, I. Human CYP2B6: Expression, inducibility and catalytic activities. *Pharmacogenetics* **1999**, *9*, 295–306. [CrossRef]
37. Haehner, B.D.; Gorski, J.C.; Vandenbranden, M.; Wrighton, S.A.; Janardan, S.K.; Watkins, P.B.; Hall, S.D. Bimodal distribution of renal cytochrome P450 3A activity in humans. *Mol. Pharmacol.* **1996**, *50*, 52–59.
38. Hukkanen, J.; Pelkonen, O.; Hakkola, J.; Raunio, H. Expression and regulation of xenobiotic-metabolizing cytochrome P450 (CYP) enzymes in human lung. *Crit. Rev. Toxicol.* **2002**, *32*, 391–411. [CrossRef]
39. Raunio, H.; Hakkola, J.; Hukkanen, J.; Lassila, A.; Päivärinta, K.; Pelkonen, O.; Anttila, S.; Piipari, R.; Boobis, A.; Edwards, R.J. Expression of xenobiotic-metabolizing CYPs in human pulmonary tissue. *Exp. Toxicol. Pathol.* **1999**, *51*, 412–417. [CrossRef]
40. Shimada, T.; Yamazaki, H.; Mimura, M.; Wakamiya, N.; Ueng, Y.F.; Guengerich, F.P.; Inui, Y. Characterization of microsomal cytochrome P450 enzymes involved in the oxidation of xenobiotic chemicals in human fetal liver and adult lungs. *Drug Metab. Dispos.* **1996**, *24*, 515–522.
41. Wei, C.; Caccavale, R.J.; Kehoe, J.J.; Thomas, P.E.; Iba, M.M. CYP1A2 is expressed along with CYP1A1 in the human lung. *Cancer Lett.* **2001**, *171*, 113–120. [CrossRef]
42. Willey, J.C.; Coy, E.L.; Frampton, M.W.; Torres, A.; Apostolakos, M.J.; Hoehn, G.; Schuermann, W.H.; Thilly, W.G.; Olson, D.E.; Hammersley, J.R.; et al. Quantitative RT-PCR measurement of cytochromes p450 1A1, 1B1, and 2B7, microsomal epoxide hydrolase, and NADPH oxidoreductase expression in lung cells of smokers and nonsmokers. *Am. J. Respir. Cell Mol. Biol.* **1997**, *17*, 114–124. [CrossRef] [PubMed]
43. Macé, K.; Bowman, E.D.; Vautravers, P.; Shields, P.G.; Harris, C.C.; Pfeifer, A.M. Characterisation of xenobiotic-metabolising enzyme expression in human bronchial mucosa and peripheral lung tissues. *Eur. J. Cancer* **1998**, *34*, 914–920. [CrossRef]
44. Guidice, J.M.; Marez, D.; Sabbagh, N.; Legrand-Andreoletti, M.; Spire, C.; Alcaïde, E.; Lafitte, J.J.; Broly, F. Evidence for CYP2D6 expression in human lung. *Biochem. Biophys. Res. Commun.* **1997**, *241*, 79–85. [CrossRef]
45. Thum, T.; Borlak, J. Gene expression in distinct regions of the heart. *Lancet* **2000**, *355*, 979–983. [CrossRef]
46. Bièche, I.; Narjoz, C.; Asselah, T.; Vacher, S.; Marcellin, P.; Lidereau, R.; Beaune, P.; de Waziers, I. Reverse transcriptase-PCR quantification of mRNA levels from cytochrome (CYP)1, CYP2 and CYP3 families in 22 different human tissues. *Pharmacogenet. Genomics* **2007**, *17*, 731–742. [CrossRef]
47. Minamiyama, Y.; Takemura, S.; Akiyama, T.; Imaoka, S.; Inoue, M.; Funae, Y.; Okada, S. Isoforms of cytochrome P450 on organic nitrate-derived nitric oxide release in human heart vessels. *FEBS Lett.* **1999**, *452*, 165–169. [CrossRef]
48. Delozier, T.C.; Kissling, G.E.; Coulter, S.J.; Dai, D.; Foley, J.F.; Bradbury, J.A.; Murphy, E.; Steenbergen, C.; Zeldin, D.C.; Goldstein, J.A. Detection of human CYP2C8, CYP2C9, and CYP2J2 in cardiovascular tissues. *Drug Metab. Dispos.* **2007**, *35*, 682–688. [CrossRef]
49. Frömel, T.; Kohlstedt, K.; Popp, R.; Yin, X.; Awwad, K.; Barbosa-Sicard, E.; Thomas, A.C.; Lieberz, R.; Mayr, M.; Fleming, I. Cytochrome P4502S1: A novel monocyte/macrophage fatty acid epoxygenase in human atherosclerotic plaques. *Basic Res. Cardiol.* **2013**, *108*, 319. [CrossRef]
50. Ande, A.; McArthur, C.; Ayuk, L.; Awasom, C.; Achu, P.N.; Njinda, A.; Sinha, N.; Rao, P.S.S.; Agudelo, M.; Nookala, A.R.; et al. Effect of mild-to-moderate smoking on viral load, cytokines, oxidative stress, and cytochrome P450 enzymes in HIV-infected individuals. *PLoS ONE* **2015**, *10*, e0122402. [CrossRef]

51. Rao, P.; Ande, A.; Sinha, N.; Kumar, A.; Kumar, S. Effects of Cigarette Smoke Condensate on Oxidative Stress, Apoptotic Cell Death, and HIV Replication in Human Monocytic Cells. *PLoS ONE* **2016**, *11*, e0155791. [CrossRef]
52. Nagai, F.; Hiyoshi, Y.; Sugimachi, K.; Tamura, H.-O. Cytochrome P450 (CYP) expression in human myeloblastic and lymphoid cell lines. *Biol. Pharm. Bull.* **2002**, *25*, 383–385. [CrossRef]
53. Haining, R.L.; Nichols-Haining, M. Cytochrome P450-catalyzed pathways in human brain: Metabolism meets pharmacology or old drugs with new mechanism of action? *Pharmacol. Ther.* **2007**, *113*, 537–545. [CrossRef] [PubMed]
54. Seden, K.; Dickinson, L.; Khoo, S.; Back, D. Grapefruit-drug interactions. *Drugs* **2010**, *70*, 2373–2407. [CrossRef] [PubMed]
55. Zheng, S.; Tasnif, Y.; Hebert, M.F.; Davis, C.L.; Shitara, Y.; Calamia, J.C.; Lin, Y.S.; Shen, D.D.; Thummel, K.E. Measurement and compartmental modeling of the effect of CYP3A5 gene variation on systemic and intrarenal tacrolimus disposition. *Clin. Pharmacol. Ther.* **2012**, *92*, 737–745. [CrossRef] [PubMed]
56. McCune, J.S.; Risler, L.J.; Phillips, B.R.; Thummel, K.E.; Blough, D.; Shen, D.D. Contribution of CYP3A5 to hepatic and renal ifosfamide N-dechloroethylation. *Drug Metab. Dispos.* **2005**, *33*, 1074–1081. [CrossRef]
57. Elfaki, I.; Mir, R.; Almutairi, F.M.; Duhier, F.M.A. Cytochrome P450: Polymorphisms and Roles in Cancer, Diabetes and Atherosclerosis. *Asian Pac. J. Cancer Prev.* **2018**, *19*, 2057–2070.
58. Ranjit, S.; Sinha, N.; Kodidela, S.; Kumar, S. Benzo(a)pyrene in Cigarette Smoke Enhances HIV-1 Replication through NF-κB Activation via CYP-Mediated Oxidative Stress Pathway. *Sci. Rep.* **2018**, *8*, 10394. [CrossRef]
59. Tyndale, R.F.; Sellers, E.M. Variable CYP2A6-mediated nicotine metabolism alters smoking behavior and risk. *Drug Metab. Dispos.* **2001**, *29*, 548–552.
60. Kubota, T.; Nakajima-Taniguchi, C.; Fukuda, T.; Funamoto, M.; Maeda, M.; Tange, E.; Ueki, R.; Kawashima, K.; Hara, H.; Fujio, Y.; et al. CYP2A6 polymorphisms are associated with nicotine dependence and influence withdrawal symptoms in smoking cessation. *Pharmacogenom. J.* **2006**, *6*, 115–119. [CrossRef]
61. Zhang, Y.; El-Sikhry, H.; Chaudhary, K.R.; Batchu, S.N.; Shayeganpour, A.; Jukar, T.O.; Bradbury, J.A.; Graves, J.P.; DeGraff, L.M.; Myers, P.; et al. Overexpression of CYP2J2 provides protection against doxorubicin-induced cardiotoxicity. *Am. J. Physiol. Heart Circ. Physiol.* **2009**, *297*, H37–H46. [CrossRef] [PubMed]
62. Wang, J.-F.; Yang, Y.; Sullivan, M.F.; Min, J.; Cai, J.; Zeldin, D.C.; Xiao, Y.-F.; Morgan, J.P. Induction of cardiac cytochrome p450 in cocaine-treated mice. *Exp. Biol. Med.* **2002**, *227*, 182–188. [CrossRef] [PubMed]
63. Jakubzick, C.V.; Randolph, G.J.; Henson, P.M. Monocyte differentiation and antigen-presenting functions. *Nat. Rev. Immunol.* **2017**, *17*, 349–362. [CrossRef] [PubMed]
64. Kurotaki, D.; Sasaki, H.; Tamura, T. Transcriptional control of monocyte and macrophage development. *Int. Immunol.* **2017**, *29*, 97–107. [CrossRef] [PubMed]
65. McMillan, D.M.; Tyndale, R.F. CYP-mediated drug metabolism in the brain impacts drug response. *Pharmacol. Ther.* **2018**, *184*, 189–200. [CrossRef] [PubMed]
66. Miksys, S.; Tyndale, R.F. Cytochrome P450-mediated drug metabolism in the brain. *J. Psychiatry Neurosci.* **2013**, *38*, 152–163. [CrossRef] [PubMed]
67. Dutheil, F.; Beaune, P.; Loriot, M.-A. Xenobiotic metabolizing enzymes in the central nervous system: Contribution of cytochrome P450 enzymes in normal and pathological human brain. *Biochimie* **2008**, *90*, 426–436. [CrossRef]
68. Ferguson, C.S.; Tyndale, R.F. Cytochrome P450 enzymes in the brain: Emerging evidence of biological significance. *Trends Pharmacol. Sci.* **2011**, *32*, 708–714. [CrossRef]
69. Howard, L.A.; Miksys, S.; Hoffmann, E.; Mash, D.; Tyndale, R.F. Brain CYP2E1 is induced by nicotine and ethanol in rat and is higher in smokers and alcoholics. *Br. J. Pharmacol.* **2003**, *138*, 1376–1386. [CrossRef]
70. Meyer, R.P.; Gehlhaus, M.; Knoth, R.; Volk, B. Expression and function of cytochrome p450 in brain drug metabolism. *Curr. Drug Metab.* **2007**, *8*, 297–306. [CrossRef]
71. Dauchy, S.; Dutheil, F.; Weaver, R.J.; Chassoux, F.; Daumas-Duport, C.; Couraud, P.-O.; Scherrmann, J.-M.; De Waziers, I.; Declèves, X. ABC transporters, cytochromes P450 and their main transcription factors: Expression at the human blood-brain barrier. *J. Neurochem.* **2008**, *107*, 1518–1528. [CrossRef]
72. Falero-Perez, J.; Sorenson, C.M.; Sheibani, N. Cyp1b1-deficient retinal astrocytes are more proliferative and migratory and are protected from oxidative stress and inflammation. *Am. J. Physiol. Cell Physiol.* **2019**, *316*, C767–C781. [CrossRef]

73. Navarro-Mabarak, C.; Camacho-Carranza, R.; Espinosa-Aguirre, J.J. Cytochrome P450 in the central nervous system as a therapeutic target in neurodegenerative diseases. *Drug Metab. Rev.* **2018**, *50*, 95–108. [CrossRef]
74. Alexaki, A.; Liu, Y.; Wigdahl, B. Cellular reservoirs of HIV-1 and their role in viral persistence. *Curr. HIV Res.* **2008**, *6*, 388–400. [CrossRef]
75. Burman, W.; Orr, L. Carbamazepine toxicity after starting combination antiretroviral therapy including ritonavir and efavirenz. *AIDS* **2000**, *14*, 2793–2794. [CrossRef]
76. Miksys, S.; Rao, Y.; Hoffmann, E.; Mash, D.C.; Tyndale, R.F. Regional and cellular expression of CYP2D6 in human brain: Higher levels in alcoholics. *J. Neurochem.* **2002**, *82*, 1376–1387. [CrossRef]
77. Miksys, S.; Tyndale, R.F. Nicotine induces brain CYP enzymes: Relevance to Parkinson's disease. *J. Neural Transm. Suppl.* **2006**, 177–180.
78. Miksys, S.; Lerman, C.; Shields, P.G.; Mash, D.C.; Tyndale, R.F. Smoking, alcoholism and genetic polymorphisms alter CYP2B6 levels in human brain. *Neuropharmacology* **2003**, *45*, 122–132. [CrossRef]
79. Mann, A.; Miksys, S.; Lee, A.; Mash, D.C.; Tyndale, R.F. Induction of the drug metabolizing enzyme CYP2D in monkey brain by chronic nicotine treatment. *Neuropharmacology* **2008**, *55*, 1147–1155. [CrossRef]
80. De Toro, J.; Herschlik, L.; Waldner, C.; Mongini, C. Emerging roles of exosomes in normal and pathological conditions: New insights for diagnosis and therapeutic applications. *Front. Immunol.* **2015**, *6*, 203. [CrossRef]
81. Hessvik, N.P.; Llorente, A. Current knowledge on exosome biogenesis and release. *Cell. Mol. Life Sci.* **2018**, *75*, 193–208. [CrossRef]
82. Simons, M.; Raposo, G. Exosomes–vesicular carriers for intercellular communication. *Curr. Opin. Cell Biol.* **2009**, *21*, 575–581. [CrossRef]
83. Conde-Vancells, J.; Gonzalez, E.; Lu, S.C.; Mato, J.M.; Falcon-Perez, J.M. Overview of extracellular microvesicles in drug metabolism. *Expert Opin. Drug Metab. Toxicol.* **2010**, *6*, 543–554. [CrossRef]
84. Palomo, L.; Mleczko, J.E.; Azkargorta, M.; Conde-Vancells, J.; González, E.; Elortza, F.; Royo, F.; Falcon-Perez, J.M. Abundance of Cytochromes in Hepatic Extracellular Vesicles Is Altered by Drugs Related with Drug-Induced Liver Injury. *Hepatol. Commun.* **2018**, *2*, 1064–1079. [CrossRef]
85. Royo, F.; Falcon-Perez, J.M. Liver extracellular vesicles in health and disease. *J. Extracell Vesicles* **2012**, *1*. [CrossRef]
86. Rowland, A.; Ruanglertboon, W.; van Dyk, M.; Wijayakumara, D.; Wood, L.S.; Meech, R.; Mackenzie, P.I.; Rodrigues, A.D.; Marshall, J.-C.; Sorich, M.J. Plasma extracellular nanovesicle (exosome)-derived biomarkers for drug metabolism pathways: A novel approach to characterize variability in drug exposure. *Br. J. Clin. Pharmacol.* **2019**, *85*, 216–226. [CrossRef]
87. Rahman, M.A.; Kodidela, S.; Sinha, N.; Haque, S.; Shukla, P.K.; Rao, R.; Kumar, S. Plasma exosomes exacerbate alcohol- and acetaminophen-induced toxicity via CYP2E1 pathway. *Sci. Rep.* **2019**, *9*, 6571. [CrossRef]
88. Cho, Y.-E.; Mezey, E.; Hardwick, J.P.; Salem, N.; Clemens, D.L.; Song, B.-J. Increased ethanol-inducible cytochrome P450-2E1 and cytochrome P450 isoforms in exosomes of alcohol-exposed rodents and patients with alcoholism through oxidative and endoplasmic reticulum stress. *Hepatol. Commun.* **2017**, *1*, 675–690. [CrossRef]
89. Cho, Y.-E.; Song, B.-J.; Akbar, M.; Baek, M.-C. Extracellular vesicles as potential biomarkers for alcohol- and drug-induced liver injury and their therapeutic applications. *Pharmacol. Ther.* **2018**, *187*, 180–194. [CrossRef]
90. Conde-Vancells, J.; Rodriguez-Suarez, E.; Embade, N.; Gil, D.; Matthiesen, R.; Valle, M.; Elortza, F.; Lu, S.C.; Mato, J.M.; Falcon-Perez, J.M. Characterization and comprehensive proteome profiling of exosomes secreted by hepatocytes. *J. Proteome Res.* **2008**, *7*, 5157–5166. [CrossRef]
91. Rodríguez-Suárez, E.; Gonzalez, E.; Hughes, C.; Conde-Vancells, J.; Rudella, A.; Royo, F.; Palomo, L.; Elortza, F.; Lu, S.C.; Mato, J.M.; et al. Quantitative proteomic analysis of hepatocyte-secreted extracellular vesicles reveals candidate markers for liver toxicity. *J. Proteomics* **2014**, *103*, 227–240. [CrossRef]
92. Haque, S.; Sinha, N.; Ranjit, S.; Midde, N.M.; Kashanchi, F.; Kumar, S. Monocyte-derived exosomes upon exposure to cigarette smoke condensate alter their characteristics and show protective effect against cytotoxicity and HIV-1 replication. *Sci. Rep.* **2017**, *7*, 16120. [CrossRef]
93. Rahman, M.A.; Patters, B.J.; Kodidela, S.; Kumar, S. Extracellular Vesicles: Intercellular Mediators in Alcohol-Induced Pathologies. *J. Neuroimmune Pharmacol.* **2019**. [CrossRef]
94. Verma, V.K.; Li, H.; Wang, R.; Hirsova, P.; Mushref, M.; Liu, Y.; Cao, S.; Contreras, P.C.; Malhi, H.; Kamath, P.S.; et al. Alcohol stimulates macrophage activation through caspase-dependent hepatocyte derived release of CD40L containing extracellular vesicles. *J. Hepatol.* **2016**, *64*, 651–660. [CrossRef]

95. Ranjit, S.; Patters, B.J.; Gerth, K.A.; Haque, S.; Choudhary, S.; Kumar, S. Potential neuroprotective role of astroglial exosomes against smoking-induced oxidative stress and HIV-1 replication in the central nervous system. *Expert Opin. Ther. Targets* **2018**, *22*, 703–714. [CrossRef]
96. Ryu, A.-R.; Kim, D.H.; Kim, E.; Lee, M.Y. The Potential Roles of Extracellular Vesicles in Cigarette Smoke-Associated Diseases. *Oxid. Med. Cell Longev.* **2018**, *2018*, 4692081. [CrossRef]
97. Batrakova, E.V.; Kim, M.S. Using exosomes, naturally-equipped nanocarriers, for drug delivery. *J. Control. Release* **2015**, *219*, 396–405. [CrossRef]
98. Giannini, E.G.; Testa, R.; Savarino, V. Liver enzyme alteration: A guide for clinicians. *CMAJ* **2005**, *172*, 367–379. [CrossRef]
99. Xu, J.; Ma, H.-Y.; Liang, S.; Sun, M.; Karin, G.; Koyama, Y.; Hu, R.; Quehenberger, O.; Davidson, N.O.; Dennis, E.A.; et al. The role of human cytochrome P450 2E1 in liver inflammation and fibrosis. *Hepatol. Commun.* **2017**, *1*, 1043–1057. [CrossRef]
100. Yang, X.; Weng, Z.; Mendrick, D.L.; Shi, Q. Circulating extracellular vesicles as a potential source of new biomarkers of drug-induced liver injury. *Toxicol. Lett.* **2014**, *225*, 401–406. [CrossRef]
101. Teschke, R. Microsomal Ethanol-Oxidizing System: Success Over 50 Years and an Encouraging Future. *Alcohol. Clin. Exp. Res.* **2019**, *43*, 386–400. [CrossRef]
102. Port, J.L.; Yamaguchi, K.; Du, B.; De Lorenzo, M.; Chang, M.; Heerdt, P.M.; Kopelovich, L.; Marcus, C.B.; Altorki, N.K.; Subbaramaiah, K.; et al. Tobacco smoke induces CYP1B1 in the aerodigestive tract. *Carcinogenesis* **2004**, *25*, 2275–2281. [CrossRef]
103. Kung, S.H.; Lund, S.; Murarka, A.; McPhee, D.; Paddon, C.J. Approaches and Recent Developments for the Commercial Production of Semi-synthetic Artemisinin. *Front Plant Sci.* **2018**, *9*, 87. [CrossRef]
104. Endo, A.; Kuroda, M.; Tsujita, Y. ML-236A, ML-236B, and ML-236C, new inhibitors of cholesterogenesis produced by Penicillium citrinium. *J. Antibiot.* **1976**, *29*, 1346–1348. [CrossRef]
105. McLean, K.J.; Hans, M.; Meijrink, B.; van Scheppingen, W.B.; Vollebregt, A.; Tee, K.L.; van der Laan, J.-M.; Leys, D.; Munro, A.W.; van den Berg, M.A. Single-step fermentative production of the cholesterol-lowering drug pravastatin via reprogramming of Penicillium chrysogenum. *Proc. Natl. Acad. Sci. USA* **2015**, *112*, 2847–2852. [CrossRef]
106. Boukouris, S.; Mathivanan, S. Exosomes in bodily fluids are a highly stable resource of disease biomarkers. *Proteomics Clin. Appl.* **2015**, *9*, 358–367. [CrossRef]
107. Sil, S.; Dagur, R.S.; Liao, K.; Peeples, E.S.; Hu, G.; Periyasamy, P.; Buch, S. Strategies for the use of Extracellular Vesicles for the Delivery of Therapeutics. *J. Neuroimmune Pharmacol.* **2019**. [CrossRef]
108. Familtseva, A.; Jeremic, N.; Tyagi, S.C. Exosomes: Cell-created drug delivery systems. *Mol. Cell. Biochem.* **2019**, *459*, 1–6. [CrossRef]
109. Vredenburg, G.; den Braver-Sewradj, S.; van Vugt-Lussenburg, B.M.A.; Vermeulen, N.P.E.; Commandeur, J.N.M.; Vos, J.C. Activation of the anticancer drugs cyclophosphamide and ifosfamide by cytochrome P450 BM3 mutants. *Toxicol. Lett.* **2015**, *232*, 182–192. [CrossRef]
110. Chen, L.; Waxman, D.J. Cytochrome P450 gene-directed enzyme prodrug therapy (GDEPT) for cancer. *Curr. Pharm. Des.* **2002**, *8*, 1405–1416. [CrossRef]
111. Zhang, J.; Kale, V.; Chen, M. Gene-directed enzyme prodrug therapy. *AAPS J.* **2015**, *17*, 102–110. [CrossRef]
112. Satoh, J.; Kino, Y.; Asahina, N.; Takitani, M.; Miyoshi, J.; Ishida, T.; Saito, Y. TMEM119 marks a subset of microglia in the human brain. *Neuropathology* **2016**, *36*, 39–49. [CrossRef]
113. Kodidela, S.; Pradhan, S.C.; Dubashi, B.; Basu, D. Interethnic differences in single and haplotype structures of folylpolyglutamate synthase and gamma-glutamyl hydrolase variants and their influence on disease susceptibility to acute lymphoblastic leukemia in the Indian population: An exploratory study. *Ind. J. Med. Paediatr. Oncol.* **2018**, *39*, 331. [CrossRef]
114. Thakkar, D.N.; Kodidela, S.; Sandhiya, S.; Dubashi, B.; Dkhar, S.A. A Polymorphism Located Near PMAIP1/Noxa Gene Influences Susceptibility to Hodgkin Lymphoma Development in South India. *Asian Pac. J. Cancer Prev.* **2017**, *18*, 2477–2483.
115. Zhou, S.-F.; Liu, J.-P.; Chowbay, B. Polymorphism of human cytochrome P450 enzymes and its clinical impact. *Drug Metab. Rev.* **2009**, *41*, 89–295. [CrossRef]
116. Kodidela, S.; Suresh Chandra, P.; Dubashi, B. Pharmacogenetics of methotrexate in acute lymphoblastic leukaemia: Why still at the bench level? *Eur. J. Clin. Pharmacol.* **2014**, *70*, 253–260. [CrossRef]

117. Kodidela, S.; Pradhan, S.C.; Dubashi, B.; Basu, D. Influence of dihydrofolate reductase gene polymorphisms rs408626 (-317A>G) and rs442767 (-680C>A) on the outcome of methotrexate-based maintenance therapy in South Indian patients with acute lymphoblastic leukemia. *Eur. J. Clin. Pharmacol.* **2015**, *71*, 1349–1358. [CrossRef]
118. LLerena, A.; Naranjo, M.E.G.; Rodrigues-Soares, F.; Penas-LLedó, E.M.; Fariñas, H.; Tarazona-Santos, E. Interethnic variability of CYP2D6 alleles and of predicted and measured metabolic phenotypes across world populations. *Expert Opin. Drug Metab. Toxicol.* **2014**, *10*, 1569–1583. [CrossRef]
119. Zhou, S.-F. Polymorphism of human cytochrome P450 2D6 and its clinical significance: Part I. *Clin. Pharmacokinet* **2009**, *48*, 689–723. [CrossRef]
120. Langman, L.; van Gelder, T.; van Schaik, R.H.N. Chapter 5—Pharmacogenomics aspect of immunosuppressant therapy. In *Personalized Immunosuppression in Transplantation*; Oellerich, M., Dasgupta, A., Eds.; Elsevier: San Diego, CA, USA, 2016; pp. 109–124. ISBN 978-0-12-800885-0.
121. Lai, C.P.; Mardini, O.; Ericsson, M.; Prabhakar, S.; Maguire, C.; Chen, J.W.; Tannous, B.A.; Breakefield, X.O. Dynamic biodistribution of extracellular vesicles in vivo using a multimodal imaging reporter. *ACS Nano* **2014**, *8*, 483–494. [CrossRef]
122. EV-TRACK Consortium; Van Deun, J.; Mestdagh, P.; Agostinis, P.; Akay, Ö.; Anand, S.; Anckaert, J.; Martinez, Z.A.; Baetens, T.; Beghein, E.; et al. EV-TRACK: Transparent reporting and centralizing knowledge in extracellular vesicle research. *Nat. Methods* **2017**, *14*, 228–232. [CrossRef]
123. Lötvall, J.; Hill, A.F.; Hochberg, F.; Buzás, E.I.; Di Vizio, D.; Gardiner, C.; Gho, Y.S.; Kurochkin, I.V.; Mathivanan, S.; Quesenberry, P.; et al. Minimal experimental requirements for definition of extracellular vesicles and their functions: A position statement from the International Society for Extracellular Vesicles. *J. Extracell. Vesicles* **2014**, *3*, 26913. [CrossRef]
124. Suk, J.S.; Xu, Q.; Kim, N.; Hanes, J.; Ensign, L.M. PEGylation as a strategy for improving nanoparticle-based drug and gene delivery. *Adv. Drug Deliv. Rev.* **2016**, *99*, 28–51. [CrossRef]
125. Matsumoto, A.; Takahashi, Y.; Nishikawa, M.; Sano, K.; Morishita, M.; Charoenviriyakul, C.; Saji, H.; Takakura, Y. Role of Phosphatidylserine-Derived Negative Surface Charges in the Recognition and Uptake of Intravenously Injected B16BL6-Derived Exosomes by Macrophages. *J. Pharm. Sci.* **2017**, *106*, 168–175. [CrossRef]
126. Kamerkar, S.; LeBleu, V.S.; Sugimoto, H.; Yang, S.; Ruivo, C.F.; Melo, S.A.; Lee, J.J.; Kalluri, R. Exosomes facilitate therapeutic targeting of oncogenic KRAS in pancreatic cancer. *Nature* **2017**, *546*, 498–503. [CrossRef]

© 2019 by the authors. Licensee MDPI, Basel, Switzerland. This article is an open access article distributed under the terms and conditions of the Creative Commons Attribution (CC BY) license (http://creativecommons.org/licenses/by/4.0/).

Article

Comparative Analyses of Cytochrome P450s and Those Associated with Secondary Metabolism in *Bacillus* Species

Bongumusa Comfort Mthethwa [1], Wanping Chen [2], Mathula Lancelot Ngwenya [1], Abidemi Paul Kappo [1], Puleng Rosinah Syed [3], Rajshekhar Karpoormath [3], Jae-Hyuk Yu [4], David R. Nelson [5,*] and Khajamohiddin Syed [1,*]

1. Department of Biochemistry and Microbiology, Faculty of Science and Agriculture, University of Zululand, KwaDlangezwa 3886, South Africa; 07bcomfort@gmail.com (B.C.M.); NgwenyaM@unizulu.ac.za (M.L.N.); KappoA@unizulu.ac.za (A.P.K.)
2. College of Food Science and Technology, Huazhong Agricultural University, Wuhan 430070, China; chenwanping@mail.hzau.edu.cn
3. Department of Pharmaceutical Chemistry, College of Health Sciences, University of KwaZulu-Natal, Durban 4000, South Africa; prosinah@gmail.com (P.R.S.); Karpoormath@ukzn.ac.za (R.K.)
4. Department of Bacteriology, University of Wisconsin-Madison, 3155 MSB, 1550 Linden Drive, Madison, WI 53706, USA; jyu1@wisc.edu
5. Department of Microbiology, Immunology and Biochemistry, University of Tennessee Health Science Center, Memphis, TN 38163, USA
* Correspondence: drnelson1@gmail.com (D.R.N.); khajamohiddinsyed@gmail.com (K.S.)

Received: 25 September 2018; Accepted: 16 October 2018; Published: 16 November 2018

Abstract: Cytochrome P450 monooxygenases (CYPs/P450s) are among the most catalytically-diverse enzymes, capable of performing enzymatic reactions with chemo-, regio-, and stereo-selectivity. Our understanding of P450s' role in secondary metabolite biosynthesis is becoming broader. Among bacteria, *Bacillus* species are known to produce secondary metabolites, and recent studies have revealed the presence of secondary metabolite biosynthetic gene clusters (BGCs) in these species. However, a comprehensive comparative analysis of P450s and P450s involved in the synthesis of secondary metabolites in *Bacillus* species has not been reported. This study intends to address these two research gaps. *In silico* analysis of P450s in 128 *Bacillus* species revealed the presence of 507 P450s that can be grouped into 13 P450 families and 28 subfamilies. No P450 family was found to be conserved in *Bacillus* species. *Bacillus* species were found to have lower numbers of P450s, P450 families and subfamilies, and a lower P450 diversity percentage compared to mycobacterial species. This study revealed that a large number of P450s (112 P450s) are part of different secondary metabolite BGCs, and also identified an association between a specific P450 family and secondary metabolite BGCs in *Bacillus* species. This study opened new vistas for further characterization of secondary metabolite BGCs, especially P450s in *Bacillus* species.

Keywords: Antibiotics; *Bacillus*; biosynthetic gene clusters; comparative analysis; cytochrome P450 monooxygenase; *Mycobacterium*; P450 diversity percentage; P450 profiling; secondary metabolites

1. Introduction

Cytochrome P450 monooxygenases, also known as CYPs/P450s, are undoubtedly among the most catalytically-diverse enzymes, performing enzymatic reactions with chemo-, regio- and stereo-selectivity [1–6]. The catalytic diversity combined with chemo-, regio- and stereo-specific oxidation of substrates exerted by P450s are used in diverse biotechnological applications ranging from drug discovery to bioethanol production and synthesis of different secondary metabolites [7–12].

P450s are heme-thiolate proteins ubiquitously found in species belonging to different biological kingdoms, including non-living entities such as viruses [13,14]. In bacteria, P450s have been found to play a key role in enzymatic reactions, leading to the biosynthesis of physiological compounds or the biodegradation of xenobiotics [9,11,15,16].

P450s' role in the synthesis of a diverse array of secondary metabolites has been thoroughly reviewed [8,12]. Secondary metabolites are natural products that are widely-used in human and veterinary medicine, agriculture, and manufacturing, and are known to mediate a variety of microbe-host and microbe-microbe interactions [17]. P450s were found to play a key role in the synthesis of different secondary metabolites, including terpenes, alkaloids, shikimates, polyketides, and peptides [12]. The coding sequences (genes) of enzymes involved in the synthesis of different secondary metabolites, including P450s, were found to be part of gene clusters named biosynthetic gene clusters (BGCs) [17]. Bacterial species have been found to have more than 1000 different types of BGCs involved in the synthesis of known and unknown secondary metabolites [17].

Among bacteria, species belonging to the genus *Bacillus* are ubiquitously present in the biosphere, and are well known for their distinct features with one common characteristic, i.e., making dormant endospores during unfavorable growth conditions [18,19]. Applications of *Bacillus* species across different spectra have been well explored in the industrial, agricultural, and ecological fields, and by academics, against a backdrop of being a well-known human pathogen [18,19]. Comprehensive *in silico* studies detailing *Bacillus* species' ability to produce different secondary metabolites and different types of BGCs have frequently been reported [20,21]. Analysis of 1566 *Bacillus* species' genomes revealed the presence of 20,000 BGCs, most of which were found to produce known secondary metabolites that play a key role in the physiology and development of *Bacillus* species [21]. The study by Grubbs et al. [21] also reported that secondary metabolite alkylpyrones play a key role in inhibiting spore development in *Bacillus* species.

Despite comprehensive analysis of *Bacillus* species' secondary metabolite BGCs, P450s that are part of different BGCs have not been reported. Analysis of P450s in *Bacillus* species date back to 2009, when the authors performed a comparative analysis of P450s in 29 *Bacillus* species, and identified a few P450s belonging to a limited number of P450 families such as CYP102, CYP106, CYP107, CYP109, CYP134, CYP152, and CYP197 [22]. Among the P450 families identified in *Bacillus* species, the CYP102 P450 family has a special place in P450 research, being one of the most extensively-studied bacterial P450s, for its structural, functional, and biotechnological potential. Even the *Bacillus* species *B. megaterium* has become very famous owing to the identification of CYP102 P450 from this organism [23,24]. *In silico* comparative analysis of P450s in bacterial species is gaining momentum. Recently, a comprehensive comparative analysis of P450s in 60 mycobacterial species has been reported; the authors identified that mycobacteria possess a large number of P450s, and that different mycobacterial categories have characteristic P450 families that can be used as biomarkers to identify different mycobacterial species [25].

The current trend of whole-genome sequencing of organisms resulted in the genome sequencing of a large number of *Bacillus* species genomes. Quite a large number of *Bacillus* species genome sequences are available for public use at Kyoto Encyclopedia of Genes and Genomes—GenomeNet (KEGG) [26]. This gives us the opportunity to perform comprehensive a comparative analysis of P450s in *Bacillus* species as per international P450 nomenclature committee rules [27–29], and to identify P450s involved in the synthesis of different secondary metabolites. Here, we report genome data mining, annotation, phylogenetic and comparative analysis of P450s in 128 *Bacillus* species, including identification of P450s involved in the synthesis of different secondary metabolites. This study also reports comparative analysis of P450s between the genera *Bacillus* and *Mycobacterium*. Last but not least, a previous study reporting BGCs in *Bacillus* species did not clearly indicate BGCs on genomic DNA (gDNA) and plasmid DNA [21]; thus, in this study, gDNA and plasmid DNAs were individually subjected to BGC analysis.

2. Results and Discussion

2.1. Bacillus Species Have the Lowest Number of P450s

Genome data-mining and annotation of P450s in 128 *Bacillus* species revealed the presence of the lowest number of P450s in their genomes (Figures 1 and 2). In total, 507 P450s were found in 114 *Bacillus* species, where 14 species did not have P450s in their genomes (Figures 1 and 2). On average, four P450s were found in *Bacillus* species, with the highest number, 11, found in *Bacillus* subtilis subsp. *spizizenii* TU-B-10. The number of P450s found in *Bacillus* species is very low compared to mycobacterial species (60 species); the latter species have, on average, 35 P450s in their genomes [25]. The P450 count in *Bacillus* species, apart from *B. subtilis* subsp. *spizizenii* TU-B-10, which has 16, is as follows: 9 in 2 species, 8 in 4 species, 7 in 20 species, 6 in 21 species, 5 in 10 species, 4 in 9 species, 3 in 28 species, 2 in 7 species, and a single in 14 species (Figures 1 and 2). This indicates that most *Bacillus* species (28 species) have three P450s in their genomes. P450s identified in each *Bacillus* species and respective P450 sequences were presented in Supplementary datasets 1 and 2.

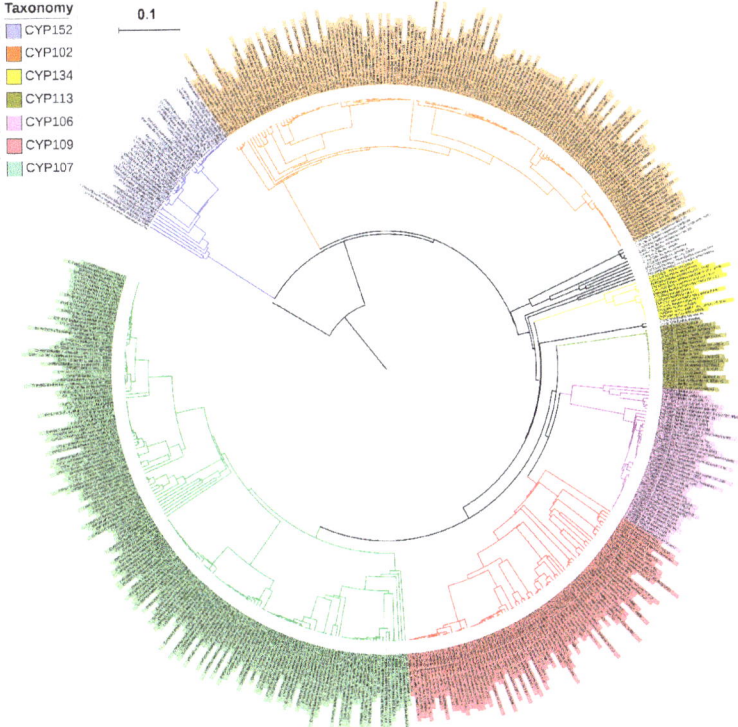

Figure 1. Phylogenetic analysis of *Bacillus* species P450s. Dominant P450 families were indicated in different colors. CYP51B1 from *Mycobacterium tuberculosis* H37Rv is used as an outgroup. A high-resolution phylogenetic tree is provided in the supplementary Figure S1.

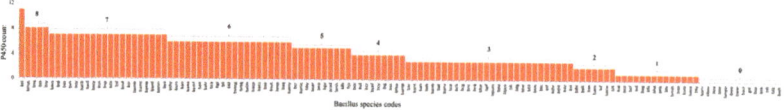

Figure 2. Comparative analysis of P450s in *Bacillus* species. The numbers next to bars indicate the number of P450s in *Bacillus* species. *Bacillus* species' names with respect to their codes can be found in Supplementary Dataset 1.

2.2. Bacillus Species Have the Lowest Number of P450 Families and Subfamilies'

As per the International P450 Nomenclature Committee rules [27–29], all 507 P450s found in 114 *Bacillus* species can be grouped into 13 P450 families and 28 subfamilies (Figures 3 and 4). Phylogenetic analysis of *Bacillus* P450s revealed P450s belonging to the same family grouped together, suggesting that the annotation of P450s in this study is accurate (Figure 1). The number of P450 families and subfamilies found in *Bacillus* species is lower compared to mycobacterial species (60 species), which have 77 P450 families and 132 subfamilies [25]. Because of the presence of the lowest number of P450 families, the P450 diversity percentage in *Bacillus* species was found to be lowest (3.9%) compared to mycobacterial species (72%) [25]. Among 13 P450 families, the CYP107 P450 family has the highest number of P450s (165 P450s) contributing 31.5% of 507 P450s (Figure 3), followed by CYP102 (143 P450s), CYP109 (79 P450s), CYP106 (40 P450s), CYP152 (36 P450s), CYP113 (18 P450s), CYP134 (13 P450s), CYP1756 (4 P450s), CYP1221 (3 P450s), CYP1179 and CYP223 (2 P450s), CYP1341 and CYP197 (single P450s) (Figure 3).

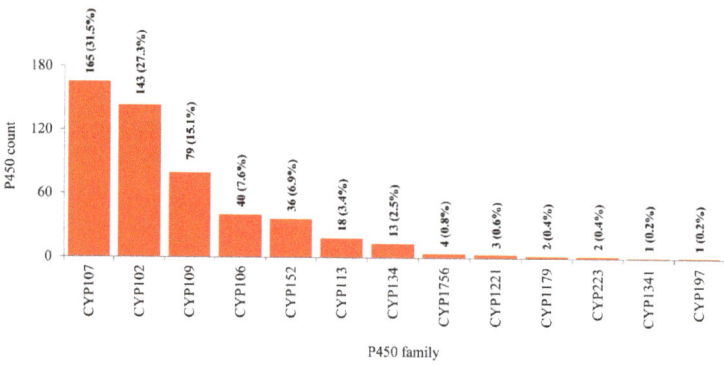

Figure 3. Comparative analysis of P450 families in *Bacillus* species. The numbers next to the family bar indicate the total number of P450s and percentage contribution (parenthesis) by a respective family to the total number of P450s. The data on the number of P450s in each P450 family in *Bacillus* species is presented in Table S1.

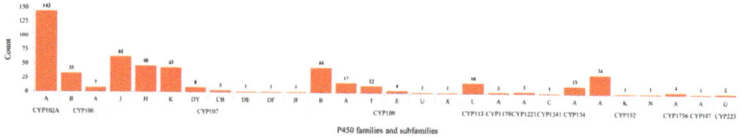

Figure 4. Comparative analysis of P450 subfamilies in *Bacillus* species. The numbers next to bars indicate the total number of members in a particular subfamily. Data on the number of P450s in each P450 subfamily in *Bacillus* species is presented in Table S1.

P450 subfamily analysis revealed that most P450 families have a single subfamily (Figure 4). Among P450 families, CYP107 has the highest number of P450 subfamilies (eight subfamilies) followed by CYP109 (six subfamilies), CYP152 (three subfamilies), and CYP106 (two subfamilies) (Figure 4).

The remaining nine P450 families, CYP102, CYP113, CYP1179, CYP1221, CYP1341, CYP134, CYP1756, CYP197, and CYP223, all have a single subfamily (Figure 4). It is interesting to note that the CYP102 P450 family, despite having the second largest number of P450s, has a single subfamily "A". Analysis of P450 subfamily profiles revealed that specific subfamilies are dominant in a particular family (Figure 4). Subfamily "J" is dominant in the CYP107 family, subfamily "B" is dominant in the CYP109 family and subfamily "A" is dominant in the CYP152 family (Figure 4). Analysis of P450 family profiles revealed that no P450 family is conserved across *Bacillus* species (Figure 5). Most *Bacillus* species have CYP102, CYP107, CYP109, CYP106, and CYP152 P450 families (Figure 5). CYP197, CYP223, and CYP1341 are present in a single *Bacillus* species (Supplementary Dataset 1).

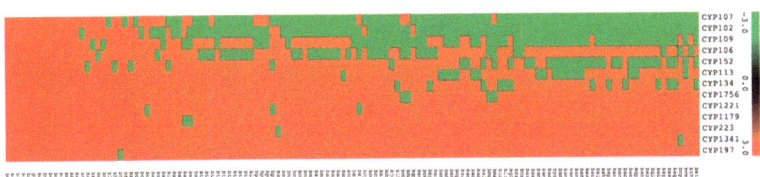

Figure 5. Heatmap of presence or absence of cytochrome P450 families in 128 species of *Bacillus*. The data have been represented as −3 for family presence (green) and 3 for family absence (red). A hundred and twenty-eight *Bacillus* species form the horizontal axis and CYP family numbers form the vertical axis. The respective data used in the generation of Figure 5 is presented in Supplementary Dataset 3.

2.3. Bacillus Species Have the Lowest Number of Secondary Metabolite BGCs

In order to identify P450s involved in the biosynthesis of secondary metabolites, the gDNA and plasmid DNA of each *Bacillus* species (Table S2) has been subjected to secondary metabolite BGCs analysis using anti-SMASH [30]. In total, 203 plasmids were identified in 60 *Bacillus* species (Table S2). Analysis of 128 *Bacillus* species genomes revealed the presence of 1098 and 26 secondary metabolite BGCs on gDNA and plasmid DNA, respectively (Figure 6 and Table S3).

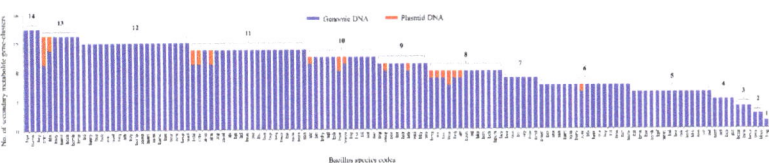

Figure 6. Comparative analysis of secondary metabolite BGCs in 128 *Bacillus* species (gDNA and plasmid DNA). Numbers next to bars indicate the number of secondary metabolite BGCs. Detailed analysis of secondary metabolite BGCs in each species is presented in Table S3.

The number of secondary metabolite BGCs varied from a maximum of 14 to one in *Bacillus* species gDNA. Interestingly, among 203 plasmid DNAs from 60 *Bacillus* species (Table S3), only 21 plasmid DNAs from 18 *Bacillus* species were found to have secondary metabolite BGCs (Figure 6 and Table S3). The number of secondary metabolites BGCs on plasmid DNAs varied from a maximum of four to one (Figure 6 and Table S3).

Analysis of types of secondary metabolite BGCs revealed the presence of 33 and 10 types of BGCs on gDNA and plasmid DNAs in *Bacillus* species (Figure 7 and Supplementary Dataset 4). The types of BGCs in individual *Bacillus* species varied from a maximum of 10 types to one (Figure 7). Among types of BGCs, Nonribosomal peptides secondary metabolite (Nrps) BGCs were dominant in *Bacillus* species, both on gDNA and plasmid DNAs (Figure 7).

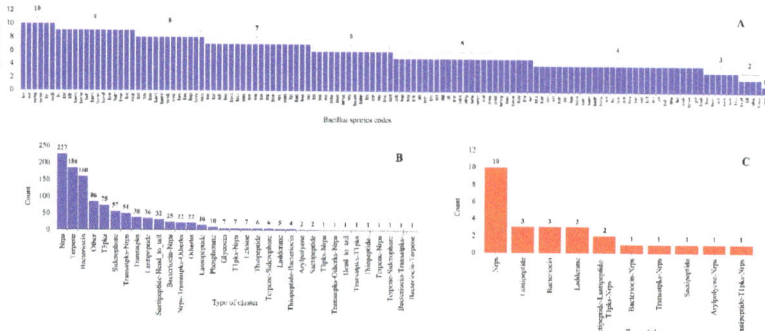

Figure 7. Comparative analysis of types of secondary metabolite BGCs in *Bacillus* species. (**A**) The number of types of secondary metabolite BGCs in *Bacillus* species. (**B,C**). Comparative analysis of types of secondary metabolite BGCs on gDNA and plasmid DNAs. Standard abbreviations representing secondary metabolite BGCs as indicated in anti-SMASH [30] were used in the figure.

Analysis of types of BGCs on gDNA and plasmid DNAs revealed the presence of seven types of common BGCs between gDNA and plasmid DNAs (Figure 7). This suggests that these plasmids might be involved in horizontal gene transfer of different BGCs among *Bacillus* species. It is important to note that horizontal gene transfer of BGCs is a common phenomenon among *Bacillus* species [21]. Interestingly, three distinct types of secondary metabolite gene clusters, namely Sactipeptide-Lantipeptide-T1pks-Nrps, Arylpolyene-Nrps, and Lantipeptide-T1pks-Nrps, were only identified on plasmid DNAs (Figure 7).

2.4. Large Number of P450s Found to Be Part of Secondary Metabolites BGCs in Bacillus Species

Among 507 P450s identified in 128 *Bacillus* species, 112 P450s (22%) from 50 *Bacillus* species were found to be part of secondary metabolite BGCs (Table 1). Among 13 P450 families, only seven families, namely CYP107, CYP113, CYP134, CYP152, CYP102, CYP109, and CYP1179, were found to be part of different secondary metabolite BGCs (Figure 8). P450 subfamily level analysis revealed that P450s belonging to the subfamilies H and K in the CYP107 family were part of secondary metabolite BGCs, despite subfamily J being dominant in that family (Figure 4). In the CYP152 family, P450s belonging to subfamily A were found to be part of the secondary metabolite BGCs. Analysis of P450s involving secondary metabolite biosynthesis revealed that P450s belonging to the CYP107 family are dominant by 61% (68 P450s) of all P450s (112 P450s) involved in secondary metabolite BGCs, followed by CYP113, CYP152, CYP102, CYP109, and CYP1179 (Figure 8). It is interesting to note that these P450 families are highly-populated in *Bacillus* species (Figures 3 and 5). This further supports the previous hypothesis that species populate specific P450s if they are useful in their adaptation to certain ecological niches or useful in their physiology [31–35]. Considering the large number of P450s, their widespread nature, and part in secondary metabolite BGCs, it can be hypothesized that the P450s belonging to the CYP107, CYP102, CYP109, CYP152, and CYP113 families play a key role in *Bacillus* species' physiology, including synthesis of different secondary metabolites. Despite secondary metabolite BGCs being found on plasmid DNA, no P450 was found to be part of these clusters. Analysis of association between P450 families and secondary metabolite BGCs revealed that CYP107 family P450s were mostly associated with BGCs Nrps-Transatpks-Otherks and Transatpks-Nrps; CYP113 family P450s are associated with Transatpks BGC, and CYP134 family P450s are associated with other, a putative gene cluster (Table 1).

Table 1. Identification of P450s that are involved in secondary metabolite BGCs in *Bacillus* species. BGCs in each species and P450 identified as part of a particular cluster are presented in Supplementary Dataset 5.

Species Name	Cluster Number	Type of BGCs	P450 Name
Bacillus subtilis subsp. *subtilis* 168	4	Nrps-Transatpks-Otherks	CYP134A1
	10	Other	CYP134A1
Bacillus subtilis subsp. *subtilis* RO-NN-1	3	Nrps-Transatpks-Otherks	CYP107K1
Bacillus subtilis subsp. *subtilis* BSP1	8	Transatpks-Otherks-Nrps	CYP107K1
Bacillus subtilis subsp. *subtilis* 6051-HGW	4	Nrps-Transatpks-Otherks	CYP107K1
	10	Other	CYP107K1
Bacillus subtilis subsp. *subtilis* BAB-1	4	Nrps-Transatpks-Otherks	CYP107K1
Bacillus subtilis subsp. *subtilis* AG1839	4	Nrps-Transatpks-Otherks	CYP107K1
	10	Other	CYP107K1
Bacillus subtilis subsp. *subtilis* JH642	4	Nrps-Transatpks-Otherks	CYP107K1
	10	Other	CYP134A1
Bacillus subtilis subsp. *subtilis* OH 131.1	1	Lantipeptide	CYP152A1
	4	Nrps-Transatpks-Otherks	CYP107K1
	9	Other	CYP134A1
Bacillus subtilis subsp. *spizizenii* W23	1	Phosphonate	CYP152A1
	4	Nrps-Transatpks-Otherks	CYP107K1
	10	Other	CYP134A1
Bacillus subtilis subsp. *spizizenii* TU-B-10	3	Nrps-Transatpks-Otherks	CYP107K1
	9	Other	CYP134A1
Bacillus subtilis BSn5	4	Other	CYP102A48
	8	Lantipeptide	CYP152A1
	11	Nrps-Transatpks-Otherks	CYP107K1
Bacillus subtilis QB928	4	Nrps-Transatpks-Otherks	CYP107K1
	10	Other	CYP134A1
Bacillus subtilis XF-1	4	Nrps-Transatpks-Otherks	CYP107K1
Bacillus subtilis PY79	4	Nrps-Transatpks-Otherks	CYP107K1
	9	Other	CYP134A1
Bacillus licheniformis ATCC 14580	7	Other	CYP134A5
Bacillus licheniformis DSM 13 = ATCC 14580	7	Other	CYP134A5
Bacillus paralicheniformis	10	Other	CYP134A5
Bacillus velezensis FZB42	5	Transatpks-Nrps	CYP107K3
	6	Transatpks-Nrps	CYP107H4
	9	Transatpks	CYP113L1
Bacillus velezensis CAU B946	5	Transatpks-Nrps	CYP107K3
	6	Transatpks-Nrps	CYP107H4
	9	Transatpks	CYP113L1
Bacillus velezensis YAU B9601-Y2	5	Transatpks	CYP107K3
	6	Transatpks-Nrps	CYP107K3
	7	Transatpks-Nrps	CYP107H4
	10	Transatpks	CYP113L1
Bacillus velezensis AS43.3	6	Transatpks	CYP107K3
	7	Transatpks-Nrps	CYP113L1
	10	Transatpks	CYP113L1
Bacillus velezensis UCMB5036	5	Transatpks-Nrps	CYP107K3
	6	Transatpks-Nrps	CYP107H4
	9	Bacteriocin-Nrps	CYP113L1
Bacillus velezensis UCMB5033	6	Transatpks-Nrps	CYP107K3
	7	Transatpks-Nrps	CYP107H4
	10	Transatpks	CYP113L1

Table 1. Cont.

Species Name	Cluster Number	Type of BGCs	P450 Name
Bacillus velezensis UCMB5113	7	Transatpks-Nrps	CYP107K3
	8	Transatpks-Nrps	CYP107H4
	11	Transatpks	CYP113L1
Bacillus velezensis NAU-B3	3	Transatpks	CYP113L1
	6	Transatpks-Nrps	CYP107H4
	7	Transatpks-Nrps	CYP107K3
Bacillus velezensis TrigoCor1448	5	Transatpks-Nrps	CYP107K3
	6	Transatpks-Nrps	CYP107H4
Bacillus velezensis SQR9	6	Transatpks-Nrps	CYP107K3
	7	Transatpks-Nrps	CYP107H4
	10	Transatpks	CYP113L1
Bacillus velezensis	6	Transatpks-Nrps	CYP107K3
	7	Transatpks-Nrps	CYP107H4
	10	Transatpks	CYP113L1
Bacillus amyloliquefaciens DSM 7	5	Transatpks-Nrps	CYP107K3
	6	Transatpks-Nrps	CYP107H2
Bacillus amyloliquefaciens TA208	7	Transatpks-Nrps	CYP107H2
	8	Transatpks-Nrps	CYP107K3
Bacillus amyloliquefaciens LL3	5	Transatpks-Nrps	CYP107K3
	6	Transatpks-Nrps	CYP107H2
Bacillus amyloliquefaciens XH7	7	Transatpks-Nrps	CYP107H2
	8	Transatpks-Nrps	CYP107K3
Bacillus amyloliquefaciens Y2	6	Transatpks-Nrps	CYP107K3
	7	Transatpks-Nrps	CYP107H4
	10	Transatpks	CYP113L1
Bacillus amyloliquefaciens IT-45	3	Transatpks	CYP113L1
	6	Transatpks-Nrps	CYP107H4
	7	Transatpks-Nrps	CYP107K3
Bacillus amyloliquefaciens CC178	6	Transatpks-Nrps	CYP107H4
	9	Transatpks	CYP113L1
Bacillus amyloliquefaciens LFB112	7	Transatpks-Nrps	CYP107K3
	8	Transatpks-Nrps	CYP107H4
	11	Transatpks	CYP113L1
Bacillus atrophaeus 1942	3	Nrps-Transatpks-Otherks	CYP107K2
	10	Nrps	CYP152A9
Bacillus atrophaeus NRS 1221A	3	Nrps-Transatpks-Otherks	CYP107K2
	10	Nrps	CYP152A9
Bacillus vallismortis	6	Transatpks-Nrps	CYP107K3
	7	Transatpks-Nrps	CYP107H4
	10	Transatpks	CYP113L1
Bacillus pumilus SH-B9	8	Nrps	CYP109B6
Bacillus sp. JS	4	Nrps-Transatpks-Otherks	CYP107K1
Bacillus sp. Pc3	1	Bacteriocin-Transatpks-Nrps	CYP107H4
	2	Transatpks-Nrps	CYP107K3
	10	Transatpks	CYP113L1
Bacillus sp. BH072	8	Transatpks-Nrps	CYP107K3
	9	Transatpks-Nrps	CYP107H4
	12	Transatpks	CYP107H4
Bacillus sp. YP1	4	Nrps-Transatpks-Otherks	CYP107K1
Bacillus sp. BS34A	4	Nrps-Transatpks-Otherks	CYP107K1
	10	Other	CYP134A1
Bacillus sp. LM 4-2	3	Nrps-Transatpks-Otherks	CYP107K1
	7	Other	CYP102A48
	9	Other	CYP134A1

Table 1. Cont.

Species Name	Cluster Number	Type of BGCs	P450 Name
Bacillus gibsonii	1	Nrps-Transatpks-Otherks	CYP107K1
	6	Other	CYP134A1
	9	Lantipeptide	CYP152A1
Bacillus xiamenensis	2	Nrps	CYP1179A4
Bacillus altitudinis	2	Nrps	CYP1179A4
	8	Nrps	CYP109B5
Bacillus sp. SDLI1	3	Transatpks-Nrps	CYP107H4
	4	Transatpks-Nrps	CYP107K3
	11	Transatpks	CYP113L1

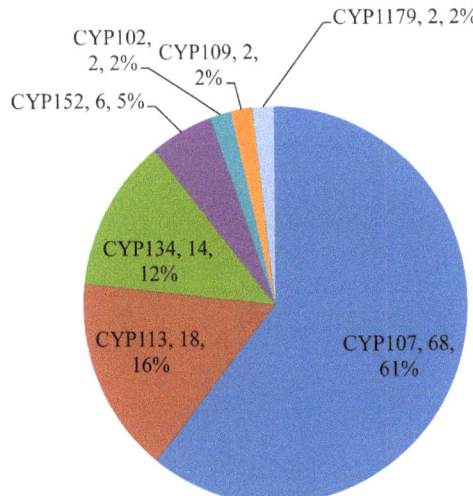

Figure 8. Comparative analysis of P450 families involved in secondary metabolite biosynthesis. The P450 family name, number of P450s and their percentage of the total number of 112 P450s are presented in the figure.

2.5. Bacillus P450s Indeed Involved in the Synthesis of Secondary Metabolites

Based on *in silico* analysis (in this study), seven P450 families, namely CYP107, CYP113, CYP134, CYP152, CYP102, CYP109, and CYP1179, were identified as part of secondary metabolite BGCs in *Bacillus* species (Figure 8). Functional data available for some P450s confirms that the predicted P450s, in this study, are indeed involved in biosynthesis of different secondary metabolites, and some of the P450 families, such as CYP105, CYP107, and CYP109, have been found to display highly-diverse functions [9,12,36]. CYP102A1 from *B. megaterium* [24,37,38] and CYP152A1 from *B. subtilis* [39,40] were found to be fatty acid hydroxylases. P450s belonging to the CYP106, CYP107, CYP109, and CYP134 families were found to hydroxylate different steroids, albeit with different substrate specificities [22]. CYP134A1 is involved in the synthesis of pulcherriminic acid, a natural product [41], and CYP107H1 (P450 biol) is involved in the synthesis of polyketides [42]. Based on functionally characterized homolog P450s from other organisms, CYP105, CYP107, and CYP109 family P450s have been found to be associated with the degradation and biotransformation of a diverse array of xenobiotics and secondary metabolites [36,43,44]. CYP113 P450s are involved in the biosynthesis of secondary metabolites such as erythromycin [45,46] and tylosin [47,48]. Despite CYP102 and CYP152 P450s being found in secondary metabolite BGCs (in this study), their role in secondary metabolites biosynthesis has not been yet elucidated.

3. Materials and Methods

3.1. Species and Database

In total, 128 *Bacillus* species genomes available for public use at KEGG (https://www.genome.jp/kegg-bin/show_organism?category=Bacillus) were used in this study (Table S4). *Bacillus* species used in this study, along with their names, species codes, and individual genome database links, were presented in Table S4.

3.2. Genome Data Mining and Annotation of P450s

P450 mining in *Bacillus* species was carried out following the methods described elsewhere [25]. Briefly, the whole proteome of *Bacillus* species was downloaded from the databases listed in Table S4, and subjected to the NCBI Batch Web CD-Search Tool (http://www.ncbi.nlm.nih.gov/Structure/bwrpsb/bwrpsb.cgi). Proteins that belong to a P450 superfamily were selected and based on the International P450 Nomenclature Committee rule; proteins with >40% identity and >55% identity were grouped under the same family and subfamily, respectively [27–29]. Proteins with less than 40% identity were assigned to a new P450 family.

3.3. Phylogenetic Analysis of P450s

The phylogenetic tree of *Bacillus* species P450s was built as described elsewhere [25], with slight modifications. Briefly, the *Bacillus* P450s protein sequences along with the outgroup *M. tuberculosis* CYP51B1 (Rv0764c) protein were aligned by MAFFT v6.864 [49], embedded on the Trex web server [50]. Then, the alignments were automatically subjected to tree inferring and optimization by the Trex web server. Finally, the best-inferred trees were visualized, colored, and generated by iTOL (http://itol.embl.de/about.cgi) [51].

3.4. P450 Diversity Percentage Analysis

P450 diversity percentage analysis was carried out as described elsewhere [25,34]. Briefly, the P450 diversity percentage in *Bacillus* species was measured as a percentage contribution of the number of P450 families in the total number of P450s.

3.5. Generation of P450 Profile Heat-Maps

The presence or absence of P450s in *Bacillus* species was shown with heat-maps generated using P450 family data. The data was represented as −3 for family presence (green) and 3 for family absence (red). A tab-delimited file was imported into Mev (Multi-experiment viewer) [52]. Hierarchical clustering using a Euclidean distance metric was used to cluster the data. A hundred and twenty-eight *Bacillus* species formed the horizontal axis (see Supplementary dataset 3 for codes) and CYP family numbers formed the vertical axis.

3.6. Secondary Metabolite BGCs Analysis

Individual *Bacillus* species genome ID and plasmids IDs from the various species databases (Table S2) were submitted to anti-SMASH [30] for identification of secondary metabolite BGCs. Results were downloaded both in the form of gene cluster sequences and Excel spreadsheets representing species-wise cluster information, and finally, P450s that are part of a specific gene cluster were identified. Standard gene cluster abbreviation terminology available at anti-SMASH database [30] was maintained in this study.

3.7. Comparative Analysis of P450s

Mycobacterial P450s were retrieved from a published article [25] and used for comparative analysis with *Bacillus* species P450s. P450 families and subfamilies and the P450 diversity percentage were compared between the genera *Mycobacterium* and *Bacillus*.

4. Conclusions

Comparative analysis of P450s in bacterial species is gaining momentum and the availability of a large number of bacterial genome sequences is fueling this process. This study is an attempt to perform a comprehensive comparative analysis of P450s and to identify the P450s involved in secondary metabolite synthesis in *Bacillus* species. Future work involves understanding the role of different *Bacillus* P450s, identified in this study, in the synthesis of various secondary metabolites.

Supplementary Materials: Supplementary materials can be found at http://www.mdpi.com/1422-0067/19/11/3623/s1.

Author Contributions: K.S. conceived and designed the experiments; all the authors were involved in performing the experiments, analysis of the data and writing of the manuscript. All the authors reviewed and approved the manuscript.

Funding: Khajamohiddin Syed expresses sincere gratitude to the University of Zululand Research Committee for funding (Grant No. C686) and to the National Research Foundation (NRF), South Africa for a research grant (Grant No. 114159). Puleng Rosinah Syed thanks NRF, South Africa for DST-NRF Innovation Master's Scholarship (Grant No. 114575). Abidemi Paul Kappo is grateful to South African Medical Research Council (SAMRC) for research grant (Grant No. PC57009).

Acknowledgments: The authors want to thank Barbara Bradley, Pretoria, South Africa for English language editing.

Conflicts of Interest: The authors declare no conflict of interest. The founding sponsors had no role in the design of the study, the collection, analysis, or interpretation of data, the writing of the manuscript or the decision to publish the results.

References

1. Sono, M.; Roach, M.P.; Coulter, E.D.; Dawson, J.H. Heme-containing oxygenases. *Chem. Rev.* **1996**, *96*, 2841–2888. [CrossRef] [PubMed]
2. Bernhardt, R. Cytochromes P450 as versatile biocatalysts. *J. Biotechnol.* **2006**, *124*, 128–145. [CrossRef] [PubMed]
3. Isin, E.M.; Guengerich, F.P. Complex reactions catalyzed by cytochrome P450 enzymes. *Biochim. Biophys. Acta (BBA) Gen. Subj.* **2007**, *1770*, 314–329. [CrossRef] [PubMed]
4. Fasan, R. Tuning P450 enzymes as oxidation catalysts. *ACS Catal.* **2012**, *2*, 647–666. [CrossRef]
5. Syed, K.; Porollo, A.; Lam, Y.W.; Grimmett, P.E.; Yadav, J.S. CYP63A2, a catalytically versatile fungal P450 monooxygenase capable of oxidizing higher-molecular-weight polycyclic aromatic hydrocarbons, alkylphenols, and alkanes. *Appl. Environ. Microbiol.* **2013**, *79*, 2692–2702. [CrossRef] [PubMed]
6. Le-Huu, P.; Heidt, T.; Claasen, B.; Laschat, S.; Urlacher, V.B. Chemo-, regio-, and stereoselective oxidation of the monocyclic diterpenoid β-cembrenediol by P450 BM3. *ACS Catal.* **2015**, *5*, 1772–1780. [CrossRef]
7. Syed, K.; Yadav, J.S. P450 monooxygenases (P450ome) of the model white rot fungus *Phanerochaete chrysosporium*. *Crit. Rev. Microbiol.* **2012**, *38*, 339–363. [CrossRef] [PubMed]
8. Podust, L.M.; Sherman, D.H. Diversity of P450 enzymes in the biosynthesis of natural products. *Nat. Prod. Rep.* **2012**, *29*, 1251–1266. [CrossRef] [PubMed]
9. McLean, K.J.; Leys, D.; Munro, A.W. Microbial cytochrome P450s. In *Cytochrome P450: Structure, Mechanism, and Biochemistry*, 4th ed.; Montellano, P.R.O., Ed.; Springer: Basel, Switzerland, 2015; pp. 261–407, ISBN 978-3-319-12108-6.
10. Guengerich, F.P. Human cytochrome P450 enzymes. In *Cytochrome P450: Structure, Mechanism, and Biochemistry*, 4th ed.; Montellano, P.R.O., Ed.; Springer: Basel, Switzerland, 2015; pp. 523–785, ISBN 978-3-319-12108-6.

11. Girhard, M.; Bakkes, P.J.; Mahmoud, O.; Urlacher, V.B. P450 Biotechnology. In *Cytochrome P450: Structure, Mechanism, and Biochemistry*, 4th ed.; Montellano, P.R.O., Ed.; Springer: Basel, Switzerland, 2015; pp. 451–520, ISBN 978-3-319-12108-6.
12. Greule, A.; Stok, J.E.; De Voss, J.J.; Cryle, M.J. Unrivalled diversity: The many roles and reactions of bacterial cytochromes P450 in secondary metabolism. *Nat. Prod. Rep.* **2018**, *35*, 757–791. [CrossRef] [PubMed]
13. Lamb, D.C.; Lei, L.; Warrilow, A.G.; Lepesheva, G.I.; Mullins, J.G.; Waterman, M.R.; Kelly, S.L. The first virally encoded cytochrome p450. *J. Virol.* **2009**, *83*, 8266–8269. [CrossRef] [PubMed]
14. Nelson, D.R. Cytochrome P450 diversity in the tree of life. *Biochim. Biophys. Acta (BBA) Proteins Proteom.* **2018**, *1866*, 141–154. [CrossRef] [PubMed]
15. Urlacher, V.B.; Eiben, S. Cytochrome P450 monooxygenases: Perspectives for synthetic application. *Trends Biochem. Sci.* **2006**, *24*, 324–330. [CrossRef] [PubMed]
16. Urlacher, V.B.; Girhard, M. Cytochrome P450 monooxygenases: An update on perspectives for synthetic application. *Trends Biochem. Sci.* **2012**, *30*, 26–36. [CrossRef] [PubMed]
17. Cimermancic, P.; Medema, M.H.; Claesen, J.; Kurita, K.; Brown, L.C.W.; Mavrommatis, K.; Pati, A.; Godfrey, P.A.; Koehrsen, M.; Clardy, J.; et al. Insights into secondary metabolism from a global analysis of prokaryotic biosynthetic gene clusters. *Cell* **2014**, *158*, 412–421. [CrossRef] [PubMed]
18. Zeigler, D.R.; Perkins, J.B. The genus *Bacillus*. In *Practical Handbook of Microbiology*, 2nd ed.; Goldman, E., Green, L.H., Eds.; CRC Press, Taylor & Francis Group: Boca Raton, FL, USA, 2009; pp. 309–337, ISBN 978-0-8493-9365-5. [CrossRef]
19. Graumann, P. *Bacillus: Cellular and Molecular Biology*, 2nd ed.; Caister Academic Press: Haverhill, UK, 2012; ISBN 978-1-904455-97-4.
20. Zhao, X.; Kuipers, O.P. Identification and classification of known and putative antimicrobial compounds produced by a wide variety of Bacillales species. *BMC Genom.* **2016**, *17*, 882. [CrossRef] [PubMed]
21. Grubbs, K.J.; Bleich, R.M.; Santa Maria, K.C.; Allen, S.E.; Farag, S.; Team, A.; Shank, E.A.; Bowers, A.A. Large-scale bioinformatics analysis of *Bacillus* genomes uncovers conserved roles of natural products in Bacterial physiology. *MSystems* **2017**, *2*, e00040-17. [CrossRef] [PubMed]
22. Furuya, T.; Shibata, D.; Kino, K. Phylogenetic analysis of *Bacillus* P450 monooxygenases and evaluation of their activity towards steroids. *Steroid* **2009**, *74*, 906–912. [CrossRef] [PubMed]
23. Ruettinger, R.T.; Wen, L.P.; Fulco, A.J. Coding nucleotide, 5′ regulatory, and deduced amino acid sequences of P-450$_{BM-3}$, a single peptide cytochrome P-450: NADPH-P-450 reductase from *Bacillus megaterium*. *J. Biol. Chem.* **1989**, *264*, 10987–10995. [PubMed]
24. Munro, A.W.; Leys, D.G.; McLean, K.J.; Marshall, K.R.; Ost, T.W.; Daff, S.; Miles, C.S.; Chapman, S.K.; Lysek, D.A.; Moser, C.C.; et al. P450 BM3: The very model of a modern flavocytochrome. *Trends Biochem. Sci.* **2002**, *27*, 250–257. [CrossRef]
25. Parvez, M.; Qhanya, L.B.; Mthakathi, N.T.; Kgosiemang, I.K.R.; Bamal, H.D.; Pagadala, N.S.; Xie, T.; Yang, H.; Chen, H.; Theron, C.W.; et al. Molecular evolutionary dynamics of cytochrome P450 monooxygenases across kingdoms: Special focus on mycobacterial P450s. *Sci. Rep.* **2016**, *6*, 33099. [CrossRef] [PubMed]
26. Kanehisa, M.; Sato, Y.; Kawashima, M.; Furumichi, M.; Tanabe, M. KEGG as a reference resource for gene and protein annotation. *Nucleic Acids Res.* **2015**, *44*, D457–D462. [CrossRef] [PubMed]
27. Nelson, D.R.; Kamataki, T.; Waxman, D.J.; Guengerich, F.P.; Estabrook, R.W.; Feyereisen, R.; Gonzalez, F.J.; Coon, M.J.; Gunsalus, I.C.; Gotoh, O.; et al. The P450 superfamily: Update on new sequences, gene mapping, accession numbers, early trivial names of enzymes, and nomenclature. *DNA Cell Biol.* **1993**, *12*, 1–51. [CrossRef] [PubMed]
28. Nelson, D.R. Cytochrome P450 nomenclature. *Methods Mol. Biol.* **1998**, *107*, 15–24. [CrossRef] [PubMed]
29. Nelson, D.R. Cytochrome P450 nomenclature, 2004. *Methods Mol. Biol.* **2006**, *320*, 1–10. [CrossRef] [PubMed]
30. Weber, T.; Blin, K.; Duddela, S.; Krug, D.; Kim, H.U.; Bruccoleri, R.; Lee, S.Y.; Fischbach, M.A.; Müller, R.; Wohlleben, W.; et al. AntiSMASH 3.0—A comprehensive resource for the genome mining of biosynthetic gene clusters. *Nucleic Acids Res.* **2015**, *43*, W237–W243. [CrossRef] [PubMed]
31. Syed, K.; Shale, K.; Pagadala, N.S.; Tuszynski, J. Systematic identification and evolutionary analysis of catalytically versatile cytochrome P450 monooxygenase families enriched in model basidiomycete fungi. *PLoS ONE* **2014**, *9*, e86③383. [CrossRef] [PubMed]
32. Kgosiemang, I.K.R.; Mashele, S.S.; Syed, K. Comparative genomics and evolutionary analysis of cytochrome P450 monooxygenases in fungal subphylum *Saccharomycotina*. *J. Pure Appl. Microbiol.* **2014**, *8*, 291–302.

33. Qhanya, L.B.; Matowane, G.; Chen, W.; Sun, Y.; Letsimo, E.M.; Parvez, M.; Yu, J.H.; Mashele, S.S.; Syed, K. Genome-wide annotation and comparative analysis of cytochrome P450 monooxygenases in Basidiomycete biotrophic plant pathogens. *PLoS ONE* **2015**, *10*, e0142100. [CrossRef] [PubMed]
34. Ngwenya, M.L.; Chen, W.; Basson, A.K.; Shandu, J.S.; Yu, J.H.; Nelson, D.R.; Syed, K. Blooming of unusual cytochrome P450s by tandem duplication in the pathogenic fungus *Conidiobolus coronatus*. *Int. J. Mol. Sci.* **2018**, *19*, 1711. [CrossRef] [PubMed]
35. Sello, M.M.; Jafta, N.; Nelson, D.R.; Chen, W.; Yu, J.H.; Parvez, M.; Kgosiemang, I.K.R.; Monyaki, R.; Raselemane, S.C.; Qhanya, L.B.; et al. Diversity and evolution of cytochrome P450 monooxygenases in Oomycetes. *Sci. Rep.* **2015**, *5*, 11572. [CrossRef] [PubMed]
36. Moody, S.C.; Loveridge, E.J. CYP 105—Diverse structures, functions and roles in an intriguing family of enzymes in *Streptomyces*. *J. Appl. Microbiol.* **2014**, *117*, 1549–1563. [CrossRef] [PubMed]
37. Li, H.; Poulos, T.L. The structure of the cytochrome p450BM-3 haem domain complexed with the fatty acid substrate, palmitoleic acid. *Nat. Struct. Biol.* **1997**, *4*, 140–146. [CrossRef] [PubMed]
38. Noble, M.A.; Miles, C.S.; Reid, G.A.; Chapman, S.K.; Munro, A.W. Catalytic properties of key active site mutants of flavocytochrome P-450 BM3. *Biochem. Soc. Trans.* **1999**, *27*, A44. [CrossRef]
39. Lee, D.S.; Yamada, A.; Matsunaga, I.; Ichihara, K.; Adachi, S.I.; Park, S.Y.; Shiro, Y. Crystallization and preliminary X-ray diffraction analysis of fatty-acid hydroxylase cytochrome P450BSβ from *Bacillus subtilis*. *Acta Crystallogr. D Struct. Biol.* **2002**, *58*, 687–689. [CrossRef]
40. Lee, D.S.; Yamada, A.; Sugimoto, H.; Matsunaga, I.; Ogura, H.; Ichihara, K.; Adachi, S.I.; Park, S.Y.; Shiro, Y. Substrate recognition and molecular mechanism of fatty acid hydroxylation by cytochrome P450 from *Bacillus subtilis*: Crystallographic, spectroscopic and mutational studies. *J. Biol. Chem.* **2003**, *278*, 9761–9767. [CrossRef] [PubMed]
41. Cryle, M.J.; Bell, S.G.; Schlichting, I. Structural and biochemical characterization of the cytochrome P450 CypX (CYP134A1) from *Bacillus subtilis*: A cyclo-L-leucyl-L-leucyl dipeptide oxidase. *Biochemistry* **2010**, *49*, 7282–7296. [CrossRef] [PubMed]
42. Cryle, M.J.; Schlichting, I. Structural insights from a P450 carrier protein complex reveal how specificity is achieved in the P450BioI ACP complex. *Proc. Natl. Acad. Sci. USA* **2008**, *105*, 15696–15701. [CrossRef] [PubMed]
43. Li, Z.Z.; Li, X.F.; Yang, W.; Dong, X.; Yu, J.; Zhu, S.L.; Li, M.; Xie, L.; Tong, W.Y. Identification and functional analysis of cytochrome P450 complement in *Streptomyces virginiae* IBL14. *BMC Genom.* **2013**, *14*, 130. [CrossRef] [PubMed]
44. Zhang, A.; Zhang, T.; Hall, E.A.; Hutchinson, S.; Cryle, M.J.; Wong, L.L.; Zhou, W.; Bell, S.G. The crystal structure of the versatile cytochrome P450 enzyme CYP109B1 from *Bacillus subtilis*. *Mol. Biosyst.* **2015**, *11*, 869–881. [CrossRef] [PubMed]
45. Savino, C.; Montemiglio, L.C.; Sciara, G.; Miele, A.E.; Kendrew, S.G.; Jemth, P.; Gianni, S.; Vallone, B. Investigating the structural plasticity of a cytochrome P450: Three dimensional structures of P450 EryK and binding to its physiological substrate. *J. Biol. Chem.* **2009**. [CrossRef] [PubMed]
46. Montemiglio, L.C.; Gianni, S.; Vallone, B.; Savino, C. Azole drugs trap cytochrome P450 EryK in alternative conformational states. *Biochemistry* **2010**, *49*, 9199–9206. [CrossRef] [PubMed]
47. Merson-Davies, L.A.; Cundiiffe, E. Analysis of five tyiosin biosynthetic genes from the tyllBA region of the *Streptomyces fradiae* genome. *Mol. Microbiol.* **1994**, *13*, 349–355. [CrossRef] [PubMed]
48. Fouces, R.; Mellado, E.; Díez, B.; Barredo, J.L. The tylosin biosynthetic cluster from *Streptomyces fradiae*: Genetic organization of the left region. *Microbiology* **1999**, *145*, 855–868. [CrossRef] [PubMed]
49. Katoh, K.; Kuma, K.; Toh, H.; Miyata, T. MAFFT version 5: Improvement in accuracy of multiple sequence alignment. *Nucleic Acids Res.* **2005**, *33*, 511–518. [CrossRef] [PubMed]
50. Boc, A.; Diallo, A.B.; Makarenkov, V. T-REX: A web server for inferring, validating and visualizing phylogenetic trees and networks. *Nucleic Acids Res.* **2012**, *40*, W573–W579. [CrossRef] [PubMed]

51. Letunic, I.; Bork, P. Interactive Tree of Life (iTOL) v3: An online tool for the display and annotation of phylogenetic and other trees. *Nucleic Acids Res.* **2016**, *44*, W242–W245. [CrossRef] [PubMed]
52. Saeed, A.I.; Sharov, V.; White, J.; Li, J.; Liang, W.; Bhagabati, N.; Braisted, J.; Klapa, M.; Currier, T.; Thiagarajan, M.; et al. TM4: A free, open-source system for microarray data management and analysis. *Biotechniques* **2003**, *34*, 374–378. [CrossRef] [PubMed]

© 2018 by the authors. Licensee MDPI, Basel, Switzerland. This article is an open access article distributed under the terms and conditions of the Creative Commons Attribution (CC BY) license (http://creativecommons.org/licenses/by/4.0/).

Article

Cytochrome P450 Monooxygenase CYP139 Family Involved in the Synthesis of Secondary Metabolites in 824 Mycobacterial Species

Puleng Rosinah Syed [1], Wanping Chen [2], David R. Nelson [3], Abidemi Paul Kappo [4], Jae-Hyuk Yu [5,6], Rajshekhar Karpoormath [1,*] and Khajamohiddin Syed [4,*]

1. Department of Pharmaceutical Chemistry, College of Health Sciences, University of KwaZulu-Natal, Durban 4000, South Africa; prosinah@gmail.com
2. College of Food Science and Technology, Huazhong Agricultural University, Wuhan 430070, China; chenwanping@mail.hzau.edu.cn
3. Department of Microbiology, Immunology and Biochemistry, University of Tennessee Health Science Center, Memphis, TN 38163, USA; drnelson1@gmail.com
4. Department of Biochemistry and Microbiology, Faculty of Science and Agriculture, University of Zululand, KwaDlangezwa 3886, South Africa; KappoA@unizulu.ac.za
5. Department of Bacteriology, University of Wisconsin-Madison, 3155 MSB, 1550 Linden Drive, Madison, WI 53706, USA; jyu1@wisc.edu
6. Department of Systems Biotechnology, Konkuk University, Seoul 05029, Korea
* Correspondence: Karpoormath@ukzn.ac.za (R.K.); khajamohiddinsyed@gmail.com (K.S.)

Received: 1 April 2019; Accepted: 11 May 2019; Published: 31 May 2019

Abstract: Tuberculosis (TB) is one of the top infectious diseases causing numerous human deaths in the world. Despite enormous efforts, the physiology of the causative agent, *Mycobacterium tuberculosis*, is poorly understood. To contribute to better understanding the physiological capacity of these microbes, we have carried out extensive in silico analyses of the 1111 mycobacterial species genomes focusing on revealing the role of the orphan cytochrome P450 monooxygenase (CYP) CYP139 family. We have found that CYP139 members are present in 894 species belonging to three mycobacterial groups: *M. tuberculosis* complex (850-species), *Mycobacterium avium* complex (34-species), and non-tuberculosis mycobacteria (10-species), with all CYP139 members belonging to the subfamily "A". CYP139 members have unique amino acid patterns at the CXG motif. Amino acid conservation analysis placed this family in the 8th among CYP families belonging to different biological domains and kingdoms. Biosynthetic gene cluster analyses have revealed that 92% of CYP139As might be associated with producing different secondary metabolites. Such enhanced secondary metabolic potentials with the involvement of CYP139A members might have provided mycobacterial species with advantageous traits in diverse niches competing with other microbial or viral agents, and might help these microbes infect hosts by interfering with the hosts' metabolism and immune system.

Keywords: biosynthetic gene clusters; cytochrome P450 monooxygenase; CYP139A1; genome data mining; host metabolism; *Mycobacterium tuberculosis*; polyketides; secondary metabolites; tuberculosis

---

1. Introduction

Tuberculosis (TB), a prehistoric disease, remains one of the top 10 causes of death and the leading cause from a single infectious agent, *Mycobacterium tuberculosis*, despite global efforts in disease control programs during the past 20 years [1]. TB is a global disease, found in every country in the world [1]. It became mankind's oldest and worst enemy owing to its widespread nature across the world and developing resistance to known and available drugs [1]. In 2017, 10 million people developed TB, and an estimated 1.3 million deaths among human immunodeficiency virus (HIV)-negative people and an additional 300,000 deaths from TB among HIV-positive people occurred [1]. The latest data from the Statistics South Africa show that TB is one of the top killers in South Africa [2], suggesting an urgent need to understand *M. tuberculosis* physiology to be able to come up with novel drugs and drug targets.

Despite living in the most advanced medicine era, TB remains a major threat to human health [1]. After 21 years of *M. tuberculosis* genome sequencing [3], to date its physiology is poorly understood and many proteins remain orphans. Genome sequencing analysis of *M. tuberculosis* H37Rv revealed the presence of 20 cytochrome P450 monooxygenases (CYPs/P450s) in its genome [3]. P450s are mixed function oxidoreductases ubiquitously distributed across the biological kingdoms [4]. P450s are well known for their role in essential cellular anabolic and catabolic processes.

Among 20 P450s, to date, the role of only six *M. tuberculosis* H37Rv P450s in its physiology have been elucidated [5]. CYP51B1, highly conserved P450 family across microbes, has been found to catalyse the 14α-demethylation of lanosterol [6–8]; CYP121A1 catalyses oxidative crosslinking of the two tyrosines in a cyclodipeptide [9]; CYP125A1 and CYP142A1 catalyse the 26-hydroxylation of cholesterol and cholest-4-en-3-one [10,11]; CYP124A1 catalyses the terminal hydroxylation of methyl-branched hydrocarbons such as those of phytanic acid and farnesol [12], cholesterol and related sterols [10,13], and vitamin D_3 and CYP128A1 is involved in oxidation of menaquinone MK9 [14].

Among *M. tuberculosis* H37Rv P450s, the *CYP139A1* gene was found downstream of polyketide synthase genes (*pks10*, *pks7*, *pks8*, *pks17*, *pks9* and *pks11*) and situated next to macrolide transport protein [15,16]. Two of the polyketide synthases, *pks7* and *pks8*, were found to be essential for the survival of *M. tuberculosis* [17,18]. Polyketide synthases along with other genes were found to be part of biosynthetic gene clusters (BGCs). As per Medema et al. [19], a BGC can be defined as a physically clustered group of two or more genes in a particular genome that together encode a biosynthetic pathway for the production of a specialised metabolite (including its chemical variants). Bacteria, fungi and plants are known to possess different types of BGCs producing a variety of secondary metabolites that are beneficial to humans. Among the genes that are part of a BGC, P450s play a key role in contributing to the diversity of a secondary metabolite owing to their regio and stereo-specific oxidation [20]. Recently, comprehensive comparative analysis of P450s and those associated with secondary metabolism revealed a large number of P450s involved in the production of secondary metabolites in different bacterial species [21,22].

Based on *CYP139A1* location, this P450 is assumed to be involved in oxidative tailoring of the macrolide structure. In the latest study, involving comprehensive comparative analysis of P450s in bacterial species belonging to the genera *Mycobacterium* and *Streptomyces*, CYP139 P450s were found to be dominantly located in different secondary metabolite BGCs [22]. This strongly indicates that CYP139 P450s are possibly involved in the synthesis of secondary metabolites. This study is aimed at using an in silico approach to unravel the CYP139 P450 family's role in mycobacterial species physiology.

2. Results and Discussion

2.1. CYP139 P450s Are Present Only in Certain Mycobacterial Category Species

Comprehensive comparative analysis of CYP139 P450s in 1111 mycobacterial species belonging to six different categories (Table S1) revealed that CYP139 P450s are present in 894 mycobacterial species belonging to three categories, namely the *Mycobacterium tuberculosis* complex (MTBC), *M. avium* complex (MAV) and non-tuberculosis mycobacteria (NTM) (Figure 1 and Table S2). This phenomenon of identifying CYP139 P450s only in these three mycobacterial categories was also observed previously when 60 mycobacterial species were analysed [23]. Results from this study, which involved such a large data set, not only supported, but also confirmed that mycobacterial species belonging to categories such as *Mycobacterium* causing leprosy (MCL), Saprophytes (SAP) and the *Mycobacterium chelonae-abscessus* complex (MCAC) do not have CYP139 P450s in their genomes, as seen in Figure 1. Interestingly, not all mycobacterial species of MTBC, NTM and MAC categories have CYP139 P450 (Figure 1). Among 956 mycobacterial species, only 850 mycobacterial species of MTBC have CYP139 P450; 10 of 14 and 34 of 57 mycobacterial species of NTM and MAC, respectively, have this P450 (Figure 1 and Table S2). A detailed analysis of CYP139 P450s along with species names and protein ID is presented in Table S2 and the CYP139 P450 sequences are presented in Supplementary Dataset 1.

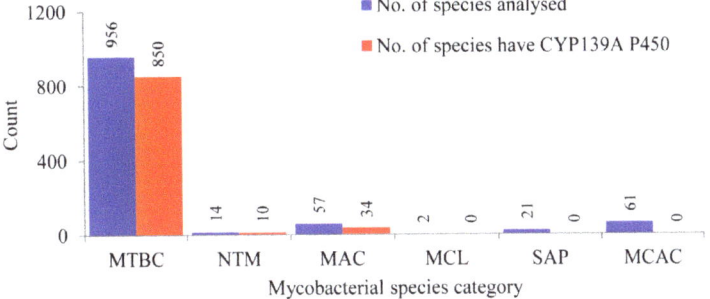

Figure 1. Comparative analysis of CYP139A P450s in species belonging to six different mycobacterial categories. Abbreviations: MTBC, *Mycobacterium tuberculosis* complex; MAV, *M. avium* complex; NTM, non-tuberculosis mycobacteria; MCL, *Mycobacterium* causing leprosy; SAP, Saprophytes and MCAC, *Mycobacterium chelonae-abscessus* complex. Information on mycobacterial species and CYP139A P450s is presented in Supplementary Tables S1 and S2, respectively.

Analysis of *CYP139* P450s in the genomes of mycobacterial species revealed that only a single copy of the *CYP139* P450 gene is present in all mycobacterial species (Table S2). Furthermore, P450 subfamily analysis revealed that all CYP139 P450s found in 894 mycobacterial species belong to the subfamily "A" (Figure 2). Phylogenetic analysis of CYP139A P450s revealed that CYP139A P450s grouped per their mycobacterial category, indicating after speciation CYP139A P450s were subjected to amino acid changes specific to their category (Figure 1), similar to what was observed for other P450s described elsewhere [23,24]. However, four CYP139A P450s belonging to *M. genavense* ATCC 51234 and *Mycobacterium sp.* JDM601 of NTM and *Mycobacterium sp.* UM CSW and *M. avium avium* Env 77 of MAC were aligned separately, suggesting that these CYP139A P450s had deviated from their counterparts (Figure 2). Percentage identity among CYP139 P450s further confirmed that CYP139A P450s from these species have a low percentage identity with their counterparts (Supplementary Dataset 2). CYP139A P450s of *Mycobacterium sp.* UM CSW and *M. avium avium* Env 77 have an average of ~77% and ~63% identity, whereas CYP139A P450s of *M. genavense* ATCC 51234 and *Mycobacterium sp.* JDM601 have an average of 75% and 60% with their counterparts (Supplementary Dataset 2) suggesting these P450s have been subjected to significant amino acid changes. The phenomenon of P450s not grouping with their counterpart species was also observed in fungal species, where CYP53D1 has been subjected to extensive amino acid changes [24], the same as what was observed for the four CYP139A P450s identified in this study. Determining the effect of these amino acid changes on functional specificity of four CYP139A P450s, if any, will be interesting future work.

2.2. CYP139 P450 Family Ranked among Top 10 P450 Families

Ranking of P450 families belonging to different biological kingdoms, based on the number of conserved amino acids in their protein sequence, placed the CYP139 P450 family in the twelfth rank [23,25]. While ranking the CYP139 P450 family, only 54 CYP139A P450s were used [23,25]. Identification of quite a large number of CYP139A P450s in this study necessitated re-analysis of the ranking of this P450 family. In order to identify the conservation rank, CYP139A P450s were subjected to PROfile Multiple Alignment with Local Structures and 3D constraints (PROMALS3D) [26] analysis (Supplementary Dataset 3). PROMALS3D analysis revealed the presence of 165 amino acids invariantly conserved in CYP139 P450s (Table 1). Comparative analysis with other P450 families from different biological kingdoms revealed that the CYP139 P450 family now occupies the eighth rank compared to the twelfth rank as assigned previously (Table 1).

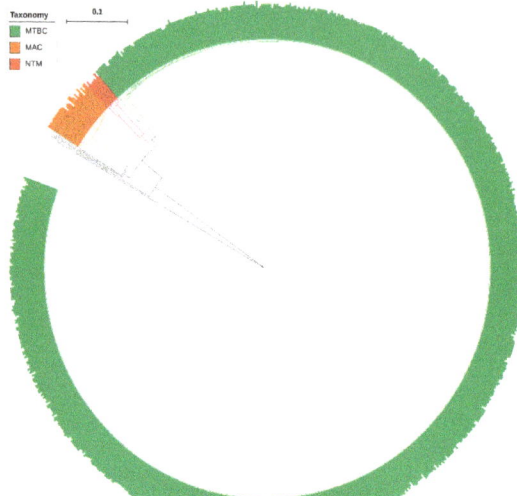

Figure 2. Phylogenetic analysis of CYP139A P450s. Different mycobacterial categories were indicated in different colours. CYP51B1 from *Mycobacterium tuberculosis* H37Rv is used as an outgroup. Abbreviations: MTBC, *Mycobacterium tuberculosis* complex; MAV, *M. avium* complex; NTM, non-tuberculosis mycobacteria. A high-resolution phylogenetic tree is provided in Supplementary Figure S1.

Table 1. Comparative amino acid conservation analysis of CYP139 P450 family with top 10 ranked P450 families [23,25]. The conservation index score is obtained as described in the section on materials and methods, following the procedure described elsewhere [27]. The conservation score (5–9) obtained via PROMALS3D is presented in the table, where the number 9 indicates invariantly conserved amino acids in P450 members. P450 families were arranged from the highest to the lowest number of amino acids conserved. CYP139 P450 family is indicated in bold.

P450 Family	Number of Member P450s	Kingdom	PROMALS3D Conservation Index					Rank (Highest to Lowest Conservation)
			5	6	7	8	9	
CYP141	29	Bacteria	0	0	0	0	389	1
CYP51	50	Bacteria	11	102	0	0	264	2
CYP137	38	Bacteria	145	0	0	0	251	3
CYP121	34	Bacteria	0	0	0	0	233	4
CYP132	39	Bacteria	175	0	0	0	217	5
CYP5619	23	Stramenopila (oomycetes)	118	38	170	0	199	6
CYP124	71	Bacteria	52	35	59	0	170	7
CYP139	**894**	**Bacteria**	**0**	**127**	**0**	**0**	**165**	**8 (formerly 12)**
CYP188	67	Bacteria	62	0	100	0	141	9
CYP123	74	Bacteria	62	0	82	0	137	10

2.3. CYP139 Family Has Unique Amino Acid Patterns at CXG Motif

In a study by Syed and Mashele [28], analysis of the P450 signature motifs, EXXR and CXG, among different P450 families led to the discovery of amino acid patterns characteristic of a P450 family. The authors proposed that "during the divergence of P450 families from a common ancestor, these amino acids patterns evolved and are retained in each P450 family as a signature of that family" [28]. However, in that study, the CYP139 P450 family is not included. Furthermore, identification of a large number of CYP139A P450s, in this study, gives us an opportunity to identify CYP139 P450 family characteristic amino acid patterns at EXXR and CXG motifs, if any.

Analysis of EXXR and CXG motifs in 894 CYP139A P450s revealed that the CYP139 P450 family EXXR domain is absolutely conserved with amino acid patterns E-T-L-R, whereas, eight amino acids are invariantly conserved in CXG motifs with amino acid patterns of F-S-G(96%)/A(4%)-G-L-H-R-C-I(96%)/V(4%)-G (Figure 3). It is interesting to note that the CYP139 P450 family EXXR motif amino acid pattern absolutely matched with the CYP5 family [28] and amino acid patterns at the CXG motif were unique and not matched with any P450 families described in the literature [25,28,29]. The CYP139 P450 family amino acid patterns at the EXXR and CXG motifs further strongly support the above hypothesis proposed by Syed and Mashele [28].

Figure 3. Analysis of amino acid patterns at the EXXR and CXG motif in CYP139 P450 family. In total 894 CYP139 P450 sequences were analysed for EXXR and CXG signature sequences.

2.4. Most CYP139A P450s Are Part of Secondary Metabolite Biosynthetic Gene Clusters

Analysis of CYP139A P450s as part of secondary metabolite BGCs in mycobacterial species revealed that most of the CYP139A P450s are part of different BGCs (Figure 4A and Table S2). Among 894 CYP139A P450s, 824 CYP139A P450s (92%) were found to be part of secondary metabolic BGCs (Figure 4A). This means 70 CYP139A P450s were not found to be part of any secondary metabolite BGCs. Comparison of CYP139A P450s that are part of BGCs in three categories revealed that most of the CYP139A P450s in MTBC and NTM species were part of BGCs, compared to species of MAC, where fewer than half of CYP139A P450s were part of secondary metabolite BGCs (Figure 4B).

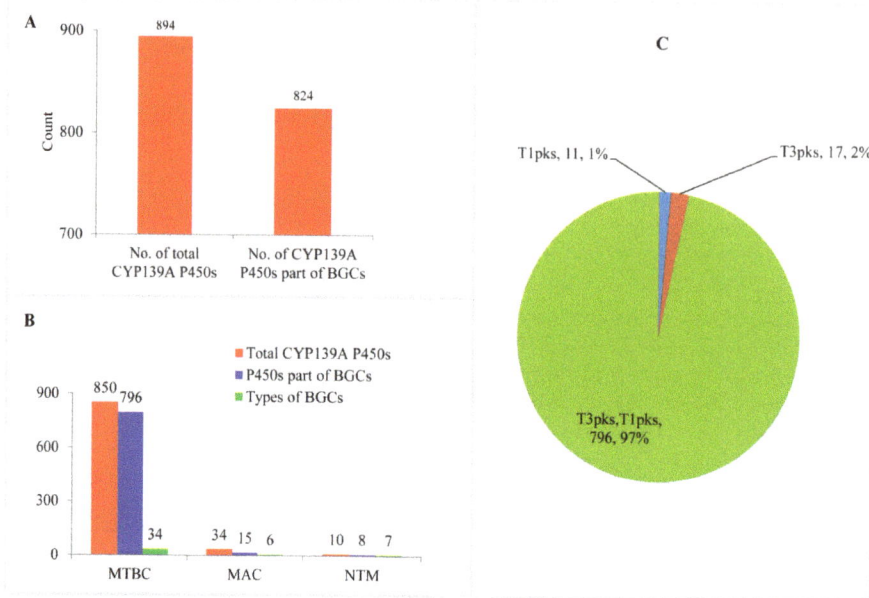

Figure 4. CYP139A P450s secondary metabolite BGCs analysis in mycobacterial species. (**A**) Analysis of CYP139A P450s that are part of BGCs. (**B**) Comparative analysis of CYP139A P450s that are part of BGCs and types of BGCs in different mycobacterial categories. Abbreviations: MTBC, *Mycobacterium tuberculosis* complex; MAV, *M. avium* complex; NTM, non-tuberculosis mycobacteria. (**C**) Comparative analysis of CYP139A P450 cluster types. The type of cluster and the number of CYP139A P450s and their percentage in the total number of P450s were presented in the figure. Abbreviation: T1pks, Type 1 polyketide synthase; T2pks, Type 2 polyketide synthase.

Analysis of secondary metabolite BGCs revealed that CYP139A P450s were part of only three different cluster types (Figure 4C and Table S2). Among three different cluster types, CYP139A P450s were found to be present dominantly as part of Type 3-Type 1 polyketide synthase (T3PKS-T1PKS) (97%) compared to T3 PKS (2%) and T1 PKS (1%) (Figure 4C and Table S2). There were 796 CYP139A P450s found to be part of T3PKS-T1PKS, followed by 17 and 11 CYP139 P450s found to be part of T3 PKS and T1 PKS, respectively (Figure 4C and Table S2). Analysis of gene clusters revealed that 824 CYP139A P450s were part of 39 different gene clusters (Figure 4). There were 34 CYP139A P450 gene clusters found in MTBC species, followed by seven gene clusters in NTM species and six gene clusters in MAC (Figure 4B). Among different gene clusters, ML-449 was dominant, with 349 CYP139A P450s followed by methylated alkyl-resorcinol/methylated acyl-phloroglucinol (MAR/MAP) with 104 CYP139A P450s, Nystatin with 74 CYP139A P450s and Jerangolid with 55 CYP139A P450s (Figure 5). Among 39 gene clusters only 11 gene clusters were found to have 10 or more CYP139A P450s (Figure 5). Analysis of DNA sequence percentage identity between CYP139A P450 gene clusters compared to known gene clusters revealed that some of the gene clusters have 100% identity, such as Leucanicidin, MAR/MAP and Micromonolactam (Figure 5), indicating CYP139A P450s are indeed involved in the synthesis of these secondary metabolites.

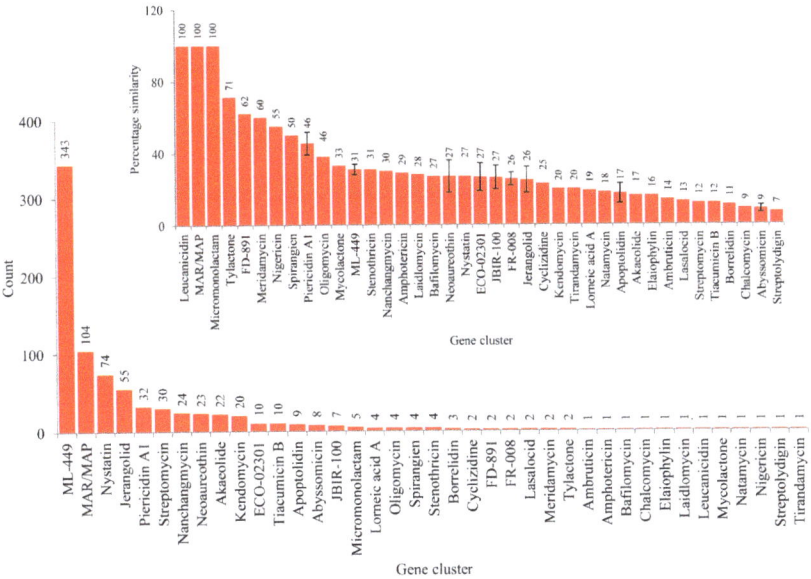

Figure 5. CYP139A P450 gene cluster analysis in mycobacterial species. The 39 gene clusters were presented with their standard abbreviated names as per anti-SMASH. The number next to bars represents the number of CYP139A P450s that is part of that gene cluster. The inset figure shows the percentage identity of CYP139A P450 gene clusters to the known gene clusters available at anti-SMASH. The number next to the bars represents the percentage identity. For some gene clusters the percentage identity is represented with standard deviation (indicated with bars).

2.5. CYP139A P450s Involved in the Synthesis of Secondary Metabolites in Mycobacterial Species

Comprehensive comparative analysis of CYP139A P450s secondary BGCs in mycobacterial species revealed that CYP139A P450s are indeed involved in the synthesis of different secondary metabolites, as 92% of CYP139A P450s were found to be part of secondary metabolite BGCs (Figures 4 and 5 and Table S2). To understand the role of CYP139A P450s in mycobacterial species' physiology well, a functional comparison of CYP139A P450s gene clusters' homolog secondary metabolites was carried out (Table 2). As shown in Table 2, it is clear that CYP139A P450s are involved in the production of chemicals that have antibacterial, antifungal, antiviral and antitumor properties. Interestingly, some of these metabolites in fact showed antimycobacterial activity (Table 2). This indicates that CYP139A P450s are possibly helping mycobacterial species to kill other bacteria, including other mycobacterial species, thus gaining the upper hand in the niche area for their survival. It is interesting to note that CYP139A P450s are present only in MTBC, NTM and MAC categories, but not present in SAP, MCAC or MCL. This necessitates understanding its role in mycobacterial species when they are surviving in hosts such as humans or other animals. In this direction, analysis of some secondary metabolite functions pointed out that some secondary metabolites are certainly helping mycobacterial species to survive in their hosts. For example, MAR/MAP BGC products are found to be part of the cell envelope in *M. marinum*, possibly complicating its access to host immune system or drug actions [30]; Akaeolide has cytotoxic activity against fibroblasts, suggesting it may play a role in tissue weakening in the host [31]; JBIR-100 exhibits cytotoxic activities and inhibition of proton pumps such as vacuolar-type ATPases (V-ATPases) activities and is thus linked with an increasing number of diseases such as osteopetrosis, male infertility and renal acidosis [32,33]. Lorneic acid A inhibits phosphodiesterase PDE5 blocking the degradation of cGMP [34] and thus it might be playing a role in pulmonary hypertension. Meridamycin has been found to bind FK506-binding proteins (FKBP12) [35]. FKBP12 proteins play a key role in regulating fundamental aspects of cell biology and have been found to be critical in mice survival [36]. Nigericin inhibits the Golgi functions in eukaryotic cells and is a well-known activator of the NLRP3 inflammasome [37–39], indicating bacterial infection. One secondary metabolite, namely mycolactone, a lipid-like toxin with cytotoxic, immunosuppressive and tissue necrosis activity, has been shown to be involved in the development of Buruli ulcer by *M. ulcerans* [40].

Table 2. Functional analysis of homolog CYP139A P450 gene clusters.

Gene Cluster	Function	Reference
ML-449	Macrolactam antifungal-antibiotic production.	[41]
MAR/MAP	Synthesis of methylated alkyl-resorcinol and methylated acyl-phloroglucinol products found to be part of cell envelope in *M. marinum*.	[30]
Nystatin	Polyene antifungal antibiotic.	[42]
Jerangolid	Antifungal polyketide.	[43]
Piericidin A1	A member of α-pyridone antibiotics, exhibits various biological activities such as antimicrobial, antifungal, and antitumour properties and possesses potent respiration-inhibitory activity against insects owing to its competitive binding capacity to mitochondrial complex I.	[44]
Streptomycin	Antibiotic used to treat bacterial infections, including tuberculosis.	[45]
Nanchangmycin	A polyether ionophore antibiotic produced by *Streptomyces nanchangensis* NS3226 that has insecticidal and in vitro antibacterial properties. Nanchangmycin exhibits antiviral properties against the Zika virus.	[46–48]
Neoaureothin	Neoaureothin is an unusual chain-extended analog of aureothin. It was first reported as a co-metabolite of neoantimycin in *Streptomyces orinoci*. It has been reported to have anti-HIV and antifungal activity.	[49]
Akaeolide	A carbocyclic polyketide with moderate antimicrobial activity and cytotoxicity to rat fibroblasts.	[31]
Kendomycin	Macrolide antibiotic with antibacterial activity.	[50]
ECO-02301	Antifungal agent.	[51]
Tiacumicin B	Macrolide antibiotic, which is used for the treatment of *Clostridium difficile* infections.	[52,53]
Apoptolidin	Macrolide antibiotic well known as apoptosis inducer and inhibitor of F0F1-ATPase. It is a promising new therapeutic lead that exhibits remarkable selectivity against cancer cells relative to normal cells.	[54–56]
Abyssomicin	A novel spirotetronate polyketide Class I antimicrobial. The biological activity of abyssomicins includes their antimicrobial activity against Gram-positive bacteria and mycobacteria, antitumour properties, latent HIV reactivator, anti-HIV and HIV replication inducer properties	[57]
JBIR-100	A new 16-membered tetraene macrolide from the *Streptomyces* species. Its structure is identical to TS155-2, which is an inhibitor of the thrombin-induced calcium influx. It exhibits cytotoxic and V-ATPases inhibition activities. V-ATPases are ubiquitous proton pumps present in the endomembrane system of all eukaryotic cells and in the plasma membranes of many animal cells that have been correlated with an increasing number of diseases such as osteoporosis, male infertility and renal acidosis.	[32,33]
Micromonolactam	A new polyene macrolactam antibiotic	[58]
Lorneic acid A	It has a fatty acid-like structure in which a benzene ring is embedded. It inhibits phosphodiesterases (PDE) with selectivity toward PDE5, thus, blocking the degradation of cGMP and having a possible linkage to pulmonary hypertension	[34]
Leucanicidin	A potent nematocide and insecticide macrolide	[59]
Oligomycin	A natural antibiotic that inhibits mitochondrial ATP synthase, thus affecting the electron transport chain.	[60]
Spirangien	Highly cytotoxic and antifungal spiroketal	[61]
Stenothricin	A peptide antibiotic inhibiting bacterial cell wall synthesis	[62]
Borrelidin	A small molecule nitrile-containing macrolide, which is an inhibitor of bacterial and eukaryal threonyl-tRNA synthetase. It exhibits among others antibacterial and anti-angiogenesis activities, suppresses growth and induces apoptosis in malignant acute lymphoblastic leukemia cells.	[63,64]
FD-891	Profoundly blocked both perforin- and FasL-dependent cytotoxicity by cytotoxic T lymphocytes—immunosuppressive.	[65]
FR-008	Macrolide antibiotic with antifungal activity.	[66]
Meridamycin	A 27-membered macrolide that acts as non-immunosuppressive FK506-binding proteins (FKBP12) ligand.	[35]
Ambruticin	Antifungal polyketide	[67]
Nigericin	Nigericin acts as an H^+, K^+, Pb^{2+} ionophore. Most commonly it is an antiporter of H^+ and K^+. In the past nigericin was used as an antibiotic active against Gram-positive bacteria. It inhibits Golgi functions in eukaryotic cells. Its ability to induce K^+ efflux also makes it a potent activator of the NLRP3 inflammasome.	[37–39]
Mycolactone	Lipid-like toxin with cytotoxic, immunosuppressive and tissue necrosis activity. It plays a key role in the development of Buruli ulcer by *M. ulcerans*.	[40]

3. Materials and Methods

3.1. Mycobacterial Species and Genome Databases

In total, 1111 mycobacterial species genomes that are available for public use (as of 12 June 2018) at Integrated Microbial Genomes & Microbiomes (IMG/M) [68] were used in the study (Table S1). Mycobacterial species used in the study, along with their name, genome ID and individual genome database links, were presented in Table S1.

3.2. Genome Data Mining and Annotation of CYP139 P450s

The *M. tuberculosis* H37Rv CYP139A1 (Rv1666c) P450 sequence has been blasted with the default settings against individual mycobacterial species genomes at IMG/M [68]. However, each time, only 20 mycobacterial species were selected for BLAST analysis. The hit proteins with more than 40% identity were selected and then subjected to BLAST analysis at the P450 BLAST server (https://ksyed.weebly.com/p450-blast.html) to identify the homolog P450. Hit proteins were then grouped into families and subfamilies based on the International Cytochrome P450 Nomenclature criteria, i.e., P450s showing >40% identity were assigned to the same P450 family and P450s that showed >55% identity were grouped under the same P450 subfamily [69–71]. Protein with more than 90% identity considered as ortholog and assigned the same subfamily number.

3.3. Phylogenetic Analysis of CYP139A P450s

The phylogenetic tree of CYP139 family members was built with *M. tuberculosis* CYP51B1 (Rv0764c) protein as outgroup. First, the protein sequences were aligned by MAFFT v6.864 [72], embedded on the Trex web serve [73]. Then, the alignments were automatically subjected to infer the best tree by the Trex web server with its embedded weighting procedure. Finally, the tree was visualised and colored by iTOL (http://itol.embl.de/about.cgi) [74].

3.4. Analysis of Homology and Amino Acid Conservation

Analysis of percentage identity among CYP139A P450s from species belonging to MAC and NTM categories was carried out as described elsewhere [23,29]. Briefly, the percentage identity between CYP139 P450s was determined using the Clustal Omega [75]. The Clustal Omega percentage identity matrix was downloaded and pasted into an Excel sheet by converting the text into a column option.

Amino acid conservation among CYP139A P450s was carried out following the method described elsewhere [23,25,29]. Briefly, CYP139 P450s were subjected to PROMALS3D [26] to identify invariantly conserved amino acids [27]. The conservation index follows numbers from 5–9, where 9 is the invariantly conserved amino acid across the sequences. The total number of conserved residues indicated by number 9 was recorded. The conserved nature of the CYP139 family was compared to other P450 families from different biological kingdoms, as reported elsewhere [23,25].

3.5. Generation of EXXR and CXG Sequence Logo

CYP139 P450 family EXXR and CXG sequence logos were generated following the method described elsewhere [25,28,29]. Briefly, CYP139 P450 sequences were aligned using ClustalW multiple alignments using MEGA7 [76]. After sequence alignment the EXXR and CXG region amino acids (4 and 10 amino acids, respectively), were selected and entered in the WebLogo program (http://weblogo.berkeley.edu/logo.cgi). As a selection parameter, the image format was selected as PNG (bitmap) at 300 dpi resolution. The percentage predominance of amino acids at particular positions is calculated considering the total number of amino acids as 100%. The generated EXXR and CXG logos were used for analysis and compared to the different P450 family EXXR and CXG logos that have been published and are available to the public [25,28,29].

3.6. Identification of CYP139 P450 Secondary Metabolite BGCs

BGCs listed on the IMG/M [68] website for each of the mycobacterial species were manually searched for the presence of CYP139 P450s using the protein ID. The BGCs that have CYP139 P450 were selected for further study. The listed BGCs at IMG/M are general [68] and in order to identify the specific type of BGCs, the selected BGCs genome sequences were subjected to secondary metabolite BGCs analysis, as described elsewhere [21]. Briefly, the individual BGC genome sequences downloaded from IMG/M [68] were submitted to anti-SMASH [77]. The type of BGC, percentage similarity to a known cluster and the cluster name were noted. Standard BGC abbreviation terminology developed by anti-SMASH [77] was used in the study.

4. Conclusions

The advancement of genome sequencing and bioinformatics tools helps significantly in understanding the role of orphan proteins in organisms. This study is an attempt to utilize the availability of quite a large number of mycobacterial species genome sequences and different bioinformatics tools to understand the role of the orphan CYP139 family in mycobacterial species. This study revealed that the CYP139 family indeed plays a role in the synthesis of secondary metabolites in mycobacterial species. Based on the functions of homolog CYP139 P450 gene clusters' secondary metabolites, it can be assumed that these metabolites indeed help mycobacterial species to survive in the host, being part of the cell envelope and inhibiting fibroblast, thus causing tissue weakening and causing ulcers via tissue necrosis. The metabolites that exhibit antibacterial (including antimycobacterial), antifungal and antiviral properties certainly help mycobacterial species to gain the upper hand in the niche area compared to those agents. It would be interesting to determine the roles of CYP139A P450s that are not part of gene clusters. Predictions made in the study are based on the functions of homolog secondary metabolites. However, wet laboratory biosynthesis and functional analysis of secondary metabolites should be carried out to understand the role of these metabolites in mycobacterial physiology. Study results can be used as a reference for future experimental studies.

Supplementary Materials: Supplementary materials can be found at http://www.mdpi.com/1422-0067/20/11/2690/s1.

Author Contributions: Conceptualization, R.K. and K.S.; data curation, P.R.S., W.C., D.R.N., A.P.K., J.-H.Y., R.K. and K.S.; formal analysis, P.R.S., W.C., D.R.N., A.P.K., J.-H.Y., R.K. and K.S.; funding acquisition, R.K., A.P.K. and K.S.; investigation, P.R.S., W.C., D.R.N., A.P.K., J.-H.Y., R.K. and K.S.; methodology, P.R.S, W.C., D.R.N., A.P.K., J.-H.Y., R.K. and K.S.; project administration, R.K. and K.S.; resources, R.K., A.P.K. and K.S.; supervision, R.K. and K.S.; validation, P.R.S., W.C., D.R.N., A.P.K., J.-H.Y., R.K. and K.S.; visualization, P.R.S., R.K. and K.S.; writing—original draft, P.R.S., W.C., D.R.N., A.P.K., J.-H.Y., R.K. and K.S.; writing—review and editing, P.R.S., R.K. and K.S. **Funding:**

Khajamohiddin Syed expresses sincere gratitude to the University of Zululand Research Committee for funding (Grant No. C686) and to the National Research Foundation (NRF), South Africa for a research grant (Grant No. 114159). Puleng Rosinah Syed thanks the NRF, South Africa for a DST-NRF Innovation Master's Scholarship (Grant No. 114575). Rajshekhar Karpoormath (RK) and Puleng Rosinah Syed are grateful to the Discipline of Pharmaceutical Sciences, College of Health Sciences, University of KwaZulu-Natal, Durban, South Africa for providing access to necessary facilities. RK is also thankful to the NRF, South Africa for the research grants (Grant No. 103728 and 112079). Abidemi Paul Kappo is grateful to the South African Medical Research Council (SAMRC) for a research grant (Grant No. PC57009).

Acknowledgments: The authors want to thank Barbara Bradley, Pretoria, South Africa for English language editing.

Conflicts of Interest: The authors declare no conflict of interest. The funders had no role in the design of the study; in the collection, analyses, or interpretation of data, in the writing of the manuscript, or in the decision to publish the results.

References

1. World Health Organization (WHO). Global Tuberculosis Report 2018. Available online: https://www.who.int/tb/publications/global_report/en/ (accessed on 22 March 2019).

2. *Mortality and Causes of Death in South Africa, 2016: Findings from Death Notification*; Statistics South Africa: Pretoria, South Africa, 2018. Available online: http://www.statssa.gov.za/publications/P03093/P030932016.pdf (accessed on 22 March 2019).
3. Cole, S.T.; Brosch, R.; Parkhill, J.; Garnier, T.; Churcher, C.; Harris, D.; Gordon, S.V.; Eiglmeier, K.; Gas, S.; Barry, C.E.; et al. Deciphering the biology of *Mycobacterium tuberculosis* from the complete genome sequence. *Nature* **1998**, *393*, 537–544. [CrossRef] [PubMed]
4. Nelson, D.R. Cytochrome P450 diversity in the tree of life. *Biochim. Biophys. Acta Proteins Proteom.* **2018**, *1866*, 141–154. [CrossRef]
5. Ortiz de Montellano, P.R. Potential drug targets in the *Mycobacterium tuberculosis* cytochrome P450 system. *J. Inorg. Biochem.* **2018**, *180*, 235–245. [CrossRef] [PubMed]
6. Bellamine, A.; Mangla, A.T.; Dennis, A.L.; Nes, W.D.; Waterman, M.R. Structural requirements for substrate recognition of *Mycobacterium tuberculosis* 14 alpha-demethylase: Implications for sterol biosynthesis. *J. Lipid Res.* **2001**, *42*, 128–136. [PubMed]
7. Bellamine, A.; Mangla, A.T.; Nes, W.D.; Waterman, M.R. Characterization and catalytic properties of the sterol 14alpha-demethylase from *Mycobacterium tuberculosis*. *Proc. Natl. Acad. Sci. USA* **1999**, *96*, 8937–8942. [CrossRef] [PubMed]
8. McLean, K.J.; Warman, A.J.; Seward, H.E.; Marshall, K.R.; Girvan, H.M.; Cheesman, M.R.; Waterman, M.R.; Munro, A.W. Biophysical characterization of the sterol demethylase P450 from *Mycobacterium tuberculosis*, its cognate ferredoxin, and their interactions. *Biochemistry* **2006**, *45*, 8427–8443. [CrossRef]
9. Belin, P.; Le Du, M.H.; Fielding, A.; Lequin, O.; Jacquet, M.; Charbonnier, J.B.; Lecoq, A.; Thai, R.; Courcon, M.; Masson, C.; et al. Identification and structural basis of the reaction catalyzed by CYP121, an essential cytochrome P450 in *Mycobacterium tuberculosis*. *Proc. Natl. Acad. Sci. USA* **2009**, *106*, 7426–7431. [CrossRef]
10. Johnston, J.B.; Ouellet, H.; Ortiz de Montellano, P.R. Functional redundancy of steroid C26-monooxygenase activity in *Mycobacterium tuberculosis* revealed by biochemical and genetic analyses. *J. Biol. Chem.* **2010**, *285*, 36352–36360. [CrossRef] [PubMed]
11. Driscoll, M.D.; McLean, K.J.; Levy, C.; Mast, N.; Pikuleva, I.A.; Lafite, P.; Rigby, S.E.; Leys, D.; Munro, A.W. Structural and biochemical characterization of *Mycobacterium tuberculosis* CYP142: Evidence for multiple cholesterol 27-hydroxylase activities in a human pathogen. *J. Biol. Chem.* **2010**, *285*, 38270–38282. [CrossRef] [PubMed]
12. Johnston, J.B.; Kells, P.M.; Podust, L.M.; Ortiz de Montellano, P.R. Biochemical and structural characterization of CYP124: A methyl-branched lipid omega-hydroxylase from *Mycobacterium tuberculosis*. *Proc. Natl. Acad. Sci. USA* **2009**, *106*, 20687–20692. [CrossRef] [PubMed]
13. Johnston, J.B.; Singh, A.A.; Clary, A.A.; Chen, C.K.; Hayes, P.Y.; Chow, S.; De Voss, J.J.; Ortiz de Montellano, P.R. Substrate analog studies of the omega-regiospecificity of *Mycobacterium tuberculosis* cholesterol metabolizing cytochrome P450 enzymes CYP124A1, CYP125A1 and CYP142A1. *Bioorg. Med. Chem.* **2012**, *20*, 4064–4081. [CrossRef] [PubMed]
14. Sogi, K.M.; Holsclaw, C.M.; Fragiadakis, G.K.; Nomura, D.K.; Leary, J.A.; Bertozzi, C.R. Biosynthesis and regulation of sulfomenaquinone, a metabolite associated with virulence in *Mycobacterium tuberculosis*. *ACS Infect. Dis.* **2016**, *2*, 800–806. [CrossRef]
15. McLean, K.J.; Munro, A.W. Structural biology and biochemistry of cytochrome P450 systems in *Mycobacterium tuberculosis*. *Drug Metab. Rev.* **2008**, *40*, 427–446. [CrossRef] [PubMed]
16. Ouellet, H.; Johnston, J.B.; Ortiz de Montellano, P.R. The *Mycobacterium tuberculosis* cytochrome P450 system. *Arch Biochem. Biophys.* **2010**, *493*, 82–95. [CrossRef] [PubMed]
17. Griffin, J.E.; Gawronski, J.D.; Dejesus, M.A.; Ioerger, T.R.; Akerley, B.J.; Sassetti, C.M. High-resolution phenotypic profiling defines genes essential for mycobacterial growth and cholesterol catabolism. *PLoS Pathog.* **2011**, *7*, e1002251. [CrossRef]
18. Sassetti, C.M.; Boyd, D.H.; Rubin, E.J. Genes required for mycobacterial growth defined by high density mutagenesis. *Mol. Microbiol.* **2003**, *48*, 77–84. [CrossRef] [PubMed]
19. Medema, M.H.; Kottmann, R.; Yilmaz, P.; Cummings, M.; Biggins, J.B.; Blin, K.; de Bruijn, I.; Chooi, Y.H.; Claesen, J.; Coates, R.C.; et al. Minimum information about a biosybnthetic gene cluster. *Nat. Chem. Biol.* **2015**, *11*, 625–631. [CrossRef]
20. Greule, A.; Stok, J.E.; De Voss, J.J.; Cryle, M.J. Unrivalled diversity: The many roles and reactions of bacterial cytochromes P450 in secondary metabolism. *Nat. Prod. Rep.* **2018**, *35*, 757–791. [CrossRef] [PubMed]

21. Mthethwa, B.C.; Chen, W.; Ngwenya, M.L.; Kappo, A.P.; Syed, P.R.; Karpoormath, R.; Yu, J.H.; Nelson, D.R.; Syed, K. Comparative analyses of cytochrome P450s and those associated with secondary metabolism in *Bacillus* species. *Int. J. Mol. Sci.* **2018**, *19*, 3623. [CrossRef]
22. Senate, L.M.; Tjatji, M.P.; Pillay, K.; Chen, W.; Zondo, N.M.; Syed, P.R.; Mnguni, F.C.; Chiliza, Z.E.; Bamal, H.D.; Karpoormath, R.; et al. Similarities, variations, and evolution of cytochrome P450s in *Streptomyces* versus *Mycobacterium*. *Sci. Rep.* **2019**, *9*, 3962. [CrossRef]
23. Parvez, M.; Qhanya, L.B.; Mthakathi, N.T.; Kgosiemang, I.K.; Bamal, H.D.; Pagadala, N.S.; Xie, T.; Yang, H.; Chen, H.; Theron, C.W.; et al. Molecular evolutionary dynamics of cytochrome P450 monooxygenases across kingdoms: Special focus on mycobacterial P450s. *Sci. Rep.* **2016**, *6*, 33099. [CrossRef] [PubMed]
24. Jawallapersand, P.; Mashele, S.S.; Kovacic, L.; Stojan, J.; Komel, R.; Pakala, S.B.; Krasevec, N.; Syed, K. Cytochrome P450 monooxygenase CYP53 family in fungi: Comparative structural and evolutionary analysis and its role as a common alternative anti-fungal drug target. *PLoS ONE* **2014**, *9*, e107209. [CrossRef] [PubMed]
25. Bamal, H.D.; Chen, W.; Mashele, S.S.; Nelson, D.R.; Kappo, A.P.; Mosa, R.A.; Yu, J.H.; Tuszynski, J.A.; Syed, K. Comparative analyses and structural insights of the novel cytochrome P450 fusion protein family CYP5619 in Oomycetes. *Sci. Rep.* **2018**, *8*, 6597. [CrossRef] [PubMed]
26. Pei, J.; Kim, B.H.; Grishin, N.V. PROMALS3D: A tool for multiple protein sequence and structure alignments. *Nucleic Acids Res.* **2008**, *36*, 2295–2300. [CrossRef]
27. Pei, J.; Grishin, N.V. AL2CO: Calculation of positional conservation in a protein sequence alignment. *Bioinformatics (Oxf. Engl.)* **2001**, *17*, 700–712. [CrossRef]
28. Syed, K.; Mashele, S.S. Comparative analysis of P450 signature motifs EXXR and CXG in the large and diverse kingdom of fungi: Identification of evolutionarily conserved amino acid patterns characteristic of P450 family. *PLoS ONE* **2014**, *9*, e95616. [CrossRef]
29. Sello, M.M.; Jafta, N.; Nelson, D.R.; Chen, W.; Yu, J.H.; Parvez, M.; Kgosiemang, I.K.; Monyaki, R.; Raselemane, S.C.; Qhanya, L.B.; et al. Diversity and evolution of cytochrome P450 monooxygenases in Oomycetes. *Sci. Rep.* **2015**, *5*, 11572. [CrossRef]
30. Parvez, A.; Giri, S.; Giri, G.R.; Kumari, M.; Bisht, R.; Saxena, P. Novel Type III polyketide synthases biosynthesize methylated polyketides in *Mycobacterium marinum*. *Sci. Rep.* **2018**, *8*, 6529. [CrossRef] [PubMed]
31. Zhou, T.; Komaki, H.; Ichikawa, N.; Hosoyama, A.; Sato, S.; Igarashi, Y. Biosynthesis of akaeolide and lorneic acids and annotation of type I polyketide synthase gene clusters in the genome of *Streptomyces* sp. NPS554. *Mar. Drugs* **2015**, *13*, 581–596. [CrossRef]
32. Ueda, J.Y.; Hashimoto, J.; Yamamura, H.; Hayakawa, M.; Takagi, M.; Shin-ya, K. A new 16-membered tetraene macrolide JBIR-100 from a newly identified *Streptomyces* species. *J. Antibiot.* **2010**, *63*, 627–629. [CrossRef]
33. Huss, M.; Wieczorek, H. Inhibitors of V-ATPases: Old and new players. *J. Exp. Biol.* **2009**, *212*, 341–346. [CrossRef] [PubMed]
34. Iwata, F.; Sato, S.; Mukai, T.; Yamada, S.; Takeo, J.; Abe, A.; Okita, T.; Kawahara, H. Lorneic acids, trialkyl-substituted aromatic acids from a marine-derived actinomycete. *J. Nat. Prod.* **2009**, *72*, 2046–2048. [CrossRef] [PubMed]
35. Salituro, G.M.; Zink, D.L.; Dahl, A.; Nielsen, J.; Wu, E.; Huang, L.; Kastner, C.; Dumont, F.J. Meridamycin: A novel nonimmunosuppressive FKBP12 ligand from *Streptomyces hygroscopicus*. *Tetrahedron Lett.* **1995**, *36*, 997–1000. [CrossRef]
36. Aghdasi, B.; Ye, K.; Resnick, A.; Huang, A.; Ha, H.C.; Guo, X.; Dawson, T.M.; Dawson, V.L.; Snyder, S.H. FKBP12, the 12-kDa FK506-binding protein, is a physiologic regulator of the cell cycle. *Proc. Natl. Acad. Sci. USA* **2001**, *98*, 2425–2430. [CrossRef]
37. Wawrocki, S.; Druszczynska, M. Inflammasomes in Mycobacterium tuberculosis-Driven Immunity. *Can. J. Infect. Dis. Med Microbiol. J. Can. Mal. Infect. Microbiol. Med* **2017**, *2017*, 2309478. [CrossRef] [PubMed]
38. Rao, S.P.; Alonso, S.; Rand, L.; Dick, T.; Pethe, K. The protonmotive force is required for maintaining ATP homeostasis and viability of hypoxic, nonreplicating *Mycobacterium tuberculosis*. *Proc. Natl. Acad. Sci. USA* **2008**, *105*, 11945–11950. [CrossRef]
39. Katsnelson, M.A.; Rucker, L.G.; Russo, H.M.; Dubyak, G.R. K+ efflux agonists induce NLRP3 inflammasome activation independently of Ca2+ signaling. *J. Immunol. (Baltim. Md. 1950)* **2015**, *194*, 3937–3952. [CrossRef] [PubMed]

40. Sarfo, F.S.; Phillips, R.; Wansbrough-Jones, M.; Simmonds, R.E. Recent advances: Role of mycolactone in the pathogenesis and monitoring of *Mycobacterium ulcerans* infection/Buruli ulcer disease. *Cell. Microbiol.* **2016**, *18*, 17–29. [CrossRef] [PubMed]
41. Jorgensen, H.; Degnes, K.F.; Dikiy, A.; Fjaervik, E.; Klinkenberg, G.; Zotchev, S.B. Insights into the evolution of macrolactam biosynthesis through cloning and comparative analysis of the biosynthetic gene cluster for a novel macrocyclic lactam, ML-449. *Appl. Environ. Microbiol.* **2010**, *76*, 283–293. [CrossRef]
42. Brautaset, T.; Sekurova, O.N.; Sletta, H.; Ellingsen, T.E.; StrLm, A.R.; Valla, S.; Zotchev, S.B. Biosynthesis of the polyene antifungal antibiotic nystatin in *Streptomyces noursei* ATCC 11455: Analysis of the gene cluster and deduction of the biosynthetic pathway. *Chem. Biol.* **2000**, *7*, 395–403. [CrossRef]
43. Julien, B.; Tian, Z.Q.; Reid, R.; Reeves, C.D. Analysis of the ambruticin and jerangolid gene clusters of *Sorangium cellulosum* reveals unusual mechanisms of polyketide biosynthesis. *Chem. Biol.* **2006**, *13*, 1277–1286. [CrossRef] [PubMed]
44. Li, Y.; Kong, L.; Shen, J.; Wang, Q.; Liu, Q.; Yang, W.; Deng, Z.; You, D. Characterization of the positive SARP family regulator PieR for improving piericidin A1 production in *Streptomyces piomogeues* var. Hangzhouwanensis. *Synth. Syst. Biotechnol.* **2019**, *4*, 16–24. [CrossRef]
45. Distler, J.; Ebert, A.; Mansouri, K.; Pissowotzki, K.; Stockmann, M.; Piepersberg, W. Gene cluster for streptomycin biosynthesis in *Streptomyces griseus*: Nucleotide sequence of three genes and analysis of transcriptional activity. *Nucleic Acids Res.* **1987**, *15*, 8041–8056. [CrossRef] [PubMed]
46. Sun, Y.; Zhou, X.; Dong, H.; Tu, G.; Wang, M.; Wang, B.; Deng, Z. A complete gene cluster from *Streptomyces nanchangensis* NS3226 encoding biosynthesis of the polyether ionophore nanchangmycin. *Chem. Biol.* **2003**, *10*, 431–441. [CrossRef]
47. Rausch, K.; Hackett, B.A.; Weinbren, N.L.; Reeder, S.M.; Sadovsky, Y.; Hunter, C.A.; Schultz, D.C.; Coyne, C.B.; Cherry, S. Screening bioactives reveals nanchangmycin as a broad spectrum antiviral active against Zika virus. *Cell Rep.* **2017**, *18*, 804–815. [CrossRef] [PubMed]
48. Liu, T.; Lin, X.; Zhou, X.; Deng, Z.; Cane, D.E. Mechanism of thioesterase-catalyzed chain release in the biosynthesis of the polyether antibiotic nanchangmycin. *Chem. Biol.* **2008**, *15*, 449–458. [CrossRef] [PubMed]
49. Cassinelli, G.; Grein, A.; Orezzi, P.; Pennella, P.; Sanfilippo, A. New antibiotics produced by *Streptoverticillium orinoci*, n. sp. *Arch. Fur Mikrobiol.* **1967**, *55*, 358–368. [CrossRef]
50. Elnakady, Y.A.; Chatterjee, I.; Bischoff, M.; Rohde, M.; Josten, M.; Sahl, H.G.; Herrmann, M.; Muller, R. Investigations to the antibacterial mechanism of action of kendomycin. *PLoS ONE* **2016**, *11*, e0146165. [CrossRef]
51. McAlpine, J.B.; Bachmann, B.O.; Piraee, M.; Tremblay, S.; Alarco, A.M.; Zazopoulos, E.; Farnet, C.M. Microbial genomics as a guide to drug discovery and structural elucidation: ECO-02301, a novel antifungal agent, as an example. *J. Nat. Prod.* **2005**, *68*, 493–496. [CrossRef] [PubMed]
52. Xiao, Y.; Li, S.; Niu, S.; Ma, L.; Zhang, G.; Zhang, H.; Zhang, G.; Ju, J.; Zhang, C. Characterization of tiacumicin B biosynthetic gene cluster affording diversified tiacumicin analogues and revealing a tailoring dihalogenase. *J. Am. Chem. Soc.* **2011**, *133*, 1092–1105. [CrossRef]
53. Glaus, F.; Altmann, K.H. Total synthesis of the tiacumicin B (lipiarmycin A3/fidaxomicin) aglycone. *Angew. Chem. (Int. Ed. Engl.)* **2015**, *54*, 1937–1940. [CrossRef]
54. Salomon, A.R.; Voehringer, D.W.; Herzenberg, L.A.; Khosla, C. Apoptolidin, a selective cytotoxic agent, is an inhibitor of F0F1-ATPase. *Chem. Biol.* **2001**, *8*, 71–80. [CrossRef]
55. Wender, P.A.; Sukopp, M.; Longcore, K. Apoptolidins B and C: Isolation, structure determination, and biological activity. *Org. Lett.* **2005**, *7*, 3025–3028. [CrossRef]
56. Kim, J.W.; Adachi, H.; Shin-ya, K.; Hayakawa, Y.; Seto, H. Apoptolidin, a new apoptosis inducer in transformed cells from *Nocardiopsis* sp. *J. Antibiot.* **1997**, *50*, 628–630. [CrossRef]
57. Sadaka, C.; Ellsworth, E.; Hansen, P.R.; Ewin, R.; Damborg, P.; Watts, J.L. Review on abyssomicins: Inhibitors of the chorismate pathway and folate biosynthesis. *Molecules (Basel Switz.)* **2018**, *23*, 1371. [CrossRef]
58. Skellam, E.J.; Stewart, A.K.; Strangman, W.K.; Wright, J.L. Identification of micromonolactam, a new polyene macrocyclic lactam from two marine *Micromonospora* strains using chemical and molecular methods: Clarification of the biosynthetic pathway from a glutamate starter unit. *J. Antibiot.* **2013**, *66*, 431–441. [CrossRef]

59. Isogai, A.; Sakuda, S.; Matsumoto, S.; Ogura, M.; Furihata, K.; Seto, H.; Suzuki, A. The structure of leucanicidin, a novel insecticidal macrolide produced by *Streptomyces halstedii*. *Agric. Biol. Chem.* **1984**, *48*, 1379–1381. [CrossRef]
60. Symersky, J.; Osowski, D.; Walters, D.E.; Mueller, D.M. Oligomycin frames a common drug-binding site in the ATP synthase. *Proc. Natl. Acad. Sci. USA* **2012**, *109*, 13961–13965. [CrossRef] [PubMed]
61. Niggemann, J.; Bedorf, N.; Flörke, U.; Steinmetz, H.; Gerth, K.; Reichenbach, H.; Höfle, G. Spirangien A and B, highly cytotoxic and antifungal spiroketals from the myxobacterium *Sorangium cellulosum*: Isolation, structure elucidation and chemical modifications. *Eur. J. Org. Chem.* **2005**, *2005*, 5013–5018. [CrossRef]
62. Hasenbohler, A.; Kneifel, H.; Konig, W.A.; Zahner, H.; Zeiler, H.J. Metabolic products of microorganisms. 134. Stenothricin, a new inhibitor of the bacterial cell wall synthesis (author's transl). *Arch. Microbiol.* **1974**, *99*, 307–321.
63. Habibi, D.; Ogloff, N.; Jalili, R.B.; Yost, A.; Weng, A.P.; Ghahary, A.; Ong, C.J. Borrelidin, a small molecule nitrile-containing macrolide inhibitor of threonyl-tRNA synthetase, is a potent inducer of apoptosis in acute lymphoblastic leukemia. *Investig. New Drugs* **2012**, *30*, 1361–1370. [CrossRef]
64. Olano, C.; Wilkinson, B.; Sanchez, C.; Moss, S.J.; Sheridan, R.; Math, V.; Weston, A.J.; Brana, A.F.; Martin, C.J.; Oliynyk, M.; et al. Biosynthesis of the angiogenesis inhibitor borrelidin by *Streptomyces parvulus* Tu4055: Cluster analysis and assignment of functions. *Chem. Biol.* **2004**, *11*, 87–97. [PubMed]
65. Kataoka, T.; Yamada, A.; Bando, M.; Honma, T.; Mizoue, K.; Nagai, K. FD-891, a structural analogue of concanamycin A that does not affect vacuolar acidification or perforin activity, yet potently prevents cytotoxic T lymphocyte-mediated cytotoxicity through the blockage of conjugate formation. *Immunology* **2000**, *100*, 170–177. [CrossRef]
66. Zhou, Y.; Li, J.; Zhu, J.; Chen, S.; Bai, L.; Zhou, X.; Wu, H.; Deng, Z. Incomplete beta-ketone processing as a mechanism for polyene structural variation in the FR-008/candicidin complex. *Chem. Biol.* **2008**, *15*, 629–638. [CrossRef] [PubMed]
67. Vetcher, L.; Menzella, H.G.; Kudo, T.; Motoyama, T.; Katz, L. The antifungal polyketide ambruticin targets the HOG pathway. *Antimicrob. Agents Chemother.* **2007**, *51*, 3734–3736. [CrossRef]
68. Chen, I.A.; Chu, K.; Palaniappan, K.; Pillay, M.; Ratner, A.; Huang, J.; Huntemann, M.; Varghese, N.; White, J.R.; Seshadri, R.; et al. IMG/M v.5.0: An integrated data management and comparative analysis system for microbial genomes and microbiomes. *Nucleic Acids Res.* **2019**, *47*, D666–D677. [CrossRef] [PubMed]
69. Nelson, D.R. Cytochrome P450 nomenclature. *Methods Mol. Biol. (Clifton N.J.)* **1998**, *107*, 15–24.
70. Nelson, D.R. Cytochrome P450 nomenclature, 2004. *Methods Mol. Biol. (Clifton N.J.)* **2006**, *320*, 1–10.
71. Nelson, D.R.; Kamataki, T.; Waxman, D.J.; Guengerich, F.P.; Estabrook, R.W.; Feyereisen, R.; Gonzalez, F.J.; Coon, M.J.; Gunsalus, I.C.; Gotoh, O.; et al. The P450 superfamily: Update on new sequences, gene mapping, accession numbers, early trivial names of enzymes, and nomenclature. *DNA Cell Biol.* **1993**, *12*, 1–51. [CrossRef] [PubMed]
72. Katoh, K.; Kuma, K.; Toh, H.; Miyata, T. MAFFT version 5: Improvement in accuracy of multiple sequence alignment. *Nucleic Acids Res.* **2005**, *33*, 511–518. [CrossRef] [PubMed]
73. Boc, A.; Diallo, A.B.; Makarenkov, V. T-REX: A web server for inferring, validating and visualizing phylogenetic trees and networks. *Nucleic Acids Res.* **2012**, *40*, W573–W579. [CrossRef] [PubMed]
74. Letunic, I.; Bork, P. Interactive tree of life (iTOL) v3: An online tool for the display and annotation of phylogenetic and other trees. *Nucleic Acids Res.* **2016**, *44*, W242–W245. [CrossRef]
75. McWilliam, H.; Li, W.; Uludag, M.; Squizzato, S.; Park, Y.M.; Buso, N.; Cowley, A.P.; Lopez, R. Analysis tool web services from the EMBL-EBI. *Nucleic Acids Res.* **2013**, *41*, W597–W600. [CrossRef] [PubMed]
76. Kumar, S.; Stecher, G.; Tamura, K. MEGA7: Molecular evolutionary genetics analysis version 7.0 for Bigger Datasets. *Mol. Biol. Evol.* **2016**, *33*, 1870–1874. [CrossRef] [PubMed]
77. Blin, K.; Pascal Andreu, V.; de Los Santos, E.L.C.; Del Carratore, F.; Lee, S.Y.; Medema, M.H.; Weber, T. The antiSMASH database version 2: A comprehensive resource on secondary metabolite biosynthetic gene clusters. *Nucleic Acids Res.* **2019**, *47*, D625–D630. [CrossRef] [PubMed]

© 2019 by the authors. Licensee MDPI, Basel, Switzerland. This article is an open access article distributed under the terms and conditions of the Creative Commons Attribution (CC BY) license (http://creativecommons.org/licenses/by/4.0/).

Article

Distribution and Diversity of Cytochrome P450 Monooxygenases in the Fungal Class *Tremellomycetes*

Olufunmilayo Olukemi Akapo [1], Tiara Padayachee [1], Wanping Chen [2], Abidemi Paul Kappo [1], Jae-Hyuk Yu [3,4], David R. Nelson [5,*] and Khajamohiddin Syed [1,*]

1. Department of Biochemistry and Microbiology, Faculty of Science and Agriculture, University of Zululand, KwaDlangezwa 3886, South Africa; akapoolufunmilayo@gmail.com (O.O.A.); teez07padayachee@gmail.com (T.P.); KappoA@unizulu.ac.za (A.P.K.)
2. College of Food Science and Technology, Huazhong Agricultural University, Wuhan 430070, China; chenwanping@mail.hzau.edu.cn
3. Department of Bacteriology, University of Wisconsin-Madison, 3155 MSB, 1550 Linden Drive, Madison, WI 53706, USA; jyu1@wisc.edu
4. Department of Systems Biotechnology, Konkuk University, Seoul 05029, Korea
5. Department of Microbiology, Immunology and Biochemistry, University of Tennessee Health Science Center, Memphis, TN 38163, USA
* Correspondence: drnelson1@gmail.com (D.R.N.); khajamohiddinsyed@gmail.com (K.S.)

Received: 20 April 2019; Accepted: 30 May 2019; Published: 13 June 2019

Abstract: *Tremellomycetes*, a fungal class in the subphylum *Agaricomycotina*, contain well-known opportunistic and emerging human pathogens. The azole drug fluconazole, used in the treatment of diseases caused by some species of *Tremellomycetes*, inhibits cytochrome P450 monooxygenase CYP51, an enzyme that converts lanosterol into an essential component of the fungal cell membrane ergosterol. Studies indicate that mutations and over-expression of CYP51 in species of *Tremellomycetes* are one of the reasons for fluconazole resistance. Moreover, the novel drug, VT-1129, that is in the pipeline is reported to exert its effect by binding and inhibiting CYP51. Despite the importance of CYPs, the CYP repertoire in species of *Tremellomycetes* has not been reported to date. This study intends to address this research gap. Comprehensive genome-wide CYP analysis revealed the presence of 203 CYPs (excluding 16 pseudo-CYPs) in 23 species of *Tremellomycetes* that can be grouped into 38 CYP families and 72 CYP subfamilies. Twenty-three CYP families are new and three CYP families (CYP5139, CYP51 and CYP61) were conserved across 23 species of *Tremellomycetes*. Pathogenic cryptococcal species have 50% fewer CYP genes than non-pathogenic species. The results of this study will serve as reference for future annotation and characterization of CYPs in species of *Tremellomycetes*.

Keywords: *cryptococcus*; *cryptococcus neoformans*; cytochrome P450 monooxygenase; CYP51; fungal pathogens; genome data-mining; human pathogens; CYP diversity analysis; *tremellomycetes*; *trichosporon*

1. Introduction

Cryptococcosis is a fungal infectious disease ubiquitously distributed around the world [1]. Two fungal species, *Cryptococcus neoformans* and *C. gattii*, are the main infectious agents causing cryptococcal meningitis in both immunocompetent and immunocompromised humans [1–4]. This disease is the major cause of morbidity and mortality among people living with advanced HIV and annually accounts for 15% of all HIV-related deaths globally [5,6]. The burden of HIV-associated cryptococcal disease in Sub-Saharan Africa is alarming, as 73% of deaths in the world are reported in this region [5,6]. Apart from these opportunistic pathogens, the genus *Cryptococcus* contains species with biotechnological potential (Table 1). Among the cryptococcal species, *C. amylolentus* is closely related to the pathogenic *C. neoformans* and is extensively used for comparative studies to identify the pathogenic traits in *C. neoformans* [7].

The genus *Cryptococcus* belongs to *Tremellomycetes*, a fungal class in the subphylum *Agaricomycotina*, which contains organisms adapted to different niches and/or having different lifestyles (Table 1). Some of the organisms in this class are now regarded as emerging opportunistic human pathogens and some species are adapted to extreme ecological niches, such as cold regions (Table 1). Despite being fungi, *Naematella encephala* and *Tremella mesenterica* Fries exhibit fungal parasitism. The diverse lifestyles or characteristics of some species of *Tremellomycetes* are summarized in Table 1.

Table 1. Some species of *Tremellomycetes* and their well-known characteristics.

Species Name	Information	References
Cryptococcus neoformans	*C. neoformans* causes meningitis in immunocompromised and apparently in immunocompetent humans. This organism is considered a major opportunistic pathogen and a leading cause of mortality in patients infected with HIV.	[2]
Cryptococcus gattii	*C. gattii* causes respiratory (pneumonia) and neurological (meningoencephalitis) diseases in humans and animals and it can infect immunocompetent hosts.	[3,4]
Cryptococcus terricola JCM 24523	*C. terricola* is oleaginous yeast and has been suggested as a candidate for the consolidated bioprocessing of hydrocarbon chemicals. It has the ability to accumulate unsaturated 18 carbon chain length fatty acids, with additional minor contributions of saturated 18 carbon and 16 carbon fatty acids.	[8–10]
Cryptococcus curvatus	*C. curvatus* is oleaginous yeast capable of accumulating 18 carbon chain length fatty acids while growing on low or negative cost feedstock. Thus, it is a potential candidate for the use in industrial fermentation processes. In a rare case *C. curvatus* was found to be involved in peritonitis associated with gastric lymphoma.	[8,11,12]
Naganishia vishniacii (formerly known as *Cryptococcus vishniacii*)	*N. vishniacii* is psychrophilic yeast adapted to live in extreme conditions, such as low-temperature oligotrophic deserts. It also has the ability to grow in a low-nutrient environment, without added vitamins.	[8,13,14]
Cryptococcus wieringae	This is associated with pectin hydrolysis during the dew-wetting process of flax and found at the beginning of grape wine fermentation.	[8,15]
Cryptococcus amylolentus CBS 6273	*C. amylolentus* is the most closely known related species of the pathogenic *Cryptococcus* species complex, and is non-pathogenic.	[7,16]
Kockovaella imperatae NRRL Y-17943	*K. imperatae* is a non-pathogenic fungus used in the analysis of widespread adenine N6-methylation of active genes in fungal species.	[17]
Naematella encephela UCDFST	It is a parasite of another fungus, *Stereum sanguinolentum*. This fungus' genome sequencing was carried out for the analysis of widespread adenine N6-methylation of active genes in fungal species.	[17]

Table 1. Cont.

Species Name	Information	References
Trichosporon asahii	Some species belonging to the genus *Trichosporon* are considered emerging opportunistic human pathogens and are the third most commonly isolated non-*Candida* yeasts from humans. They live in soil and are adapted to colonize the skin, gastrointestinal, respiratory and urinary tracts of humans. *T. asahii* is the most important species causing disseminated disease in immunocompromised patients, while the inhalation of *T. asahii* spores is the most important cause of summer-type hypersensitivity pneumonitis in healthy individuals. Some *Trichosporon* species have also emerged as rare but frequently fatal pathogens causing disseminated infections (trichosporonosis) in immunocompromised individuals and intensive care unit patients.	[18–20]
Trichosporon oleaginosus IBC0246	*T. oleaginosus* is oleaginous yeast with the ability to accumulate lipids equivalent to biosynthetic kerosene, and thus is a biotechnologically valuable player for the generation of environmentally friendly (carbon-neutral) energy by converting agro-industrial waste to fuel (biodiesel).	[8,21]
Tremella mesenterica Fries	It is a parasite of crust fungus of the genus *Peniophora* and has a false appearance, as if it were growing on wood. Whereas in fact, it grows on the crust of fungal mycelium.	[22]

In countering cryptococcosis, three classes of antifungal agents are available: polyenes (such as amphotericin B), azoles (such as fluconazole) and the pyrimidine analogue to flucytosine [1]. The gold standard induction treatment includes giving amphotericin B along with flucytosine [23]. However, this combination therapy has substantial side effects and the need for intravenous medications poses a problem, as these are not readily available in developing countries, which are most affected by cryptococcosis [24]. To overcome this problem, a combination of fluconazole along with flucytosine has been recommended after initial therapy with amphotericin B and flucytosine [1,23].

Fluconazole binds to the fungal cytochrome P450 monooxygenase (CYP/P450) enzyme 14α-demethylase, named CYP51, which converts lanosterol into ergosterol, an essential component of the fungal cell membrane [25]. *C. neoformans* also has *CYP51* and quite a number of studies have indicated that the development of drug resistance to fluconazole is due to the mutations in the *CYP51* gene and to the elevated levels of CYP51 in cryptococcal species [26–30]. In addition to *C. neoformans*, drug resistance in other species of *Tremellomycetes* has also been reported owing to mutations in *CYP51* [31,32]. Recent studies have demonstrated that the new anti-cryptococcosis drug named VT-1129 that is in the pipeline strongly binds and inhibits CYP51 of *C. neoformans* and *C. gattii* [33–35].

Despite the importance of CYPs as drug targets, to date, the CYP repertoire in cryptococcal species or in other species of *Tremellomycetes* has not been elucidated. A few studies reported the CYP contingent of *C. neoformans* and *T. mesenterica* Fries with the purpose of comparing the CYP profiles with wood-degrading fungi [22,36,37]. Thus, in this study we present a comparative analysis of CYPs in species of *Tremellomycetes*.

2. Results and Discussion

2.1. Pathogenic Cryptococcal Species Have Few CYPs in Their Genomes

Genome-wide data mining of CYPs in 16 cryptococcal species revealed the presence of 112 CYPs in their genomes (Figure 1). *C. curvatus* and *C. terricola* have the highest number of CYPs (16 CYPs each), and *C. gattii* VGIV IND107 has the lowest number of CYPs (Figure 1). An interesting pattern was observed when comparing the CYP count among cryptococcal species. Almost 50% fewer CYPs were found in pathogenic cryptococcal species compared to non-pathogenic cryptococcal species (Figure 1). This suggests that adaptation to survive in a host (mainly animals) that has a rich source of simple nutrients might have led to the loss of CYPs. The same phenomenon was observed in fungal species belonging to the subphylum *Saccharomycotina*, where species lost a considerable number of CYPs owing to their adaptation to simpler carbon sources [38].

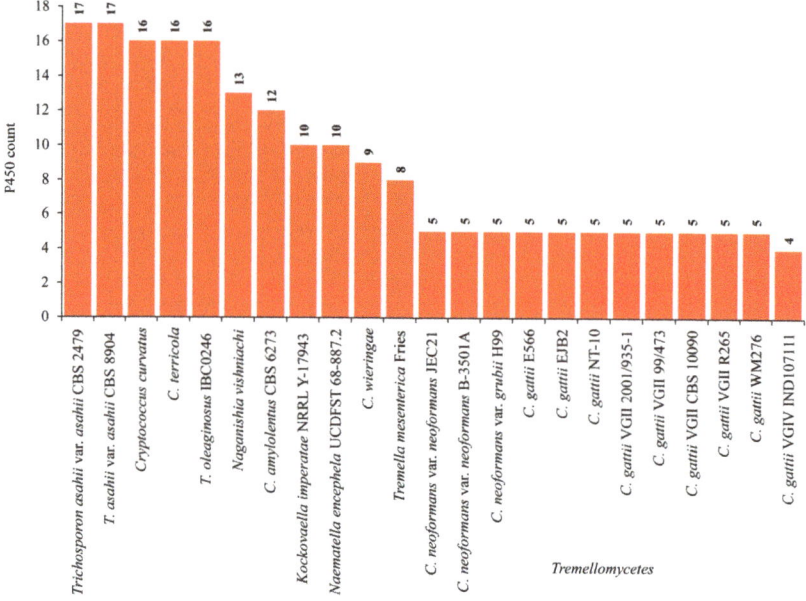

Figure 1. Comparative analysis of CYPs in the species of *Tremellomycetes*.

The comparison of cryptococcal species' CYP count with other species belonging to the same subphylum *Agaricomycotina*, especially the well-studied wood-degrading fungi, is not logical, since the wood-degrading species have quite a large number of CYPs in their genomes [22]. As cryptococcal species fall under *Tremellomycetes*, in this study, a comprehensive comparative analysis of CYPs in *Tremellomycetes* was carried out (Figure 1). As shown in Figure 1, the comparison of CYPs among species of *Tremellomycetes* indicated that pathogenic cryptococcal species have a lower number of CYPs compared to other species of *Tremellomycetes*. Fungal parasites such as *T. mesenterica* Fries and *N. encephela* have eight and 10 CYPs in their genomes, somewhat lower than non-pathogens. It is interesting to note that the species belonging to the genus *Trichosporon* have the highest number of CYPs in their genomes, both pathogenic and non-pathogenic (Figure 1). It is well-known that most of the species belonging to this genus are considered commensals of the human skin and gastrointestinal tract, and these species are now increasingly causing superficial and invasive diseases in immunocompromised individuals and intensive care unit patients [18,39]. This indicates that these organisms have a long way to go to adapt better, similar to the cryptococcal species, and thus, in this process they may lose CYPs as well.

2.2. New CYP Families Were Found in Tremellomycetes

A total of 203 CYPs were found in 23 species of *Tremellomycetes* (Figure 2 and Supplementary Dataset 1). Sixteen CYPs were found to be pseudo/false positives, as they lack one of the CYP characteristic motifs and/or short fragments (listed in Supplementary Dataset 1). Thus, these CYPs were not included in the study. The annotation of CYPs as per International P450 Nomenclature Committee rules [40–42] in combination with phylogenetic analysis (Figure 2) revealed that 203 *Tremellomycetes* CYPs could be grouped into 38 CYP families and 72 CYP subfamilies (Figure 2 and Supplementary Dataset 2, sheet 1). Phylogenetic analysis of CYPs is critical in assigning the CYP family and subfamily for the CYPs that have a borderline percentage identity of around 40–41% (for a family) and 55–56% (for a subfamily) with the named fungal CYPs.

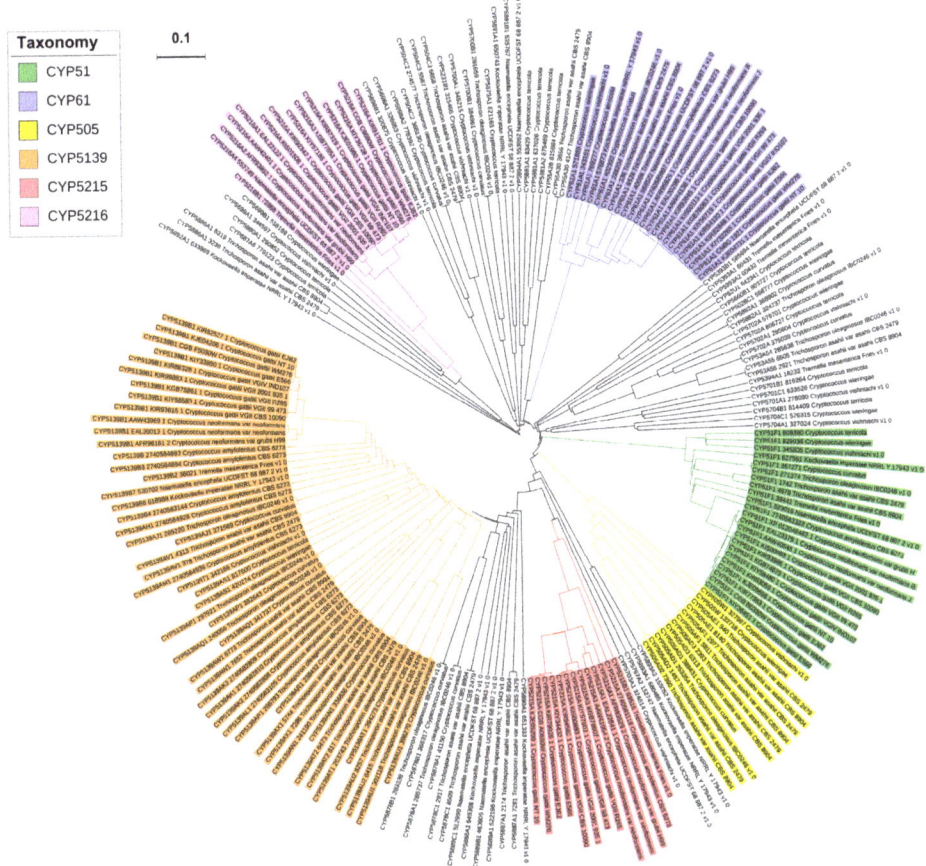

Figure 2. Phylogenetic analysis of CYPs from the species of *Tremellomycetes*. CYP families that are highly populated in species of *Tremellomycetes* are indicated in different colors. A high-resolution phylogenetic tree is provided in Supplementary Figure S1.

A total of 23 new CYP families, named CYP5215, CYP5216, CYP5393, CYP5394, CYP5698-CYP5702, CYP5878-CYP5882, and CYP5886-CYP5894A1, were identified in species of *Tremellomycetes*. *Kockovaella imperatae* NRRL Y-17943 have the highest number of new CYP families (six CYP families: CYP5888-CYP5893) followed by *Naganishia vishniacii* (five new CYP families: CYP5698-CYP5702), *C. terricola* (three new CYP families: CYP5879-CYP5881), *T. mesenterica* Fries (two new CYP families: CYP5393 and CYP5394), *T. oleaginosus* IBC0246 (CYP5878 and CYP5882) and two new CYP families (CYP5886 and CYP5887) were found in *T. asahii* var. *asahii* CBS 2479 and *T. asahii* var. *asahii* CBS 8904. Three species of *Tremellomycetes* have only one new CYP family: *C. neoformans* var. *grubii* H99 (CYP5215), *C. neoformans* var. *neoformans* B-3501A (CYP5216), and *N. encephela* UCDFST 68-887.2 (CYP5894).

2.3. Four CYP Families Are Conserved in Pathogenic Cryptococcal Species

CYP family-level comparative analysis revealed that among 38 CYP families, the CYP5139 family was found to be dominant in species of *Tremellomycetes* with 51 members, following the CYP51 and CYP61 families each with 23 members, the CYP5216 family with 14 members, the CYP5215 family with 13 members, and the CYP505 family with 12 members (Figure 3). Analysis of CYP family conservation revealed that three CYP families, namely CYP5139, CYP51, and CYP61, are conserved in all 23 species of *Tremellomycetes* (Figure 4). CYP family comparison among pathogenic cryptococcal species revealed conservation of two more CYP families, CYP5215 and CY5216, in all species except *C. gattii* VGIV IND107, which does not have CYP5215 (Figure 4). These two CYP families are also present in fungal parasites, *T. mesenterica* Fries (both families) and *N. encephela* UCDFST 68-887.2 (only CYP5216 family), and non-pathogenic *C. amylolentus* CBS 6273 (Figure 4). The CYP family CYP5231 found in *N. vishniacii* is also present in the fungal species, *Melampsora laricis-populina* and *Puccinia graminis*, belonging to the class *Pucciniomycotina*, where this family is bloomed in both species [36]. The presence of the CYP5126 family only in pathogenic or parasitic *Tremellomycetes* indicates that this CYP family might be playing a role in the adaptation of these organisms to their host. The analysis of CYP subfamilies revealed that the CYP5139 family has 17 CYP subfamilies, indicating the blooming of members in this family. The same was observed for quite a number of CYP families in other fungi [36,37].

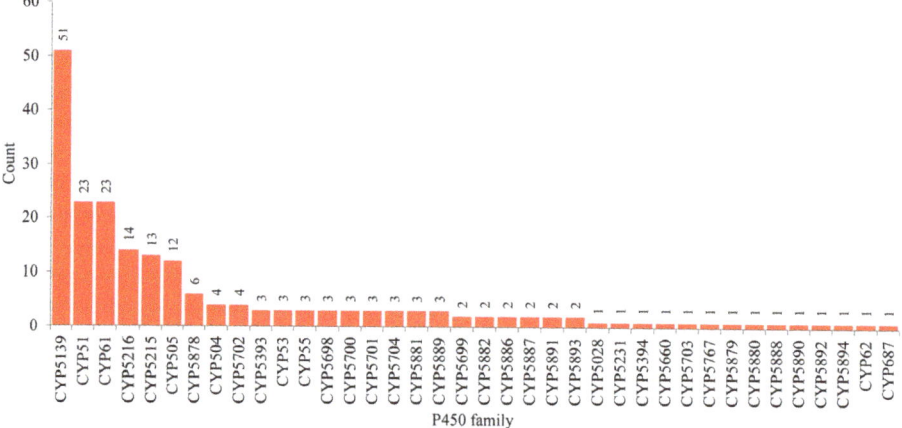

Figure 3. The CYP family-level comparative analysis in the species of *Tremellomycetes*. The numbers next to the family bar indicate the total number of CYPs. The data on the number of CYPs in each CYP family, along with subfamilies, are presented in Supplementary Dataset 2, sheet 1.

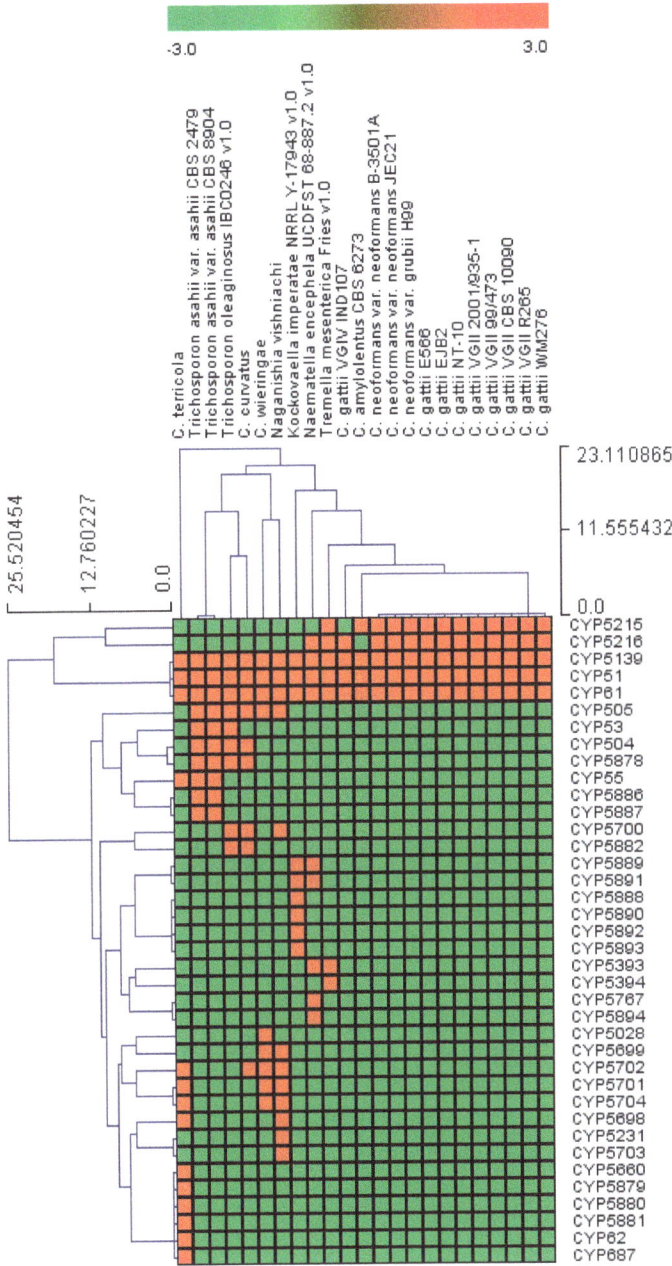

Figure 4. Heat map representing the presence or absence of cytochrome P450 families in 23 species of *Tremellomycetes*. The data have been represented as −3 for gene absence (green) and 3 for gene presence (red). Twenty-three species of *Tremellomycetes* form the horizontal axis and CYP families form the vertical axis. The data used in the generation of Figure 4 are presented in Supplementary Dataset 2, sheet 2.

2.4. Pathogenic Cryptococcal Species Have the Highest CYP Diversity

CYP diversity analysis revealed that pathogenic cryptococcal species, along with non-pathogenic *C. wieringae* and the fungal parasite *N. encephela* UCDFST 68-887.2, have 100% CYP diversity in their genomes (Figure 5 and Supplementary Dataset 2, sheet 3). *Tremellomycetes* such as *C. curvatus*, *C. amylolentus* CBS 6273 and *T. asahii* var. *asahii* strains had the lowest CYP diversity percentage. This is due to the blooming of CYP5139 members in their genome (Supplementary Dataset 2, sheet 1). The highest CYP diversity observed in pathogenic cryptococcal species is perfectly matched with species belonging to the fungal subphylum *Saccharomycotina* [38]. One commonality can be found between the species belonging to *Tremellomycetes* and *Saccharomycotina*: It can be assumed that some species of *Tremellomycetes* lost CYPs, compared to their counterparts, which may be due to the adaptation to use simple carbon sources present in the host, as observed for species of *Saccharomycotina*, where the loss of CYPs in response to the adaptation to use simpler carbon sources was observed [38].

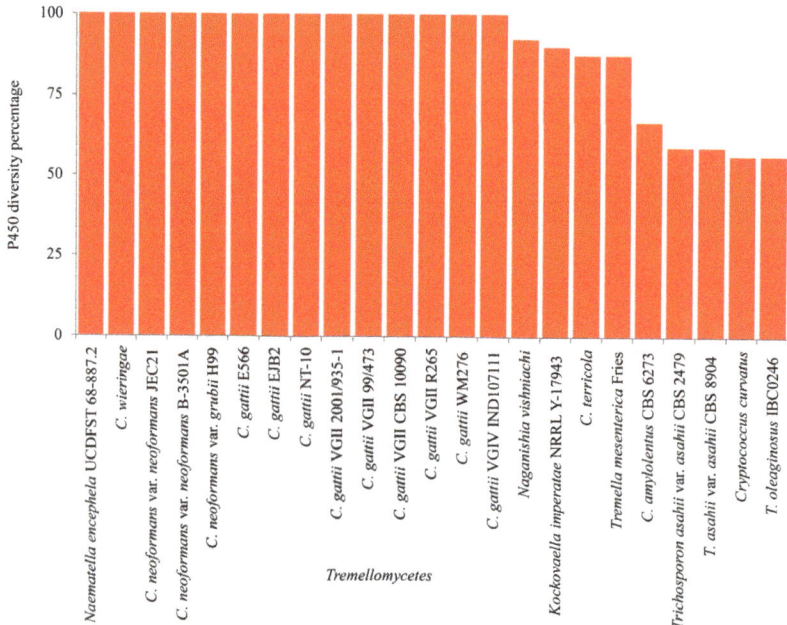

Figure 5. CYP diversity percentage analysis in *Tremellomycetes*.

2.5. Most of the CYPs from the Species of Tremellomycetes Are Orphans with No Known Function

Among CYPs from the species of *Tremellomycetes*, CYP51F1 of *C. neoformans* has been shown to be involved in 14α-demethylation of lanosterol [30] and the *CYP51F1* gene was cloned from *T. asahii* ATCC MYA-1296 = OMU239 = TIMM4014 [27]. Apart from CYP51F1, some of the CYPs' functions can be predicted based on characterized homolog CYPs. CYP61 family members are involved in membrane ergosterol biosynthesis where they catalyze C-22 sterol desaturase activity [43]. CYP505 family members are involved in oxidation of fatty acids [44]. CYP504 family members are involved in conversion of phenyl acetate to 2-hydroxyphenylacetate [45]. CYP53 family members are involved in detoxification of toxic molecules, including benzoate and its derived compounds [46–48]. The primary function of CYP53 is the conversion of benzoate to *para*-hydroxy-benzoate. A study reported that CYP53 could be an alternative antifungal drug target in view of its critical role in fungal organisms [49]. It is interesting to note that this CYP family is only present in three species belonging to the genus *Trichosporon* (Figure 4 and Supplementary Dataset 2, sheet 1). CYP55 family members are involved in the reduction of nitric oxide (NO) to nitrous oxide (N_2O) [50,51]. It is interesting to note that in addition to CYP53, CYP55 family members are also found in two species belonging to the genus *Trichosporon* (Figure 4 and Supplementary Dataset 2, sheet 1). Apart from the CYP families listed above, the rest of the CYPs found in species of *Tremellomycetes* are orphans.

3. Materials and Methods

3.1. Species and Databases

Twenty-three species of *Tremellomycetes* were used in the study (Table 2). Different databases, such as NCBI (https://www.ncbi.nlm.nih.gov/), FungiDB [52] and Joint Genome Institute MycoCosm portal [8], were browsed for cryptococcal species genomes. All the species' genomes used in the study have been published and are available for public use, with the exception of two species, *C. wieringae* and *N. vishniacii*, and permission to use these two species' CYPs was obtained from the principal investigators listed at the respective species' databases at the Joint Genome Institute MycoCosm portal [8].

Table 2. Species of *Tremellomycetes* used in the study. The respective genome databases used for analysis of CYPs, along with the reference articles, are listed in the table. Abbreviations: NCBI, National Center for Biotechnology Information and JGI, Joint Genome Institute.

Species Name	Database	Reference
Cryptococcus gattii VGII R265		
Cryptococcus gattii NT-10		
Cryptococcus gattii VGII 99/473		
Cryptococcus gattii E566		
Cryptococcus gattii VGII 2001/935-1	NCBI	[3,4]
Cryptococcus gattii VGIV IND107		
Cryptococcus gattii VGII CBS 10090		
Cryptococcus gattii VGII 2001/935-1		
Cryptococcus gattii EJB2		
Cryptococcus gattii WM276	NCBI	[3]
Cryptococcus terricola JCM 24523 v1.0	JGI	[9]
Cryptococcus curvatus ATCC 20509	JGI	[11]
Naganishia vishniacii v1.0 (formerly known as *Cryptococcus vishniacii*)	JGI	
Cryptococcus wieringae	JGI	
Cryptococcus neoformans var. *neoformans* B-3501A	NCBI	[2]
Cryptococcus neoformans var. *neoformans* JEC21	NCBI	[2]
Cryptococcus amylolentus CBS 6273	JGI	[7,16]
Kockovaella imperatae NRRL Y-17943 v1.0	JGI	[17]
Naematella encephela UCDFST 68-887.2 v1.0	JGI	[17]
Trichosporon asahii var. *asahii* CBS 2479	JGI	[19]
Trichosporon asahii var. *asahii* CBS 8904	JGI	[20]
Trichosporon oleaginosus IBC0246 v1.0	JGI	[21]
Tremella mesenterica Fries v1.0	JGI	[22]
Cryptococcus neoformans var. *grubii* H99	NCBI	[2]

3.2. CYP Mining and Annotation

Genome mining for CYPs and subsequent annotation was carried out following the protocol described elsewhere [36,38,53,54]. Briefly, proteomes cryptococcal species from different genome databases, listed in Section 3.1, were downloaded and subjected to an NCBI conserved domain search [55] to classify the proteins into different subfamilies. The proteins grouped under a CYP superfamily were selected and subjected to blast analysis against fungal CYPs [42] to identify homolog CYPs. Based on percentage identity with a named homolog CYP, the hit proteins were then assigned to different CYP families and CYP subfamilies, following the International P450 Nomenclature Committee rule, that is, >40% identity for a family and >55% identity for a subfamily [40,41]. CYPs that had less than 40% identity with named fungal CYPs [42] were assigned to a new family. For each species, CYPs from different databases were compared and duplicate CYPs were removed from the final CYP count.

3.3. Phylogenetic Analysis of CYPs

Phylogenetic analysis of CYPs was carried out following the procedure described elsewhere [36,53,56,57]. Briefly, first, the protein sequences were aligned by MAFFT v6.864 [58], and embedded on the T-REX web server [59]. Then, the alignments were automatically subjected to tree inferring and optimization by the T-REX web server. Finally, the best-inferred trees were visualized and colored by iTOL (http://itol.embl.de/about.cgi) [60].

3.4. Generation of CYP Profile Heat-Maps

The presence or absence of CYPs in species of *Tremellomycetes* was shown with heat-maps generated using CYP family data following the method described elsewhere [54,61]. The data were represented as −3 for gene absence (green) and 3 for gene presence (red). A tab-delimited file was imported into multi-experiment viewer (MeV) [62]. Hierarchical clustering using a Euclidean distance metric was used to cluster the data. Twenty-three species of *Tremellomycetes* form the horizontal axis and 38 CYP families form the vertical axis.

3.5. CYP Diversity Percentage Analysis

CYP diversity percentage analysis was carried out as described elsewhere [53,57,63,64]. Briefly, the CYP diversity percentage in species of *Tremellomycetes* was measured as a percentage contribution of the number of CYP families in the total number of CYPs.

3.6. Functional Prediction of CYPs

Literature was searched for characterized CYPs from species of *Tremellomycetes*, if any. Furthermore, functional prediction of CYPs was carried out based on the characterized homolog CYPs from different fungal organisms. CYP family level functional prediction was presented in the article.

4. Conclusions

Infections caused by human pathogenic species of *Tremellomycetes* are regarded as neglected diseases. Research on unraveling the infectious fungal pathogens' physiology and development of novel drugs against these pathogens is seldom done because of the lack of a lucrative market. However, cryptococcal meningitis remains a huge killer among people living with HIV in Sub-Saharan Africa and some of the species in the genus *Trichosporon* are now emerging human pathogens. This study's results provide insight into the CYP enzymes in the species of *Tremellomycetes*. This study revealed that cryptococcal species have almost 50% fewer CYP genes than their non-pathogenic counterparts and furthermore have the highest CYP diversity. Four CYP families were found to be conserved in pathogenic *Cryptococcus* species, indicating their important role in these pathogens. Interestingly, the CYP5139 family was bloomed with 17 CYP subfamilies in species of *Tremellomycetes*, indicating its possible key role in the physiology of these organisms. This study serves as a reference for future annotation of CYPs and has opened new vistas for the characterization of CYPs in the species of *Tremellomycetes*.

Supplementary Materials: Supplementary materials can be found at http://www.mdpi.com/1422-0067/20/12/2889/s1.

Author Contributions: Conceptualization, K.S.; methodology, O.O.A., T.P., W.C., A.P.K., J.-H.Y., D.R.N. and K.S.; validation, O.O.A., T.P., W.C., A.P.K., J.-H.Y., D.R.N. and K.S.; formal analysis, O.O.A., T.P., W.C., A.P.K., J.-H.Y., D.R.N. and K.S.; investigation, O.O.A., T.P., W.C., A.P.K., J.-H.Y., D.R.N. and K.S.; resources, K.S. and A.P.K.; data curation, O.O.A., T.P., W.C., A.P.K., J.-H.Y., D.R.N. and K.S.; writing—original draft preparation, O.O.A., W.C., J.-H.Y., D.R.N. and K.S.; writing—K.S.; visualization, O.O.A., T.P., W.C., D.R.N. and K.S.; supervision, K.S.; project administration, K.S.; funding acquisition, A.P.K. and K.S.

Funding: Khajamohiddin Syed expresses sincere gratitude to the University of Zululand Research Committee for funding (Grant No. C686) and to the National Research Foundation (NRF), South Africa for a research grant (Grant No. 114159). Tiara Padayachee is grateful to the NRF, South Africa for an honors bursary. Abidemi Paul Kappo is grateful to the South African Medical Research Council (SAMRC) for a research grant (Grant No. PC57009).

Acknowledgments: The authors want to thank Laurie Connell, School of Marine Sciences, The University of Maine, Maine, USA and Joey Spatafora, Dept. Botany & Plant Pathology, Oregon State University, Oregon, USA for permission to use *Naganishia vishniacii* (formerly *Cryptococcus vishniacii*) and *Cryptococcus wieringae* genome data for CYP analysis. The authors also want to thank Barbara Bradley, Pretoria, South Africa for English language editing.

Conflicts of Interest: The authors declare no conflict of interest. The funders had no role in the design of the study, in the collection, analyses, or interpretation of data, in the writing of the manuscript, or in the decision to publish the results.

References

1. May, R.C.; Stone, N.R.; Wiesner, D.L.; Bicanic, T.; Nielsen, K. *Cryptococcus*: from environmental saprophyte to global pathogen. *Nat. Rev. Microbiol.* **2016**, *14*, 106–117. [CrossRef] [PubMed]
2. Loftus, B.J.; Fung, E.; Roncaglia, P.; Rowley, D.; Amedeo, P.; Bruno, D.; Vamathevan, J.; Miranda, M.; Anderson, I.J.; Fraser, J.A.; et al. The genome of the basidiomycetous yeast and human pathogen *Cryptococcus neoformans*. *Science* **2005**, *307*, 1321–1324. [CrossRef] [PubMed]
3. D'Souza, C.A.; Kronstad, J.W.; Taylor, G.; Warren, R.; Yuen, M.; Hu, G.; Jung, W.H.; Sham, A.; Kidd, S.E.; Tangen, K.; et al. Genome variation in *Cryptococcus gattii*, an emerging pathogen of immunocompetent hosts. *MBio* **2011**, *2*. [CrossRef] [PubMed]
4. Farrer, R.A.; Desjardins, C.A.; Sakthikumar, S.; Gujja, S.; Saif, S.; Zeng, Q.; Chen, Y.; Voelz, K.; Heitman, J.; May, R.C.; et al. Genome evolution and innovation across the four major lineages of *Cryptococcus gattii*. *MBio* **2015**, *6*. [CrossRef] [PubMed]
5. World Health Organization (WHO). Cryptococcal Disease: What's New and Important. Available online: https://www.who.int/hiv/mediacentre/news/cryptococcal-disease-key-messages/en/ (accessed on 8 April 2019).
6. Rajasingham, R.; Smith, R.M.; Park, B.J.; Jarvis, J.N.; Govender, N.P.; Chiller, T.M.; Denning, D.W.; Loyse, A.; Boulware, D.R. Global burden of disease of HIV-associated cryptococcal meningitis: An updated analysis. *Lancet Infect. Dis.* **2017**, *17*, 873–881. [CrossRef]
7. Passer, A.R.; Coelho, M.A.; Billmyre, R.B.; Nowrousian, M.; Mittelbach, M.; Yurkov, A.M.; Averette, A.F.; Cuomo, C.A.; Sun, S.; Heitman, J. Genetic and genomic analyses reveal boundaries between species closely related to *Cryptococcus* pathogens. *bioRxiv* **2019**. [CrossRef]
8. Kuo, A.; Salamov, A.; Korzeniewski, F.; Nordberg, H.; Shabalov, I.; Dubchak, I.; Otillar, R.; Riley, R.; Ohm, R.; Nikitin, R.; et al. MycoCosm portal: gearing up for 1000 fungal genomes. *Nucleic Acids Res.* **2013**, *42*, D699–D704.
9. Close, D.; Ojumu, J.; Zhang, G. Draft genome sequence of *Cryptococcus terricola* JCM 24523, an oleaginous yeast capable of expressing exogenous DNA. *Genome Announc.* **2016**, *4*. [CrossRef]
10. Tanimura, A.; Takashima, M.; Sugita, T.; Endoh, R.; Kikukawa, M.; Yamaguchi, S.; Sakuradani, E.; Ogawa, J.; Ohkuma, M.; Shima, J. *Cryptococcus terricola* is a promising oleaginous yeast for biodiesel production from starch through consolidated bioprocessing. *Sci. Rep.* **2014**, *4*, 4776. [CrossRef]
11. Close, D.; Ojumu, J. Draft Genome sequence of the oleaginous yeast *Cryptococcus curvatus* ATCC 20509. *Genome Announc.* **2016**, *4*. [CrossRef]

12. Nowicka, J.; Nawrot, U.; Haus, O.; Kuliczkowski, K.; Fonteyne, P.-A.; Nolard, N. *Cryptococcus curvatus* in peritoneal fluid of gastric lymphoma patient with complex chromosome aberrations—Case report. *Med. Mycol. Mikol.* **2007**, *14*, 285–287.
13. Vishniac, H.S.; Hempfling, W.P. *Cryptococcus vishniacii* sp. nov. an Antarctic yeast. *Int. J. Syst. Evol. Microbiol.* **1979**, *29*, 153–158. [CrossRef]
14. Schmidt, S.K.; Vimercati, L.; Darcy, J.L.; Arán, P.; Gendron, E.M.; Solon, A.J.; Porazinska, D.; Dorador, C. A Naganishia in high places: functioning populations or dormant cells from the atmosphere? *Mycology* **2017**, *8*, 153–163. [CrossRef] [PubMed]
15. Milanović, V.; Comitini, F.; Ciani, M. Grape berry yeast communities: influence of fungicide treatments. *Int. J. Food Microbiol.* **2013**, *161*, 240–246. [CrossRef] [PubMed]
16. Sun, S.; Yadav, V.; Billmyre, R.B.; Cuomo, C.A.; Nowrousian, M.; Wang, L.; Souciet, J.L.; Boekhout, T.; Porcel, B.; Wincker, P.; et al. Fungal genome and mating system transitions facilitated by chromosomal translocations involving intercentromeric recombination. *PLoS Biol.* **2017**, *15*, e2002527. [CrossRef] [PubMed]
17. Mondo, S.J.; Dannebaum, R.O.; Kuo, R.C.; Louie, K.B.; Bewick, A.J.; LaButti, K.; Haridas, S.; Kuo, A.; Salamov, A.; Ahrendt, S.R.; et al. Widespread adenine N6-methylation of active genes in fungi. *Nat. Genet.* **2017**, *49*, 964–968. [CrossRef] [PubMed]
18. Davies, G.E.; Thornton, C.R. Differentiation of the emerging human pathogens *Trichosporon asahii* and *Trichosporon asteroides* from other pathogenic yeasts and moulds by using species-specific monoclonal antibodies. *PLoS ONE* **2014**, *9*, e84789. [CrossRef] [PubMed]
19. Yang, R.; Ao, J.; Wang, W.; Song, K.; Li, R.; Wang, D. Disseminated trichosporonosis in China. *Mycoses* **2003**, *46*, 519–523. [CrossRef]
20. Yang, R.Y.; Li, H.T.; Zhu, H.; Zhou, G.P.; Wang, M.; Wang, L. Genome sequence of the *Trichosporon asahii* environmental strain CBS 8904. *Eukaryot. Cell* **2012**, *11*, 1586–1587. [CrossRef]
21. Kourist, R.; Bracharz, F.; Lorenzen, J.; Kracht, O.N.; Chovatia, M.; Daum, C.; Deshpande, S.; Lipzen, A.; Nolan, M.; Ohm, R.A.; et al. Genomics and transcriptomics analyses of the oil-accumulating basidiomycete yeast *Trichosporon oleaginosus*: Insights into substrate utilization and alternative evolutionary trajectories of fungal mating systems. *MBio* **2015**, *6*, e00918. [CrossRef]
22. Floudas, D.; Binder, M.; Riley, R.; Barry, K.; Blanchette, R.A.; Henrissat, B.; Martínez, A.T.; Otillar, R.; Spatafora, J.W.; Yadav, J.S. The Paleozoic origin of enzymatic lignin decomposition reconstructed from 31 fungal genomes. *Science* **2012**, *336*, 1715–1719. [CrossRef] [PubMed]
23. Perfect, J.R.; Dismukes, W.E.; Dromer, F.; Goldman, D.L.; Graybill, J.R.; Hamill, R.J.; Harrison, T.S.; Larsen, R.A.; Lortholary, O.; Nguyen, M.H.; et al. Clinical practice guidelines for the management of cryptococcal disease: 2010 update by the infectious diseases society of america. *Clin. Infect. Dis.* **2010**, *50*, 291–322. [CrossRef] [PubMed]
24. Hanson, K.E.; Catania, J.; Alexander, B.D.; Perfect, J.R. Drug Resistance in Cryptococcosis. In *Antimicrobial Drug Resistance: Clinical and Epidemiological Aspects*; Mayers, D.L., Sobel, J.D., Ouellette, M., Kaye, K.S., Marchaim, D., Eds.; Springer International Publishing: Cham, Switzerland, 2017; Volume 2, pp. 1119–1140.
25. Kelly, S.L.; Kelly, D.E. Microbial cytochromes P450: biodiversity and biotechnology. Where do cytochromes P450 come from, what do they do and what can they do for us? *Philos. Trans. R. Soc. Lond. B Biol. Sci.* **2013**, *368*. [CrossRef] [PubMed]
26. Lamb, D.C.; Corran, A.; Baldwin, B.C.; Kwon-Chung, J.; Kelly, S.L. Resistant P45051A1 activity in azole antifungal tolerant *Cryptococcus neoformans* from AIDS patients. *FEBS Lett.* **1995**, *368*, 326–330. [CrossRef]
27. Sionov, E.; Chang, Y.C.; Garraffo, H.M.; Dolan, M.A.; Ghannoum, M.A.; Kwon-Chung, K.J. Identification of a *Cryptococcus neoformans* cytochrome P450 lanosterol 14alpha-demethylase (Erg11) residue critical for differential susceptibility between fluconazole/voriconazole and itraconazole/posaconazole. *Antimicrob. Agents Chemother.* **2012**, *56*, 1162–1169. [CrossRef] [PubMed]
28. Rodero, L.; Mellado, E.; Rodriguez, A.C.; Salve, A.; Guelfand, L.; Cahn, P.; Cuenca-Estrella, M.; Davel, G.; Rodriguez-Tudela, J.L. G484S amino acid substitution in lanosterol 14-alpha demethylase (ERG11) is related to fluconazole resistance in a recurrent *Cryptococcus neoformans* clinical isolate. *Antimicrob. Agents Chemother.* **2003**, *47*, 3653–3656. [CrossRef] [PubMed]
29. Ngamskulrungroj, P.; Chang, Y.; Hansen, B.; Bugge, C.; Fischer, E.; Kwon-Chung, K.J. Characterization of the chromosome 4 genes that affect fluconazole-induced disomy formation in *Cryptococcus neoformans*. *PLoS ONE* **2012**, *7*, e33022. [CrossRef]

30. Revankar, S.G.; Fu, J.; Rinaldi, M.G.; Kelly, S.L.; Kelly, D.E.; Lamb, D.C.; Keller, S.M.; Wickes, B.L. Cloning and characterization of the lanosterol 14alpha-demethylase (ERG11) gene in *Cryptococcus neoformans*. *Biochem. Biophys. Res. Commun.* **2004**, *324*, 719–728. [CrossRef]

31. Kushima, H.; Tokimatsu, I.; Ishii, H.; Kawano, R.; Shirai, R.; Kishi, K.; Hiramatsu, K.; Kadota, J. Cloning of the lanosterol 14-alpha-demethylase (ERG11) gene in *Trichosporon asahii*: A possible association between G453R amino acid substitution and azole resistance in *T. asahii*. *FEMS Yeast Res.* **2012**, *12*, 662–667. [CrossRef]

32. Kushima, H.; Tokimatsu, I.; Ishii, H.; Kawano, R.; Watanabe, K.; Kadota, J.I. A new amino acid substitution at G150S in lanosterol 14-alpha eemethylase (Erg11 protein) in multi-azole-resistant *Trichosporon asahii*. *Med. Mycol. J.* **2017**, *58*, E23–E28. [CrossRef]

33. Lockhart, S.R.; Fothergill, A.W.; Iqbal, N.; Bolden, C.B.; Grossman, N.T.; Garvey, E.P.; Brand, S.R.; Hoekstra, W.J.; Schotzinger, R.J.; Ottinger, E.; et al. The investigational fungal Cyp51 inhibitor VT-1129 demonstrates potent in vitro activity against *Cryptococcus neoformans* and *Cryptococcus gattii*. *Antimicrob. Agents Chemother.* **2016**, *60*, 2528–2531. [CrossRef] [PubMed]

34. Warrilow, A.G.; Parker, J.E.; Price, C.L.; Nes, W.D.; Garvey, E.P.; Hoekstra, W.J.; Schotzinger, R.J.; Kelly, D.E.; Kelly, S.L. The investigational drug VT-1129 is a highly potent inhibitor of *Cryptococcus* species CYP51 but only weakly inhibits the human enzyme. *Antimicrob. Agents Chemother.* **2016**, *60*, 4530–4538. [CrossRef]

35. Nielsen, K.; Vedula, P.; Smith, K.D.; Meya, D.B.; Garvey, E.P.; Hoekstra, W.J.; Schotzinger, R.J.; Boulware, D.R. Activity of VT-1129 against *Cryptococcus neoformans* clinical isolates with high fluconazole MICs. *Med. Mycol.* **2017**, *55*, 453–456. [PubMed]

36. Qhanya, L.B.; Matowane, G.; Chen, W.; Sun, Y.; Letsimo, E.M.; Parvez, M.; Yu, J.H.; Mashele, S.S.; Syed, K. Genome-wide annotation and comparative analysis of cytochrome P450 monooxygenases in Basidiomycete biotrophic plant pathogens. *PLoS ONE* **2015**, *10*, e0142100. [CrossRef] [PubMed]

37. Syed, K.; Shale, K.; Pagadala, N.S.; Tuszynski, J. Systematic identification and evolutionary analysis of catalytically versatile cytochrome P450 monooxygenase families enriched in model basidiomycete fungi. *PLoS ONE* **2014**, *9*, e86683. [CrossRef] [PubMed]

38. Kgosiemang, I.K.R.; Syed, K.; Mashele, S.S. Comparative genomics and evolutionary analysis of cytochrome P450 monooxygenases in fungal subphylum *Saccharomycotina*. *J. Pure Appl. Microbiol.* **2014**, *8*, 291–302.

39. Duarte-Oliveira, C.; Rodrigues, F.; Goncalves, S.M.; Goldman, G.H.; Carvalho, A.; Cunha, C. The cell biology of the *Trichosporon*-host interaction. *Front. Cell. Infect. Microbiol.* **2017**, *7*, 118. [CrossRef]

40. Nelson, D.R. Cytochrome P450 nomenclature. *Methods Mol. Biol.* **1998**, *107*, 15–24. [PubMed]

41. Nelson, D.R. Cytochrome P450 nomenclature, 2004. *Methods Mol. Biol.* **2006**, *320*, 1–10. [PubMed]

42. Nelson, D.R. The cytochrome p450 homepage. *Hum. Genom.* **2009**, *4*, 59–65.

43. Kelly, S.L.; Lamb, D.C.; Baldwin, B.C.; Corran, A.J.; Kelly, D.E. Characterization of *Saccharomyces cerevisiae* CYP61, sterol delta22-desaturase, and inhibition by azole antifungal agents. *J. Biol. Chem.* **1997**, *272*, 9986–9988. [CrossRef]

44. Nakayama, N.; Takemae, A.; Shoun, H. Cytochrome P450foxy, a catalytically self-sufficient fatty acid hydroxylase of the fungus *Fusarium oxysporum*. *J. Biochem.* **1996**, *119*, 435–440. [CrossRef]

45. Mingot, J.M.; Penalva, M.A.; Fernandez-Canon, J.M. Disruption of *phacA*, an *Aspergillus nidulans* gene encoding a novel cytochrome P450 monooxygenase catalyzing phenylacetate 2-hydroxylation, results in penicillin overproduction. *J. Biol. Chem.* **1999**, *274*, 14545–14550. [CrossRef]

46. Faber, B.W.; van Gorcom, R.F.; Duine, J.A. Purification and characterization of benzoate-para-hydroxylase, a cytochrome P450 (CYP53A1), from *Aspergillus niger*. *Arch. Biochem. Biophys.* **2001**, *394*, 245–254. [CrossRef]

47. Matsuzaki, F.; Wariishi, H. Molecular characterization of cytochrome P450 catalyzing hydroxylation of benzoates from the white-rot fungus *Phanerochaete chrysosporium*. *Biochem. Biophys. Res. Commun.* **2005**, *334*, 1184–1190. [CrossRef]

48. Durairaj, P.; Jung, E.; Park, H.H.; Kim, B.G.; Yun, H. Comparative functional characterization of a novel benzoate hydroxylase cytochrome P450 of *Fusarium oxysporum*. *Enzym. Microb. Technol.* **2015**, *70*, 58–65. [CrossRef]

49. Jawallapersand, P.; Mashele, S.S.; Kovacic, L.; Stojan, J.; Komel, R.; Pakala, S.B.; Krasevec, N.; Syed, K. Cytochrome P450 monooxygenase CYP53 family in fungi: Comparative structural and evolutionary analysis and its role as a common alternative anti-fungal drug target. *PLoS ONE* **2014**, *9*, e107209. [CrossRef]

50. Nakahara, K.; Tanimoto, T.; Hatano, K.; Usuda, K.; Shoun, H. Cytochrome P-450 55A1 (P-450dNIR) acts as nitric oxide reductase employing NADH as the direct electron donor. *J. Biol. Chem.* **1993**, *268*, 8350–8355.

51. Shoun, H.; Fushinobu, S.; Jiang, L.; Kim, S.W.; Wakagi, T. Fungal denitrification and nitric oxide reductase cytochrome P450nor. *Philos. Trans. R. Soc. Lond. B Biol. Sci.* **2012**, *367*, 1186–1194. [CrossRef]
52. Basenko, E.Y.; Pulman, J.A.; Shanmugasundram, A.; Harb, O.S.; Crouch, K.; Starns, D.; Warrenfeltz, S.; Aurrecoechea, C.; Stoeckert, C.J., Jr.; Kissinger, J.C.; et al. FungiDB: An integrated bioinformatic resource for fungi and Oomycetes. *J. Fungi* **2018**, *4*, 39. [CrossRef]
53. Matowane, R.G.; Wieteska, L.; Bamal, H.D.; Kgosiemang, I.K.R.; Van Wyk, M.; Manume, N.A.; Abdalla, S.M.H.; Mashele, S.S.; Gront, D.; Syed, K. *In silico* analysis of cytochrome P450 monooxygenases in chronic granulomatous infectious fungus *Sporothrix schenckii*: Special focus on CYP51. *Biochim. Biophys. Acta Proteins Proteom.* **2018**, *1866*, 166–177. [CrossRef]
54. Ngwenya, M.L.; Chen, W.; Basson, A.K.; Shandu, J.S.; Yu, J.H.; Nelson, D.R.; Syed, K. Blooming of unusual cytochrome P450s by tandem duplication in the pathogenic fungus *Conidiobolus coronatus*. *Int. J. Mol. Sci.* **2018**, *19*, 1711. [CrossRef]
55. Marchler-Bauer, A.; Bo, Y.; Han, L.; He, J.; Lanczycki, C.J.; Lu, S.; Chitsaz, F.; Derbyshire, M.K.; Geer, R.C.; Gonzales, N.R.; et al. CDD/SPARCLE: functional classification of proteins *via* subfamily domain architectures. *Nucleic Acids Res.* **2017**, *45*, D200–D203. [CrossRef]
56. Bamal, H.D.; Chen, W.; Mashele, S.S.; Nelson, D.R.; Kappo, A.P.; Mosa, R.A.; Yu, J.H.; Tuszynski, J.A.; Syed, K. Comparative analyses and structural insights of the novel cytochrome P450 fusion protein family CYP5619 in Oomycetes. *Sci. Rep.* **2018**, *8*, 6597. [CrossRef]
57. Senate, L.M.; Tjatji, M.P.; Pillay, K.; Chen, W.; Zondo, N.M.; Syed, P.R.; Mnguni, F.C.; Chiliza, Z.E.; Bamal, H.D.; Karpoormath, R.; et al. Similarities, variations, and evolution of cytochrome P450s in *Streptomyces* versus *Mycobacterium*. *Sci. Rep.* **2019**, *9*, 3962. [CrossRef]
58. Katoh, K.; Kuma, K.; Toh, H.; Miyata, T. MAFFT version 5: improvement in accuracy of multiple sequence alignment. *Nucleic Acids Res.* **2005**, *33*, 511–518. [CrossRef]
59. Boc, A.; Diallo, A.B.; Makarenkov, V. T-REX: A web server for inferring, validating and visualizing phylogenetic trees and networks. *Nucleic Acids Res.* **2012**, *40*, W573–W579. [CrossRef]
60. Letunic, I.; Bork, P. Interactive tree of life (iTOL) v3: An online tool for the display and annotation of phylogenetic and other trees. *Nucleic Acids Res.* **2016**, *44*, W242–W245. [CrossRef]
61. Mthethwa, B.C.; Chen, W.; Ngwenya, M.L.; Kappo, A.P.; Syed, P.R.; Karpoormath, R.; Yu, J.H.; Nelson, D.R.; Syed, K. Comparative analyses of cytochrome P450s and those associated with secondary metabolism in *Bacillus* species. *Int. J. Mol. Sci.* **2018**, *19*, 3623. [CrossRef]
62. Saeed, A.I.; Sharov, V.; White, J.; Li, J.; Liang, W.; Bhagabati, N.; Braisted, J.; Klapa, M.; Currier, T.; Thiagarajan, M.; et al. TM4: A free, open-source system for microarray data management and analysis. *BioTechniques* **2003**, *34*, 374–378. [CrossRef]
63. Parvez, M.; Qhanya, L.B.; Mthakathi, N.T.; Kgosiemang, I.K.; Bamal, H.D.; Pagadala, N.S.; Xie, T.; Yang, H.; Chen, H.; Theron, C.W.; et al. Molecular evolutionary dynamics of cytochrome P450 monooxygenases across kingdoms: Special focus on mycobacterial P450s. *Sci. Rep.* **2016**, *6*, 33099. [CrossRef]
64. Sello, M.M.; Jafta, N.; Nelson, D.R.; Chen, W.; Yu, J.H.; Parvez, M.; Kgosiemang, I.K.; Monyaki, R.; Raselemane, S.C.; Qhanya, L.B.; et al. Diversity and evolution of cytochrome P450 monooxygenases in Oomycetes. *Sci. Rep.* **2015**, *5*, 11572. [CrossRef]

© 2019 by the authors. Licensee MDPI, Basel, Switzerland. This article is an open access article distributed under the terms and conditions of the Creative Commons Attribution (CC BY) license (http://creativecommons.org/licenses/by/4.0/).

Article

Comprehensive Analyses of Cytochrome P450 Monooxygenases and Secondary Metabolite Biosynthetic Gene Clusters in *Cyanobacteria*

Makhosazana Jabulile Khumalo [1], Nomfundo Nzuza [1], Tiara Padayachee [1], Wanping Chen [2], Jae-Hyuk Yu [3,4], David R. Nelson [5,*] and Khajamohiddin Syed [1,*]

1. Department of Biochemistry and Microbiology, Faculty of Science and Agriculture, University of Zululand, KwaDlangezwa 3886, South Africa; khosietens@gmail.com (M.J.K.); nomfundonzuza11@gmail.com (N.N.); teez07padayachee@gmail.com (T.P.)
2. Department of Molecular Microbiology and Genetics, University of Göttingen, 37077 Göttingen, Germany; chenwanping1@foxmail.com
3. Department of Bacteriology, University of Wisconsin-Madison, 3155 MSB, 1550 Linden Drive, Madison, WI 53706, USA; jyu1@wisc.edu
4. Department of Systems Biotechnology, Konkuk University, Seoul 05029, Korea
5. Department of Microbiology, Immunology and Biochemistry, University of Tennessee Health Science Center, Memphis, TN 38163, USA
* Correspondence: dnelson@uthsc.edu (D.R.N.); khajamohiddinsyed@gmail.com (K.S.)

Received: 22 December 2019; Accepted: 15 January 2020; Published: 19 January 2020

Abstract: The prokaryotic phylum *Cyanobacteria* are some of the oldest known photosynthetic organisms responsible for the oxygenation of the earth. Cyanobacterial species have been recognised as a prosperous source of bioactive secondary metabolites with antibacterial, antiviral, antifungal and/or anticancer activities. Cytochrome P450 monooxygenases (CYPs/P450s) contribute to the production and diversity of various secondary metabolites. To better understand the metabolic potential of cyanobacterial species, we have carried out comprehensive analyses of P450s, predicted secondary metabolite biosynthetic gene clusters (BGCs), and P450s located in secondary metabolite BGCs. Analysis of the genomes of 114 cyanobacterial species identified 341 P450s in 88 species, belonging to 36 families and 79 subfamilies. In total, 770 secondary metabolite BGCs were found in 103 cyanobacterial species. Only 8% of P450s were found to be part of BGCs. Comparative analyses with other bacteria *Bacillus*, *Streptomyces* and mycobacterial species have revealed a lower number of P450s and BGCs and a percentage of P450s forming part of BGCs in cyanobacterial species. A mathematical formula presented in this study revealed that cyanobacterial species have the highest gene-cluster diversity percentage compared to *Bacillus* and mycobacterial species, indicating that these diverse gene clusters are destined to produce different types of secondary metabolites. The study provides fundamental knowledge of P450s and those associated with secondary metabolism in cyanobacterial species, which may illuminate their value for the pharmaceutical and cosmetics industries.

Keywords: cytochromes P450 monooxygenases; secondary metabolites; *Cyanobacteria*; biosynthetic gene clusters; gene-cluster diversity percentage; mathematical formula; phylogenetic analysis

1. Introduction

Cyanobacteria are thought to be some of the oldest known photosynthetic organisms that played a major role in the evolution of life by contributing to the oxygenation of the earth's atmosphere [1–5]. These Gram-negative photosynthetic prokaryotes also played a significant role in nitrogen and carbon cycles as they are able to fix atmospheric nitrogen and carbon. It is also well known that cyanobacterial species are responsible for the evolution of plants on earth owing to their endosymbiotic lifestyle, which led to the development of light-harvesting organelles in plants [6,7]. Cyanobacterial species can be found in diverse terrestrial habitats, ranging from the oceans to fresh-water bodies, soil to desert rocks, and extreme environments such as Antarctic dry valleys, Arctic and thermophilic lakes [8,9].

To survive in a wide range of environments, cyanobacterial species produce diverse natural products comprising both primary and secondary metabolites belonging to the group's non-ribosomal proteins, polyketides, and terpenes and alkaloids. These products have varying activities of anticancer, antiviral, and ultraviolet-protective activities, as well as hepatotoxicity and neurotoxicity [10,11]. Many cyanobacterial species are used as model organisms to understand fundamental processes such as photosynthesis, nitrogen fixation, and circadian rhythm [12–14]. Owing to their amenability to gene manipulation, these organisms have been genetically modified/engineered for the production of valuable human compounds [15,16]. Despite all their highly beneficial characteristics, cyanobacterial species are also known to cause cyanobacterial blooms in water, resulting in the production of toxins that are harmful to humans, wild and domestic animals [17]. An overview of some of the cyanobacterial species' beneficial and harmful characteristics is presented in Table 1.

Table 1. Some of the cyanobacterial species well-known characteristics.

Species Name	Well Known for	Reference(s)
Acaryochloris marina	Species were isolated from the marine environment and can produce chlorophyll d as primary photosynthetic pigment that is able to use far-red light for photosynthesis.	[18]
Anabaena sp. WA102	Filamentous nitrogen-fixing cyanobacterium that often form blooms in eutrophic water bodies and able to produce a range of neurotoxic secondary metabolites.	[19]
Synechocystis sp. PCC 6803	Species found in fresh water and capable of both phototrophic and heterotrophic growth; owing to this ability, it was one of the most highly studied cyanobacterium for these characteristics. This species lost its nitrogen-fixing ability.	[20]
Synechococcus elongatus PCC 6301	Unicellular, rod-shaped, fresh-water living, obligate photoautotrophic organism that has long been used as a model organism for photosynthesis research.	[21]
Synechococcus sp. WH8102	Widely found in marine water across the world. It is well known for its oligotrophic nature, as it can utilize nitrogen and phosphorus sources. It also developed strategies to conserve limited iron stores by using nickel and cobalt in some enzymes. Species belonging to the genus Synchococcus are considered generalist compared to Prochlorococcus species, as they are nutritionally versatile and adapted to different ecological niches. These species developed a unique type of swimming motility, as they propel in the absence of any demonstrable external organelle.	[22]
Thermosynechococcus elongates	This species is unique among cyanobacterial species, as it grows in hot springs and has an optimal growth temperature of 55 °C.	[23]
Cyanobium sp. NIES-981	This species is used for standard inhibition tests for toxicants in water, as it fulfills the criteria provided by the Organization for Economic Co-operation and Development test guidelines.	[24]
Dactylococcopsis salina	Gas-vacuolate cyanobacterium isolated from Solar Lake, a stratified heliothermal saline pool in Sinai.	[25]
Chamaesiphon minutus	It is an epiphyte of fresh water red alga Paralemanea catenata (Rhodophyta).	[26]
Leptolyngbya sp. NIES-3755	Species belonging to this genus are found in various environments ranging from soil and fresh water to hypogean sites. This species was isolated from the soil at the Toyohashi University of Technology, Japan.	[27–29]
Halomicronema hongdechloris	It is the first cyanobacterium to be identified that produces chlorophyll f and is isolated from a stromatolite in the World Heritage site of Shark Bay, Western Australia.	[30,31]
Pseudanabaena sp. ABRG5-3	It is a semifilamentous, non-heterocystous cyanobacterium isolated from a pond in Japan.	[32]
Prochlorococcus marinus subsp. marinus CCMP1375	Among species of the Prochlorococcus genus, this cyanobacterium is extreme as it can grow at very low light levels in the ocean. Species belonging to this genus are the smallest known oxygen-evolving autotrophs and dominate the tropical and subtropical oceanic phytoplankton community. Species in this genus are adapted to different light levels in the ocean.	[33]
Geminocystis sp. NIES-3709	Fresh water living cyanobacterium capable of accumulating large amounts of phycoerythrin, light-harvesting antenna proteins, compared to Geminocystis sp. NIES-3708.	[34]
Microcystis aeruginosa	Species belonging to this genus are the most representative of toxic bloom-forming cyanobacteria in eutrophic waters. M. aeruginosa is well-known for its toxicity by producing various toxic small polypeptides, including microcystin and cyanopeptolin.	[35]
Cyanobacterium sp. Strain HL-69	It is isolated from the magnesium sulfate-dominated hypersaline Hot Lake in northern Washington.	[36]
Crocosphaera watsonii	Nitrogen-fixing cyanobacterium found in oligotrophic oceans adapted to iron and phosphorus limitation.	[37]
Crocosphaera subtropica	Unicellular cyanobacteria capable of fixing atmospheric dinitrogen (diazotroph) in marine environments, like filamentous cyanobacterial species.	[38]
Trichodesmium erythraeum	Filamentous cyanobacterium known as the primary producer and supplier of new nitrogen through its ability to fix atmospheric dinitrogen (diazotroph) in tropical and subtropical oceans.	[39]
Arthrospira (Spirulina) platensis	Economically important cyanobacterium, an important source of nutrition and medicinal value. This species is consumed as a source of protein around the world.	[40]
Planktothrix agardhii	Cyanobacterium forming bloom in eutrophic water and capable of producing toxins.	[41]

Table 1. Cont.

Species Name	Well Known for	Reference(s)
Moorea producens	Prolific secondary metabolite producing filamentous tropical marine cyanobacterium. One-fifth of its genome is devoted to the production of secondary metabolites.	[42]
Gloeobacter violaceus	Ancient cyanobacterium that lacks thylakoid membranes.	[43]
Nostoc sp. PCC 7120	Filamentous cyanobacterium capable of fixing atmospheric dinitrogen (diazotroph).	[44]
Nostoc punctiforme	A facultative heterotroph symbiotic cyanobacterium capable of establishing symbiosis with Anthoceros punctatus.	[45]
Nostoc azollae 0708	A nitrogen-fixing endosymbiont of water fern Azolla filiculoides Lam.	[46]
Anabaena sp. strain 90	Hepatotoxic bloom-forming cyanobacterium with 5% of its genome devoted to synthesis of small peptides that are toxic to animals.	[47]
Calothrix strain 336/3	Industrially relevant cyanobacterium capable of producing higher levels of hydrogen (biofuel) compared to N. punctiforme PCC 73102 and Nostoc (Anabaena) sp. strain PCC 7120.	[48]
Fischerella sp. NIES-3754	Cyanobacterium isolated from hot spring in Japan with potential to have thermoresistant optogenetic tools.	[49]
Nodularia spumigena UHCC 0039	Cyanobacterium responsible for Baltic sea brackish water cyanobacterial blooms producing toxins.	[50]

Currently, world-wide research continues to identify novel secondary metabolites with potential biotechnological value from cyanobacterial species. These secondary metabolites are usually produced by a set of genes that are clustered in an organism, and these clusters are known as secondary metabolite biosynthetic gene clusters (BGCs) [51]. Among different genes involved in the production of secondary metabolites, P450s occupy a special place, as these enzymes contribute to the diversity of secondary metabolites due to regio- and stereo-specific oxidation of substrates [52,53]. Recent studies in different bacterial populations belonging to the genera *Bacillus* [54], *Streptomyces* and *Mycobacterium* [55,56] revealed the presence of a large number of P450s in secondary metabolite BGCs. A paper published in 2010 [11] reported some P450s present in cyanobacterial species where the authors performed blast analysis using P450s from *Nostoc* sp. strain PCC 7120 [57] and *Synechocystis* sp. PCC 6803 [58]. However, to date, a comprehensive comparative analysis of P450s and those associated with secondary metabolism in cyanobacterial species has not been reported. The availability of a large number of cyanobacterial species genomes gives us an opportunity to address these issues. In this study, genome data-mining, annotation and phylogenetic analysis of P450s in 114 cyanobacterial species were performed. The study also reports a comparative analysis of secondary metabolite BGCs in cyanobacterial species and identification of P450s that are part of the different BGCs. Furthermore, a comparative analysis of P450s and secondary metabolite key features of cyanobacterial species with other bacterial species belonging to the genera *Bacillus*, *Streptomyces* and *Mycobacterium* is presented.

2. Results and Discussion

2.1. Cyanobacterial Species Have Lowest Number of P450s

Analysis of 41 cyanobacterial genera and species belonging to the unclassified *Cyanobacteria* (Table S1) revealed the presence of P450s in all cyanobacterial genera except in the genera *Thermosynechococcus*, *Atelocyanobacterium* and *Crocosphaera* (Figure 1). This may be due to the lowest number of species genomes being available for analysis in these genera. Furthermore, two species in the unclassified *Cyanobacteria* group did not have P450s (Figure 1). Analysis of P450s at species level revealed that among 114 species, 88 species had P450s and 26 species did not have P450s in their genomes (Figure 1), indicating that most of the species in different genera had P450s.

In total, 341 P450s were found in 88 cyanobacterial species (Figures 2 and 3, Table S2). The analysis also revealed the presence of 13 P450-fragments and 15 P450 false positives in different cyanobacterial species (Table S2). A list of P450s, P450-fragments, and P450 false positives identified in cyanobacterial species, along with their sequences, is presented in Supplementary Dataset 1. The occurrence of P450-fragments and P450 false positives in organisms is quite common and these sequences were not taken for further analysis. Among cyanobacterial species, *Rivularia* sp. PCC 7116 has the highest number of P450s (16 P450s), followed by *Nostocales cyanobacterium* HT-58-2 (13 P450s), and *Nostoc flagelliforme* (12 P450s) (Figure 3 and Table S2). Most of the cyanobacterial species have a single P450 in their genome (Figure 3). Comparative analysis with other bacterial species revealed that cyanobacterial species have the lowest number of P450s compared to *Bacillus* species, *Streptomyces* and mycobacterial species (Table 2). Cyanobacterial species have an average of three P450s in their genomes compared to an average of four P450s in *Bacillus* species [54], 30 P450s in mycobacterial species [59], and 34 P450s in *Streptomyces* species [55] (Table 2).

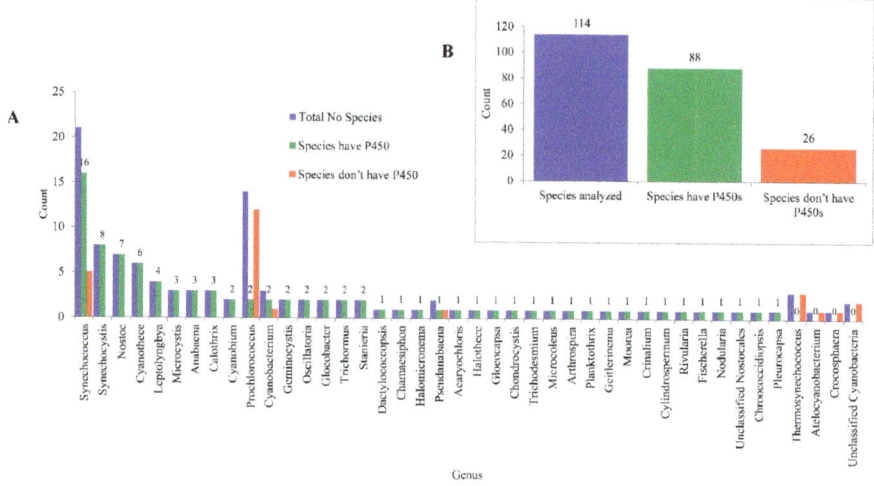

Figure 1. Analysis of P450s in *Cyanobacteria*. P450s were analyzed at both the genus level (**A**) and species level (**B**). The numbers next to the bars indicate the number of species. In Panel A, only species numbers for the species that have P450s are presented. Detailed information is presented in Tables S1 and S2.

Table 2. Comparative analysis of key features of P450s in different bacterial species.

	Cyanobacterial Species	*Bacillus* Species	Mycobacterial Species	*Streptomyces* Species
Total No. of Species Analyzed	114	128	60	48
No. of P450s	341	507	1784	1625
No. of Families	36	13	77	144
No. of Subfamilies	79	28	132	377
Dominant P450 family	CYP110	CYP107	CYP125	CYP107
No. of BGCs *	770	1098	898	1461
Types of BGCs	73	33	18	159
No. of P450s Part of BGCs	27	112	204	554
Average No. of P450s	3	4	30	34
P450 Diversity Percentage	0.09	0.02	0.07	0.18
Average No. of BGCs	7	9	15	30
Gene Cluster Diversity Percentage	0.08	0.02	0.03	0.23
Percentage of P450s Part of BGCs	8	22	11	34
Reference	This work	[54]	[55,59]	[55]

Note: * 103 cyanobacterial species gave results with anti-SMASH (antibiotics & Secondary Metabolite Analysis Shell). Eleven species genomes did not give results. Detailed information on gene clusters is presented in Table S2.

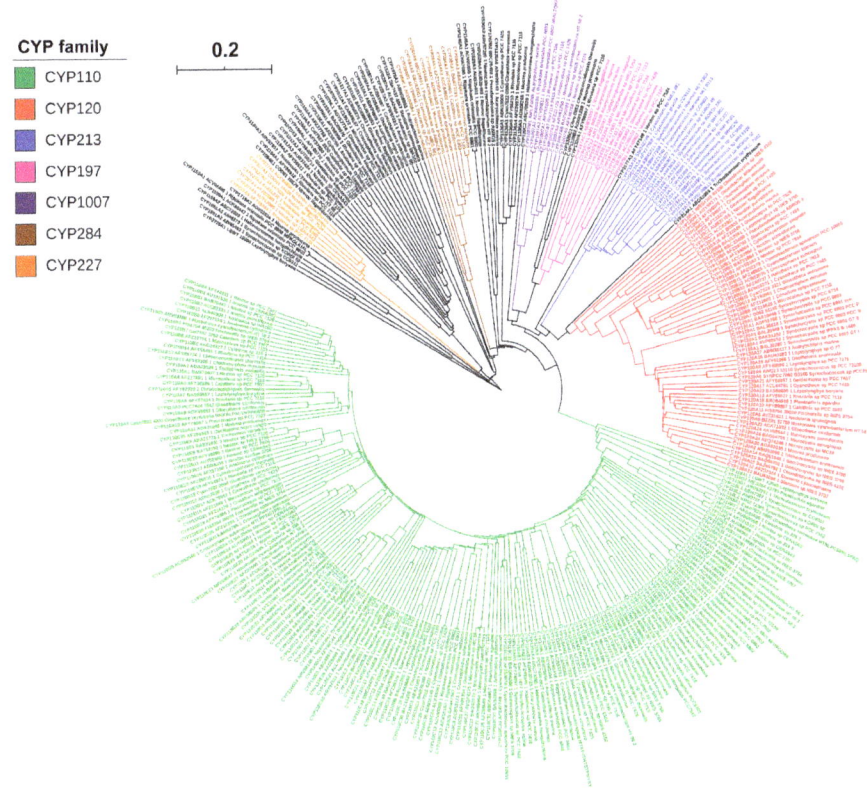

Figure 2. Phylogenetic analysis of cyanobacterial species P450s. Dominant P450 families are indicated in different colours. A high-resolution phylogenetic tree is provided as Supplementary Dataset 2.

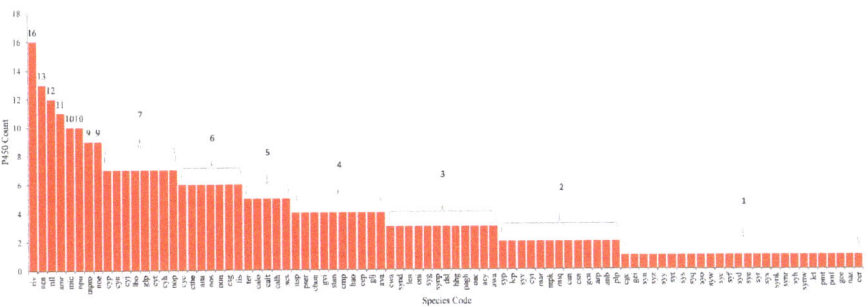

Figure 3. Comparative analysis of P450s in cyanobacterial species. The numbers next to bars indicate the number of P450s in each species. The species names with respect to their codes can be found in Table S2.

2.2. CYP110 is the Dominant P450 Family in Cyanobacterial Species

As per the International P450 Nomenclature Committee rules [60–62], 341 P450s found in 88 cyanobacterial species can be grouped into 36 P450 families and 79 P450 subfamilies (Figure 4 and Table S3). Phylogenetic analysis of cyanobacterial species P450s revealed grouping of P450s belonging to the same family together on the tree (Figure 2), indicating the correct assignment of P450 families and subfamilies. Among 36 P450 families, CYP110 has the highest number of P450s (176 P450s), followed by CYP120 (59 P450s), CYP213 (16 P450s) and P450 families, CYP197 and CYP284, which have 11 P450s in each (Figure 4). Comparative analysis of P450 families among different bacterial species revealed that different bacterial species have different dominant P450 families (Table 1). CYP110 is the dominant P450 family in cyanobacterial species, whereas CYP125 is dominant in mycobacterial species and CYP107 in *Bacillus* species and *Streptomyces* species (Table 2). The molecular basis for the blooming (P450 family with many members) [63] of certain P450 families is attributed to the species habitat and lifestyle [55]. In addition to this, P450 subfamily-level blooming was observed in cyanobacterial species, where some subfamilies were found to be dominant in a particular family (Table S3). For example, among 26 subfamilies found in the CYP110 family, subfamilies C (34 P450s), E (29 P450s), D (26 P450s) and A (15 P450s) are dominant; among six subfamilies found in CYP120 family, subfamilies A (35 P450s) and B (16 P450s) are dominant (Table S3), indicating the subfamily-level blooming of P450s in cyanobacterial species.

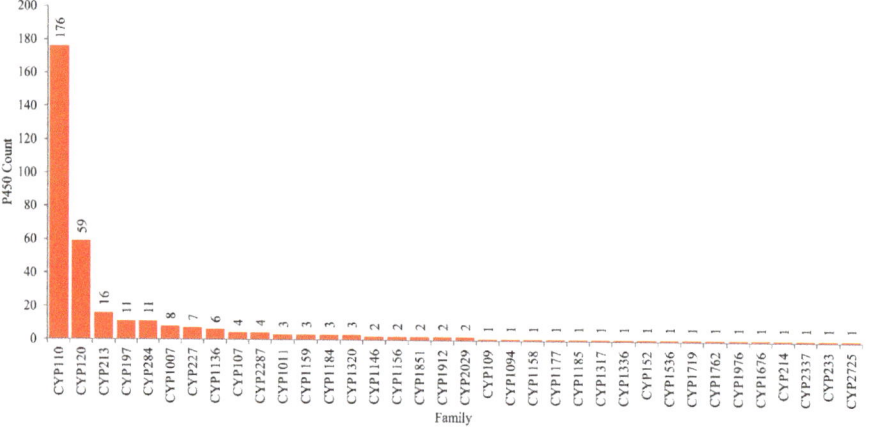

Figure 4. Family level comparative analysis of P450s in the species of *Cyanobacteria*. The numbers next to the family bar indicate the number of P450s. The data on the number of P450 families, along with subfamilies, are presented in Supplementary Table S3.

Apart from five P450 families, the remaining 31 P450 families in cyanobacterial species have a single-digit number of members (Figure 4). In fact, 17 P450 families have a single P450, indicating high P450 family diversity in cyanobacterial species. This was further confirmed when the P450 diversity percentage was compared among different bacterial species (Table 2). The P450 diversity percentage in cyanobacterial species was found to be highest compared to *Bacillus* species and mycobacterial species and almost 50% lower compared to *Streptomyces* species (Table 2). The highest P450 diversity observed for cyanobacterial species indicates that these P450s might have diverse roles, as was observed for *Streptomyces* species [55]. However, future functional analysis of cyanobacterial species P450s will provide more evidence on this observation.

P450 family conservation analysis revealed that none of the 36 families were conserved in 88 cyanobacterial species (Figure 5). The P450 profile heat-map revealed that the P450 families CYP110 and CYP120 were found to be a co-presence in most of the cyanobacterial species (Figure 5). Non-conservation of P450 families was also observed in *Bacillus* species [54], but in the *Streptomyces* [55] and mycobacterial species [59] quite a large number of P450 families were found to be conserved.

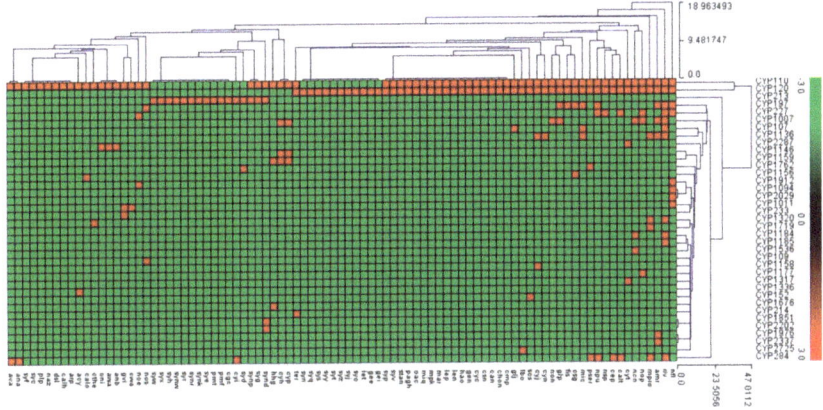

Figure 5. Heat-map of the presence/absence of P450 families in 88 cyanobacterial species. The data is represented as 3 for family presence (**red**) and −3 for family absence (**green**). Eighty-nine cyanobacterial species form the horizontal axis and P450 family numbers form the vertical axis. A detailed table showing P450 family profiles in each of the cyanobacterial species is presented in Supplementary Dataset 3.

2.3. Cyanobacterial Species Have Lowest Secondary Metabolite Biosynthetic Gene Clusters

A total of 770 secondary metabolite BGCs were found in 103 cyanobacterial species (Table S2). Species-wise comparative analysis revealed that 29 cyanobacterial species have at least two-digit numbers of secondary metabolites BGCs in their genomes (Figure 6). Among cyanobacterial species, *Cylindrospermum stagnale* (Csg) and *Prochlorococcus marinus* MIT 9303 (Pmf) have the highest number of secondary metabolite BGCs (23 BGCs in each), followed by *Nostoc* sp. CENA543 (Noe) (18 BGCs) and 17 BGCs each in *Nostocales cyanobacterium* HT-58-2 (Ncn) and *P. marinus* MIT 9313 (Pmt) (Figure 6). Comparative analysis of secondary metabolite BGCs revealed that cyanobacterial species have the lowest number of secondary metabolite BGCs in their genomes compared to *Bacillus* species, mycobacterial species and *Streptomyces* species (Table 2). On average, seven secondary metabolite BGCs were found in cyanobacterial species compared to nine in *Bacillus* species, 15 in mycobacterial species and 30 in *Streptomyces* species (Table 2). This indicates that *Streptomyces* species dominate the production of secondary metabolites and this is the reason why more than 80% of the antibiotics that are in use today are in fact sourced from *Streptomyces* species [64].

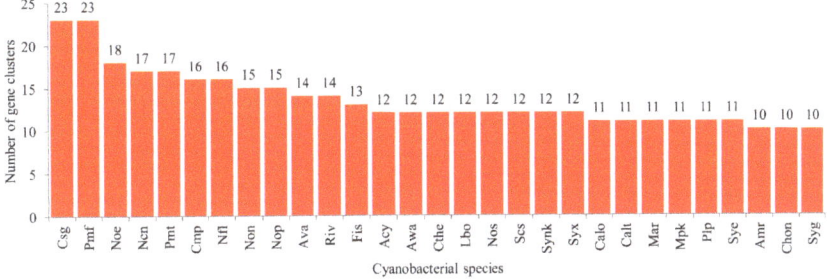

Figure 6. Comparative analysis of secondary metabolite biosynthetic gene clusters (BGCs) in cyanobacterial species. Species with a two-digit number of secondary metabolite BGCs are shown in the figure. The species names with respect to their codes can be found in Table S1. Detailed information on each of the species' secondary metabolite BGCs is presented in Table S2.

2.4. Cyanobacterial Species Has Highest Gene Cluster Diversity Percentage Compared to Bacillus and Mycobacterial Species

Analysis of types of gene clusters in 103 cyanobacterial species revealed the presence of 73 different types of secondary metabolite BGCs (Figure 7 and Table S4). Among secondary metabolite BGCs, terpene BGC is dominant (235 clusters), followed by bacteriocin (183 clusters) and non-ribosomal peptides (NRPS) (64 clusters) (Figure 7). Forty types of BGCs have only a single gene cluster, indicating the highest diversity in types of gene clusters in cyanobacterial species (Figure 7). Comparative analysis of types of BGCs among different bacterial species revealed that cyanobacterial species have the highest number of types of BGCs compared to *Bacillus* and mycobacterial species, but the lowest compared to *Streptomyces* species (Table 1).

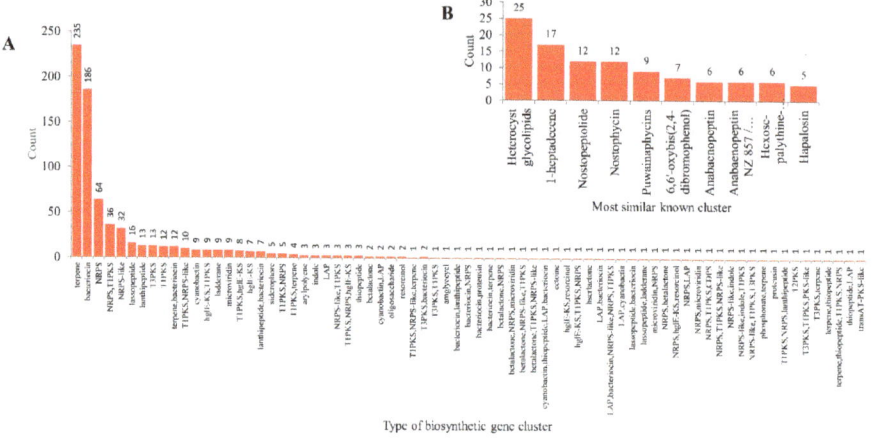

Figure 7. Comparative analysis of types of secondary metabolite biosynthetic gene clusters (BGCs) in 103 cyanobacterial species (**A**) and most similar known clusters (**B**). Standard abbreviations representing secondary metabolite BGCs as indicated in anti-SMASH (antibiotics & Secondary Metabolite Analysis Shell) [65] were used in the figure. Detailed information is presented in Supplementary Table S4.

In order to measure accurate BGC diversity among different bacterial species, we have developed a new equation, similar to the one we developed for P450 diversity percentage calculation [55], with some modification. The formula below will nullify the number of species used and will give an accurate gene cluster diversity percentage comparison between different populations.

$$\text{Gene cluster diversity percentage} = \frac{100 \times \text{Total number of types of clusters}}{\text{Total number of clusters} \times \text{number of species}}$$

Based on the above formula, the gene cluster diversity percentage in cyanobacterial species was found to be four times higher compared to *Bacillus* and mycobacterial species (Table 2). This indicates that despite cyanobacterial species having the lowest number of gene clusters, these clusters are diverse and destined to produce different types of secondary metabolites. This was evident when looking into the most similar known clusters where, among 770 clusters, only 228 clusters showed similarity to the 79 best known clusters (Figure 7 and Table S4). Among the known similar clusters, only four most similar known clusters are dominant, with 25 (heterocyst glycolipids) 17 (1-heptadecene) and 12 (Nostopeptolide and Nostophycin) (Figure 7 and Table S4). A detailed analysis on most similar known clusters is presented in Supplementary Table S4. The remaining 542 BGCs have no similar known clusters, indicating that these BGCs might encode novel secondary metabolites, possibly with potential biotechnological value.

2.5. Few Cyanobacterial Species P450s Found to be Part of Secondary Metabolite Biosynthetic Gene Clusters

Analysis of P450s that are part of different secondary metabolite BGCs revealed that only a few P450s were part of secondary metabolite BGCs in cyanobacterial species compared to *Bacillus*, mycobacterial and *Streptomyces* species (Tables 2 and 3). Only 8% of P450s are part of BGCs in cyanobacterial species compared to other bacterial species, where 22% (*Bacillus* species), 11% (mycobacterial species) and 34% (*Streptomyces* species) of P450s were found to be part of BGCs (Table 2). Among 341 P450s only 27 P450s were found to be part of secondary metabolite BGCs in cyanobacterial species, indicating that cyanobacterial species P450s might play a major role in their primary metabolism. The 27 P450s that are part of BGCs belong to six P450 families (Table 3). P450s belonging to the CYP110 family are dominantly present in BGCs (17 P450s—63%), followed by CYP213 (4 P450s—15%), CYP120 (3 P450s—11%) and a single member found in P450 families CYP1011, CYP1185, and CYP197 (Table 3). A point to be noted is that the CYP110 family is dominantly present in cyanobacterial species, indicating its requirement for the production of secondary metabolites, as the same phenomenon was observed where dominant P450 families were found to be part of BGCs in *Bacillus*, mycobacterial and *Streptomyces* species [54,55]. The 27 P450s were found to be part of 10 types of clusters, where nine P450s were found to be part of an NRPS, Type I PKS (polyketide synthase), followed by five P450s that were part of terpene and three P450s that were part of bacteriocin (Table 2). P450s found in each of the clusters and most similar known clusters were presented in Table 3. Analysis of the most similar known clusters revealed that CYP110AH1 from *Synechococcus* sp. PCC 7502 is certainly involved in the production of anabaenopeptin NZ 857/nostamide, as this P450 NRPS cluster showed 100% similarity to the gene cluster that produces the metabolite (Table 3). Apart from this match, the percentage similarity to most known clusters is very low and thus the metabolites produced by different gene clusters cannot be predicted.

2.6. Cyanobacterial Species P450s Functions and Features Resemblance to Eukaryotic P450s

Functional analysis of a few cyanobacterial species P450s revealed that these P450s have some unusual catalytic diversity and resemblance to eukaryotic P450s. Based on our study, it can safely be predicted that the 27 cyanobacterial species P450s listed in Table 3 are involved in the production of different secondary metabolites. Some of the cyanobacterial species P450 functions against different compounds were elucidated. However, the biological relevance of these reactions is not clear. CYP120A1 from *Synechocystis* sp. PCC 6803 was found to be the first non-animal retinoic acid hydroxylase [66]. This P450-catalyzed reaction represents a novel modification of retinoids compared to vertebrate CYP26 family P450s. CYP120A1 hydroxylated all-*trans*-retinoic acid at C16 or C17 positons and converted *cis*-retinoic acids (9-*cis*-retinoic acid and 13-*cis*-retinoic acid), retinal, 3(R)-OH-retinal, retinol, β-apo-13-carotenone (C_{18}) and β-apo-14'-carotenal (C_{22}) resulted in the formation of the corresponding hydroxyl derivatives [66]. CYP120A1 had the highest preference for all-*trans* substrates compared to *cis*- substrates. Among the compounds analysed, CYP120A1 had the highest activity of all-*trans*-retinoic acids, followed by β-apo-13-carotenone (C_{18}) [66]. CYP110C1 from *Nostoc* sp. PCC7120 was found to be germacrene A hydroxylase involved in sesquiterpene biosynthesis [57,67]. CYP110C1 converted germacrene A into two different products and the predominant product of this reaction was identified as 1,2,3,5,6,7,8,8aoctahydro-6-isopropenyl-4,8a-dimethylnaphth-1-ol [67]. CYP110E1 from the *Nostoc* sp. strain PCC 7120 was found to be flavone synthase, the first prokaryotic P450 with such activity [68]. CYP110E1 hydroxylated naringenin and (hydroxyl) flavanones into apigenin and (hydroxyl) flavones [68]. CYP110E1 also hydroxylated different compounds such as sesquiterpenes (zerumbone), drugs (ibuprofen and flurbiprofen), and aryl compounds (1-methoxy and 1-ethoxy naphthalene) into novel compounds that are usually difficult to synthesize chemically [68]. CYP110A1 from *Nostoc* sp. PCC 7120 was predicted to be a fatty acid ω-hydroxylase as the purified P450 binds to long-chain saturated and unsaturated fatty acids [69]. Unlike other prokaryotic P450s, CYP110A1 was found to be associated with membrane fraction, indicating its close resemblance to eukarotic P450s [69].

Table 3. Comparative analysis of P450s that are associated with secondary metabolites biosynthetic gene clusters (BGCs). Types of clusters, most similar known cluster and similarity were obtained by submitting individual P450 clusters to anti-SMASH (antibiotics & Secondary Metabolite Analysis Shell) [65]. Standard abbreviations representing type of clusters as indicated in anti-SMASH [65] were used in the table.

P450 Names	Type of Clusters	Most Similar Known Cluster	Similarity
CYP213A8	T3PKS	Xenocyloins	25%
CYP213A5	bacteriocin		
CYP213A6	T3PKS	Colicin V	2%
CYP110AH1	NRPS	Anabaenopeptin NZ 857/nostamide A	100%
CYP213A3	bacteriocin		
CYP120C2	T2PKS	Ambiguine	6%
CYP110K6	NRPS		
CYP120A21	bacteriocin		
CYP110Q3	NRPS, T1PKS	Hapalosin	40%
CYP110C17	terpene		
CYP110C29	NRPS, T1PKS	Nostophycin	27%
CYP1011G1	NRPS, T1PKS	Crocacin	23%
CYP110AP1	terpene		
CYP110AT1	NRPS, T1PKS	Hapalosin	40%
CYP110Q4	NRPS, T1PKS	Hapalosin	40%
CYP110C21	NRPS-like	Anacyclamide	14%
CYP197E3	NRPS, T1PKS	Cryptophycin	37%
CYP110AG1	terpene	Hectochlorin	25%
CYP110E29	terpene, thiopeptide, T1PKS, NRPS	Nostophycin	27%
CYP110E18	terpene, thiopeptide, T1PKS, NRPS	Nostophycin	27%
CYP110C21	terpene		
CYP110Q4	NRPS, T1PKS	Puwainaphycins	40%
CYP110AT1	NRPS, T1PKS	Puwainaphycins	40%
CYP120A13	ladderane		
CYP1185A1	lassopeptide, bacteriocin		
CYP110Q2	NRPS, T1PKS	Hapalosin	40%
CYP110C14	terpene	6,6'-oxybis(2,4-dibromophenol)	14%

P450s' role in the synthesis of carotenoids, light-harvesting pigments, in cyanobacterial species will help in addressing an evolutionary link between these species and plants since cyanobacterial species considered as precursors of chloroplasts in plants [6,7]. It is well-known that members of the CYP97 P450 family are conserved across plant taxa [70] and involved in the synthesis of carotenoids in plants [71,72]. Furthermore, research in this direction will also help in identifying the biological relevance of P450s in cyanobacterial species.

3. Materials and Methods

3.1. Species and Databases

In this study, 114 cyanobacterial species belonging to 41 genera and unclassified *Cyanobacteria* that are available for public use at Kyoto Encyclopaedia of Genes and Genomes (KEGG) [73] were used. Detailed information on different genera and species used in this study is presented in Table S1.

3.2. Genome Data Mining and Annotation of P450s

Genome data mining of P450s and their annotation was carried out following the procedure described in the literature [54,55]. Briefly, individual cyanobacterial species proteomes were downloaded from KEGG and subjected to NCBI Batch Web CD-Search Tool analysis [74]. After analysis, the results were analyzed and proteins that belong to the P450 superfamily were selected. The selected proteins were subjected to BLAST analysis at http://bioshell.pl/p450/blast_only.html as part of the P450 page at http://www.p450.unizulu.ac.za/. Based on the percentage identity to the named homolog P450s, the proteins were then annotated (assigning P450 family and subfamily), following the International P450 Nomenclature Committee rule, i.e., sequences with >40% identity were assigned to the same family as named homolog P450 and sequences with >55% identity to the same subfamily as named homolog P450 [60–62]. Proteins with less than 40% identity to a named homolog P450 were assigned to a new P450 family.

3.3. Phylogenetic Tree Construction of Cyanobacterial Species P450s

The phylogenetic tree of cyanobacterial species P450s was built using the methodology described elsewhere [54–56]. Firstly, the alignment of cyanobacterial species P450s protein sequences was performed by the MAFFT v6.864 program available at the Trex web server [75] (http://www.trex.uqam.ca/index.php?action=mafft). Then, the alignments were automatically subjected to tree inferring and optimisation by the Trex web server [76]. Briefly, the server inferred the trees with different algorithms, including maximum likelihood, maximum parsimony, neighbor joining, in the library, and searched out the best phylogenetic tree in the least-squares sense. Finally, the best-inferred tree was envisioned and coloured using iTOL [77] (https://itol.embl.de/).

3.4. Generation of P450 Profile Heat-Maps

P450 profile heat-maps were generated following the method reported in the literature [56,78,79]. The presence or absence of P450s in cyanobacterial species was shown with heat-maps generated using P450 family data. The data were represented as 3 for family presence (red) and −3 for family absence (green). A tab-delimited file was imported into Mev 4.9 (multi-experiment viewer) [80]. Hierarchical clustering using a Euclidean distance metric was used to cluster the data. Eight-nine cyanobacterial species formed the horizontal axis and P450 families formed the vertical axis.

3.5. Secondary Metabolite Biosynthetic Gene Clusters Analysis

Secondary metabolite BGCs analysis in cyanobacterial species was carried out following the method as mentioned previously [54,55]. Briefly, individual cyanobacterial species genome IDs (Table S1) were submitted to anti-SMASH (antibiotics & Secondary Metabolite Analysis Shell) [65] for identification of secondary metabolite BGCs. The gene cluster information generated by anti-SMASH is analyzed for the presence of P450s by manually mining the cluster sequences. Information on the type of cluster, most similar known cluster and percentage similarity to a known cluster is also noted and presented in table format. Among 114 cyanobacterial species, 11 species' genome IDs did not pass through anti-SMASH analysis. Thus, in this study, secondary metabolite BGCs data for 103 species is presented. Lists of species that are not part of the secondary metabolite cluster analysis are presented in Table S5.

3.6. Data Analysis

P450 diversity percentage analysis in cyanobacterial species was carried out following the method described elsewhere [55]. P450 diversity percentage is calculated using the formula: P450 diversity percentage = 100 × Total number of P450 families/Total number of P450s × Number of species. The average number of P450s was calculated using the formula: Average number of P450s = Number of P450s/Number of species. The average number of BGCs was calculated using the formula: Average number of BGCs = Total number of BGCs/Number of species. A new formula was developed in order to calculate gene cluster diversity percentage and is described in Section 2.4.

4. Conclusions

Research on harnessing the biotechnological potential of cyanobacterial species is gaining momentum. In this direction, this study is an attempt to provide a complete picture of P450 enzymes in different cyanobacterial species as these enzymes are the key players in primary and secondary metabolism of organisms, including the production of different secondary metabolites. Furthermore, providing names for P450s as per International P450 Nomenclature Committee rules enables researchers to make use of the cyanobacterial species P450 names presented in the study. A limited amount of cyanobacterial species P450s functional analysis revealed that cyanobacterial species P450s are unique in terms of their catalytic activity and they show high resemblance to eukaryotic P450s. Unravelling the role of P450s in carotenoid synthesis is necessary to understand their biological relevance in cyanobacterial species and also to address the evolutionary link between these species and plants since cyanobacterial species are considered as precursors of chloroplasts in plants. The mathematical formula presented in this study will enable researchers to conduct accurate comparison of secondary metabolite biosynthetic gene cluster diversity among different organisms. The highest gene cluster diversity observed for cyanobacterial species compared to species belonging to the genera *Bacillus* and *Mycobacterium* and the fact that a large number of biosynthetic gene clusters have no similar known clusters indicate that these gene clusters might encode novel secondary metabolites with new biological properties whose potential needs to be explored for the food, cosmetic and pharmaceutical industries.

Supplementary Materials: Supplementary materials can be found at http://www.mdpi.com/1422-0067/21/2/656/s1.

Author Contributions: Conceptualization, K.S.; data curation, M.J.K., N.N., T.P., W.C., J.-H.Y., D.R.N. and K.S.; formal analysis, M.J.K., N.N., T.P., W.C., J.-H.Y., D.R.N. and K.S.; funding acquisition, K.S.; investigation, M.J.K., N.N., T.P., W.C., J.-H.Y., D.R.N. and K.S.; methodology, M.J.K., N.N., T.P., W.C., J.-H.Y., D.R.N. and K.S.; project administration, K.S.; resources, K.S.; supervision, K.S.; validation, M.J.K., N.N., T.P., W.C., J.-H.Y., D.R.N. and K.S.; visualization, M.J.K., N.N., T.P., W.C., J.-H.Y., D.R.N. and K.S.; writing—original draft, M.J.K., N.N., T.P., W.C., J-H.Y., D.R.N. and K.S.; writing—review and editing, W.C., J.-H.Y., D.R.N. and K.S. All authors have read and agreed to the published version of the manuscript.

Funding: Khajamohiddin Syed expresses sincere gratitude to the University of Zululand Research Committee for funding (Grant No. C686) and to the National Research Foundation (NRF), South Africa for a research grant (Grant No. 114159). Honours students, Makhosazana Jabulile Khumalo (HBG181018376835), Nomfundo Nzuza (MND190626451135) and Tiara Padayachee (MND190619448759) thank the NRF, South Africa for honours bursaries.

Acknowledgments: The authors want to thank Barbara Bradley, Pretoria, South Africa for English language editing.

Conflicts of Interest: The authors declare no conflict of interest. The funders had no role in the design of the study, in the collection, analyses, or interpretation of data, in the writing of the manuscript, or in the decision to publish the results.

References

1. Bekker, A.; Holland, H.; Wang, P.-L.; Rumble Iii, D.; Stein, H.; Hannah, J.; Coetzee, L.; Beukes, N. Dating the rise of atmospheric oxygen. *Nature* **2004**, *427*, 117–120. [CrossRef]
2. Kump, L.R. The rise of atmospheric oxygen. *Nature* **2008**, *451*, 277–278. [CrossRef]
3. Shih, P.M. Cyanobacterial evolution: Fresh insight into ancient questions. *Curr. Biol.* **2015**, *25*, R192–R193. [CrossRef]
4. Flores, F.G. *The Cyanobacteria: Molecular Biology, Genomics, and Evolution*; Horizon Scientific Press: Heatherset, UK, 2008.
5. Buick, R. The antiquity of oxygenic photosynthesis: Evidence from stromatolites in sulphate-deficient Archaean lakes. *Science* **1992**, *255*, 74–77. [CrossRef]
6. Raven, J.A.; Allen, J.F. Genomics and chloroplast evolution: What did cyanobacteria do for plants? *Genome Biol.* **2003**, *4*, 209. [CrossRef]
7. Willis, K.; McElwain, J. *The Evolution of Plants*; Oxford University Press: Oxford, UK, 2014.
8. Gaysina, L.A.; Saraf, A.; Singh, P. Cyanobacteria in Diverse Habitats. In *Cyanobacteria*; Elsevier: Amsterdam, The Netherlands, 2019; pp. 1–28.
9. Dvořák, P.; Casamatta, D.A.; Hašler, P.; Jahodářová, E.; Norwich, A.R.; Poulíčková, A. Diversity of the cyanobacteria. In *Modern Topics in the Phototrophic Prokaryotes*; Springer: Berlin/Heidelberg, Germany, 2017; pp. 3–46.
10. Kultschar, B.; Llewellyn, C. Secondary Metabolites in Cyanobacteria. *Second. Metab. Sources Appl.* **2018**, *23*. [CrossRef]

11. Robert, F.O.; Pandhal, J.; Wright, P.C. Exploiting cyanobacterial P450 pathways. *Curr. Opin. Microbiol.* **2010**, *13*, 301–306. [CrossRef]
12. Cohen, S.E.; Golden, S.S. Circadian rhythms in cyanobacteria. *Microbiol. Mol. Biol. Rev.* **2015**, *79*, 373–385. [CrossRef] [PubMed]
13. Jensen, P.E.; Leister, D. Cyanobacteria as an experimental platform for modifying bacterial and plant photosynthesis. *Front. Bioeng. Biotechnol.* **2014**, *2*, 7. [CrossRef] [PubMed]
14. Bothe, H.; Schmitz, O.; Yates, M.G.; Newton, W.E. Nitrogen fixation and hydrogen metabolism in cyanobacteria. *Microbiol. Mol. Biol. Rev.* **2010**, *74*, 529–551. [CrossRef] [PubMed]
15. Santos-Merino, M.; Singh, A.K.; Ducat, D.C. New applications of synthetic biology tools for cyanobacterial metabolic engineering. *Front. Bioeng. Biotechnol.* **2019**, *7*. [CrossRef]
16. Wilde, A.; Dienst, D. Tools for genetic manipulation of cyanobacteria. In *Bioenergetic Processes of Cyanobacteria*; Springer: Berlin/Heidelberg, Germany, 2011; pp. 685–703.
17. Dittmann, E.; Fewer, D.P.; Neilan, B.A. Cyanobacterial toxins: Biosynthetic routes and evolutionary roots. *FEMS Microbiol. Rev.* **2013**, *37*, 23–43. [CrossRef] [PubMed]
18. Swingley, W.D.; Chen, M.; Cheung, P.C.; Conrad, A.L.; Dejesa, L.C.; Hao, J.; Honchak, B.M.; Karbach, L.E.; Kurdoglu, A.; Lahiri, S.; et al. Niche adaptation and genome expansion in the chlorophyll d-producing cyanobacterium *Acaryochloris marina*. *Proc. Natl. Acad. Sci. USA* **2008**, *105*, 2005–2010. [CrossRef] [PubMed]
19. Brown, N.M.; Mueller, R.S.; Shepardson, J.W.; Landry, Z.C.; Morre, J.T.; Maier, C.S.; Hardy, F.J.; Dreher, T.W. Structural and functional analysis of the finished genome of the recently isolated toxic *Anabaena* sp. WA102. *BMC Genom.* **2016**, *17*, 457. [CrossRef] [PubMed]
20. Kaneko, T.; Tanaka, A.; Sato, S.; Kotani, H.; Sazuka, T.; Miyajima, N.; Sugiura, M.; Tabata, S. Sequence analysis of the genome of the unicellular cyanobacterium *Synechocystis* sp. strain PCC6803. I. Sequence features in the 1 Mb region from map positions 64% to 92% of the genome. *DNA Res. Int. J. Rapid Publ. Rep. Genes Genomes* **1995**, *2*, 153–166, 191–198.
21. Sugita, C.; Ogata, K.; Shikata, M.; Jikuya, H.; Takano, J.; Furumichi, M.; Kanehisa, M.; Omata, T.; Sugiura, M.; Sugita, M. Complete nucleotide sequence of the freshwater unicellular cyanobacterium *Synechococcus elongatus* PCC 6301 chromosome: Gene content and organization. *Photosynth. Res.* **2007**, *93*, 55–67. [CrossRef]
22. Palenik, B.; Brahamsha, B.; Larimer, F.W.; Land, M.; Hauser, L.; Chain, P.; Lamerdin, J.; Regala, W.; Allen, E.E.; McCarren, J.; et al. The genome of a motile marine *Synechococcus*. *Nature* **2003**, *424*, 1037–1042. [CrossRef]
23. Nakamura, Y.; Kaneko, T.; Sato, S.; Ikeuchi, M.; Katoh, H.; Sasamoto, S.; Watanabe, A.; Iriguchi, M.; Kawashima, K.; Kimura, T.; et al. Complete genome structure of the thermophilic cyanobacterium *Thermosynechococcus elongatus* BP-1. *DNA Res. Int. J. Rapid Publ. Rep. Genes Genomes* **2002**, *9*, 123–130.
24. Yamaguchi, H.; Shimura, Y.; Suzuki, S.; Yamagishi, T.; Tatarazako, N.; Kawachi, M. Complete Genome Sequence of Cyanobium sp. NIES-981, a Marine Strain Potentially Useful for Ecotoxicological Bioassays. *Genome Announc.* **2016**, *4*. [CrossRef]
25. Walsby, A.; Van Rijn, J.; Cohen, Y. The biology of a new gas-vacuolate cyanobacterium, *Dactylococcopsis salina* sp. nov., in Solar Lake. *Proc. R. Soc. Lond. Ser. B Biol. Sci.* **1983**, *217*, 417–447.
26. Kučera, P.; Uher, B.; Komárek, O. Epiphytic cyanophytes Xenococcus kerneri and Chamaesiphon minutus on the freshwater red alga *Paralemanea catenata* (Rhodophyta). *Biologia* **2006**, *61*, 11–13. [CrossRef]
27. Hirose, Y.; Fujisawa, T.; Ohtsubo, Y.; Katayama, M.; Misawa, N.; Wakazuki, S.; Shimura, Y.; Nakamura, Y.; Kawachi, M.; Yoshikawa, H.; et al. Complete Genome Sequence of Cyanobacterium *Leptolyngbya* sp. NIES-3755. *Genome Announc.* **2016**, *4*. [CrossRef] [PubMed]
28. Shimura, Y.; Hirose, Y.; Misawa, N.; Osana, Y.; Katoh, H.; Yamaguchi, H.; Kawachi, M. Comparison of the terrestrial cyanobacterium Leptolyngbya sp. NIES-2104 and the freshwater *Leptolyngbya boryana* PCC 6306 genomes. *DNA Res. Int. J. Rapid Publ. Rep. Genes Genomes* **2015**, *22*, 403–412.
29. Bruno, L.; Billi, D.; Bellezza, S.; Albertano, P. Cytomorphological and genetic characterization of troglobitic Leptolyngbya strains isolated from Roman hypogea. *Appl. Environ. Microbiol.* **2009**, *75*, 608–617. [CrossRef] [PubMed]
30. Chen, M.; Hernandez-Prieto, M.A.; Loughlin, P.C.; Li, Y.; Willows, R.D. Genome and proteome of the chlorophyll f-producing cyanobacterium *Halomicronema hongdechloris*: Adaptative proteomic shifts under different light conditions. *BMC Genom.* **2019**, *20*, 207. [CrossRef]
31. Chen, M.; Schliep, M.; Willows, R.D.; Cai, Z.L.; Neilan, B.A.; Scheer, H. A red-shifted chlorophyll. *Science* **2010**, *329*, 1318–1319. [CrossRef]

32. Tajima, N.; Kanesaki, Y.; Sato, S.; Yoshikawa, H.; Maruyama, F.; Kurokawa, K.; Ohta, H.; Nishizawa, T.; Asayama, M.; Sato, N. Complete Genome Sequence of the Nonheterocystous Cyanobacterium *Pseudanabaena* sp. ABRG5-3. *Genome Announc.* **2018**, *6*. [CrossRef]
33. Dufresne, A.; Salanoubat, M.; Partensky, F.; Artiguenave, F.; Axmann, I.M.; Barbe, V.; Duprat, S.; Galperin, M.Y.; Koonin, E.V.; Le Gall, F.; et al. Genome sequence of the cyanobacterium *Prochlorococcus marinus* SS120, a nearly minimal oxyphototrophic genome. *Proc. Natl. Acad. Sci. USA* **2003**, *100*, 10020–10025. [CrossRef]
34. Hirose, Y.; Katayama, M.; Ohtsubo, Y.; Misawa, N.; Iioka, E.; Suda, W.; Oshima, K.; Hanaoka, M.; Tanaka, K.; Eki, T.; et al. Complete Genome Sequence of Cyanobacterium *Geminocystis* sp. Strain NIES-3709, Which Harbors a Phycoerythrin-Rich Phycobilisome. *Genome Announc.* **2015**, *3*. [CrossRef]
35. Kaneko, T.; Nakajima, N.; Okamoto, S.; Suzuki, I.; Tanabe, Y.; Tamaoki, M.; Nakamura, Y.; Kasai, F.; Watanabe, A.; Kawashima, K.; et al. Complete genomic structure of the bloom-forming toxic cyanobacterium *Microcystis aeruginosa* NIES-843. *DNA Res. Int. J. Rapid Publ. Rep. Genes Genomes* **2007**, *14*, 247–256. [CrossRef]
36. Mobberley, J.M.; Romine, M.F.; Cole, J.K.; Maezato, Y.; Lindemann, S.R.; Nelson, W.C. Draft Genome Sequence of *Cyanobacterium* sp. Strain HL-69, Isolated from a Benthic Microbial Mat from a Magnesium Sulfate-Dominated Hypersaline Lake. *Genome Announc.* **2018**, *6*. [CrossRef] [PubMed]
37. Bench, S.R.; Heller, P.; Frank, I.; Arciniega, M.; Shilova, I.N.; Zehr, J.P. Whole genome comparison of six *Crocosphaera watsonii* strains with differing phenotypes. *J. Phycol.* **2013**, *49*, 786–801. [CrossRef] [PubMed]
38. Welsh, E.A.; Liberton, M.; Stockel, J.; Loh, T.; Elvitigala, T.; Wang, C.; Wollam, A.; Fulton, R.S.; Clifton, S.W.; Jacobs, J.M.; et al. The genome of Cyanothece 51142, a unicellular diazotrophic cyanobacterium important in the marine nitrogen cycle. *Proc. Natl. Acad. Sci. USA* **2008**, *105*, 15094–15099. [CrossRef] [PubMed]
39. Capone, D.G.; Burns, J.A.; Montoya, J.P.; Subramaniam, A.; Mahaffey, C.; Gunderson, T.; Michaels, A.F.; Carpenter, E.J. Nitrogen fixation by *Trichodesmium* spp.: An important source of new nitrogen to the tropical and subtropical North Atlantic Ocean. *Glob. Biogeochem. Cycles* **2005**, *19*. [CrossRef]
40. Fujisawa, T.; Narikawa, R.; Okamoto, S.; Ehira, S.; Yoshimura, H.; Suzuki, I.; Masuda, T.; Mochimaru, M.; Takaichi, S.; Awai, K.; et al. Genomic structure of an economically important cyanobacterium, *Arthrospira* (*Spirulina*) *platensis* NIES-39. *DNA Res. Int. J. Rapid Publ. Rep. Genes Genomes* **2010**, *17*, 85–103. [CrossRef]
41. Churro, C.; Azevedo, J.; Vasconcelos, V.; Silva, A. Detection of a *Planktothrix agardhii* Bloom in Portuguese Marine Coastal Waters. *Toxins* **2017**, *9*, 391. [CrossRef]
42. Leao, T.; Castelao, G.; Korobeynikov, A.; Monroe, E.A.; Podell, S.; Glukhov, E.; Allen, E.E.; Gerwick, W.H.; Gerwick, L. Comparative genomics uncovers the prolific and distinctive metabolic potential of the cyanobacterial genus *Moorea*. *Proc. Natl. Acad. Sci. USA* **2017**, *114*, 3198–3203. [CrossRef]
43. Nakamura, Y.; Kaneko, T.; Sato, S.; Mimuro, M.; Miyashita, H.; Tsuchiya, T.; Sasamoto, S.; Watanabe, A.; Kawashima, K.; Kishida, Y.; et al. Complete genome structure of *Gloeobacter violaceus* PCC 7421, a cyanobacterium that lacks thylakoids. *Dna Res. Int. J. Rapid Publ. Rep. Genes Genomes* **2003**, *10*, 137–145. [CrossRef]
44. Kaneko, T.; Nakamura, Y.; Wolk, C.P.; Kuritz, T.; Sasamoto, S.; Watanabe, A.; Iriguchi, M.; Ishikawa, A.; Kawashima, K.; Kimura, T.; et al. Complete genomic sequence of the filamentous nitrogen-fixing cyanobacterium *Anabaena* sp. strain PCC 7120. *DNA Res. Int. J. Rapid Publ. Rep. Genes Genomes* **2001**, *8*, 205–213, 227–253.
45. Ekman, M.; Picossi, S.; Campbell, E.L.; Meeks, J.C.; Flores, E. A Nostoc punctiforme sugar transporter necessary to establish a Cyanobacterium-plant symbiosis. *Plant Physiol.* **2013**, *161*, 1984–1992. [CrossRef]
46. Ran, L.; Larsson, J.; Vigil-Stenman, T.; Nylander, J.A.; Ininbergs, K.; Zheng, W.W.; Lapidus, A.; Lowry, S.; Haselkorn, R.; Bergman, B. Genome erosion in a nitrogen-fixing vertically transmitted endosymbiotic multicellular cyanobacterium. *PLoS ONE* **2010**, *5*. [CrossRef]
47. Wang, H.; Sivonen, K.; Rouhiainen, L.; Fewer, D.P.; Lyra, C.; Rantala-Ylinen, A.; Vestola, J.; Jokela, J.; Rantasarkka, K.; Li, Z.; et al. Genome-derived insights into the biology of the hepatotoxic bloom-forming cyanobacterium *Anabaena* sp. strain 90. *BMC Genom.* **2012**, *13*, 613. [CrossRef]
48. Isojarvi, J.; Shunmugam, S.; Sivonen, K.; Allahverdiyeva, Y.; Aro, E.M.; Battchikova, N. Draft genome sequence of *Calothrix* strain 336/3, a novel h2-producing cyanobacterium isolated from a finnish lake. *Genome Announc.* **2015**, *3*. [CrossRef]
49. Hirose, Y.; Fujisawa, T.; Ohtsubo, Y.; Katayama, M.; Misawa, N.; Wakazuki, S.; Shimura, Y.; Nakamura, Y.; Kawachi, M.; Yoshikawa, H.; et al. Complete genome sequence of cyanobacterium *Fischerella* sp. NIES-3754, providing thermoresistant optogenetic tools. *J. Biotechnol.* **2016**, *220*, 45–46. [CrossRef]

50. Teikari, J.E.; Hou, S.; Wahlsten, M.; Hess, W.R.; Sivonen, K. Comparative Genomics of the Baltic Sea Toxic Cyanobacteria *Nodularia spumigena* UHCC 0039 and Its Response to Varying Salinity. *Front. Microbiol.* **2018**, *9*, 356. [CrossRef]
51. Cimermancic, P.; Medema, M.H.; Claesen, J.; Kurita, K.; Wieland Brown, L.C.; Mavrommatis, K.; Pati, A.; Godfrey, P.A.; Koehrsen, M.; Clardy, J.; et al. Insights into secondary metabolism from a global analysis of prokaryotic biosynthetic gene clusters. *Cell* **2014**, *158*, 412–421. [CrossRef]
52. Podust, L.M.; Sherman, D.H. Diversity of P450 enzymes in the biosynthesis of natural products. *Nat. Prod. Rep.* **2012**, *29*, 1251–1266. [CrossRef]
53. Greule, A.; Stok, J.E.; De Voss, J.J.; Cryle, M.J. Unrivalled diversity: The many roles and reactions of bacterial cytochromes P450 in secondary metabolism. *Nat. Prod. Rep.* **2018**, *35*, 757–791. [CrossRef]
54. Mthethwa, B.; Chen, W.; Ngwenya, M.; Kappo, A.; Syed, P.; Karpoormath, R.; Yu, J.-H.; Nelson, D.; Syed, K. Comparative analyses of cytochrome P450s and those associated with secondary metabolism in *Bacillus* species. *Int. J. Mol. Sci.* **2018**, *19*, 3623. [CrossRef]
55. Senate, L.M.; Tjatji, M.P.; Pillay, K.; Chen, W.; Zondo, N.M.; Syed, P.R.; Mnguni, F.C.; Chiliza, Z.E.; Bamal, H.D.; Karpoormath, R. Similarities, variations, and evolution of cytochrome P450s in *Streptomyces* versus *Mycobacterium*. *Sci. Rep.* **2019**, *9*, 3962. [CrossRef]
56. Syed, P.R.; Chen, W.; Nelson, D.R.; Kappo, A.P.; Yu, J.-H.; Karpoormath, R.; Syed, K. Cytochrome P450 Monooxygenase CYP139 Family Involved in the Synthesis of Secondary Metabolites in 824 Mycobacterial Species. *Int. J. Mol. Sci.* **2019**, *20*, 2690. [CrossRef] [PubMed]
57. Agger, S.A.; Lopez-Gallego, F.; Hoye, T.R.; Schmidt-Dannert, C. Identification of sesquiterpene synthases from *Nostoc punctiforme* PCC 73102 and *Nostoc* sp. strain PCC 7120. *J. Bacteriol.* **2008**, *190*, 6084–6096. [CrossRef] [PubMed]
58. Kuhnel, K.; Ke, N.; Cryle, M.J.; Sligar, S.G.; Schuler, M.A.; Schlichting, I. Crystal structures of substrate-free and retinoic acid-bound cyanobacterial cytochrome P450 CYP120A1. *Biochemistry* **2008**, *47*, 6552–6559. [CrossRef] [PubMed]
59. Parvez, M.; Qhanya, L.B.; Mthakathi, N.T.; Kgosiemang, I.K.R.; Bamal, H.D.; Pagadala, N.S.; Xie, T.; Yang, H.; Chen, H.; Theron, C.W. Molecular evolutionary dynamics of cytochrome P450 monooxygenases across kingdoms: Special focus on mycobacterial P450s. *Sci. Rep.* **2016**, *6*. [CrossRef]
60. Nelson, D.R. Cytochrome P450 nomenclature. *Methods Mol. Biol. (Clifton N. J.)* **1998**, *107*, 15–24.
61. Nelson, D.R. Cytochrome P450 nomenclature, 2004. *Methods Mol. Biol. (Clifton N. J.)* **2006**, *320*, 1–10.
62. Nelson, D.R. The cytochrome p450 homepage. *Hum. Genom.* **2009**, *4*, 59–65. [CrossRef]
63. Feyereisen, R. Insect CYP genes and P450 enzymes. In *Insect Molecular Biology and Biochemistry*; Elsevier: Amsterdam, The Netherlands, 2012; pp. 236–316.
64. De Lima Procópio, R.E.; da Silva, I.R.; Martins, M.K.; de Azevedo, J.L.; de Araújo, J.M. Antibiotics produced by *Streptomyces*. *Braz. J. Infect. Dis.* **2012**, *16*, 466–471. [CrossRef]
65. Blin, K.; Shaw, S.; Steinke, K.; Villebro, R.; Ziemert, N.; Lee, S.Y.; Medema, M.H.; Weber, T. antiSMASH 5.0: Updates to the secondary metabolite genome mining pipeline. *Nucleic Acids Res.* **2019**, *47*, W81–W87. [CrossRef]
66. Alder, A.; Bigler, P.; Werck-Reichhart, D.; Al-Babili, S. In vitro characterization of *Synechocystis* CYP120A1 revealed the first nonanimal retinoic acid hydroxylase. *FEBS J.* **2009**, *276*, 5416–5431. [CrossRef]
67. Harada, H.; Shindo, K.; Iki, K.; Teraoka, A.; Okamoto, S.; Yu, F.; Hattan, J.; Utsumi, R.; Misawa, N. Efficient functional analysis system for cyanobacterial or plant cytochromes P450 involved in sesquiterpene biosynthesis. *Appl. Microbiol. Biotechnol.* **2011**, *90*, 467–476. [CrossRef] [PubMed]
68. Makino, T.; Otomatsu, T.; Shindo, K.; Kitamura, E.; Sandmann, G.; Harada, H.; Misawa, N. Biocatalytic synthesis of flavones and hydroxyl-small molecules by recombinant *Escherichia coli* cells expressing the cyanobacterial CYP110E1 gene. *Microb. Cell Factories* **2012**, *11*, 95. [CrossRef] [PubMed]
69. Torres, S.; Fjetland, C.R.; Lammers, P.J. Alkane-induced expression, substrate binding profile, and immunolocalization of a cytochrome P450 encoded on the *nifD* excision element of Anabaena 7120. *BMC Microbiol* **2005**, *5*, 16. [CrossRef] [PubMed]
70. Nelson, D.; Werck-Reichhart, D. A P450-centric view of plant evolution. *Plant J.* **2011**, *66*, 194–211. [CrossRef] [PubMed]
71. Tian, L.; DellaPenna, D. Progress in understanding the origin and functions of carotenoid hydroxylases in plants. *Arch. Biochem. Biophys.* **2004**, *430*, 22–29. [CrossRef]

72. Kim, J.; Smith, J.J.; Tian, L.; DellaPenna, D. The evolution and function of carotenoid hydroxylases in Arabidopsis. *Plant Cell Physiol.* **2009**, *50*, 463–479. [CrossRef]
73. Kanehisa, M.; Sato, Y.; Furumichi, M.; Morishima, K.; Tanabe, M. New approach for understanding genome variations in KEGG. *Nucleic Acids Res.* **2018**, *47*, D590–D595. [CrossRef]
74. Marchler-Bauer, A.; Bo, Y.; Han, L.; He, J.; Lanczycki, C.J.; Lu, S.; Chitsaz, F.; Derbyshire, M.K.; Geer, R.C.; Gonzales, N.R. CDD/SPARCLE: Functional classification of proteins via subfamily domain architectures. *Nucleic Acids Res.* **2016**, *45*, D200–D203. [CrossRef]
75. Katoh, K.; Kuma, K.-i.; Toh, H.; Miyata, T. MAFFT version 5: Improvement in accuracy of multiple sequence alignment. *Nucleic Acids Res.* **2005**, *33*, 511–518. [CrossRef]
76. Boc, A.; Diallo, A.B.; Makarenkov, V. T-REX: A web server for inferring, validating and visualizing phylogenetic trees and networks. *Nucleic Acids Res.* **2012**, *40*, W573–W579. [CrossRef]
77. Letunic, I.; Doerks, T.; Bork, P. SMART 7: Recent updates to the protein domain annotation resource. *Nucleic Acids Res.* **2011**, *40*, D302–D305. [CrossRef] [PubMed]
78. Ngwenya, M.; Chen, W.; Basson, A.; Shandu, J.; Yu, J.-H.; Nelson, D.; Syed, K. Blooming of unusual cytochrome P450s by tandem duplication in the pathogenic fungus *Conidiobolus coronatus*. *Int. J. Mol. Sci.* **2018**, *19*, 1711. [CrossRef]
79. Akapo, O.O.; Padayachee, T.; Chen, W.; Kappo, A.P.; Yu, J.-H.; Nelson, D.R.; Syed, K. Distribution and Diversity of Cytochrome P450 Monooxygenases in the Fungal Class *Tremellomycetes*. *Int. J. Mol. Sci.* **2019**, *20*, 2889. [CrossRef] [PubMed]
80. Saeed, A.; Sharov, V.; White, J.; Li, J.; Liang, W.; Bhagabati, N.; Braisted, J.; Klapa, M.; Currier, T.; Thiagarajan, M. TM4: A free, open-source system for microarray data management and analysis. *Biotechniques* **2003**, *34*, 374–378. [CrossRef] [PubMed]

© 2020 by the authors. Licensee MDPI, Basel, Switzerland. This article is an open access article distributed under the terms and conditions of the Creative Commons Attribution (CC BY) license (http://creativecommons.org/licenses/by/4.0/).

Article

More P450s Are Involved in Secondary Metabolite Biosynthesis in *Streptomyces* Compared to *Bacillus*, *Cyanobacteria*, and *Mycobacterium*

Fanele Cabangile Mnguni [1], Tiara Padayachee [1], Wanping Chen [2], Dominik Gront [3], Jae-Hyuk Yu [4,5], David R. Nelson [6,*] and Khajamohiddin Syed [1,*]

1. Department of Biochemistry and Microbiology, Faculty of Science and Agriculture, University of Zululand, KwaDlangezwa 3886, South Africa; fanelemngun@gmail.com (F.C.M.); teez07padayachee@gmail.com (T.P.)
2. Department of Molecular Microbiology and Genetics, University of Göttingen, 37077 Göttingen, Germany; chenwanping1@foxmail.com
3. Faculty of Chemistry, Biological and Chemical Research Center, University of Warsaw, Pasteura 1, 02-093 Warsaw, Poland; dgront@gmail.com
4. Department of Bacteriology, University of Wisconsin-Madison, 3155 MSB, 1550 Linden Drive, Madison, WI 53706, USA; jyu1@wisc.edu
5. Department of Systems Biotechnology, Konkuk University, Seoul 05029, Korea
6. Department of Microbiology, Immunology and Biochemistry, University of Tennessee Health Science Center, Memphis, TN 38163, USA
* Correspondence: drnelson1@gmail.com (D.R.N.); khajamohiddinsyed@gmail.com (K.S.)

Received: 18 January 2020; Accepted: 13 February 2020; Published: 7 July 2020

Abstract: Unraveling the role of cytochrome P450 monooxygenases (CYPs/P450s), heme-thiolate proteins present in living and non-living entities, in secondary metabolite synthesis is gaining momentum. In this direction, in this study, we analyzed the genomes of 203 *Streptomyces* species for P450s and unraveled their association with secondary metabolism. Our analyses revealed the presence of 5460 P450s, grouped into 253 families and 698 subfamilies. The CYP107 family was found to be conserved and highly populated in *Streptomyces* and *Bacillus* species, indicating its key role in the synthesis of secondary metabolites. *Streptomyces* species had a higher number of P450s than *Bacillus* and cyanobacterial species. The average number of secondary metabolite biosynthetic gene clusters (BGCs) and the number of P450s located in BGCs were higher in *Streptomyces* species than in *Bacillus*, mycobacterial, and cyanobacterial species, corroborating the superior capacity of *Streptomyces* species for generating diverse secondary metabolites. Functional analysis *via* data mining confirmed that many *Streptomyces* P450s are involved in the biosynthesis of secondary metabolites. This study was the first of its kind to conduct a comparative analysis of P450s in such a large number (203) of *Streptomyces* species, revealing the P450s' association with secondary metabolite synthesis in *Streptomyces* species. Future studies should include the selection of *Streptomyces* species with a higher number of P450s and BGCs and explore the biotechnological value of secondary metabolites they produce.

Keywords: *Streptomyces*; *Mycobacterium*; *Bacillus*; *Cyanobacteria*; cytochrome P450 monooxygenases; secondary metabolites; biosynthetic gene clusters; terpenes; polyketides; P450 blooming; non-ribosomal peptides

1. Introduction

Cytochrome P450 monooxygenases (CYPs/P450s) are biotechnologically valuable enzymes [1]. P450s have heme (protoporphyrin IX), an iron(III)-containing porphyrin, as a prosthetic group in their structure [2]. Because of the presence of this prosthetic group, these enzymes absorb wavelengths at 450 nm; thus, the name P450s has been assigned to these proteins [3–6]. Since their identification, a large number of P450s have been identified in almost all living organisms [7] and, surprisingly, in non-living entities as well [8]. The regio- and stereo-specific catalytic nature of these enzymes makes them essential for the survival of some organisms, and these enzymes are thus good drug targets in the case of pathogenic organisms [9–13]. Application of these enzymes in all fields of research continues, and excellent success has been achieved in using them for the production of substances valuable to humans or as drug targets or in drug metabolism, as reported previously [1]. One of the applications of P450s currently being explored is their role in the production of secondary metabolites, compounds with potential biotechnological value, owing to their stereo- and regio-specific enzymatic activity, which contributes to the diversity of secondary metabolites [14–16].

Unlike other enzymes, P450 enzymes have a typical nomenclature system established by the International P450 Nomenclature Committee [17–19]. According to the committee's rules, P450s begin with the prefix "CYP" for cytochrome P450 monooxygenase, followed by an Arabic numeral which designates the family, a capital letter designating the subfamily, and an Arabic numeral designating the individual P450 in a family. The annotation of P450s (assigning family and subfamily) follows a rule that all P450s with > 40% identity belong to the same family and all P450s with > 55% identity belong to the same subfamily [17–19]. Worldwide, researchers follow this P450 nomenclature system. The nomenclature of P450s is also be verified by phylogenetic analysis to enable their correct annotation, as phylogenetic-based annotation could detect similarity cues beyond a simple percentage identity cutoff, as mentioned elsewhere [20].

The continued genomic rush has resulted in genome sequencing of a large number of species belonging to all biological kingdoms. P450s in the newly sequenced species need to be annotated as per the International P450 Nomenclature Committee rules [17–19] to enable researchers to use the same names for functional and evolutionary analysis of P450s. For this reason, large numbers of P450s have recently been annotated in bacterial species belonging to the genera *Mycobacterium* [21], *Bacillus* [22], *Streptomyces* [20], and *Cyanobacteria* [23]. These studies have revealed numerous P450s involved in the synthesis of different types of secondary metabolites. This type of *in silico* study is highly important for identifying unique P450s that can be drug-targeted and for P450 evolutionary analysis, as the P450 profiles in species have been found to be characteristic of species' lifestyle [11,20,22,24–27].

Among bacterial species, *Streptomyces* species are well-known for producing over two-thirds of the clinically useful antibiotics in the world [28]. Because of this importance, *Streptomyces* species have been subjected to exhaustive secondary metabolite production studies [29,30]. *Streptomyces* P450s play a key role in the production of different secondary metabolites; their contribution to secondary metabolite diversity and applications in drug metabolism have been reviewed extensively [15,16,31–33]. In the latest study, comprehensive comparative analysis of P450 and secondary metabolite biosynthetic gene clusters (BGCs) in 48 *Streptomyces* species was elucidated [20]. The study revealed the presence of novel P450s in *Streptomyces* species and numerous P450s forming parts of secondary metabolite BGCs [20]. The study results indicated that lifestyle or ecological niches play a key role in the evolution of P450 profiles in species belonging to the genera *Streptomyces* and *Mycobacterium* [20].

To date, a large number of *Streptomyces* species genomes have been sequenced and are available for public use. This provided an opportunity to annotate P450s in these species to analyze and compare their profiles among different bacterial species, including the identification and comparative analysis of P450s involved in the production of secondary metabolites. This study thus aimed to perform genome data mining, annotation, and phylogenetic analysis of P450s in 155 newly available *Streptomyces* species genomes. It also included the identification and comparative analysis of P450s that are parts of secondary metabolite BGCs among bacterial species belonging to the genera *Streptomyces*, *Bacillus*, *Mycobacterium*, and *Cyanobacteria*, as the species belonging to these genera are known to have P450s and to produce secondary metabolites.

2. Results and Discussion

2.1. Streptomyces Species Have Large Number of P450s

Genome-wide data mining and annotation of P450s in 203 *Streptomyces* species (Supplementary Table S1) revealed the presence of 5460 P450s in their genomes (Figure 1, Table 1, and Supplementary Dataset 1). The P450 count in the *Streptomyces* species ranged from 10 to 69 P450s, with an average of 27 P450s. Apart from the complete P450 sequences, pseudo-P450s (6 hit proteins), P450-fragments (114 hit proteins), P450-derived glycosyltransferase activator proteins (22 hit proteins), and P450 false-positive hits (2 hit proteins) were also found in some *Streptomyces* species (Supplementary Table S2). The presence of these types of P450 hit proteins in species is common and, because of the nature of these proteins, they were not included in the study for further analysis. Among *Streptomyces* species, *Streptomyces albulus* ZPM was found to have the highest number of P450s in its genome (69 P450s) followed by *S. clavuligerus* (65 P450s); the lowest number of P450s was found in *Streptomyces* sp. CNT372 and *S. somaliensis* DSM 40738 (10 P450s each) (Figure 1 and Table 1). Analysis of the most prevalent number of P450s revealed that 19 P450s was the prevalent number in *Streptomyces* species (Table 1). The average number of P450s in *Streptomyces* species was found to be higher than in *Bacillus* species [22] and cyanobacterial species [23], and almost the same as in mycobacterial species [21] (Table 2). A point to be noted is that the number of species greatly influences the average number of P450s and, thus, the higher the number of species in the analysis, the better and more accurate the results, as mentioned elsewhere [20,23]. This is the reason *Streptomyces* species showed a slightly lower average number of P450s in their genomes compared to mycobacterial species, since only 60 species were employed in the study [21]. Thus, future annotation of P450s in more mycobacterial species will provide accurate insights into this aspect.

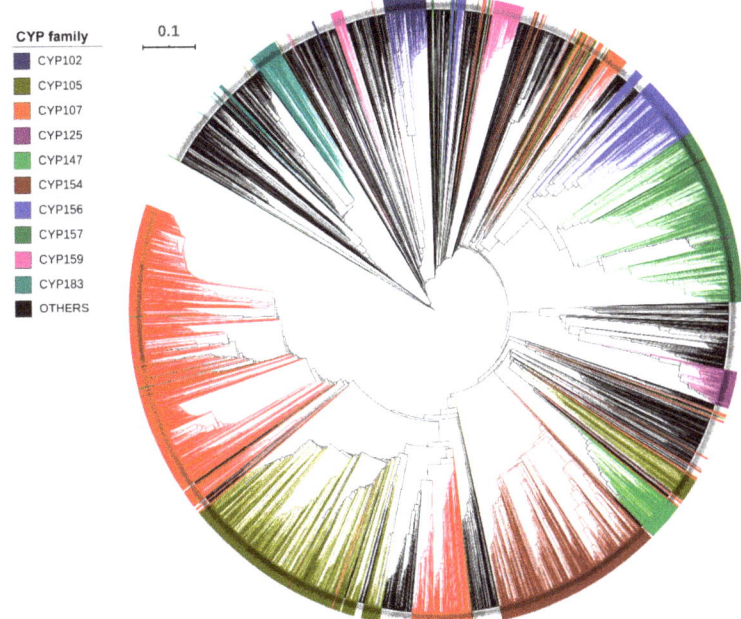

Figure 1. Phylogenetic analysis of *Streptomyces* P450s. In total, 5 460 P450s were used to construct the tree and the dominant P450 families are highlighted in different colors and indicated in the figure. A high-resolution phylogenetic tree is provided in Supplementary Dataset 2.

Table 1. Genome-wide data mining and annotation of P450s in 203 *Streptomyces* species.

Species Name	P450s	No. F	No. SF	Species Name	P450s	No. F	No. SF
Streptomyces sp. Tu6071	22	13	20	*Streptomyces* sp. CNT372	10	8	10
Streptomyces purpureus KA281, ATCC 21405	22	17	20	*Streptomyces* sp. CNS606	16	9	14
Streptomyces sp. W007	28	12	24	*Streptomyces* sp. 303MFCol5.2	23	14	22
Streptomyces sp. TAA486-18	18	12	17	*Streptomyces acidiscabies* 84-104	47	22	44
Streptomyces lysosuperificus ATCC 31396	25	19	24	*Streptomyces roseosporus* NRRL 11379	19	10	16
Streptomyces sp. PVA 94-07	20	7	18	*Streptomyces* sp. OspMP-M45	19	9	19
Streptomyces sp. SPB78	20	12	20	*Streptomyces* sp. AmelKG-A3	19	9	19
Streptomyces canus 299MFChir4.1	28	17	27	*Streptomyces* sp. S4	19	9	19
Streptomyces sp. FxanaA7	30	15	29	*Streptomyces* sp. SM8	18	8	16
Streptomyces sulphureus DSM 40104	26	13	25	*Streptomyces* sp. LaPpAH-199	26	11	21
Streptomyces sp. MspMP-M5	44	20	41	*Streptomyces* sp. 140Col2.1E	22	9	17
Streptomyces coeliflavus ZG0656	17	12	16	*Streptomyces* sp. DvalAA-21	24	10	22
Streptomyces pristinaespiralis ATCC 25486	18	11	17	*Streptomyces* sp. CNT371	17	13	17
Streptomyces sp. LaPpAH-201	19	8	19	*Streptomyces somaliensis* DSM 40738	10	8	9
Streptomyces albulus CCRC 11814	64	26	50	*Streptomyces* sp. 351MFTsu5.1	22	11	22
Streptomyces viridochromogenes DSM 40736	24	15	24	*Streptomyces* sp. DvalAA-83	24	10	22
Streptomyces sp. LaPpAH-95	24	9	22	*Streptomyces* sp. AmelKG-F2B	24	17	23
Streptomyces mirabilis YR139	42	26	41	*Streptomyces* sp. CNT302	26	13	22
Streptomyces sp. AA1529	26	15	24	*Streptomyces olindensis* DAUFPE 5622	26	14	22
Streptomyces atratus OK008	15	10	14	*Streptomyces* sp. CNY243	17	14	16
Streptomyces sp. PsTaAH-130	36	21	32	*Streptomyces* sp. AA0539	19	10	19
Streptomyces sp. CNT318	27	15	24	*Streptomyces atratus* OK807	31	13	27
Streptomyces sp. CNH099	16	12	16	*Streptomyces* sp. CNS335	16	13	17
Streptomyces sp. CNH287	16	12	16	*Streptomyces* sp. FxanaC1	27	15	24
Streptomyces sp. MnatMP-M77	32	14	27	*Streptomyces* sp. WMMB 322	19	11	17
Streptomyces zinciresistens K42	19	11	18	*Streptomyces* sp. TOR3209	20	13	19
Streptomyces sp. SolWspMP-so12th	22	11	19	*Streptomyces* sp. AmelKG-E11A	24	15	22
Streptomyces sp. GXT6	13	8	11	*Streptomyces* sp. PP-C42	16	6	14
Streptomyces roseosporus NRRL 15998	19	10	16	*Streptomyces* sp. DpondAA-E10	25	10	22
Streptomyces sp. LaPpAH-108	24	12	23	*Streptomyces* sp. HPH0547	32	18	32
Streptomyces aurantiacus JA 4570	30	20	30	*Streptomyces* sp. DpondAA-A50	25	10	22
Streptomyces hygroscopicus ATCC 53653	57	21	49	*Streptomyces* sp. TAA040	15	10	15
Streptomyces sp. Tu 6176	30	15	26	*Streptomyces* sp. PgraA7	23	10	20
Streptomyces ghanaensis ATCC 14672	35	20	34	*Streptomyces* sp. FxanaD5	15	11	15
Streptomyces sp. KhCrAH-337	26	12	22	*Streptomyces* sp. LamerLS-316	25	11	22
Streptomyces sp. LaPpAH-202	19	8	19	*Streptomyces viridochromogenes* Tue57	31	17	29
Streptomyces sp. UNC401CLCol	15	11	15	*Streptomyces* sp. GBA 94-10	20	7	18
Streptomyces sp. SirexAA-H	21	12	20	*Streptomyces* sp. CNQ-525	18	14	18
Streptomyces turgidiscabies Car8	28	20	27	*Streptomyces* sp. SceaMP-e96	41	18	36
Streptomyces sp. KhCrAH-40	26	12	22	*Streptomyces mirabilis* OK461	37	16	31
Streptomyces rimosus rimosus ATCC 10970	54	30	52	*Streptomyces* sp. LaPpAH-185	44	27	40
Streptomyces gancidicus BKS 13-15	18	11	17	*Streptomyces exfoliatus* DSMZ 41693	26	16	24
Streptomyces auratus AGR0001	35	14	33	*Streptomyces* sp. PsTaAH-137	29	16	28
Kitasatospora sp. SolWspMP-SS2h	25	15	24	*Streptomyces* sp. Amel2xE9	27	15	26
Streptomyces sp. NTK 937	17	8	17	*Streptomyces* sp. AmelKG-D3	22	11	19
Streptomyces sp. SceMP-e48	19	10	17	*Streptomyces prunicolor* NBRC 13075	44	18	39
Streptomyces sp. HmicA12	25	14	24	*Streptomyces* sp. e14	28	13	25
Streptomyces griseoaurantiacus M045	16	11	16	*Streptomyces* sp. CNX435	12	9	12
Streptomyces afghaniensis 772	28	17	29	*Streptomyces* sp. HCCB10043	17	10	14
Streptomyces sulphureus L180	19	11	19	*Streptomyces* sp. JS01	24	11	19
Streptomyces sp. KhCrAH-340	26	12	22	*Streptomyces chartreusis* NRRL 3882	29	19	26
Streptomyces sp. C	30	17	27	*Streptomyces* sp. CNY228	19	9	19
Streptomyces violaceusniger SPC6	13	8	12	*Streptomyces* sp. Amel2xB2	27	13	25
Streptomyces sp. HGB0020	23	13	22	*Streptomyces* sp. LaPpAH-165	24	9	22
Streptomyces sp. CNS615	27	15	24	*Streptomyces albulus* ZPM	68	27	51

Table 1. Cont.

Species Name	P450s	No. F	No. SF	Species Name	P450s	No. F	No. SF
Streptomyces tsukubaensis NRRL 18488	30	18	30	Streptomyces albulus NK660	64	27	50
Streptomyces vitaminophilus DSM 41686	18	10	15	Streptomyces noursei	64	26	52
Streptomyces sp. SA3_actG	21	12	20	Streptomyces violaceusniger Tu4113	50	16	42
Streptomyces bottropensis ATCC 25435 (2517572239)	31	19	30	Streptomyces bingchenggensis	49	26	44
Streptomyces sp. CNQ865	16	13	16	Streptomyces rapamycinicus	63	23	56
Streptomyces sp. CNT360	19	13	18	Streptomyces sp. 769	59	24	49
Streptomyces sp. 142MFCol3.1	27	14	24	Streptomyces hygroscopicus subsp. jinggangensis 5008	38	18	33
Streptomyces sp. ScaeMP-e122	25	11	23	Streptomyces cattleya NRRL 8058 = DSM 46488	41	21	38
Streptomyces griseoflavus Tu4000	20	15	19	Streptomyces cattleya NRRL 8057	40	20	37
Streptomyces sp. ACT-1	30	13	26	Streptomyces hygroscopicus subsp. jinggangensis TL01	37	18	33
Streptomyces sp. TAA204	18	10	16	Streptomyces avermitilis MA-4680	52	23	42
Streptomyces sp. SPB74	18	10	18	Streptomyces collinus	34	16	27
Streptomyces sp. CNQ329	13	10	13	Streptomyces lydicus A02	38	19	35
Streptomyces sp. 4F	16	11	15	Streptomyces lydicus 103	32	13	29
Streptomyces sp. KhCrAH-244	26	12	22	Streptomyces sp. Mg1	37	21	36
Streptomyces chartreusis NRRL 12338	23	15	23	Streptomyces leeuwenhoekii C34(2013)	36	17	34
Streptomyces sviceus ATCC 29083	19	12	19	Streptomyces pratensis/flavogriseus IAF 45	29	16	26
Streptomyces sp. CcalMP-8W	23	12	20	Streptomyces reticuli	47	26	43
Streptomyces sp. SS	15	11	15	Streptomyces griseus	28	13	24
Streptomyces sp. CNQ766	16	13	16	Streptomyces sp. PAMC 26508	29	16	26
Streptomyces sp. URHA0041	16	9	15	Streptomyces sp. SirexAA-E	24	10	22
Streptomyces sp. CNB091	27	14	24	Streptomyces davawensis	32	19	30
Streptomyces flavidovirens DSM 40150	24	15	23	Streptomyces cyaneogriseus	30	14	28
Streptomyces yeochonensis CN732	18	11	18	Streptomyces lincolnensis	24	15	23
Streptomyces viridosporus T7A, ATCC 39115	32	19	31	Streptomyces pristinaespiralis HCCB 10218	23	12	18
Streptomyces sp. FXJ7.023	27	12	23	Streptomyces venezuelae	23	16	21
Streptomyces mirabilis OV308	28	14	27	Streptomyces sp. CFMR 7	24	13	20
Streptomyces sp. AW19M42	27	12	24	Streptomyces vietnamensis	30	20	29
Streptomyces sp. ATexAB-D23	28	11	26	Streptomyces xiamenensis 318	19	12	19
Streptomyces sp. BoleA5	17	8	15	Streptomyces coelicolor	18	10	17
Streptomyces sp. AA4	35	17	29	Streptomyces albus J1074	18	9	18
Streptomyces sp. CNS654	27	10	22	Streptomyces ambofaciens	19	10	18
Streptomyces ipomoeae 91-03	44	26	43	Streptomyces lividans	20	10	18
Streptomyces sp. DpondAA-B6	19	9	19	Streptomyces scabiei 87.22	30	16	30
Streptomyces sp. PCS3-D2	25	18	24	Streptomyces glaucescens	18	11	17
Streptomyces sp. PRh5	57	20	51	Streptomyces albus DSM 41398	25	13	24
Streptomyces sp. CNR698	29	17	26	Streptomyces fulvissimus	19	10	16
Amycolatopsis sp. 75iv2, ATCC 39116	28	18	27	Streptomyces sp. CNQ-509	16	11	16
Streptomyces cattleya ATCC 35852	41	21	38	Streptomyces rubrolavendulae	20	12	19
Streptomyces sp. WMMB 714	21	10	18	Streptomyces clavuligerus	64	30	58
Streptomyces scabrisporus DSM 41855	37	27	36	Streptomyces griseochromogenes	46	24	40
Streptomyces sp. Ncost-T6T-1	25	14	22	Streptomyces sp. S10(2016)	20	15	20
Streptomyces sp. CNB632	16	12	16	Streptomyces globisporus	23	13	19
Streptomyces mobaraensis NBRC 13819	22	13	21	Streptomyces sp. CdTB01	26	17	25
Streptomyces sp. KhCrAH-43	26	12	22	Streptomyces parvulus	25	15	25
Streptomyces sp. PsTaAH-124	32	16	27	Streptomyces sp. SAT1	25	15	22
Streptomyces sp. Amel2xC10	15	10	15				

Abbreviations: No. F: number of P450 families; No. SF: number of P450 subfamilies.

Table 2. Comparative analysis of key features of P450s in different bacterial species.

	Streptomyces Species	Mycobacterial Species	*Bacillus* Species	Cyanobacterial Species
Total No. of species analyzed	203	60	128	114
No. of P450s	5460	1784	507	341
No. of families	253	77	13	36
No. of subfamilies	698	132	28	79
Dominant P450 family	CYP107	CYP125	CYP107	CYP110
Average no. of P450s	27	30	4	3
No. of BGCs *	4457	898	1098	770
Average no. of BGCs	31	15	9	7
No. of P450s part of BGCs	1231	204	112	27
Percentage of P450s part of BGCs	22	11	22	8
Reference	This work	[20,21]	[22]	[23]

Abbreviations: BGC: biosynthetic gene cluster. Symbol: * 103 cyanobacterial species [23] and 144 *Streptomyces* species were used for BGC analysis.

2.2. CYP107 Family Was Found to Be Dominant and Conserved in 203 Streptomyces Species

Analysis of P450 families and subfamilies in 203 *Streptomyces* species revealed that 5460 P450s could be grouped into 253 P450 families and 698 P450 subfamilies (Table 2 and Supplementary Table S3). Among *Streptomyces* species, *S. clavuligerus* had the highest number of P450 families (30) and P450 subfamilies (58) in its genome (Table 1). Although *S. rimosus rimosus* ATCC 10970 had the same number of P450 families as *S. clavuligerus*, the number of subfamilies was the third highest (52 subfamilies) (Table 1). One interesting observation is that the species with the highest number of P450s did not have the highest number of P450 families, suggesting that some of the P450 families were populated (bloomed). Blooming of P450 families is common across species, and this phenomenon has been observed in different species belonging to different biological kingdoms [24,26,34–36]. Phylogenetic analysis revealed that some of the P450 families were scattered across the evolutionary tree (Figure 1). This phenomenon was also observed previously for *Streptomyces* species P450s, and it has been hypothesized that the phylogenetic-based annotation of P450s could be detecting similarity cues beyond a simple percentage identity cutoff [20]. Analysis of P450 families in the 155 *Streptomyces* species used in this study revealed the presence of 38 new P450 families, i.e., CYP1200A1, CYP1216A1, CYP1223A1, CYP1228A1, CYP1236A1, CYP1238A1, CYP1265A1, CYP1279A1, CYP1369A1, CYP1432A1, CYP1518A1, CYP1529A1, CYP1543A1, CYP1568A1, CYP159A1, CYP1607A1, CYP1658A1, CYP1759A1, CYP1810A1, CYP1832A1, CYP1866A1, CYP1896A1, CYP1920A1, CYP1929A1, CYP1931A1, CYP1940A1, CYP1941A1, CYP1943A1, CYP1972A1, CYP1984A1, CYP1994A1, CYP2076A1, CYP2080A1, CYP2134A1, CYP2180A1, CYP2349A1, CYP2427A1, and CYP2723A1. A detailed analysis of the number of new P450 families found in different *Streptomyces* species is presented in Supplementary Table S2.

Among the P450 families, the CYP107 family was found to be dominant, with 1 235 P450s in *Streptomyces* species, followed by CYP105 with 684 P450s, CYP157 with 525 P450s, and CYP154 with 510 P450s (Figure 2 and Supplementary Table S3), indicating the possible blooming of these families in *Streptomyces* species, as observed in species belonging to different biological kingdoms [24,26,34–36]. It is interesting to note that the CYP107 family was also found to be dominant in the *Bacillus* species [22], indicating its dominant role in the synthesis of secondary metabolites in both the *Streptomyces* and *Bacillus* genera. An interesting pattern was observed when comparing subfamily diversity in the dominant P450 families (Figure 2, Table 3, and Supplementary Table S3). P450 families such as CYP107, CYP105, CYP183, and CYP113 had the highest diversity at the subfamily level, as numerous subfamilies were found in these families (Supplementary Table S3). This phenomenon of the highest diversity in P450 families being found in *Streptomyces* species is not uncommon, and this proved to be the key contributor in the production of diverse secondary metabolites in *Streptomyces* species compared to mycobacterial species [20]. Strong support for this argument is the fact that the CYP105 P450 family members in *Streptomyces* species have been shown to be involved in oxidation of numerous endogenous and exogenous compounds and in the generation of different secondary metabolites [32]. However, in contrast to the diversity at subfamily level for the P450 families CYP107, CYP105, CYP183, and CYP113, the rest of the dominant P450 families had single or double or triple subfamilies, indicating subfamily-level blooming in these P450 families (Table 3).

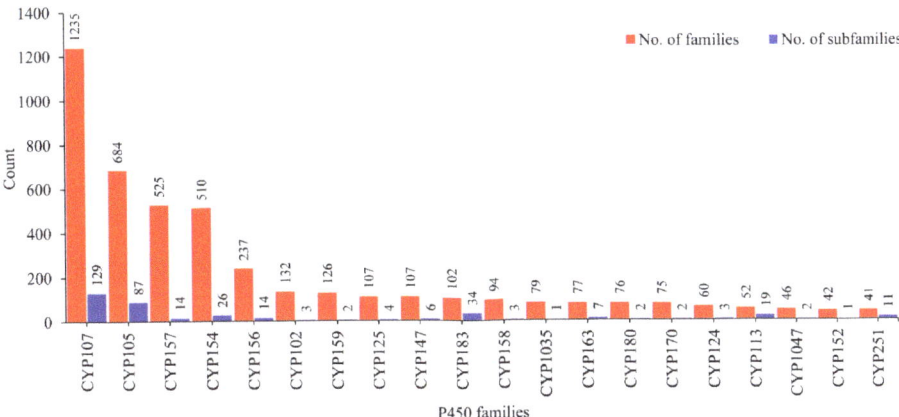

Figure 2. P450 family and subfamily analysis in 203 *Streptomyces* species. Only the dominant P450 families with more than 40 P450s are shown in the figure. Detailed data on P450 families and subfamilies are presented in Supplementary Table S3.

Table 3. P450 subfamily analysis in the dominant families in 203 *Streptomyces* species. The number of members in the dominant P450 subfamily is presented. Detailed data on different subfamilies are presented in Supplementary Table S3.

P450 Family	Dominant Subfamilies						
	A	B	C	D	E	F	G
CYP157	174		177				
CYP154	127		164	76			
CYP156		120					
CYP102		78					48
CYP159	125						
CYP125	104						
CYP147						73	
CYP158	91						
CYP1035	79						
CYP163		50					
CYP180	54						
CYP170	57						
CYP124							50
CYP1047	43						
CYP152					42		
CYP251	23						

P450 family conservation analysis revealed that the CYP107 family is conserved in all 203 *Streptomyces* species (Figure 3 and Supplementary Dataset 3). P450 families such as CYP156, CYP105, CYP154 and CYP157 are also present in the majority of the *Streptomyces* species (Figure 3 and Supplementary Dataset 3).

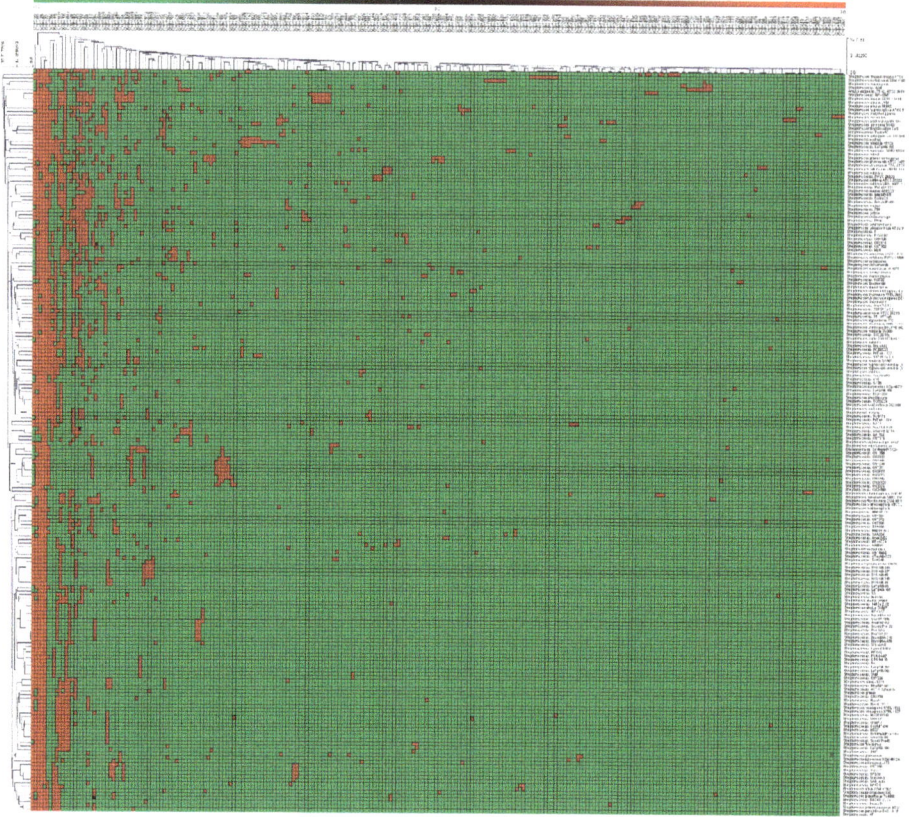

Figure 3. Heat-map of P450 family conservation analysis in *Streptomyces* species. In the heat-map, the presence and absence of P450 families are indicated in red and green colors. The horizontal axis represents P450 families and the vertical axis represents *Streptomyces* species.

2.3. Numerous P450s Involved in Secondary Metabolite Production in Streptomyces Compared to Other Bacterial Species

Analysis of 144 *Streptomyces* species' genomes revealed the presence of 4457 BGCs in their genomes (Table 2 and Supplementary Table S4). The number of BGCs found in 144 *Streptomyces* species was found to be higher than in mycobacterial, *Bacillus*, and cyanobacterial species (Table 2), indicating the superiority of the *Streptomyces* species in producing secondary metabolites; two-thirds of the antibiotics used in the world currently come from these species [28]. The average number of BGCs in *Streptomyces* species was found to be double compared to mycobacterial species and close to four times higher than that in *Bacillus* and cyanobacterial species (Table 2). Analysis of BGCs revealed that a large proportion of *Streptomyces* species' P450s are part of BGCs compared to other bacterial species; 1231 P450s in *Streptomyces* species compared to 112 in *Bacillus* species, 204 in mycobacterial species, and 27 in cyanobacterial species (Table 2). A total of 1231 P450s were found to be part of BGCs belonging to 135 P450 families (Figure 4 and Supplementary Table S5). Among 135 P450 families, P450s belonging to the CYP107 family were dominantly present in BGCs, followed by CYP105, CYP157, and CYP154 (Figure 4 and Supplementary Table S5). This clearly suggests that the P450 families that are bloomed in *Streptomyces* species are actually involved in the production of secondary metabolites. This strongly supports the proposed hypothesis that in *Streptomyces* species, P450s are evolved to generate secondary metabolites, thus helping these bacteria to thrive in their environment [20]. In order to assess the *in silico* results generated by this study, in which a large number of *Streptomyces* species P450s were predicted to be involved in secondary metabolite production, we performed an extensive literature review to identify *Streptomyces* P450s involved in the production of secondary metabolites. As shown in Table 4, a large number of P450s belonging to different P450 families, as predicted in this study, were found to be involved in the production of different secondary metabolites. This strongly supports the notion that the P450s identified as part of different BGCs in this study produce secondary metabolites.

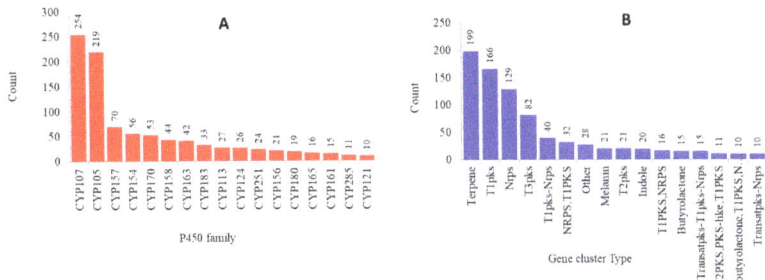

Figure 4. Analysis of P450s associated with secondary metabolite production in *Streptomyces* species. (**A**) Dominant P450 families (families with higher numbers of members) that are part of biosynthetic gene clusters (BGCs) and (**B**) dominant BGCs (present in higher numbers) containing P450s were presented in the figure. The numbers next to bars indicate the number of P450s in panel A and the number of BGCs in panel B. Detailed information is presented in Supplementary Table S5.

Analysis of P450 BGCs revealed the presence of 235 types of BGCs, where the BGC type, such as terpene, was dominant, followed by T1PKS, NRPS, and T3PKS (Figure 4 and Supplementary Table S5). A detailed analysis of P450s that are part of BGCs and types of BGCs containing P450s is presented in Supplementary Table S5. Analysis of the linkage between a particular P450 family and BGC revealed that some P450s are linked to a particular BGC (Supplementary Table S4), indicating horizontal transfer of BGCs between *Streptomyces* species. *Streptomyces* P450s such as CYP283A are linked to bacteriocin and bottromycin; CYP113K3 is linked to Bacteriocin-Nrps, CYP124G is linked to melanin, and CYP105A is linked to NRPS and butyrolactone. A point to be noted is that horizontal transfer of BGCs among different organisms is well-documented in the literature [37].

Table 4. List of *Streptomyces* species P450s involved in synthesis of secondary metabolites.

P450	Species	Function	References
CYP158A1	*Streptomyces coelicolor* A3(2)	Flaviolin biosynthesis	[38]
CYP1048A1	*Streptomyces scabiei*	Thaxtomin (phytotoxin) biosynthesis	[39]
CYP105A1	*Streptomyces griseolus*	Diterpenoids synthesis	[40]
CYP105A3 (P450sca-2)	*Streptomyces carbophilus*	Pravastatin synthesis	[41]
CYP105B28(GfsF) *	*Streptomyces graminofaciens*	Macrolide antibiotic synthesis	[42,43]
CYP105D6	*Streptomyces avermitilis*	Filipin biosynthesis	[44]
CYP105D7	*Streptomyces avermitilis*	Filipin biosynthesis	[45]
CYP105D8	*Streptomyces tubercidicus* strain I-1529	Avermectin oxidation	[32,46]

Table 4. Cont.

P450	Species	Function	References
CYP105D9	*Streptomyces* sp. JP95	Griseorhodin biosynthesis	[32,47]
CYP105F2	*Streptomyces peucetius*	Oleandomycin biosynthesis	[48,49]
CYP105H1	*Streptomyces noursei* ATCC 11455	Nystatin biosynthesis	[32]
CYP105H3	*Streptomyces natalensis*	Pimaricin biosynthesis	[32,50]
CYP105H4 (AmphN) !	*Streptomyces nodosus*	Amphotericin biosynthesis	[51,52]
CYP105H5	*Streptomyces griseus*	Candidicin biosynthesis	[32]
CYP105K1	*Streptomyces tendae* strain Tue901	Nikkomycin biosynthesis	[32,53]
CYP105K2	*Streptomyces ansochromogenes*	Nikkomycin biosynthesis	[32]
CYP105L1 (TylH1,orf7) !	*Streptomyces fradiae*	Tylosin biosynthesis	[54,55]
CYP105L4(ChmH1) *	*Streptomyces bikiniensis*	Chalcomycin biosynthesis	[56]
CYP105M1 (orf10) !	*Streptomyces clavuligerus*	Clavulanic acid antibiotic biosynthesis	[57]
CYP105N1	*Streptomyces coelicolor* A3(2)	Coelibactin siderophore biosynthesis	[58,59]
CYP105P1	*Streptomyces avermitilis*	Filipin biosynthesis	[44]
CYP105U1	*Streptomyces hygroscopicus*	Geldanamycin biosynthesis	[60]
CYP105V1	*Streptomyces* sp. HK803	Phoslactomycin biosynthesis	[32,61]
CYP105AA1	*Streptomyces tubercidicus* strain R922	Avermectin oxidation	[32,46]
CYP105AA2	*Streptomyces tubercidicus* strain I-1529	Avermectin oxidation	[32,46]
CYP107A1	*Streptomyces peucetius*	Dealkylation of 7-ethoxycoumarin	[62]
CYP107A1	*Saccharopolyspora erythraea*	Erythromycin biosynthesis	[63,64]
CYP107B (HmtN) !	*Streptomyces himastatinicus* ATCC 53653	Himastatin biosynthesis	[65]
CYP107B (HmtN)	*Streptomyces himastatinicus*	Himastatin biosynthesis	[66]
CYP107C1	*Streptomyces thermotolerans*	Carbomycin biosynthesis	[67]
CYP107E40(chmPII) *	*Streptomyces bikiniensis*	Chalcomycin biosynthesis	[56]
CYP107EE2(chmPI) *	*Streptomyces bikiniensis*	Chalcomycin biosynthesis	[56]
CYP107FH5(TamI) *	*Streptomyces* sp. 307-9	Tirandamycin biosynthesis	[68,69]
CYP107G1	*Streptomyces rapamycinicus*	Rapamycin biosynthesis	[70,71]
CYP107G1 (rapN) !	*Streptomyces hygroscopicus*	Rapamycin biosynthesis	[71,72]
CYP107L1	*Streptomyces venezuelae*	Macrolide antibioitics biosynthesis	[73]
CYP107L59(FosK) *	*Streptomyces pulveraceus*	Fostriecin biosynthesis	[74]
CYP107MD3(FosG) *	*Streptomyces pulveraceus*	Fostriecin biosynthesis	[74]
CYP107W1	*Streptomyces avermitilis*	Oligomycin A biosynthesis	[75,76]
CYP112A2	*Streptomyces rapamycinicus*	Rapamycin biosynthesis	[70,71]
CYP113A1	*Saccharopolyspora erythraea*	Erythromycin biosynthesis	[63,64]
CYP113B1 (TylI) !	*Streptomyces fradiae*	Tylosin biosynthesis	[54,55]
CYP113D3(HmtT) *	*Streptomyces himastatinicus* ATCC 53653	Himastatin biosynthesis	[65]
CYP113D3 (HmtT) *	*Streptomyces himastatinicus*	Himastatin biosynthesis	[66]
CYP113HI (HmtS) *	*Streptomyces himastatinicus*	Himastatin biosynthesis	[66]
CYP122A2 (rapJ) !	*Streptomyces hygroscopicus*	Rapamycin biosynthesis	[70,71]
CYP122A3	*Streptomyces hygroscopicus*	Rapamycin biosynthesis	[70,71]
CYP122A4 (FkbD) !	*Streptomyces tsukubaensis*	FK506 (immunosuppressant) polyketide biosynthesis	[77]
CYP129A2	*Streptomyces peucetius*	Doxorubicin biosynthesis	[78,79]
CYP129A2 (dox A) !	*Streptomyces* sp. strain C5	Doxorubicin biosynthesis	[80,81]
CYP131A2 (dnrQ) !	*Streptomyces* sp. strain C5	Doxorubicin biosynthesis	[80,81]
CYP140M1(TtnI) *	*Streptomyces griseochromogenes*	Tautomycetin biosynthesis	[82]
CYP151A (AurH) !	*Streptomyces thioluteus*	Aureothin biosynthesis	[83]
CYP154A1	*Streptomyces coelicolor* A3(2)	Polyketide synthesis and cyclization of a cellular dipentaenone	[84,85]

224

Table 4. Cont.

P450	Species	Function	References
CYP154B1	*Streptomyces fradiae*	Tylosin biosynthesis	[54,55]
CYP154C1	*Streptomyces coelicolor* A3(2)	Macrolide biosynthesis	[86]
CYP158A2	*Streptomyces coelicolor* A3(2)	Flaviolin biosynthesis	[87]
CYP161A2 (PimD) [!]	*Streptomyces natalensis*	Pimaricin biosynthesis	[88]
CYP161A3 (AmphL) [!]	*Streptomyces nodosus*	Amphotericin biosynthesis	[51]
CYP162A1	*Streptomyces tendae*	Nikkomycin biosynthesis	[53,89]
CYP163A1 (NovI) [!]	*Streptomyces spheroids*	Novobiocin biosynthesis	[90]
CYP163B3 (P450 Sky) [!]	*Streptomyces* sp. Acta 2897	Skyllamycin biosynthesis	[91]
CYP170A1	*Streptomyces coelicolor* A3(2)	Albaflavenone biosynthesis	[92]
CYP170A2	*Streptomyces avermitilis*	Albaflavenone biosynthesis	[93]
CYP170B1	*Streptomyces albus*	Albaflavenone biosynthesis	[94]
CYP171A1	*Streptomyces avermitilis*	Avermectin biosynthesis	[95,96]
CYP183A1	*Streptomyces avermitilis*	Pentalenolactone biosynthesis	[96,97]
CYP244A1 (StaN) [!]	*Streptomyces* sp tp-a0274	Rapamycin biosynthesis	[70,71]
CYP245A1 (StaP) [!]	*Streptomyces* sp tp-a0274	Rapamycin biosynthesis	[70,71]
CYP246A1	*Streptomyces scabiei*	Thaxtomin (phytotoxin) biosynthesis	[98]
CYP248A1	*Streptomyces thioluteus*	Aureothin biosynthesis	[83]

Note: For some P450s, protein notations are given in parentheses. These P450s were annotated in this study (indicated with asterisk superscript) and previously (indicated with exclamation mark) [20] by browsing the individual biosynthetic gene-cluster sequences reported in the literature. To enable readers to match the P450s with the published literature, we have provided protein notations in the parentheses. If known, the name of the secondary metabolite of which P450s are involved in production is indicated in the table.

3. Materials and Methods

3.1. Information on Streptomyces Species and Genome Database

In total, 203 *Streptomyces* species genomes (permanent and finished draft genomes) available for public use at the Joint Genome Institute Integrated Microbial Genomes and Microbiomes (JGI IMG/M) [99] and Kyoto Encyclopedia of Genes and Genomes (KEGG) [100] were used in this study. The 203 *Streptomyces* species included 48 *Streptomyces* species for which P450s and BGCs were annotated previously [20]. For these 48 species, P450 and BGCs data were retrieved from published articles and used in the study [20]. Thus, 155 *Streptomyces* species were data-mined for P450s and BGCs in this study. Information on the species used in the study is provided in Supplementary Table S1.

3.2. Genome Data Mining and Identification of P450s

Identification and annotation of P450s in *Streptomyces* species were carried out following a method described elsewhere [20–22]. Briefly, each *Streptomyces* species genome available at JGI IMG/M [99] was searched for P450s using the InterPro code "IPR001128". The hit protein sequences were then searched for the presence of P450 characteristic motifs such as EXXR and CXG [101]. Proteins having one of these motifs were considered pseudo-P450s, and proteins that were short in amino acid length and lacking both motifs as P450 fragments. Neither the pseudo-P450s nor the P450 fragments were considered for further analysis.

3.3. Allocating Family and Subfamily to P450s

The hit proteins that were collected were subjected to BLAST analysis against bacterial P450s at the website http://www.p450.unizulu.ac.za/. Based on the International P450 Nomenclature Committee rule [17–19], proteins with a percentage identity greater than 40% were assigned to the same family as named homolog P450s, and those that had greater than 55% identity were assigned to the same subfamily as named homolog P450s. Proteins that had a percentage identity less than 40% were assigned to a new family.

3.4. Streptomyces P450 Phylogenetic Analysis

Phylogenetic analysis of the *Streptomyces* P450s was carried out following the method described in the literature [102]. First, the *Streptomyces* P450 sequences were aligned using the MAFFT v6.864 program with an automatically optimized model option [103], available at the Trex web server [104]. The alignments were then automatically subjected to inference and optimization of the tree by the Trex web server with its embedded weighting procedure, and the best inferred tree was visualized and annotated by iTOL [105].

3.5. Streptomyces P450 Profile Heat-Maps

P450 profile heat-maps were generated following a method published previously [22,27] to check the presence and absence of P450s in *Streptomyces* species. Briefly, a tab-delimited file was imported into Multi-Experiment Viewer (Mev) [106] and hierarchical clustering using a Euclidean distance metric was used to cluster the data. In total, 203 *Streptomyces* species formed the vertical axis and P450 family numbers formed the horizontal axis. Data were presented as −3 for family absence (green) and 3 for family presence (red).

3.6. Identification of P450s That Are Part of Secondary Metabolite BGCs

Secondary metabolite BGCs analysis and identification of P450s that are part of these BGCs were carried out following the procedure mentioned previously [102], with slight modification. For each *Streptomyces* species genome available at JGI IMG/M, the secondary metabolite BGCs were searched for the presence of P450s. The DNA sequence of BGCs with P450s was collected and formatted to fasta format using PSPad editor (http://www.pspad.com/en/). The fasta-formatted files were then used to identify the type of cluster and most similar known clusters using the Antibiotics and Secondary Metabolite Analysis Shell (anti-SMASH) program [107]. The results obtained were recorded on Excel spreadsheets and represented as species-wise BGCs, type and similar known BGCs, percentage similarity to known BGCs, and P450s that are part of specific BGCs. Some *Streptomyces* species genome IDs did not pass through anti-SMASH analysis, and thus these species were not included in P450s analysis as part of secondary metabolite BGCs. A list of *Streptomyces* species subjected to anti-SMASH analysis is presented in Supplementary Table S4.

3.7. Data Analysis

All calculations were done following the method described in the literature [23]. The average number of P450s was calculated using the formula: Average number of P450s = Number of P450s/Number of species. The average number of BGCs was calculated using the formula: Average number of BGCs = Total number of BGCs/Number of species. The percentage of P450s that formed part of BGCs was calculated using the formula: Percentage of P450s part of BGCs = 100 × Number of P450s part of BGCs /Total number of P450s present in species. For comparative analysis of P450s and BGCs, information for bacterial species belonging to the genera *Bacillus* [22], *Mycobacterium* [21], and *Cyanobacteria* [23] was resourced from published articles.

4. Conclusions

In the last five decades, research on cytochrome P450 monooxygenases (CYPs/P450s) has mainly focused on their function and structural aspects, with little focus on evolutionary analysis, especially in microbes. The availability of a large number of microbial species genomes gives us an opportunity to focus on exploring the evolutionary aspects of P450s. Because a typical nomenclature system that has been established for P450s, each species genome needs to be data-mined and P450 proteins need to be annotated (assigning family and subfamily). In this way, researchers around the world can make use of uniform P450 names. In this study, we therefore annotated a large number of P450s in 203 *Streptomyces* species and found 38 new P450 families. Some P450 families were found to be bloomed in *Streptomyces* species even at the subfamily level. Comparative analysis of key P450 features among different bacterial species revealed that *Streptomyces* species had a greater number of P450s, more secondary metabolite BGCs, and the highest number of P450s as part of BGCs compared to the bacterial species belonging to the genera *Bacillus*, *Mycobacterium*, and *Cyanobacteria*. This further confirmed that the higher the number of P450s, the higher the secondary metabolite diversity in a species. This was true for *Streptomyces* species, as large number of P450s were found to be involved in the generation of diverse secondary metabolites. One interesting phenomenon observed was the linkage between a particular P450 family and BGC. This indicates that these BGCs were horizontally transferred among different *Streptomyces* species. This study is a good addition to the comparative analysis of P450s and BGCs among different bacterial populations. Data presented in the study will serve as a reference for further annotation of P450s in *Streptomyces* species and other bacterial species. *In silico* predicted BGCs need to be experimentally validated to assess the secondary metabolites' biological properties.

Supplementary Materials: Supplementary materials can be found at http://www.mdpi.com/1422-0067/21/13/4814/s1.

Author Contributions: Conceptualization, K.S.; data curation, F.C.M., T.P., W.C., D.G., J.-H.Y., D.R.N. and K.S.; formal analysis, F.C.M., T.P., W.C., D.G., J.-H.Y., D.R.N. and K.S.; funding acquisition, K.S.; investigation, F.C.M., T.P., W.C., J.-H.Y., D.R.N. and K.S.; methodology, F.C.M., T.P., W.C., D.G., J.-H.Y., D.R.N. and K.S.; project administration, K.S.; resources, K.S.; supervision, K.S.; validation, F.C.M., T.P., W.C., D.G., J.-H.Y., D.R.N. and K.S.; visualization, F.C.M., T.P., W.C., and K.S.; writing—original draft, F.C.M., T.P., W.C., J.-H.Y., D.R.N. and K.S.; writing—review and editing, F.C.M., T.P., W.C., J.-H.Y., D.R.N. and K.S. All authors have read and agreed to the published version of the manuscript.

Funding: The work presented in this article is part of research funded by a National Research Foundation (NRF), South Africa grant awarded to Khajamohiddin Syed (Grant No. 114159), where all the international authors involved in the study are listed as international collaborators. Fanele Cabangile Mnguni thanks the NRF, South Africa for a DST-NRF Innovation Master's Scholarship for the year 2019 (Grant No. 117171). Honours student, Tiara Padayachee, thanks the NRF, South Africa for an honours bursary (Grant No. MND190619448759). Dominik Gront was supported by the National Science Centre, Poland (Grant No. 2018/29/B/ST6/01989). Khajamohiddin Syed expresses sincere gratitude to the University of Zululand Research Committee for funding (Grant No. C686) and for the laboratory facilities.

Acknowledgments: The authors want to thank Barbara Bradley, Pretoria, South Africa for English language editing.

Conflicts of Interest: The authors declare no conflict of interest. The funders had no role in the design of the study, in the collection, analyses, or interpretation of data, in the writing of the manuscript, or in the decision to publish the results.

References

1. Urlacher, V.B.; Girhard, M. Cytochrome P450 Monooxygenases in Biotechnology and Synthetic Biology. *Trends Biotechnol.* **2019**, *37*, 882–897. [CrossRef]
2. Poulos, T.L.; Finzel, B.C.; Howard, A.J. High-resolution crystal structure of cytochrome P450cam. *J. Mol. Biol.* **1987**, *195*, 687–700. [CrossRef]
3. Garfinkel, D. Studies on pig liver microsomes. I. Enzymic and pigment composition of different microsomal fractions. *Arch. Biochem. Biophys.* **1958**, *77*, 493–509. [CrossRef]
4. Klingenberg, M. Pigments of rat liver microsomes. *Arch Biochem. Biophys.* **1958**, *75*, 376–386. [CrossRef]
5. Omura, T. Recollection of the early years of the research on cytochrome P450. *Proc. Jpn. Acad. Ser. B Phys. Biol. Sci.* **2011**, *87*, 617–640. [CrossRef] [PubMed]
6. Omura, T.; Sato, R. A new cytochrome in liver microsomes. *J. Biol. Chem.* **1962**, *237*, 1375–1376.
7. Nelson, D.R. Cytochrome P450 diversity in the tree of life. *Biochim. Biophys. Acta Proteins Proteom.* **2018**, *1866*, 141–154. [CrossRef]

8. Lamb, D.C.; Follmer, A.H.; Goldstone, J.V.; Nelson, D.R.; Warillow, A.G.; Price, C.L.; True, M.Y.; Kelly, S.L.; Poulos, T.L.; Stegeman, J.J. On the occurrence of cytochrome P450 in viruses. *Proc. Natl. Acad. Sci. USA* **2019**, *116*, 12343–12352. [CrossRef]
9. Lepesheva, G.I.; Hargrove, T.Y.; Kleshchenko, Y.; Nes, W.D.; Villalta, F.; Waterman, M.R. CYP51: A major drug target in the cytochrome P450 superfamily. *Lipids* **2008**, *43*, 1117–1125. [CrossRef]
10. Kelly, S.L.; Kelly, D.E. Microbial cytochromes P450: Biodiversity and biotechnology. Where do cytochromes P450 come from, what do they do and what can they do for us? *Philos. Trans. R. Soc. London Ser. B Biol. Sci.* **2013**, *368*, 20120476. [CrossRef]
11. Jawallapersand, P.; Mashele, S.S.; Kovacic, L.; Stojan, J.; Komel, R.; Pakala, S.B.; Krasevec, N.; Syed, K. Cytochrome P450 monooxygenase CYP53 family in fungi: Comparative structural and evolutionary analysis and its role as a common alternative anti-fungal drug target. *PLoS ONE* **2014**, *9*, e107209. [CrossRef]
12. Ziniel, P.D.; Karumudi, B.; Barnard, A.H.; Fisher, E.M.; Thatcher, G.R.; Podust, L.M.; Williams, D.L. The *Schistosoma mansoni* Cytochrome P450 (CYP3050A1) Is Essential for Worm Survival and Egg Development. *PLoS Negl. Trop. Dis.* **2015**, *9*, e0004279. [CrossRef]
13. Ortiz de Montellano, P.R. Potential drug targets in the *Mycobacterium tuberculosis* cytochrome P450 system. *J. Inorg. Biochem.* **2018**, *180*, 235–245. [CrossRef]
14. Podust, L.M.; Sherman, D.H. Diversity of P450 enzymes in the biosynthesis of natural products. *Nat. Prod. Rep.* **2012**, *29*, 1251–1266. [CrossRef]
15. Greule, A.; Stok, J.E.; De Voss, J.J.; Cryle, M.J. Unrivalled diversity: The many roles and reactions of bacterial cytochromes P450 in secondary metabolism. *Nat. Prod. Rep.* **2018**, *35*, 757–791. [CrossRef]
16. Rudolf, J.D.; Chang, C.Y.; Ma, M.; Shen, B. Cytochromes P450 for natural product biosynthesis in *Streptomyces*: Sequence, structure, and function. *Nat. Prod. Rep.* **2017**, *34*, 1141–1172. [CrossRef]
17. Nelson, D.R.; Kamataki, T.; Waxman, D.J.; Guengerich, F.P.; Estabrook, R.W.; Feyereisen, R.; Gonzalez, F.J.; Coon, M.J.; Gunsalus, I.C.; Gotoh, O.; et al. The P450 superfamily: Update on new sequences, gene mapping, accession numbers, early trivial names of enzymes, and nomenclature. *DNA Cell Biol.* **1993**, *12*, 1–51. [CrossRef]
18. Nelson, D.R. Cytochrome P450 nomenclature. *Methods Mol. Biol. (Clifton NJ)* **1998**, *107*, 15–24. [CrossRef]
19. Nelson, D.R. Cytochrome P450 nomenclature, 2004. *Methods Mol. Biol. (Clifton NJ)* **2006**, *320*, 1–10. [CrossRef]
20. Senate, L.M.; Tjatji, M.P.; Pillay, K.; Chen, W.; Zondo, N.M.; Syed, P.R.; Mnguni, F.C.; Chiliza, Z.E.; Bamal, H.D.; Karpoormath, R. Similarities, variations, and evolution of cytochrome P450s in *Streptomyces* versus *Mycobacterium*. *Sci. Rep.* **2019**, *9*, 3962. [CrossRef]
21. Parvez, M.; Qhanya, L.B.; Mthakathi, N.T.; Kgosiemang, I.K.; Bamal, H.D.; Pagadala, N.S.; Xie, T.; Yang, H.; Chen, H.; Theron, C.W.; et al. Molecular evolutionary dynamics of cytochrome P450 monooxygenases across kingdoms: Special focus on mycobacterial P450s. *Sci. Rep.* **2016**, *6*, 33099. [CrossRef]
22. Mthethwa, B.; Chen, W.; Ngwenya, M.; Kappo, A.; Syed, P.; Karpoormath, R.; Yu, J.-H.; Nelson, D.; Syed, K. Comparative analyses of cytochrome P450s and those associated with secondary metabolism in *Bacillus* species. *Int. J. Mol. Sci.* **2018**, *19*, 3623. [CrossRef]
23. Khumalo, M.J.N.; Padayachee, T.; Chen, W.; Yu, J.-H.; Nelson, D.; Syed, K. Comprehensive analyses of cytochrome P450 monooxygenases and secondary metabolite biosynthetic gene clusters in *Cyanobacteria*. *Int. J. Mol. Sci.* **2020**, *21*, 656. [CrossRef]
24. Syed, K.; Shale, K.; Pagadala, N.S.; Tuszynski, J. Systematic identification and evolutionary analysis of catalytically versatile cytochrome P450 monooxygenase families enriched in model basidiomycete fungi. *PLoS ONE* **2014**, *9*, e86683. [CrossRef]
25. Kgosiemang, I.K.R.; Syed, K.; Mashele, S.S. Comparative genomics and evolutionary analysis of cytochrome P450 monooxygenases in fungal subphylum *Saccharomycotina*. *J. Pure Appl. Microbiol.* **2014**, *8*, 12.
26. Sello, M.M.; Jafta, N.; Nelson, D.R.; Chen, W.; Yu, J.H.; Parvez, M.; Kgosiemang, I.K.; Monyaki, R.; Raselemane, S.C.; Qhanya, L.B.; et al. Diversity and evolution of cytochrome P450 monooxygenases in Oomycetes. *Sci. Rep.* **2015**, *5*, 11572. [CrossRef]
27. Akapo, O.O.; Padayachee, T.; Chen, W.; Kappo, A.P.; Yu, J.H.; Nelson, D.R.; Syed, K. Distribution and Diversity of Cytochrome P450 Monooxygenases in the Fungal Class *Tremellomycetes*. *Int. J. Mol. Sci.* **2019**, *20*, 2889. [CrossRef]
28. de Lima Procópio, R.E.; da Silva, I.R.; Martins, M.K.; de Azevedo, J.L.; de Araújo, J.M. Antibiotics produced by *Streptomyces*. *Braz. J. Infect. Dis.* **2012**, *16*, 466–471. [CrossRef]

29. Hwang, K.-S.; Kim, H.U.; Charusanti, P.; Palsson, B.Ø.; Lee, S.Y. Systems biology and biotechnology of *Streptomyces* species for the production of secondary metabolites. *Biotechnol. Adv.* **2014**, *32*, 255–268. [CrossRef]
30. Harir, M.; Bendif, H.; Bellahcene, M.; Fortas, Z.; Pogni, R. *Streptomyces* Secondary Metabolites. In *Basic Biology and Applications of Actinobacteria*; IntechOpen: London, UK, 2018; pp. 99–122. [CrossRef]
31. Cho, M.-A.; Han, S.; Lim, Y.-R.; Kim, V.; Kim, H.; Kim, D. *Streptomyces* Cytochrome P450 Enzymes and Their Roles in the Biosynthesis of Macrolide Therapeutic Agents. *Biomol. Ther.* **2019**, *27*, 127. [CrossRef]
32. Moody, S.C.; Loveridge, E.J. CYP105-diverse structures, functions and roles in an intriguing family of enzymes in *Streptomyces*. *J. Appl. Microbiol.* **2014**, *117*, 1549–1563. [CrossRef]
33. Lamb, D.C.; Waterman, M.R.; Zhao, B. *Streptomyces* cytochromes P450: Applications in drug metabolism. *Expert Opin. Drug Metab. Toxicol.* **2013**, *9*, 1279–1294. [CrossRef]
34. Feyereisen, R. Insect CYP genes and P450 enzymes. In *Insect Molecular Biology and Biochemistry*; Elsevier: Amsterdam, The Netherlands, 2012; pp. 236–316.
35. Qhanya, L.B.; Matowane, G.; Chen, W.; Sun, Y.; Letsimo, E.M.; Parvez, M.; Yu, J.H.; Mashele, S.S.; Syed, K. Genome-wide annotation and comparative analysis of cytochrome P450 monooxygenases in Basidiomycete biotrophic plant pathogens. *PLoS ONE* **2015**, *10*, e0142100. [CrossRef]
36. Ngwenya, M.L.; Chen, W.; Basson, A.K.; Shandu, J.S.; Yu, J.H.; Nelson, D.R.; Syed, K. Blooming of unusual cytochrome P450s by tandem duplication in the pathogenic fungus *Conidiobolus coronatus*. *Int. J. Mol. Sci.* **2018**, *19*, 1711. [CrossRef]
37. Tran, P.N.; Yen, M.R.; Chiang, C.Y.; Lin, H.C.; Chen, P.Y. Detecting and prioritizing biosynthetic gene clusters for bioactive compounds in bacteria and fungi. *Appl. Microbiol. Biotechnol.* **2019**, *103*, 3277–3287. [CrossRef]
38. Zhao, B.; Lamb, D.C.; Lei, L.; Kelly, S.L.; Yuan, H.; Hachey, D.L.; Waterman, M.R. Different binding modes of two flaviolin substrate molecules in cytochrome P450 158A1 (CYP158A1) compared to CYP158A2. *Biochemistry* **2007**, *46*, 8725–8733. [CrossRef]
39. Yu, F.; Li, M.; Xu, C.; Wang, Z.; Zhou, H.; Yang, M.; Chen, Y.; Tang, L.; He, J. Structural insights into the mechanism for recognizing substrate of the cytochrome P450 enzyme TxtE. *PLoS ONE* **2013**, *8*, e81526. [CrossRef]
40. Janocha, S.; Zapp, J.; Hutter, M.; Kleser, M.; Bohlmann, J.; Bernhardt, R. Resin acid conversion with CYP105A1: An enzyme with potential for the production of pharmaceutically relevant diterpenoids. *Chembiochem* **2013**, *14*, 467–473. [CrossRef]
41. Watanabe, I.; Nara, F.; Serizawa, N. Cloning, characterization and expression of the gene encoding cytochrome P-450sca-in2 from *Streptomyces carbophilus* involved in production of pravastatin, a specific HMG-CoA reductase inhibitor. *Gene* **1995**, *163*, 81–85. [CrossRef]
42. Kudo, F.; Motegi, A.; Mizoue, K.; Eguchi, T. Cloning and characterization of the biosynthetic gene cluster of 16-membered macrolide antibiotic FD-891: Involvement of a dual functional cytochrome P450 monooxygenase catalyzing epoxidation and hydroxylation. *Chembiochem* **2010**, *11*, 1574–1582. [CrossRef]
43. Kataoka, T.; Yamada, A.; Bando, M.; Honma, T.; Mizoue, K.; Nagai, K. FD-891, a structural analogue of concanamycin A that does not affect vacuolar acidification or perforin activity, yet potently prevents cytotoxic T lymphocyte-mediated cytotoxicity through the blockage of conjugate formation. *Immunology* **2000**, *100*, 170–177. [CrossRef]
44. Xu, L.H.; Fushinobu, S.; Takamatsu, S.; Wakagi, T.; Ikeda, H.; Shoun, H. Regio- and stereospecificity of filipin hydroxylation sites revealed by crystal structures of cytochrome P450 105P1 and 105D6 from *Streptomyces avermitilis*. *J. Biol. Chem.* **2010**, *285*, 16844–16853. [CrossRef] [PubMed]
45. Takamatsu, S.; Xu, L.H.; Fushinobu, S.; Shoun, H.; Komatsu, M.; Cane, D.E.; Ikeda, H. Pentalenic acid is a shunt metabolite in the biosynthesis of the pentalenolactone family of metabolites: Hydroxylation of 1-deoxypentalenic acid mediated by CYP105D7 (SAV_7469) of *Streptomyces avermitilis*. *J. Antibiot.* **2011**, *64*, 65–71. [CrossRef]
46. Jungmann, V.; Molnar, I.; Hammer, P.E.; Hill, D.S.; Zirkle, R.; Buckel, T.G.; Buckel, D.; Ligon, J.M.; Pachlatko, J.P. Biocatalytic conversion of avermectin to 4"-oxo-avermectin: Characterization of biocatalytically active bacterial strains and of cytochrome p450 monooxygenase enzymes and their genes. *Appl. Environ. Microbiol.* **2005**, *71*, 6968–6976. [CrossRef]

47. Yunt, Z.; Reinhardt, K.; Li, A.; Engeser, M.; Dahse, H.M.; Gutschow, M.; Bruhn, T.; Bringmann, G.; Piel, J. Cleavage of four carbon-carbon bonds during biosynthesis of the griseorhodin a spiroketal pharmacophore. *J. Am. Chem. Soc.* **2009**, *131*, 2297–2305. [CrossRef]
48. Rodriguez, A.M.; Olano, C.; Mendez, C.; Hutchinson, C.R.; Salas, J.A. A cytochrome P450-like gene possibly involved in oleandomycin biosynthesis by *Streptomyces antibioticus*. *FEMS Microbiol. Lett.* **1995**, *127*, 117–120. [CrossRef]
49. Shrestha, P.; Oh, T.J.; Liou, K.; Sohng, J.K. Cytochrome P450 (CYP105F2) from *Streptomyces peucetius* and its activity with oleandomycin. *Appl. Microbiol. Biotechnol.* **2008**, *79*, 555–562. [CrossRef]
50. Aparicio, J.F.; Fouces, R.; Mendes, M.V.; Olivera, N.; Martin, J.F. A complex multienzyme system encoded by five polyketide synthase genes is involved in the biosynthesis of the 26-membered polyene macrolide pimaricin in *Streptomyces natalensis*. *Chem. Biol.* **2000**, *7*, 895–905. [CrossRef]
51. Caffrey, P.; Lynch, S.; Flood, E.; Finnan, S.; Oliynyk, M. Amphotericin biosynthesis in *Streptomyces nodosus*: Deductions from analysis of polyketide synthase and late genes. *Chem. Biol.* **2001**, *8*, 713–723. [CrossRef]
52. Agarwal, P.K.; Agarwal, P.; Reddy, M.K.; Sopory, S.K. Role of DREB transcription factors in abiotic and biotic stress tolerance in plants. *Plant Cell Rep.* **2006**, *25*, 1263–1274. [CrossRef]
53. Lauer, B.; Russwurm, R.; Schwarz, W.; Kalmanczhelyi, A.; Bruntner, C.; Rosemeier, A.; Bormann, C. Molecular characterization of co-transcribed genes from *Streptomyces tendae* Tu901 involved in the biosynthesis of the peptidyl moiety and assembly of the peptidyl nucleoside antibiotic nikkomycin. *Mol. Gen. Genet. MGG* **2001**, *264*, 662–673. [CrossRef]
54. Merson-Davies, L.A.; Cundliffe, E. Analysis of five tylosin biosynthetic genes from the tyllBA region of the *Streptomyces fradiae* genome. *Mol. Microbiol.* **1994**, *13*, 349–355. [CrossRef] [PubMed]
55. Fouces, R.; Mellado, E.; Diez, B.; Barredo, J.L. The tylosin biosynthetic cluster from *Streptomyces fradiae*: Genetic organization of the left region. *Microbiology* **1999**, *145 Pt 4*, 855–868. [CrossRef] [PubMed]
56. Ward, S.L.; Hu, Z.; Schirmer, A.; Reid, R.; Revill, W.P.; Reeves, C.D.; Petrakovsky, O.V.; Dong, S.D.; Katz, L. Chalcomycin biosynthesis gene cluster from *Streptomyces bikiniensis*: Novel features of an unusual ketolide produced through expression of the chm polyketide synthase in *Streptomyces fradiae*. *Antimicrob. Agents Chemother.* **2004**, *48*, 4703–4712. [CrossRef] [PubMed]
57. Reading, C.; Cole, M. Clavulanic acid: A beta-lactamase-inhibiting beta-lactam from *Streptomyces clavuligerus*. *Antimicrob. Agents Chemother.* **1977**, *11*, 852–857. [CrossRef]
58. Lim, Y.R.; Hong, M.K.; Kim, J.K.; Doan, T.T.; Kim, D.H.; Yun, C.H.; Chun, Y.J.; Kang, L.W.; Kim, D. Crystal structure of cytochrome P450 CYP105N1 from *Streptomyces coelicolor*, an oxidase in the coelibactin siderophore biosynthetic pathway. *Arch Biochem. Biophys.* **2012**, *528*, 111–117. [CrossRef]
59. Zhao, B.; Moody, S.C.; Hider, R.C.; Lei, L.; Kelly, S.L.; Waterman, M.R.; Lamb, D.C. Structural analysis of cytochrome P450 105N1 involved in the biosynthesis of the zincophore, coelibactin. *Int. J. Mol. Sci.* **2012**, *13*, 8500–8513. [CrossRef]
60. Li, T.; Ni, S.; Jia, C.; Wang, H.; Sun, G.; Wu, L.; Gan, M.; Shan, G.; He, W.; Lin, L.; et al. Identification of 4,5-dihydro-4-hydroxygeldanamycins as shunt products of geldanamycin biosynthesis. *J. Nat. Prod.* **2012**, *75*, 1480–1484. [CrossRef]
61. Palaniappan, N.; Kim, B.S.; Sekiyama, Y.; Osada, H.; Reynolds, K.A. Enhancement and selective production of phoslactomycin B, a protein phosphatase II a inhibitor, through identification and engineering of the corresponding biosynthetic gene cluster. *J. Biol. Chem.* **2003**, *278*, 35552–35557. [CrossRef]
62. Niraula, N.P.; Kanth, B.K.; Sohng, J.K.; Oh, T.J. Hydrogen peroxide-mediated dealkylation of 7-ethoxycoumarin by cytochrome P450 (CYP107AJ1) from *Streptomyces peucetius* ATCC27952. *Enzym. Microb. Technol.* **2011**, *48*, 181–186. [CrossRef]
63. Shafiee, A.; Hutchinson, C.R. Macrolide antibiotic biosynthesis: Isolation and properties of two forms of 6-deoxyerythronolide B hydroxylase from *Saccharopolyspora erythraea* (*Streptomyces erythreus*). *Biochemistry* **1987**, *26*, 6204–6210. [CrossRef]
64. Stassi, D.; Donadio, S.; Staver, M.J.; Katz, L. Identification of a *Saccharopolyspora erythraea* gene required for the final hydroxylation step in erythromycin biosynthesis. *J. Bacteriol.* **1993**, *175*, 182–189. [CrossRef] [PubMed]
65. Zhang, H.; Chen, J.; Wang, H.; Xie, Y.; Ju, J.; Yan, Y.; Zhang, H. Structural analysis of HmtT and HmtN involved in the tailoring steps of himastatin biosynthesis. *FEBS Lett.* **2013**, *587*, 1675–1680. [CrossRef]

66. Ma, J.; Wang, Z.; Huang, H.; Luo, M.; Zuo, D.; Wang, B.; Sun, A.; Cheng, Y.Q.; Zhang, C.; Ju, J. Biosynthesis of himastatin: Assembly line and characterization of three cytochrome P450 enzymes involved in the post-tailoring oxidative steps. *Angew. Chem. (Int. Ed. Engl.)* **2011**, *50*, 7797–7802. [CrossRef]
67. Arisawa, A.; Tsunekawa, H.; Okamura, K.; Okamoto, R. Nucleotide sequence analysis of the carbomycin biosynthetic genes including the 3-O-acyltransferase gene from *Streptomyces thermotolerans*. *Biosci. Biotechnol. Biochem.* **1995**, *59*, 582–588. [CrossRef] [PubMed]
68. Carlson, J.C.; Fortman, J.L.; Anzai, Y.; Li, S.; Burr, D.A.; Sherman, D.H. Identification of the tirandamycin biosynthetic gene cluster from *Streptomyces* sp. 307-9. *Chembiochem* **2010**, *11*, 564–572. [CrossRef]
69. Carlson, J.C.; Li, S.; Gunatilleke, S.S.; Anzai, Y.; Burr, D.A.; Podust, L.M.; Sherman, D.H. Tirandamycin biosynthesis is mediated by co-dependent oxidative enzymes. *Nat. Chem.* **2011**, *3*, 628–633. [CrossRef] [PubMed]
70. Aparicio, J.F.; Molnar, I.; Schwecke, T.; Konig, A.; Haydock, S.F.; Khaw, L.E.; Staunton, J.; Leadlay, P.F. Organization of the biosynthetic gene cluster for rapamycin in *Streptomyces hygroscopicus*: Analysis of the enzymatic domains in the modular polyketide synthase. *Gene* **1996**, *169*, 9–16. [CrossRef]
71. Molnar, I.; Aparicio, J.F.; Haydock, S.F.; Khaw, L.E.; Schwecke, T.; Konig, A.; Staunton, J.; Leadlay, P.F. Organisation of the biosynthetic gene cluster for rapamycin in *Streptomyces hygroscopicus*: Analysis of genes flanking the polyketide synthase. *Gene* **1996**, *169*, 1–7. [CrossRef]
72. Huang, S.; Bjornsti, M.-A.; Houghton, P.J. Rapamycins: Mechanisms of action and cellular resistance. *Cancer Biol. Ther.* **2003**, *2*, 222–232. [CrossRef]
73. Sherman, D.H.; Li, S.; Yermalitskaya, L.V.; Kim, Y.; Smith, J.A.; Waterman, M.R.; Podust, L.M. The structural basis for substrate anchoring, active site selectivity, and product formation by P450 PikC from *Streptomyces venezuelae*. *J. Biol. Chem.* **2006**, *281*, 26289–26297. [CrossRef] [PubMed]
74. Kong, R.; Liu, X.; Su, C.; Ma, C.; Qiu, R.; Tang, L. Elucidation of the biosynthetic gene cluster and the post-PKS modification mechanism for fostriecin in *Streptomyces pulveraceus*. *Chem. Biol.* **2013**, *20*, 45–54. [CrossRef] [PubMed]
75. Han, S.; Pham, T.V.; Kim, J.H.; Lim, Y.R.; Park, H.G.; Cha, G.S.; Yun, C.H.; Chun, Y.J.; Kang, L.W.; Kim, D. Functional characterization of CYP107W1 from *Streptomyces avermitilis* and biosynthesis of macrolide oligomycin A. *Arch Biochem. Biophys.* **2015**, *575*, 1–7. [CrossRef] [PubMed]
76. Han, S.; Pham, T.V.; Kim, J.H.; Lim, Y.R.; Park, H.G.; Cha, G.S.; Yun, C.H.; Chun, Y.J.; Kang, L.W.; Kim, D. Structural Analysis of the *Streptomyces avermitilis* CYP107W1-Oligomycin A Complex and Role of the Tryptophan 178 Residue. *Mol. Cells* **2016**, *39*, 211–216. [CrossRef] [PubMed]
77. Chen, D.; Zhang, Q.; Zhang, Q.; Cen, P.; Xu, Z.; Liu, W. Improvement of FK506 production in *Streptomyces tsukubaensis* by genetic enhancement of the supply of unusual polyketide extender units via utilization of two distinct site-specific recombination systems. *Appl. Environ. Microbiol.* **2012**, *78*, 5093–5103. [CrossRef]
78. Lomovskaya, N.; Otten, S.L.; Doi-Katayama, Y.; Fonstein, L.; Liu, X.C.; Takatsu, T.; Inventi-Solari, A.; Filippini, S.; Torti, F.; Colombo, A.L.; et al. Doxorubicin overproduction in *Streptomyces peucetius*: Cloning and characterization of the dnrU ketoreductase and dnrV genes and the doxA cytochrome P-450 hydroxylase gene. *J. Bacteriol.* **1999**, *181*, 305–318. [CrossRef]
79. Madduri, K.; Hutchinson, C.R. Functional characterization and transcriptional analysis of a gene cluster governing early and late steps in daunorubicin biosynthesis in *Streptomyces peucetius*. *J. Bacteriol.* **1995**, *177*, 3879–3884. [CrossRef]
80. Dickens, M.L.; Priestley, N.D.; Strohl, W.R. In vivo and in vitro bioconversion of epsilon-rhodomycinone glycoside to doxorubicin: Functions of DauP, DauK, and DoxA. *J. Bacteriol.* **1997**, *179*, 2641–2650. [CrossRef]
81. Walczak, R.J.; Dickens, M.L.; Priestley, N.D.; Strohl, W.R. Purification, properties, and characterization of recombinant *Streptomyces* sp. strain C5 DoxA, a cytochrome P-450 catalyzing multiple steps in doxorubicin biosynthesis. *J. Bacteriol.* **1999**, *181*, 298–304. [CrossRef]
82. Li, W.; Luo, Y.; Ju, J.; Rajski, S.R.; Osada, H.; Shen, B. Characterization of the tautomycetin biosynthetic gene cluster from *Streptomyces griseochromogenes* provides new insight into dialkylmaleic anhydride biosynthesis. *J. Nat. Prod.* **2009**, *72*, 450–459. [CrossRef]
83. Zocher, G.; Richter, M.E.; Mueller, U.; Hertweck, C. Structural fine-tuning of a multifunctional cytochrome P450 monooxygenase. *J. Am. Chem. Soc.* **2011**, *133*, 2292–2302. [CrossRef] [PubMed]

84. Podust, L.M.; Bach, H.; Kim, Y.; Lamb, D.C.; Arase, M.; Sherman, D.H.; Kelly, S.L.; Waterman, M.R. Comparison of the 1.85 A structure of CYP154A1 from *Streptomyces coelicolor* A3(2) with the closely related CYP154C1 and CYPs from antibiotic biosynthetic pathways. *Protein Sci. Publ. Protein Soc.* **2004**, *13*, 255–268. [CrossRef] [PubMed]
85. Cheng, Q.; Lamb, D.C.; Kelly, S.L.; Lei, L.; Guengerich, F.P. Cyclization of a cellular dipentaenone by *Streptomyces coelicolor* cytochrome P450 154A1 without oxidation/reduction. *J. Am. Chem. Soc.* **2010**, *132*, 15173–15175. [CrossRef] [PubMed]
86. Podust, L.M.; Kim, Y.; Arase, M.; Neely, B.A.; Beck, B.J.; Bach, H.; Sherman, D.H.; Lamb, D.C.; Kelly, S.L.; Waterman, M.R. The 1.92-A structure of *Streptomyces coelicolor* A3(2) CYP154C1. A new monooxygenase that functionalizes macrolide ring systems. *J. Biol. Chem.* **2003**, *278*, 12214–12221. [CrossRef] [PubMed]
87. Zhao, B.; Bellamine, A.; Lei, L.; Waterman, M.R. The role of Ile87 of CYP158A2 in oxidative coupling reaction. *Arch Biochem. Biophys.* **2012**, *518*, 127–132. [CrossRef] [PubMed]
88. Mendes, M.V.; Anton, N.; Martin, J.F.; Aparicio, J.F. Characterization of the polyene macrolide P450 epoxidase from *Streptomyces natalensis* that converts de-epoxypimaricin into pimaricin. *Biochem. J.* **2005**, *386*, 57–62. [CrossRef]
89. Xie, Z.; Niu, G.; Li, R.; Liu, G.; Tan, H. Identification and characterization of sanH and sanI involved in the hydroxylation of pyridyl residue during nikkomycin biosynthesis in *Streptomyces ansochromogenes*. *Curr. Microbiol.* **2007**, *55*, 537–542. [CrossRef]
90. Chen, H.; Walsh, C.T. Coumarin formation in novobiocin biosynthesis: Beta-hydroxylation of the aminoacyl enzyme tyrosyl-S-NovH by a cytochrome P450 NovI. *Chem. Biol.* **2001**, *8*, 301–312. [CrossRef]
91. Uhlmann, S.; Süssmuth, R.D.; Cryle, M.J. Cytochrome p450sky interacts directly with the nonribosomal peptide synthetase to generate three amino acid precursors in skyllamycin biosynthesis. *ACS Chem. Biol.* **2013**, *8*, 2586–2596. [CrossRef]
92. Zhao, B.; Lei, L.; Vassylyev, D.G.; Lin, X.; Cane, D.E.; Kelly, S.L.; Yuan, H.; Lamb, D.C.; Waterman, M.R. Crystal structure of albaflavenone monooxygenase containing a moonlighting terpene synthase active site. *J. Biol. Chem.* **2009**, *284*, 36711–36719. [CrossRef]
93. Takamatsu, S.; Lin, X.; Nara, A.; Komatsu, M.; Cane, D.E.; Ikeda, H. Characterization of a silent sesquiterpenoid biosynthetic pathway in *Streptomyces avermitilis* controlling epi-isozizaene albaflavenone biosynthesis and isolation of a new oxidized epi-isozizaene metabolite. *Microb. Biotechnol.* **2011**, *4*, 184–191. [CrossRef] [PubMed]
94. Moody, S.C.; Zhao, B.; Lei, L.; Nelson, D.R.; Mullins, J.G.; Waterman, M.R.; Kelly, S.L.; Lamb, D.C. Investigating conservation of the albaflavenone biosynthetic pathway and CYP170 bifunctionality in streptomycetes. *FEBS J.* **2012**, *279*, 1640–1649. [CrossRef] [PubMed]
95. Ikeda, H.; Omura, S. Avermectin Biosynthesis. *Chem. Rev.* **1997**, *97*, 2591–2610. [CrossRef] [PubMed]
96. Lamb, D.C.; Ikeda, H.; Nelson, D.R.; Ishikawa, J.; Skaug, T.; Jackson, C.; Omura, S.; Waterman, M.R.; Kelly, S.L. Cytochrome p450 complement (CYPome) of the avermectin-producer *Streptomyces avermitilis* and comparison to that of *Streptomyces coelicolor* A3(2). *Biochem. Biophys. Res. Commun.* **2003**, *307*, 610–619. [CrossRef]
97. Tetzlaff, C.N.; You, Z.; Cane, D.E.; Takamatsu, S.; Omura, S.; Ikeda, H. A gene cluster for biosynthesis of the sesquiterpenoid antibiotic pentalenolactone in *Streptomyces avermitilis*. *Biochemistry* **2006**, *45*, 6179–6186. [CrossRef] [PubMed]
98. Healy, F.G.; Krasnoff, S.B.; Wach, M.; Gibson, D.M.; Loria, R. Involvement of a cytochrome P450 monooxygenase in thaxtomin A biosynthesis by *Streptomyces acidiscabies*. *J. Bacteriol.* **2002**, *184*, 2019–2029. [CrossRef]
99. Markowitz, V.M.; Chen, I.-M.A.; Palaniappan, K.; Chu, K.; Szeto, E.; Grechkin, Y.; Ratner, A.; Jacob, B.; Huang, J.; Williams, P. IMG: The integrated microbial genomes database and comparative analysis system. *Nucleic Acids Res.* **2011**, *40*, D115–D122. [CrossRef]
100. Kanehisa, M.; Sato, Y.; Furumichi, M.; Morishima, K.; Tanabe, M. New approach for understanding genome variations in KEGG. *Nucleic Acids Res.* **2019**, *47*, D590–D595. [CrossRef]
101. Syed, K.; Mashele, S.S. Comparative analysis of P450 signature motifs EXXR and CXG in the large and diverse kingdom of fungi: Identification of evolutionarily conserved amino acid patterns characteristic of P450 family. *PLoS ONE* **2014**, *9*, e95616. [CrossRef]

102. Syed, P.R.; Chen, W.; Nelson, D.R.; Kappo, A.P.; Yu, J.H.; Karpoormath, R.; Syed, K. Cytochrome P450 Monooxygenase CYP139 Family Involved in the Synthesis of Secondary Metabolites in 824 Mycobacterial Species. *Int. J. Mol. Sci.* **2019**, *20*, 2690. [CrossRef]
103. Katoh, K.; Kuma, K.-I.; Toh, H.; Miyata, T. MAFFT version 5: Improvement in accuracy of multiple sequence alignment. *Nucleic Acids Res.* **2005**, *33*, 511–518. [CrossRef] [PubMed]
104. Boc, A.; Diallo, A.B.; Makarenkov, V. T-REX: A web server for inferring, validating and visualizing phylogenetic trees and networks. *Nucleic Acids Res.* **2012**, *40*, W573–W579. [CrossRef] [PubMed]
105. Letunic, I.; Bork, P. Interactive Tree Of Life (iTOL) v4: Recent updates and new developments. *Nucleic Acids Res.* **2019**, *47*, W256–W259. [CrossRef] [PubMed]
106. Howe, E.; Holton, K.; Nair, S.; Schlauch, D.; Sinha, R.; Quackenbush, J. Mev: Multiexperiment viewer. In *Biomedical Informatics for Cancer Research*; Springer: Berlin/Heidelberg, Germany, 2010; pp. 267–277.
107. Blin, K.; Wolf, T.; Chevrette, M.G.; Lu, X.; Schwalen, C.J.; Kautsar, S.A.; Suarez Duran, H.G.; de Los Santos, E.L.C.; Kim, H.U.; Nave, M.; et al. antiSMASH 4.0-improvements in chemistry prediction and gene cluster boundary identification. *Nucleic Acids Res.* **2017**, *45*, W36–W41. [CrossRef] [PubMed]

© 2020 by the authors. Licensee MDPI, Basel, Switzerland. This article is an open access article distributed under the terms and conditions of the Creative Commons Attribution (CC BY) license (http://creativecommons.org/licenses/by/4.0/).

Article

Comparative Analysis, Structural Insights, and Substrate/Drug Interaction of CYP128A1 in *Mycobacterium tuberculosis*

Nokwanda Samantha Ngcobo [1], Zinhle Edith Chiliza [1], Wanping Chen [2], Jae-Hyuk Yu [3,4], David R. Nelson [5], Jack A. Tuszynski [6,7], Jordane Preto [8,*] and Khajamohiddin Syed [1,*]

1. Department of Biochemistry and Microbiology, Faculty of Science and Agriculture, University of Zululand, KwaDlangezwa 3886, South Africa; mskwandosamn@gmail.com (N.S.N.); zinhlechiliza01@gmail.com (Z.E.C.)
2. Department of Molecular Microbiology and Genetics, University of Göttingen, 37077 Göttingen, Germany; chenwanping1@foxmail.com
3. Department of Bacteriology, University of Wisconsin-Madison, 3155 MSB, 1550 Linden Drive, Madison, WI 53706, USA; jyu1@wisc.edu
4. Department of Systems Biotechnology, Konkuk University, Seoul 05029, Korea
5. Department of Microbiology, Immunology and Biochemistry, University of Tennessee Health Science Center, Memphis, TN 38163, USA; drnelson1@gmail.com
6. Department of Physics and Department of Oncology, University of Alberta, Edmonton, AB T6G 2E1, Canada; jack.tuszynski@gmail.com
7. Department of Mechanical and Aerospace Engineering, Politecnico di Torino, Corso Duca degli Abruzzi, 24, 10129 Torino TO, Italy
8. Université Claude Bernard Lyon 1, INSERM 1052, CNRS 5286, Centre Léon Bérard, Centre de Recherche en Cancérologie de Lyon, 69622 Lyon, France
* Correspondence: jordane.preto@gmail.com (J.P.); khajamohiddinsyed@gmail.com (K.S.)

Received: 29 April 2020; Accepted: 11 May 2020; Published: 8 July 2020

Abstract: Cytochrome P450 monooxygenases (CYPs/P450s) are well known for their role in organisms' primary and secondary metabolism. Among 20 P450s of the tuberculosis-causing *Mycobacterium tuberculosis* H37Rv, CYP128A1 is particularly important owing to its involvement in synthesizing electron transport molecules such as menaquinone-9 (MK9). This study employs different *in silico* approaches to understand CYP128 P450 family's distribution and structural aspects. Genome data-mining of 4250 mycobacterial species has revealed the presence of 2674 *CYP128* P450s in 2646 mycobacterial species belonging to six different categories. Contrast features were observed in the *CYP128* gene distribution, subfamily patterns, and characteristics of the secondary metabolite biosynthetic gene cluster (BGCs) between *M. tuberculosis complex* (MTBC) and other mycobacterial category species. In all MTBC species (except one) CYP128 P450s belong to subfamily A, whereas subfamily B is predominant in another four mycobacterial category species. Of CYP128 P450s, 78% was a part of BGCs with *CYP124A1*, or together with *CYP124A1* and *CYP121A1*. The CYP128 family ranked fifth in the conservation ranking. Unique amino acid patterns are present at the EXXR and CXG motifs. Molecular dynamic simulation studies indicate that the CYP128A1 bind to MK9 with the highest affinity compared to the azole drugs analyzed. This study provides comprehensive comparative analysis and structural insights of CYP128A1 in *M. tuberculosis*.

Keywords: cytochrome P450 monooxygenenases; CYP128A1; *Mycobacterium tuberculosis* H37Rv; tuberculosis; molecular dynamic simulations; azole drugs; menaquinone

1. Introduction

Tuberculosis (TB), caused by *Mycobacterium tuberculosis* H37Rv, remains a serious public health problem despite the existence of international TB control programs [1]. Recent data from the World Health Organization (WHO) shows that about 10 million people fell ill with TB in 2018 [1]. TB's global threat to human health has been exacerbated in recent years by the emergence of widespread multi- and extensively drug-resistant *M. tuberculosis* strains [1]. The developing countries of South-East Asia and Africa have shown high incidence rates of TB. The prevalence of the disease in these countries is mainly due to lack of basic sanitation (causing an increase in transmission of the disease), human immunodeficiency virus (HIV) infection and lack of drugs to treat the disease [1]. Recent statistics from South Africa revealed that TB is the major killer among infectious diseases, indicating that this disease is still a major challenge in the country [2].

In 1998, determination of the *M. tuberculosis* H37Rv genome sequence encouraged more investigation of new anti-tubercular drugs and the seeking of more knowledge on the complex biology of the *M. tuberculosis* bacterium [3]. This highlighted the importance of lipid metabolism in *M. tuberculosis*; novel biosynthetic pathways were found to be involved in the synthesis of compounds such as phenolphthiocerol, mycolic acids and mycocerosic acid for the complex cell wall structure of the bacterium [4]. Among the enzymes involved in lipid metabolism, cytochrome P450 monooxygenases (CYPs/P450s) in *M. tuberculosis* were found to play a key role in the metabolism of lipids [5,6]. P450s are heme-thiolate proteins found in all species across biological domains [7]. Recent studies revealed the presence of a large number of P450s in mycobacterial species and most of these P450s were found to be involved in lipid metabolism [6]. *M. tuberculosis* H37Rv has 20 P450s in its genome and some of these P450s are indeed involved in lipid metabolism [5]. Furthermore, one of the P450 genes, namely *CYP125A1*, was used as a key factor in determining the cholesterol degrading ability of mycobacterial species [8].

Among *M. tuberculosis* H37Rv P450s, CYP128A1 gained particular interest among researchers owing to its history indicating its essentiality and its physiological importance. Transposon site hybridization mutagenesis studies indicated that *CYP128A1* is essential for in vitro survival of *M. tuberculosis* H37Rv [9]. Interestingly, another study, which used a similar approach, revealed that *CYP128A1* is not essential for survival of *M. tuberculosis* CDC1551 [10]. However, this study had a backdrop of limited gene coverage in its mutant library. In Vitro *M. tuberculosis* H37Rv latency model studies including a carbon starvation model [11] and hypoxia model [12] showed up-regulation of *CYP128A1*, suggesting that this P450 has a potential role in *M. tuberculosis* latency. Until 2016, the nature of *CYP128A1* with respect to its essentiality remained a mystery. Research revealed that *CYP128A1* is non-essential for survival of *M. tuberculosis* H37Rv as the *CYP128A1* gene knock-out strain survives [13]. However, the *CYP128A1* gene knock-out strain has proven to be hyper-virulent [13], indicating this gene actually playing a role in the synthesis of a compound that acts as a negative regulator of virulence.

Heterologous expression of *CYP128A1* posed a great challenge to researchers, as the expression of this particular P450 in *Escherichia coli* has been unsuccessful [14,15]. Genomic analysis revealed that *CYP128A1* is part of an operon that consists of two other genes, *stf3* and *rv2269c* [16]. In Vivo studies using *M. smegmatis* as model strain demonstrated that CYP128A1 is involved in hydroxylation of menaquinone-9 (MK9) and is essential in the synthesis of this compound, whereas Stf3 was found to introduce the sulfate group to menaquinone-9 and *rv2269c* was found to act as a promoter [13]. The sequence of reaction is that CYP128A1 introduces the hydroxyl group into MK9, followed by the addition of the sulfate group by the Stf3 that leads to the synthesis of sulfomenaquinone [13].

Lipoquinones are electron transport molecules that are involved in the respiratory function of bacteria and mainly consist of menaquinone and ubiquinone [17]. Menaquinones (2-methyl-3-polyprenyl-1,4-naphthoquinones) especially MK9 is ubiquitous and unique to mycobacteria [17], indicating that CYP128A1 should be present in all mycobacterial species. However, to date, the distribution of *CYP128A1* in such a large number of mycobacterial species belonging to six different mycobacterial categories, i.e., *Mycobacterium tuberculosis* complex (MTBC), *M. chelonae-abscessus* complex (MCAC), *M. avium* complex (MAC), mycobacteria causing leprosy (MCL), non-tuberculosis mycobacteria (NTM) and saprophytes (SAP) is still unknown. Anti-fungal azole drugs were shown to be promising new tools to fight TB, particularly as they showed high antimycobacterial activity [18–20] and interestingly, to date, characterized *M. tuberculosis* P450s have been found to bind quite a number of azole drugs [21], leading to *M. tuberculosis* P450s becoming the main focus as novel drug targets against this pathogen [5]. The binding of azole drugs to CYP128A1 could have an undesired effect, as this could lead to possible disruption of enzyme function, a vital component in the virulence modulation of *M. tuberculosis*.

To date, genome-wide analysis of *CYP128* P450s has only been carried out in 60 mycobacterial species [22] and because of the failure of *CYP128A1* heterologous expression [14,15], analysis of binding patterns of azole drugs to CYP128A1 has not been performed. To address these research gaps, in this study, genome-wide data mining, annotation, and phylogenetic analysis of CYP128A1 were carried out in 4250 mycobacterial species. Furthermore, *in silico* analysis of binding of different azole drugs with the CYP128A1 model was assessed. The results for CYP128A1 were discussed in the context of gaining more knowledge on the role of this P450 in mycobacterial species.

2. Results and Discussion

2.1. Presence of CYP128 P450s in Five of the Six Mycobacterial Category Species

Comprehensive comparative analysis of *CYP128* P450s in 4250 mycobacterial species belonging to six different categories (MTBC, MCAC, MAC, MCL, NTM and SAP) revealed that this P450 family is present in 2646 mycobacterial species (Figures 1 and 2; Supplementary Dataset 1). Thirty-seven mycobacterial species were found to have short P450 sequences that have none or only one of the highly conserved P450 motifs, i.e., EXXR and CXG (Supplementary Dataset 1). Thus, these short sequences are not included in the study and the species are considered not to have *CYP128* P450. Among mycobacterial species that were used in this study, most of the species (99%) belonging to the MTBC category have this P450 (Figure 2). Among 2350 MTBC species only 38 species do not have *CYP128* (Figure 2). In contrast to MTBC species, most of the species belonging to another five mycobacterial categories do not have this P450 (Figure 2). CYP128 P450s were found in only 34% of NTM species, followed by 26% of MAC species, 12% of MCAC species and 18% of SAP species. The results observed in this study revealed that none of the MCL species has CYP128 P450s, which is consistent with previous observations that MCL species do not have this P450 [22]. Overall, based on the results of this study and comparison with the results reported earlier [22], we conclude that *CYP128* P450s can be found in species belonging to the five mycobacterial categories except in species of the MCL category. Most of the CYP128 P450s (81%) are 489 amino acids in length (Supplementary Dataset 1). An anomaly was observed that two CYP128 P450 sequences, one from *M. tuberculosis* BTB10-120 and the other from *M. tuberculosis* SIT745/EAI1-MYS, has 877 amino acids. However, analysis of genome sequences revealed that *CYP128* and the sulfotransferase (*stf3*) genes were present next to each other on different DNA strands, indicating an error in annotation leading to the fused protein. A single copy of the *CYP128* gene was found in almost all species, except for 25 species where two copies (22 species), three copies (two species) and four copies (one species) of this gene were found (Supplementary Dataset 1). The species with more than one copy of this P450 gene were found across four different categories, where eight species were from MTBC, 12 species from NTM, three species from MAC and two species from MCAC (Supplementary Dataset 1). CYP128 P450 sequences identified in the study are presented in Supplementary Dataset 2.

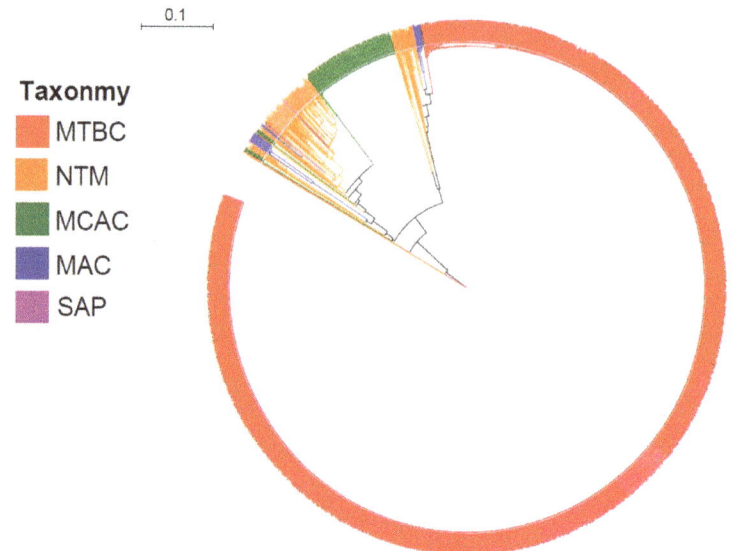

Figure 1. Phylogenetic tree of the cytochrome P450 monooxygenases (CYPs/P450s). Different mycobacterial categories are indicated indifferent colors. Abbreviations: MTBC, *Mycobacterium tuberculosis* complex; NTM, non-tuberculosis mycobacteria; MCAC, *Mycobacterium chelonae-abscessus* complex; MAC, *M. avium* complex and SAP, Saprophytes. A high-resolution phylogenetic tree is provided in Supplementary Dataset 3.

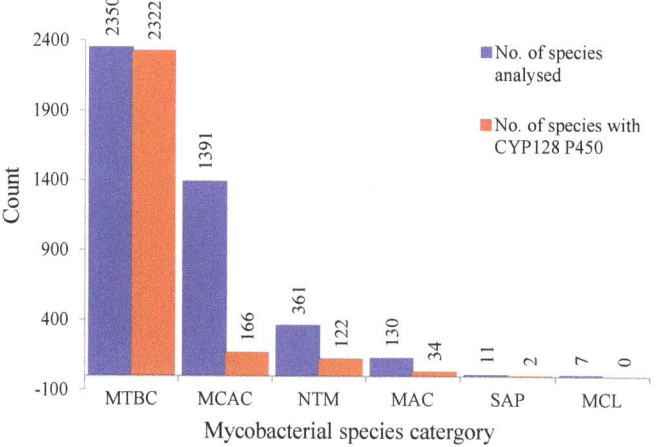

Figure 2. Comparative analysis of the CYP128 P450s in six different mycobacterial categories. Abbreviations: MTBC, *Mycobacterium tuberculosis* complex; NTM, non-tuberculosis mycobacteria; MCAC, Mycobacterium chelonae-abscessus complex; MAC, *M. avium* complex; SAP, saprophytes and MCL, mycobacteria causing leprosy. Information on mycobacterial species and CYP128 P450s is presented in Supplementary Dataset 1.

2.2. Different Mycobacterial Category Species Have Different CYP128 Subfamily Preferences

CYP128 subfamily analysis revealed the presence of a new CYP128 subfamily in mycobacterial species. Thus, at present two CYP128 subfamilies can be found in mycobacterial species, i.e., A and B

(Figure 3). Phylogenetic analysis revealed an interesting feature in CYP128 subfamilies with respect to their categories (Figure 1). In a previous study, it was observed that P450s belonging to different mycobacterial categories grouped together according to their category, indicating high conservation of P450 protein sequences after speciation into different categories [22]. In this study, the same phenomenon was observed for CYP128 subfamilies A and B, as these subfamily proteins were grouped together according to their mycobacterial categories, despite the subfamilies being clearly separated on the tree (Figure 1). Also, contrasting subfamily features were observed among different mycobacterial categories (Figure 3). Except for one species, *M. tuberculosis* XTB13-223, all species belonging to the MTBC category have CYP128 subfamily A (Figure 3). However, in contrast to MTBC species, in other mycobacterial categories subfamily B was dominantly present (Figure 3). This indicates that during evolution mycobacterial species had both subfamilies, but owing to their lifestyles or ecological niches only one subfamily was favored. This phenomenon of enriching a particular type of P450 family or subfamilies in microbial species is well known [6,22–28]. An interesting pattern of subfamilies was found in species having more than one copy of the *CYP128* gene. Eight MTBC species have two copies of the *CYP128* gene, both belonging to the same subfamily A. However, in other category species subfamilies A and B are present; especially in species belonging to the NTM category, subfamily B is populated (Supplementary Dataset 1).

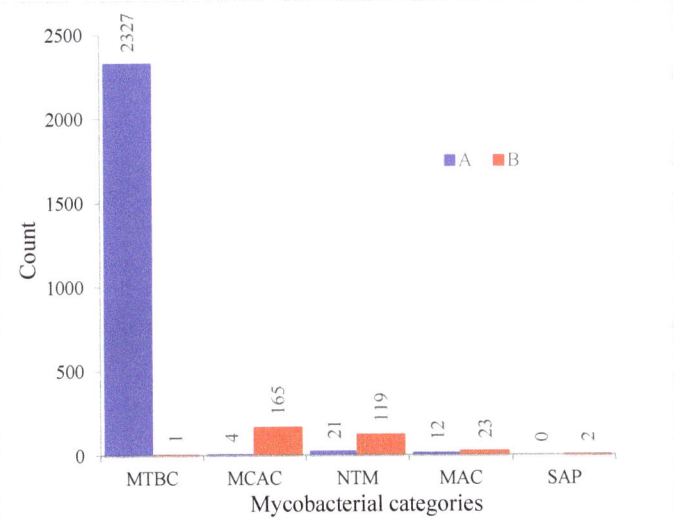

Figure 3. Comparative analysis of CYP128 P450 subfamilies in five different mycobacterial categories. Abbreviations: MTBC, *Mycobacterium tuberculosis* complex; NTM, non-tuberculosis mycobacteria; MCAC, *Mycobacterium chelonae-abscessus* complex; MAC, *M. avium* complex; and SAP, Saprophytes. Information on mycobacterial species and CYP128 P450s is presented in Supplementary Dataset 1.

2.3. CYP128 Family Ranked Fifth among P450 Families

It has been proposed that the present-day P450s are evolved from the ancient P450 CYP51 [29,30]. Evolution of different P450 families and their rate of evolution indicate that the higher the evolutionary rate, the more catalytically diverse they are [6,22,24]. Parvez and co-workers [22] proposed a ranking system for P450 families where different P450 families are given a rank based on the number of conserved amino acids and it was proposed that the higher the conservation, the less the catalytic diversity. Furthermore, a recent study revealed that better ranking of P450 families can be achieved using a larger sample size [31]. In a previous study, only 49 CYP128 P450s were used to predict the ranking [22]. Identification of quite a large number of CYP128 P450s in this study necessitates re-assessment of ranking, in order to calculate the accurate number of conserved amino acids in P450s, as it was mentioned that P450s with similar amino acid length should be used [22,25,31,32]. Thus, in this study, 2191 CYP128 P450s with amino acid lengths ranging from 479 to 489 amino acids (Supplementary Dataset 1) were used to assess the conservation ranking of the CYP128 P450 family. PROfile Multiple Alignment with Predicted Local Structures and 3D Constraints (PROMALS3D) analysis [33] revealed the presence of 217 amino acids that are invariantly conserved in CYP128 P450s (Table 1). Comparison with other P450 families from different biological kingdoms placed CYP128 family in the fifth position, compared to 23rd position previously (Table 1), indicating that this P450 family is one of the best conserved families. A complete table with updated P450 family ranking is presented in Table S1.

Table 1. Comparative amino acid conservation analysis of CYP128 P450 family with top 10 ranked families. The conservation index score (5–9) is obtained as described elsewhere [33] using PROMALS3D, where the number 9 indicates invariantly conserved amino acids in P450 members. The CYP128 family is indicated in bold.

P450 Family	Number of Member P450s	Kingdom	PROMALS3D Conservation Index					Rank (Highest to Lowest Conservation)
			5	6	7	8	9	
CYP141	29	Bacteria	0	0	0	0	389	1
CYP51	50	Bacteria	11	102	0	0	264	2
CYP137	38	Bacteria	145	0	0	0	251	3
CYP121	34	Bacteria	0	0	0	0	233	4
CYP128	**2191**	**Bacteria**	**118**	**25**	**0**	**0**	**217**	**5 (previously 23rd)**
CYP132	39	Bacteria	175	0	0	0	217	5
CYP5619	23	Stramenopila	118	38	170	0	199	6
CYP124	71	Bacteria	52	35	59	0	170	7
CYP139	894	Bacteria	0	127	0	0	165	8
CYP188	67	Bacteria	62	0	100	0	141	9
CYP123	74	Bacteria	62	0	82	0	137	10

The bold shows the position of CYP128 as it been revised from 23rd place to 5th because of this study results.

2.4. CYP128 Family Has Distinctive Amino Acid Patterns at EXXR and CXG Motif

A study of the analysis of P450 motifs EXXR and CXG revealed that all P450 families have a unique amino acid pattern that serves as a signature characteristic of that particular P450 family [34]. Subsequent studies of P450 families strongly supported this hypothesis [25,31,32]. Analysis of the EXXR and CXG P450 motifs in the CYP128 family has not been performed, and the large number of CYP128 P450s identified in this study provides an opportunity to analyze the amino acid patterns at the EXXR and CXG-motifs for this family. Analysis of the amino acids at the EXXR motif indicated the presence of E-T(84%)/Q(11%)/H(1%) -W(84%)/L(15%)-R amino acid patterns at this motif, with most of the CYP128 P450s (84%) having the E-T-W-R amino acid pattern (Figure 4). Analysis of amino acid patterns at the CXG motif revealed the presence of F-G-S(88%)/Y(12%) -G-I(88%)/V(6%)/A(5%)/P(1%) -H-L(89%)/M(11%) -C-P(87%)/I(7%)/L(6%) -G amino acid patterns, with the majority of the CYP128 P450s (86%) containing the F-G-S-G-I-H-L-C-P-G amino acid patterns (Figure 4). Amino acids patterns at the EXXR and CXG motifs of the CYP128 family were found to be unique compared to P450 families from different biological kingdoms [26,31,32,34], further supporting the hypothesis that amino acid patterns at these motifs are a signature of the P450 family [34].

Figure 4. Analysis of amino acid patterns at the EXXR and CXG motif in the CYP128 P450 family. All CYP128 P450 sequences (total of 2674 sequences) were analyzed for the EXXR and CXG signature motifs.

2.5. Most CYP128 P450s Exist in Secondary Metabolite Biosynthetic Gene Clusters

P450s are well known to play a key role in the synthesis of different secondary metabolites [35,36] and a recent study undertaking comparative analysis of secondary metabolite biosynthetic gene clusters (BGCs) between 48 *Streptomyces* species and 60 mycobacterial species revealed that *CYP128* P450s are part of a secondary metabolite BGC [6]. The *CYP128* P450 BGCs were found to have one or two other P450s in the cluster and the *CYP128* P450 is always found with CYP124A1 or together with *CYP124A1* and *CYP121A1* P450s [6]. In this study, secondary metabolite BGC analysis revealed that 78% of CYP128 P450s were found to be part of secondary metabolite BGCs; 2090 *CYP128* P450s from 2674 *CYP128* P450s were found to be part of secondary metabolite BGCs. Analysis of cluster types revealed 1994 *CYP128* P450 clusters belonging to the tRNA-dependent cyclodipeptide synthases (CDPS) cluster type, followed by 94 having no cluster types, indicating a novel BGC cluster; one belongs to Type I Polyketide synthase (T1PKS) and the last one belongs to the non-ribosomal peptide synthetase (NRPS) cluster type (Supplementary Dataset 1). In 25 species with two or more *CYP128* P450s, four species (*M. tuberculosis* KI_19771, *M. tuberculosis* 402267, *M. canettii* CIPT 140070010, and *M. canettii* CIPT 140010059) were found to have all their *CYP128* P450s as part of a cluster type, three species (*M. tuberculosis* BTB10-253, *M. tuberculosis* M1415, and *Mycobacterium* sp. GA-1199) had only one and the rest of the species had no *CYP128* P450s as part of any cluster types (Supplementary Dataset 1).

An interesting contrast pattern was observed when comparing *CYP128* P450 BGCs among different mycobacterial category species (Table 2). Most of the CYP128 P450s were found to be part of BGCs in MTBC species, whereas only a handful of CYP128 P450s were part of BGCs in species belonging to the categories MCAC, NTM, and MAC (Table 2). Furthermore, most BGCs of MTBC species have three P450s, *CYP128*, *CYP121*, and *CYP124*, followed by BGCs with *CYP128* P450 and BGCs with *CYP128* and *CYP124* P450s (Table 2). Interestingly, the remaining four mycobacterial category species BGCs have only *CYP128* P450 (Table 2). None of the BGCs from all five different mycobacterial categories has the combination of P450s, *CYP128*, and *CYP121* (Table 2).

Table 2. *CYP128* P450 gene cluster analysis in five mycobacterial category species.

Data type	Mycobacterial Category				
	MTBC	MCAC	NTM	MAC	SAP
Total number of *CYP128* P450s	2328	169	140	35	2
Number of *CYP128* P450s not part of BGCs	282	147	120	34	1
Number of *CYP128* P450s part of BGCs	2046	22	20	1	1
Number of BGCs with *CYP128*, *CYP121* and *CYP124* P450s	1908	0	0	0	0
Number of BGCs with *CYP128* P450	136	22	21	1	1
Number of BGCs with *CYP128* and *CYP121* P450s	0	0	0	0	0
Number of BGCs with *CYP128* and *CYP124* P450s	2	0	0	0	0

Abbreviations: MTBC, *Mycobacterium tuberculosis* complex; NTM, non-tuberculosis mycobacteria; MCAC, *Mycobacterium chelonae-abscessus* complex; MAC, *M. avium* complex; and SAP, saprophytes; BGCs, biosynthetic gene clusters.

2.6. CYP128A1 Protein Model Has High Affinity to Menaquinone-9

Since the expression of *CYP128A1* in heterologous hosts such as *E. coli* was found to be difficult [14,15] and at present there is no scope to generate the CYP128A1 crystal structure, in this study a 3D model of CYP128A1 was built to explore its binding affinity with MK9 and different azole drugs. The CYP128A1 model was built using the structure of Vitamin D3 hydroxylase (Vdh) from *Pseudonocardia autotrophica* as a template (PDB ID: 3VRM, 33.8% identity) [37]. The model was optimized using molecular-dynamics (MD) simulations (Figure 5) as described in the subsection on Materials and Methods. Notably, other models based on templates sharing similar coverage and identity compared to 3VRM were tested. Other structures include 1Z8P (32.4% identity) [38], 3A4G (33.5% identity) [39] and 5GNM (33.6% identity) [40]. Although some of these structures have a bit better resolution than 3VRM, they either correspond to apo structures (5GNM) or structures bound to small ligands (1Z8P and 3A4G), leaving the catalytic site "packed up" and thus inaccessible through docking simulations. This was confirmed by visual inspection together with our inability to generate binding poses in the catalytic site when docking on models based on 5GNM or 3A4G templates (results not shown). Conversely, 3VRM corresponds to the structure of Vdh mutant bound with Vitamin D3 where the catalytic site is open enough to enable the binding of ligands as big as MK9. As a result, docking simulations on this model were successful in finding binding poses for all the ligands tested.

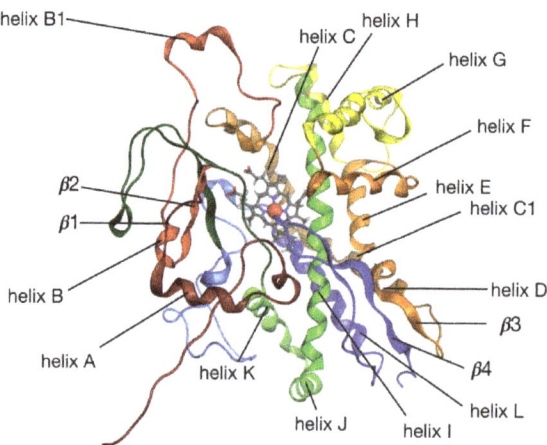

Figure 5. A 3D model of CYP128A1 with structured regions labeled. Heme and the iron atom are shown in grey and red, respectively.

Homology modeling was carried out using the Molecular Operating Environment (MOE) software as described in the Materials and Methods section. Importantly, sequence alignment did not reveal any major indel regions compared to our target sequence (see Figure S1). We reported a five-residue loop as the most significant insertion with a fairly large distance to the catalytic site (23.5 Å to the ferric ion of the HEME group) and a seven-residue gap as the most important deletion. During our protocol, 10 intermediate models were generated independently. For every model, the RMSD (root-mean-square deviation) to the mean conformation was calculated and a low standard deviation (STD) to the mean structure was obtained (STD = 2.654 A for residues in the five-residue insertion, STD = 0.102 A for residues on both sides of the seven-residue deletion, STD = 0.021 A for the whole structure), suggesting good convergence and a limited number of possible variations in the target model. Our final CYP128A1 model (Figure 5), selected based on MOE's built-in score, was used as a target in a first round of docking simulations. For each compound, the 10 top docked complexes were then equilibrated through MD (six complexes in the case of MK9, see Materials and Methods section). Next, MK9 and azole drugs were removed and re-docked independently onto each of their corresponding equilibrated receptors. In Table 3, we reported the results of our re-docking procedure with top scores (in kcal/mol) listed in column 2. The order of binding was as follows: MK9 > itraconozole > posaconazole > ketoconazole > econazole > miconzaole > voriconazole > clotrimazole > fluconazole (Table 2).

Table 3. Final results obtained after re-docking MK9 and azole drugs on equilibrated CYP128A1 models. The compounds are ranked according to their top docking score, from highest to lowest affinity (column 2). Column 3 shows the score of the first well-orientated pose with respect to the catalytic site. In the case of MK9, proper orientation means that the hydrophobic tail of MK9 is facing the ferric ion of heme, which is consistent with omega hydroxylation of the molecule reported elsewhere [13,16]. For azole drugs, the imidazole ring is facing Fe of heme and the tail should be out of the cavity. In column 3, numbers in parenthesis refer to the rank of the compound based on that score (over all compounds). In column 4, we reported the rank of the first well-oriented pose among all the poses generated for each ligand.

Compound	Top Score (kcal/mol)	Score of First Well-Oriented Pose (kcal/mol)	Rank of First Well-Oriented Pose
Menaquinone-9 (MK9)	−13.48	−12.83 (1)	7
Itraconazole	−12.57	−10.72 (3)	10
Posaconazole	−11.81	−11.81 (2)	1
Ketoconazole	−10.47	−9.84 (4)	9
Econazole	−8.24	−7.56 (7)	6
Miconazole	−8.10	−8.10 (5)	1
Voriconazole	−7.63	−7.57 (6)	2
Clotrimazole	−7.46	−7.21 (9)	5
Fluconazole	−7.38	−7.38 (8)	1

In P450 enzyme systems, substrates and azole drugs are known to coordinate with the ferric core of the heme group. In the case of MK9, coordination takes place at the end of its hydrophobic tail, consistent with omega hydroxylation of the molecule [13,16]. For azole drugs, coordination occurs via the imidazole ring. Importantly, docking of MK9 and azole drugs was performed in a non-covalent way in the present paper. Therefore, no coordination was predicted for the tested ligands, which would require more thorough investigation, especially considering changes in electronic density. However, since non-covalent binding is a critical step in the establishment of covalent interactions, we believe the present results still provide a good indication of how azoles drugs can interact with CYP128A1 and how they rank in terms of affinity.

Column 3 in Table 3 shows the score of the first well-oriented pose, i.e., where the end of the hydrocarbon tail (MK9) or the imidazole ring (azole drugs) faces the ferric core of the heme. Interestingly, considering well-oriented poses does not significantly affect the ranking of compounds. Column 4 shows the rank of the first well-oriented pose over all the poses generated for each ligand. While top-ranked poses are already well oriented in some cases (posaconazole, miconazole, voriconazole, fluconazole), well-oriented poses of other compounds such as itraconazole and ketoconazole exhibit lower ranking (10 and 9, respectively). Nonetheless, we observed that at least one pose among the top 10 is well oriented for each compound, leading to a score value similar to the top-ranked pose in each case. We depicted the first well-oriented poses of each compound in Figure 6.

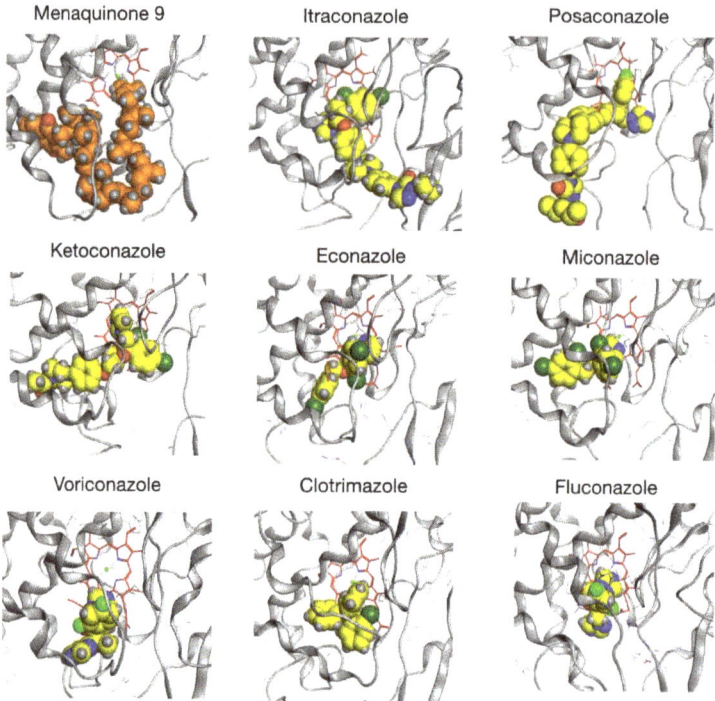

Figure 6. First well-oriented binding poses obtained for a substrate (menaquinone 9) and the azole compounds. The substrate/azole compounds are shown in orange/yellow, respectively, while the heme is shown in red and the ferric core is depicted in light green.

In summary, *in silico* results indicated that the CYP128A1 indeed binds to the substrate (MK9) with highest affinity compared to the azole drugs analyzed in the study (Table 3). Among azole drugs itraconazole, posaconazole and ketoconazole have the highest binding affinity with CYP128A1. One possible reason for the high binding affinity of these compounds compared to other azoles is that these molecules are more extended depicting the longer side chain of MK9. A point to be noted that same phenomenon of better interactions of azoles due to their more extended structures was also observed for CYP51 of *Sporothrix schenckii* [41]. CYP128A1 binding to different azole drugs is also consistent with other *M. tuberculosis* H37Rv P450s that are experimentally shown to bind azole drugs albeit with different affinities [21].

3. Materials and Methods

3.1. Species and Databases

A total of 4250 mycobacterial species genomes (permanent draft) available at the Joint Genome Institute (JGI)/Integrated Microbial Genomes and Microbes (IMG/M) database (from February 2019 to March 2019) were used in this study [42]. Species were grouped in six different mycobacterial categories, as described elsewhere [22,31]. The six categories included MTBC (2350 species), MCAC (1391 species), MAC (130 species), MCL (7 species), NTM (361 species), and SAP (11 species). Mycobacterial species along with their genome IDs and their mycobacterial categories were presented in Supplementary Dataset 1.

3.2. Genome Data Mining and Annotation of CYP128 P450s

CYP128 P450s mining in different bacterial species was carried out following the method described elsewhere [22,31]. Briefly, BLAST analysis was performed with the *M. tuberculosis* H37Rv CYP128A1 (*Rv2268c*) P450 sequence with default settings against individual mycobacterial species genomes at the IMG/M database. Considering the International Cytochrome P450 Nomenclature criteria, i.e., P450s showing >40% identity belong to the same family [43–45], all the hit proteins with more than 40% identity were selected and subjected to P450 characteristic motifs analysis as described elsewhere [34,46,47]. Proteins that showed all P450 characteristic motifs were selected for further analysis. Proteins that were short in amino acid length and had none or only one of the highly conserved P450 motifs, such as EXXR and CXG, were considered P450 fragments and not included in the study. The selected proteins were then subjected to BLAST analysis at the P450 webpage (http://www.p450.unizulu.ac.za/) to identify the named homology protein, in this case CYP128 P450. Hit proteins that showed homology to CYP128 were then selected and different subfamilies were assigned following the International Cytochrome P450 Nomenclature criteria, i.e., P450s showing >55% identity belong to the same subfamily [43–45].

3.3. Phylogenetic Analysis of CYP128 P450s

Phylogenetic analysis of CYP128 P450s was carried out following the method described elsewhere [31]. First, the protein sequences were aligned by MAFFT v6.864 embedded on the Trex web server [48]. Thereafter, the alignments were automatically subjected to tree inferring and optimization by the Trex web server [49]. Finally, the best-inferred tree was envisioned and colored using iTOL (http://itol.embl.de/about.cgi) [50].

3.4. Amino Acid Conservation Analysis

Amino acid conservation analysis of CYP128 P450s was carried out following the methods described elsewhere [22,31,32]. Briefly, 2191 CYP128 P450s with amino acid length ranging from 475–496 amino acids (Supplementary Dataset 1) were selected and subjected to PROMALS3D analysis [33]. In order to calculate the accurate number of conserved amino acids in P450s, it was mentioned that P450s with similar amino acid length should be used [22,25,31,32]. PROMALS3D analysis provided the amino acid conservation index at different protein sequence positions [33], using the numbers from 5 to 9, where 9 is the invariantly conserved amino acid. The conserved number of amino acids for each conservation index was counted and compared with other P450 families from different biological kingdoms [22,25,31,32] to determine the CYP128 P450 family conservation rank.

3.5. Generation of EXXR and CXG Sequence Logos

CYP128 P450 family EXXR and CXG sequence logos were constructed using the method described elsewhere [22,32,34]. Briefly, all CYP128 P450 sequences were aligned using ClustalW multiple alignments embedded in MEGA7 [51]. Then the EXXR and CXG region amino acids (4 and 10 amino acids, respectively), were copied and pasted in the WebLogo program (http://weblogo.berkeley.edu/logo.cgi) [52,53]. As a selection parameter, the image format was selected as PNG (bitmap) at 300 dpi resolution. The percentage predominance of amino acids at specific positions was calculated, taking into account the total number of amino acids as 100%. The generated EXXR and CXG logos were used for analysis and comparison to the different P450 family EXXR and CXG logos that have been published and are accessible to the public [22,31,32,34].

3.6. Identification of CYP128 P450 Secondary Metabolite BGCs

Secondary metabolite BGCs analysis of CYP128 P450s was carried out following the method described elsewhere [31]. Mycobacterial species BGCs listed at the IMG/M website were manually searched for the presence of CYP128 P450s using their gene ID [42]. The BGCs that contained CYP128 P450 were selected and the entire gene cluster sequence was downloaded. The listed BGCs at IMG/M are unspecific and to identify the particular BGC type, the downloaded gene cluster sequence was subjected to secondary metabolite BGC analysis using anti-SMASH [54]. The type of BGC, percentage similarity to a known cluster and the known cluster name were recorded from the anti-SMASH analysis. Standard BGC abbreviation terminology developed by anti-SMASH was used in the study.

3.7. CYP128 Homology Modeling

The Molecular Operating Environment (Chemical Computing Group) was used to build a 3D model of CYP128A1. The crystallographic structure of the P450 Vitamin D3 hydoxylase bound with vitamin D3 (VD3) (PDB ID: 3VRM) [37] was used as a template, showing 33.8% identity and 48% similarity with the targeted sequence after alignment. Homology modeling of CYP128A1 was performed by setting the number of generated models to 10 and by selecting the final model based on MOE's Generalized Born/Volume Integral (GB/VI) scoring function. During the modeling, the heme group of the template—including the ferric ion—was kept as part of the environment and included in the refinement step. The final model was eventually protonated at neutral pH and minimized using a MOE's built-in protocol.

3.8. Molecular Docking of the CYP128 Model

Non-covalent docking of menaquinone-9 (MK9), a known substrate of CYP128A1, and different azole compounds (Figure S2) was done with MOE's dock utility. Prior to this, all compounds were "washed" using MOE, i.e., we generated the most dominant protonation state of each compound at neutral pH, computed its atomic partial charges, and minimized the generated 3D structure. Docking was performed into the catalytic site of our CYP128A1 model, setting the placement method to "Triangle Matcher" and the scoring and rescoring methods to "London dG" and "GBVI/WSA dG", respectively. After docking, the ligand structures were further refined using the fixed receptor option. This refinement step entails energy minimization using the conventional Amber10:Extended Huckel Theory molecular mechanics force field to take electronic effects into account. For each compound, the top 10 complexes as identified from GBVI/WSA-dG scores were considered for further MD equilibration (see Section 3.10). In the case of MK9, since the molecule was predicted to undergo omega hydroxylation via the heme group, only poses with proper orientation (hydrophobic tail facing the heme group) were kept, resulting in six (out of 10) poses being selected.

3.9. Density Functional Theory

MD parameters—e.g., partial charges and force constants—for non-standard residues like the heme group are not provided by standard MD force fields. Hence, to run further equilibration of our docked complexes, it was necessary to generate those parameters via quantum calculations. Such calculations were performed using Gaussian 09 (g09) together with the Metal Center Parameter Builder (MCPB.py) available in the Amber16 package [55,56]. MCPB.py was applied to create the correct g09 input files by including the heme ferric cation and its nearby residues from our 3D structures. g09 was successively utilized for geometry optimization, force constant calculation and Merz-Kollman RESP charge calculation of the selected atoms. Every calculation was performed at the B3LYP/6-31G* level of theory. Finally, the MCPB.py program was re-applied to fit RESP charges and generate parameters compatible with Amber's ff14SB force field. Note that a partial charge of $0.250e$ was calculated for the ferric cation in the heme group. Keeping in mind that the partial charge of a metal ion is less than 2e, no matter how big its formal charge (oxidation state) is, our result looks reasonable.

3.10. Molecular Dynamics

While MCPB.py and g09 were used to get the correct force field parameters for the ferric core of the heme and nearby atoms, Amber's antechamber utility [55,56] was applied to generate MD parameters for MK9 and azole compounds. Amber's tleap program was applied to solvate all our docked complexes in TIP3P water and to generate the correct topology file to conduct MD simulations using the ff14SB force field. For each complex, a rectangular box with at least 10 Å distance between the box edges and the protein was considered, resulting in about 16,380 water molecules in each case. Na^+ and Cl^- ions were added to approximate 0.15 M concentration as well as to neutralize the system. Minimization, NVT, NPT, and MD production runs were all performed with Amber's pmemd utility. Minimization of each structure was carried out in two phases, both using the steepest descent and conjugate gradient methods successively. Briefly, minimization was done in 10,000 minimizations steps on hydrogens and solvent atoms only, i.e., by restraining the protein–ligand complexes. Next, a 20,000-step minimization was run without restraints. The structures were then equilibrated in the NVT ensemble during 20 ps and in the NPT ensemble during 40 ps, setting the temperature to 298 K and the pressure to 1 bar. Finally, MD production was run to relax each complex for at least 20 ns. The stability of each complex was assessed by checking if the RMSD of both the protein and the ligand reached a plateau by the end of the simulation (see Figure S3). In general, we found 20 ns sufficient to equilibrate the complexes, although in some cases, an extra 5–10 ns may have been required to equilibrate the structure fully.

3.11. Redocking of Compounds

MK9 and the azoles compounds were re-docked on their corresponding set of equilibrated structures using the MOE's dock (six structures for MK9, 10 structures for each of the azole drugs). The same options as in the first docking step were used (see Section 3.8).

4. Conclusions

The availability of quite a large number of bacterial genome sequences and different bioinformatics tools gives us an opportunity to understand the role of different genes/proteins in bacterial communities at large rather than confining the results to a single bacterium. In this study, we utilized such information and tools to understand the CYP128 P450 family profiles in different mycobacterial species and understand its structure. The study revealed interesting aspects; for example, P450 was found to be highly conserved in the MTBC species causing lung disease in humans and other animals, indicating the important function of this enzyme during the latent phase of these organisms, as this P450 was found to be expressed during such stage [11,12]. Furthermore, only 12–34% of non-MTBC species that have this P450 strongly support its important role during the latent phase of MTBC species. Contrasting CYP128 subfamily profiles between MTBC and four other different mycobacterial categories revealed by this study suggests that CYP128 P450s may play different roles in different category species, as previously observed for the CYP53 family in fungi [25]. Molecular dynamic simulation of CYP128A1 interactions with different ligands revealed that this P450 has the highest binding affinity to its substrate compared to azole drugs. Binding of azole drugs to CYP128A1 protein models further shows the complex biology of mycobacterial species, as azole drugs have been found to be effective against *Mycobacterium tuberculosis* [18–20], indicating the expression of *CYP128A1* in the latent phase [11,12], and possible binding of azole drugs during experimentation did not resulted in a hypervirulent bacterium. This may be due to the inhibition of essential P450s such as CYP121A1 and CYP125A1, leading to the death of *M. tuberculosis* H37Rv. One of the interesting observations of this study is that more than one copy of this gene was present in some species and one of these genes was found not to be part of the gene cluster. This poses a fascinating evolutionary question on the need for having more than one copy of this gene. Research on the CYP128 P450s that are not part of gene clusters would be interesting in discerning other roles of CYP128 P450s in mycobacterial species. It is also important that *CYP128* gene clusters were found to have *CYP124A1* or both *CYP124A1* and *CYP121A1*, indicating the collective efforts of these P450s in generating complex lipids in mycobacterial species, especially in MTBC species, as none of the other four mycobacterial category species has this type of combination in its *CYP128* gene clusters. Future studies should include cloning of the entire *CYP128A1* gene cluster (containing *CYP121A1* and *CYP124A1*) and analyzing the effect of the cluster molecule in *M. tuberculosis* pathogenesis.

Supplementary Materials: The following are available online at http://www.mdpi.com/1422-0067/21/14/4816/s1, Figure S1: 3VRM/CYP128A1 alignement., Figure S2: 2D structures of substrate (menaquinone 9) and azole compounds used in the study, Figure S3: RMSD *vs* time during MD simulations of MK9-CYP128A1 complexes (colors are explained in the legend of each figure), Table S1: Comparative amino acid conservation acid analysis of CYP128 P450 family with top 10 ranked families, Supplementary Dataset 1: Comprehensive comparative analyses of CYP128 P450s and those associated with secondary metabolite biosynthetic gene clusters in mycobacterial species, Supplementary Dataset 2: CYP128 P450 sequences identified and annotated in mycobacterial species, Supplementary Dataset 3: A high-resolution phylogenetic tree of CYP128 P450s.

Author Contributions: Conceptualization, K.S.; data curation, N.S.N., Z.E.C., W.C., J.-H.Y., D.R.N., J.A.T., J.P. and K.S.; formal analysis, N.S.N., Z.E.C., W.C., J.-H.Y., D.R.N., J.A.T., J.P. and K.S.; funding acquisition, K.S.; investigation, N.S.N., Z.E.C., W.C., J.-H.Y., D.R.N., J.A.T., J.P. and K.S.; methodology, N.S.N., Z.E.C., W.C., J.-H.Y., D.R.N., J.A.T., J.P. and K.S.; project administration, K.S.; resources, K.S.; supervision, K.S.; validation, N.S.N., Z.E.C., W.C., J.-H.Y., D.R.N., J.A.T., J.P. and K.S.; visualization, N.S.N., W.C., J.P. and K.S.; writing—original draft, N.S.N., W.C., J.-H.Y., D.R.N., J.P. and K.S.; writing—review and editing, N.S.N., W.C., J.-H.Y., D.R.N., J.P. and K.S. All authors have read and agreed to the published version of the manuscript.

Funding: The work presented in this article is part of a National Research Foundation (NRF), South Africa grant awarded to Khajamohiddin Syed (Grant No. 114159), where all the international authors involved in the study are listed as international collaborators. Master's study bursaries were provided for Nokwanda Samantha Ngcobo (year 2019) and Zinhle Edith Chiliza (year 2018) from the same grant. Zinhle Edith Chiliza also thanks the NRF, South Africa for a DST-NRF Innovation Master's Scholarship for the year 2019 (Grant No. 117182). Khajamohiddin Syed expresses sincere gratitude to the University of Zululand Research Committee for funding (Grant No. C686) and for the laboratory facilities.

Acknowledgments: The authors want to thank Barbara Bradley, Pretoria, South Africa for English language editing.

Conflicts of Interest: The authors declare no conflict of interest.

References

1. World Health Organization (WHO). Global Tuberculosis Report 2019. 2019. ISBN 978-92-4-156571-4. Available online: https://www.who.int/tb/publications/global_report/en/ (accessed on 7 December 2019).
2. SSA. Mortality and Causes of Death in South Africa, 2016: Findings From Death Notification; Statistics South Africa. 2018. Available online: http://www.statssa.gov.za/publications/P03093/P030932016.pdf (accessed on 22 March 2019).
3. Cole, S.; Brosch, R.; Parkhill, J.; Garnier, T.; Churcher, C.; Harris, D.; Gordon, S.; Eiglmeier, K.; Gas, S.; Barry Iii, C. Deciphering the biology of *Mycobacterium tuberculosis* from the complete genome sequence. *Nature* **1998**, *393*, 537. [CrossRef] [PubMed]
4. Ghazaei, C. Mycobacterium tuberculosis and lipids: Insights into molecular mechanisms from persistence to virulence. *J. Res. Med Sci. Off. J. Isfahan Univ. Med. Sci.* **2018**, *23*, 63. [CrossRef] [PubMed]
5. Ortiz de Montellano, P.R. Potential drug targets in the *Mycobacterium tuberculosis* cytochrome P450 system. *J. Inorg. Biochem.* **2018**, *180*, 235–245. [CrossRef]
6. Senate, L.M.; Tjatji, M.P.; Pillay, K.; Chen, W.; Zondo, N.M.; Syed, P.R.; Mnguni, F.C.; Chiliza, Z.E.; Bamal, H.D.; Karpoormath, R. Similarities, variations, and evolution of cytochrome P450s in *Streptomyces* versus *Mycobacterium*. *Sci. Rep.* **2019**, *9*, 3962. [CrossRef]
7. Nelson, D.R. Cytochrome P450 diversity in the tree of life. *Biochim. Biophys. Acta Proteins Proteom.* **2018**, *1866*, 141–154. [CrossRef]
8. Van Wyk, R.; van Wyk, M.; Mashele, S.S.; Nelson, D.R.; Syed, K. Comprehensive comparative analysis of cholesterol catabolic genes/proteins in mycobacterial species. *Int. J. Mol. Sci.* **2019**, *20*, 1032. [CrossRef]

9. Sassetti, C.M.; Rubin, E.J. Genetic requirements for mycobacterial survival during infection. *Proc. Natl. Acad. Sci. USA* **2003**, *100*, 12989–12994. [CrossRef]
10. Lamichhane, G.; Zignol, M.; Blades, N.J.; Geiman, D.E.; Dougherty, A.; Grosset, J.; Broman, K.W.; Bishai, W.R. A postgenomic method for predicting essential genes at subsaturation levels of mutagenesis: Application to *Mycobacterium tuberculosis*. *Proc. Natl. Acad. Sci. USA* **2003**, *100*, 7213–7218. [CrossRef]
11. Betts, J.C.; Lukey, P.T.; Robb, L.C.; McAdam, R.A.; Duncan, K. Evaluation of a nutrient starvation model of *Mycobacterium tuberculosis* persistence by gene and protein expression profiling. *Mol. Microbiol.* **2002**, *43*, 717–731. [CrossRef]
12. Rustad, T.R.; Harrell, M.I.; Liao, R.; Sherman, D.R. The enduring hypoxic response of *Mycobacterium tuberculosis*. *PLoS ONE* **2008**, *3*, e1502. [CrossRef]
13. Sogi, K.M.; Holsclaw, C.M.; Fragiadakis, G.K.; Nomura, D.K.; Leary, J.A.; Bertozzi, C.R. Biosynthesis and Regulation of Sulfomenaquinone, a Metabolite Associated with Virulence in *Mycobacterium tuberculosis*. *ACS Infect. Dis.* **2016**, *2*, 800–806. [CrossRef]
14. Ouellet, H.; Johnston, J.B.; de Montellano, P.R.O. The *Mycobacterium tuberculosis* cytochrome P450 system. *Arch. Biochem. Biophys.* **2010**, *493*, 82–95. [CrossRef]
15. Driscoll, M. Investigating Orphan Cytochromes P450 from *Mycobacterium tuberculosis*: The Search for Potential Drug Targets. Ph.D. Thesis, University of Manchester, Manchester, UK, 2011.
16. Holsclaw, C.M.; Sogi, K.M.; Gilmore, S.A.; Schelle, M.W.; Leavell, M.D.; Bertozzi, C.R.; Leary, J.A. Structural characterization of a novel sulfated menaquinone produced by stf3 from *Mycobacterium tuberculosis*. *ACS Chem. Biol.* **2008**, *3*, 619624. [CrossRef]
17. Dhiman, R.K.; Mahapatra, S.; Slayden, R.A.; Boyne, M.E.; Lenaerts, A.; Hinshaw, J.C.; Angala, S.K.; Chatterjee, D.; Biswas, K.; Narayanasamy, P. Menaquinone synthesis is critical for maintaining mycobacterial viability during exponential growth and recovery from non-replicating persistence. *Mol. Microbiol.* **2009**, *72*, 85–97. [CrossRef]
18. Sun, Z.; Zhang, Y. Antituberculosis activity of certain antifungal and antihelmintic drugs. *Tuber. Lung Dis.* **1999**, *79*, 319–320. [CrossRef]
19. Ahmad, Z.; Sharma, S.; Khuller, G.K. The potential of azole antifungals against latent/persistent tuberculosis. *FEMS Microbiol. Lett.* **2006**, *258*, 200–203. [CrossRef]
20. Ahmad, Z.; Sharma, S.; Khuller, G. In vitro and ex vivo antimycobacterial potential of azole drugs against *Mycobacterium tuberculosis* H37Rv. *FEMS Microbiol. Lett.* **2005**, *251*, 19–22. [CrossRef]
21. Chenge, J.T.; Duyet, L.V.; Swami, S.; McLean, K.J.; Kavanagh, M.E.; Coyne, A.G.; Rigby, S.E.; Cheesman, M.R.; Girvan, H.M.; Levy, C.W.; et al. Structural Characterization and Ligand/Inhibitor Identification Provide Functional Insights into the *Mycobacterium tuberculosis* Cytochrome P450 CYP126A1. *J. Biol. Chem.* **2017**, *292*, 1310–1329.
22. Parvez, M.; Qhanya, L.B.; Mthakathi, N.T.; Kgosiemang, I.K.; Bamal, H.D.; Pagadala, N.S.; Xie, T.; Yang, H.; Chen, H.; Theron, C.W.; et al. Molecular evolutionary dynamics of cytochrome P450 monooxygenases across kingdoms: Special focus on mycobacterial P450s. *Sci. Rep.* **2016**, *6*, 33099.
23. Mthethwa, B.; Chen, W.; Ngwenya, M.; Kappo, A.; Syed, P.; Karpoormath, R.; Yu, J.-H.; Nelson, D.; Syed, K. Comparative analyses of cytochrome P450s and those associated with secondary metabolism in Bacillus species. *Int. J. Mol. Sci.* **2018**, *19*, 3623. [CrossRef]
24. Syed, K.; Shale, K.; Pagadala, N.S.; Tuszynski, J. Systematic identification and evolutionary analysis of catalytically versatile cytochrome P450 monooxygenase families enriched in model basidiomycete fungi. *PLoS ONE* **2014**, *9*, e86683. [CrossRef] [PubMed]
25. Jawallapersand, P.; Mashele, S.S.; Kovačič, L.; Stojan, J.; Komel, R.; Pakala, S.B.; Kraševec, N.; Syed, K. Cytochrome P450 monooxygenase CYP53 family in fungi: Comparative structural and evolutionary analysis and its role as a common alternative anti-fungal drug target. *PLoS ONE* **2014**, *9*, e107209. [CrossRef] [PubMed]
26. Sello, M.M.; Jafta, N.; Nelson, D.R.; Chen, W.; Yu, J.-H.; Parvez, M.; Kgosiemang, I.K.R.; Monyaki, R.; Raselemane, S.C.; Qhanya, L.B. Diversity and evolution of cytochrome P450 monooxygenases in Oomycetes. *Sci. Rep.* **2015**, *5*, 11572. [CrossRef] [PubMed]
27. Qhanya, L.B.; Matowane, G.; Chen, W.; Sun, Y.; Letsimo, E.M.; Parvez, M.; Yu, J.-H.; Mashele, S.S.; Syed, K. Genome-wide annotation and comparative analysis of cytochrome P450 monooxygenases in Basidiomycete biotrophic plant pathogens. *PLoS ONE* **2015**, *10*, e0142100. [CrossRef] [PubMed]

28. Kgosiemang, I.K.R.; Syed, K.; Mashele, S.S. Comparative genomics and evolutionary analysis of cytochrome P450 monooxygenases in fungal subphylum *Saccharomycotina*. *J. Pure Appl. Microbiol.* **2014**, *8*, 291–302.
29. Nelson, D.R. Cytochrome P450 and the individuality of species. *Arch. Biochem Biophys* **1999**, *369*, 1–10. [CrossRef]
30. Yoshida, Y.; Aoyama, Y.; Noshiro, M.; Gotoh, O. Sterol 14-demethylase P450 (CYP51) provides a breakthrough for the discussion on the evolution of cytochrome P450 gene superfamily. *Biochem. Biophys. Res. Commun.* **2000**, *273*, 799–804. [CrossRef]
31. Syed, P.R.; Chen, W.; Nelson, D.R.; Kappo, A.P.; Yu, J.-H.; Karpoormath, R.; Syed, K. Cytochrome P450 Monooxygenase CYP139 Family Involved in the Synthesis of Secondary Metabolites in 824 Mycobacterial Species. *Int. J. Mol. Sci.* **2019**, *20*, 2690. [CrossRef]
32. Bamal, H.D.; Chen, W.; Mashele, S.S.; Nelson, D.R.; Kappo, A.P.; Mosa, R.A.; Yu, J.-H.; Tuszynski, J.A.; Syed, K. Comparative analyses and structural insights of the novel cytochrome P450 fusion protein family CYP5619 in *Oomycetes*. *Sci. Rep.* **2018**, *8*, 6597. [CrossRef]
33. Pei, J.; Kim, B.-H.; Grishin, N.V. PROMALS3D: A tool for multiple protein sequence and structure alignments. *Nucleic Acids Res.* **2008**, *36*, 2295–2300. [CrossRef]
34. Syed, K.; Mashele, S.S. Comparative analysis of P450 signature motifs EXXR and CXG in the large and diverse kingdom of fungi: Identification of evolutionarily conserved amino acid patterns characteristic of P450 family. *PLoS ONE* **2014**, *9*, e95616. [CrossRef]
35. Greule, A.; Stok, J.E.; De Voss, J.J.; Cryle, M.J. Unrivalled diversity: The many roles and reactions of bacterial cytochromes P450 in secondary metabolism. *Nat. Prod. Rep.* **2018**, *35*, 757–791. [CrossRef]
36. Podust, L.M.; Sherman, D.H. Diversity of P450 enzymes in the biosynthesis of natural products. *Nat. Prod. Rep.* **2012**, *29*, 1251–1266. [CrossRef]
37. Yasutake, Y.; Nishioka, T.; Imoto, N.; Tamura, T. A single mutation at the ferredoxin binding site of P450 Vdh enables efficient biocatalytic production of 25-hydroxyvitamin D (3). *Chembiochem* **2013**, *14*, 2284–2291. [CrossRef]
38. Nagano, S.; Cupp-Vickery, J.R.; Poulos, T.L. Crystal Structures of the Ferrous Dioxygen Complex of Wild-type Cytochrome P450eryF and Its Mutants, A245S and A245T INVESTIGATION OF THE PROTON TRANSFER SYSTEM IN P450eryF. *J. Biol. Chem.* **2005**, *280*, 22102–22107. [CrossRef]
39. Yasutake, Y.; Fujii, Y.; Nishioka, T.; Cheon, W.K.; Arisawa, A.; Tamura, T. Structural evidence for enhancement of sequential vitamin D3 hydroxylation activities by directed evolution of cytochrome P450 vitamin D3 hydroxylase. *J. Biol. Chem.* **2010**, *285*, 31193–31201. [CrossRef]
40. Yasutake, Y.; Kameda, T.; Tamura, T. Structural insights into the mechanism of the drastic changes in enzymatic activity of the cytochrome P450 vitamin D3 hydroxylase (CYP107BR1) caused by a mutation distant from the active site. *Acta Crystallogr. Sect. F Struct. Biol. Commun.* **2017**, *73*, 266–275. [CrossRef]
41. Matowane, R.G.; Wieteska, L.; Bamal, H.D.; Kgosiemang, I.K.R.; Van Wyk, M.; Manume, N.A.; Abdalla, S.M.H.; Mashele, S.S.; Gront, D.; Syed, K. In silico analysis of cytochrome P450 monooxygenases in chronic granulomatous infectious fungus *Sporothrix schenckii*: Special focus on CYP51. *Biochim. Biophys. Acta Proteins Proteom.* **2018**, *1866*, 166–177. [CrossRef]
42. Chen, I.-M.A.; Chu, K.; Palaniappan, K.; Pillay, M.; Ratner, A.; Huang, J.; Huntemann, M.; Varghese, N.; White, J.R.; Seshadri, R. IMG/M v. 5.0: An integrated data management and comparative analysis system for microbial genomes and microbiomes. *Nucleic Acids Res.* **2018**, *47*, D666–D677. [CrossRef]
43. Nelson, D.R. Cytochrome P450 Nomenclature, 2004. In *Cytochrome P450 Protocols*; Springer: Berlin, Germany, 2006; pp. 1–10.
44. Nelson, D.R. Cytochrome P450 nomenclature. In *Cytochrome P450 Protocols*; Springer: Berlin, Germany, 1998; pp. 15–24.
45. Nelson, D.R. The cytochrome p450 homepage. *Hum. Genom.* **2009**, *4*, 59. [CrossRef]
46. Sirim, D.; Widmann, M.; Wagner, F.; Pleiss, J. Prediction and analysis of the modular structure of cytochrome P450 monooxygenases. *BMC Struct. Biol.* **2010**, *10*, 34. [CrossRef]
47. Gotoh, O. Substrate recognition sites in cytochrome P450 family 2 (CYP2) proteins inferred from comparative analyses of amino acid and coding nucleotide sequences. *J. Biol. Chem.* **1992**, *267*, 83–90.
48. Katoh, K.; Kuma, K.-i.; Toh, H.; Miyata, T. MAFFT version 5: Improvement in accuracy of multiple sequence alignment. *Nucleic Acids Res.* **2005**, *33*, 511–518. [CrossRef]

49. Boc, A.; Diallo, A.B.; Makarenkov, V. T-REX: A web server for inferring, validating and visualizing phylogenetic trees and networks. *Nucleic Acids Res.* **2012**, *40*, W573–W579. [CrossRef]
50. Letunic, I.; Bork, P. Interactive tree of life (iTOL) v3: An online tool for the display and annotation of phylogenetic and other trees. *Nucleic Acids Res.* **2016**, *44*, W242–W245. [CrossRef]
51. Kumar, S.; Stecher, G.; Tamura, K. MEGA7: Molecular evolutionary genetics analysis version 7.0 for bigger datasets. *Mol. Biol. Evol.* **2016**, *33*, 1870–1874. [CrossRef]
52. Schneider, T.D.; Stephens, R.M. Sequence logos: A new way to display consensus sequences. *Nucleic Acids Res.* **1990**, *18*, 6097–6100. [CrossRef]
53. Crooks, G.E.; Hon, G.; Chandonia, J.-M.; Brenner, S.E. WebLogo: A sequence logo generator. *Genome Res.* **2004**, *14*, 1188–1190. [CrossRef]
54. Blin, K.; Pascal Andreu, V.; de los Santos, E.L.C.; Del Carratore, F.; Lee, S.Y.; Medema, M.H.; Weber, T. The antiSMASH database version 2: A comprehensive resource on secondary metabolite biosynthetic gene clusters. *Nucleic Acids Res.* **2018**, *47*, D625–D630. [CrossRef]
55. Salomon-Ferrer, R.; Case, D.A.; Walker, R.C. An overview of the Amber biomolecular simulation package. *WIREs Comput. Mol. Sci.* **2013**, *3*, 198–210. [CrossRef]
56. Case, D.; Berryman, J.; Betz, R.; Cerutti, D.; Cheatham, T., III; Darden, T.; Duke, R.; Giese, T.; Gohlke, H.; Goetz, A. *AMBER 2015*; University of California: San Francisco, CA, USA, 2015.

© 2020 by the authors. Licensee MDPI, Basel, Switzerland. This article is an open access article distributed under the terms and conditions of the Creative Commons Attribution (CC BY) license (http://creativecommons.org/licenses/by/4.0/).

Article

Transcriptome Analysis of *Sogatella furcifera* (Homoptera: Delphacidae) in Response to Sulfoxaflor and Functional Verification of Resistance-Related P450 Genes

Xue-Gui Wang [1,*,†], Yan-Wei Ruan [1,†], Chang-Wei Gong [1,†], Xin Xiang [1], Xiang Xu [2], Yu-Ming Zhang [1] and Li-Tao Shen [1]

1. Biorational Pesticide Research Lab, Sichuan Agricultural University, Chengdu 611130, China
2. Sichuan Provincial Plant Protection Station, Department of Agriculture, Chengdu 610041, China
* Correspondence: wangxuegui@sicau.edu.cn
† These authors equally contributed to this work.

Received: 18 August 2019; Accepted: 12 September 2019; Published: 15 September 2019

Abstract: The white-back planthopper (WBPH), *Sogatella furcifera*, is a major rice pest in China and in some other rice-growing countries of Asia. The extensive use of pesticides has resulted in severe resistance of *S. furcifera* to variety of chemical insecticides. Sulfoxaflor is a new diamide insecticide that acts on nicotinic acetylcholine receptors (nAChRs) in insects. The aim of this study was to explore the key genes related to the development of resistance to sulfoxaflor in *S. furcifera* and to verify their functions. Transcriptomes were compared between white-back planthoppers from a susceptible laboratory strain (Sus-Lab) and Sus-Lab screened with the sublethal LC_{25} dose of sulfoxaflor for six generations (SF-Sel). Two P450 genes (*CYP6FD1* and *CYP4FD2*) and three transcription factors (*NlE78sf*, *C2H2ZF1* and *C2H2ZF3*) with upregulated expression verified by qRT-PCR were detected in the Sus-Lab and SF-Sel strains. The functions of *CYP6FD1* and *CYP4FD2* were analyzed by RNA interference, and the relative normalized expressions of *CYP6FD1* and *CYP4FD2* in the SF-Sel population were lower than under dsGFP treatment after dsRNA injection. Moreover, the mortality rates of SF-Sel population treated with the LC_{50} concentration of sulfoxaflor after the injecting of dsRNA targeting *CYP6FD1* and *CYP4FD2* were significantly higher than in the dsGFP group from 72 h to 96 h ($p < 0.05$), and mortality in the *CYP6FD1* knockdown group was clearly higher than that of the *CYP4FD2* knockdown group. The interaction between the tertiary structures of *CYP6FD1* and *CYP4FD2* and sulfoxaflor was also predicted, and *CYP6FD1* showed a stronger metabolic ability to process sulfoxaflor. Therefore, overexpression of *CYP6FD1* and *CYP4FD2* may be one of the primary factors in the development of sulfoxaflor resistance in *S. furcifera*.

Keywords: *Sogatella furcifera*; sulfoxaflor; transcriptome; cytochrome P450 monooxygenase; RNA interference

---

1. Introduction

The white-back planthopper (WBPH), *Sogatella furcifera* (Horváth) (Homoptera: Delphacidae), is an important insect pest in rice-growing countries in Asia [1,2] that seriously affects rice yields by sucking juice from the rice stem, resulting in slow growth, yellowing, and even lodging and a phenomenon known as 'hopper burn', leading to the death of rice plants in severe cases and crop failure. In addition, recent studies have shown that the southern rice black-streaked dwarf virus (SRBSDV) can be transmitted in the main rice-growing areas of Asia, including China, Vietnam, and Japan, through the stylets of *S. furcifera* when injecting juice containing the SRBSDV into the stems of healthy rice, causing great rice yield losses. *S. furcifera* is a typical *r*-strategy pest and can reproduce rapidly when conditions are suitable [3–5]. It can also migrate long distances to adapt to environmental changes [6]. Over the past few decades, chemical pesticides have been the main strategy for controlling this pest [7]. The long-term use of insecticides against the rice planthopper has resulted in the development of resistance and population increases [8]. *S. furcifera* has now developed different levels of resistance to organophosphorus, carbamate, phenylpyrazole, neonicotinoid, pyrethroid, and insect growth regulator insecticides [9–13]. When insects are successively exposed chemicals, insecticide resistance is a natural adaptability feature and one of the most important factors is the enhanced detoxification metabolic ability by enzymes to insecticides, such as mixed-function oxidase (MFO), carboxylesterase (CarEs) and glutathione *S*-transferase (GSTs) [14], especially those for cytochrome P450 monooxygenases (P450s) activity, could play a major role in the detoxification of insecticides in a number of insect pests [15,16].

Sulfoxaflor is an insecticide produced by Dow AgroSciences (DAS) from a new chemical class of sulfoximines, which act on nicotinic acetylcholine receptors (nAChRs) in the insect nervous system [17,18], and is the first commercial agrochemical to be used for the control of a broad range of sap-feeding insect pests [19]. Although several other chemically distinct classes of insecticides (spinosyns, neonicotinoids, nereistoxin analogs) also act on nAChRs, sulfoxaflor presents some special biological characteristics, such as an absence of cross-resistance between sulfoxaflor and the other nAChR-acting insecticides, which make it highly effective against a wide range of sap-feeding insects, especially against aphids such as the green peach aphid [20]. Sulfoxaflor is also effective against insect pests that are resistant to other classes of insecticides, including many insects that are resistant to neonicotinoids [21]. In 2012, sulfoxaflor was first registered for the control of Miridae pests in cotton cultivation in Arkansas, Louisiana, and Mississippi in the United States, and in 2013, Dow AgroSciences introduced sulfoxaflor to China for the control of cotton aphids, wheat aphids, scale insect pests in citrus, and Delphacidae insect pests in rice [22]. As a relatively recently developed insecticide, there is still a risk that the efficiency of sulfoxaflor may become compromised through various mechanisms, e.g., through modification of the target site of neonicotinoid insecticides or positive sublethal effects (hormesis) in exposed individuals [23]. Thus, it is crucial to understand the resistance mechanisms of insects related to this insecticide before its widespread use in integrated pest management (IPM).

In this study, we examined toxicity and transcript profiles in a susceptible laboratory (Sus-Lab) strain of *S. furcifera* and the Sus-Lab strain in which resistance was continuously induced by treatment with the sublethal LC_{25} dose of sulfoxaflor for six generations (SF-Sel). We analyzed the genes showing upregulated expression and verified the results using qRT-PCR. Furthermore, the functions of two P450 genes, *CYP6FD1* and *CYP4FD2*, were also analyzed with RNA interference technology through the design of suitable dsRNAs, whose silencing specificity was confirmed using qRT-PCR, and mortality was determined in nymphs treated with the dsRNAs combined with sulfoxaflor. The objectives of this study were to explore the risk of the development of resistance to sulfoxaflor, to conduct a preliminary investigation for the functional verification of P450 genes induced by sulfoxaflor, and to manage the development of resistance through the design of suitable dsRNAs for the silencing of significantly upregulated genes in the future.

2. Results

2.1. Toxicity of Sulfoxaflor in S. furcifera

The LC_{25} for the Sus-Lab strain was estimated at 2.102 µg/mL, and a 2.06-fold decrease in the susceptibility level was observed in the SF-Sel strain, which presented LC_{50} value of 7.284 µg/mL for sulfoxaflor, compared with the LC_{50} value of 3.544 µg/mL for Sus-Lab.

2.2. Synergism Experiment

According to the results of the synergism experiment, when the Sus-Lab strain was treated with the synergistic agents triphenyl phosphate (TPP) acting on CarEs, Diethyl maleate (DEM) on GSTs and piperonyl butoxide (PBO) on cytochrome P450 monooxygenases (P450s), all synergistic treatments showed strong synergistic effects compared with treatment without a synergistic agent, and their toxicities were clearly enhanced (not overlapping the 95% confidence interval, 95%CI), with synergism ratio (SR) values of 1.495, 1.205 and 1.211, respectively, while as the toxicity showed no difference between the synergistic treatments (overlapping 95% CI). The toxicity of the SF-Sel strain treated with the synergistic agent PBO was significantly enlarged compared with those in the other two treatments involving TPP and DEM (not overlapping 95% CIs for the other two synergistic agents). The SR value reached 5.328 and was significantly higher than those in the treatments involving TPP and DEM, for which SR values were 2.689 and 2.283, respectively (Table 1).

Table 1. Synergistic effect of three synergists and sulfoxaflor on *S. furcifera*.

Strain	Treatment	LC_{50} (µg/mL) 95%CI	Slope ± SE	x^2 (df)	SR *
Sus-Lab	sulfoxaflor	3.544(3.287–3.804)	6.499 ± 0.722	11.226(13)	/
	sulfoxaflor plus TPP	2.371(2.051–2.700)	3.253 ± 0.418	4.864(13)	1.495
	sulfoxaflor plus DEM	2.940(2.627–3.282)	4.069 ± 0.512	4.997(13)	1.205
	sulfoxaflor plus PBO	2.927(2.549–3.212)	6.893 ± 1.199	10.213(13)	1.211
SF-Sel	sulfoxaflor	7.284(6.265–8.385)	4.101 ± 0.454	14.726(13)	/
	sulfoxaflor plus TPP	2.709(2.180–3.449)	1.972 ± 0.248	4.292(13)	2.689
	sulfoxaflor plus DEM	3.191(2.529–4.061)	2.200 ± 0.307	3.268(13)	2.283
	sulfoxaflor plus PBO	1.367(1.129–1.648)	2.829 ± 0.353	4.908(13)	5.328

* SR (synergism ratio) = LC_{50} of a strain treated with sulfoxaflor alone divided by LC_{50} of the same strain treated with sulfoxaflor plus a synergist. The synergists of TPP, DEM and PBO stand for triphenyl phosphate, Diethyl maleateand piperonyl butoxide, respectively.

2.3. Detoxification Enzyme Activity

To further determine the potential role of detoxification enzymes in the development of resistance of *S. furcifera* to sulfoxaflor, both the Sus-Lab strain, either untreated or treated with a synergistic agent (DEM, TPP or PBO), and the SF-Sel strain, either untreated or treated with a synergistic agent (DEM, TPP or PBO), were analyzed to determine the activities of CarEs, GSTs and P450s. As shown in Figure 1, the activity of CarEs in the SF-Sel strain treated with TPP was the highest (0.9616 mmol × min^{-1}× mg pro^{-1}), followed by those in the SF-Sel strain (0.8257 mmol × min^{-1}× mg pro^{-1}), the Sus-Lab strain treated with TPP (0.8187 mmol × min^{-1}× mg pro^{-1}) and the Sus-Lab strain (0.7397 mmol × min^{-1}× mg pro^{-1}). However, there was no significant difference in CarE activity under any of the treatments ($p > 0.05$). The GSTs activity of the SF-Sel strain was the greatest, at 1.0103 mmol × min^{-1}× mg pro^{-1}, which was significantly different from the other three treatments ($p < 0.05$), where the activities ranged from 0.6824 to 0.8113 mmol × min^{-1}× mg pro^{-1} and did not significantly differ from each other ($p > 0.05$). P450s activity was also the greatest in the SF-Sel strain (13.5345 nmol × min^{-1}× mg pro^{-1}, which was significantly different from those under the other treatments ($p < 0.05$)), followed by the SF-Sel strain treated with PBO (10.7384 nmol × min^{-1}× mg pro^{-1}). The P450s activities of the Sus-Lab strain and Sus-Lab strain treated with PBO were the weakest (5.6181 and 4.8470 nmol × min^{-1}× mg pro^{-1}), and both of these values were significantly different from those under the other treatments ($p < 0.05$) (Figure 1).

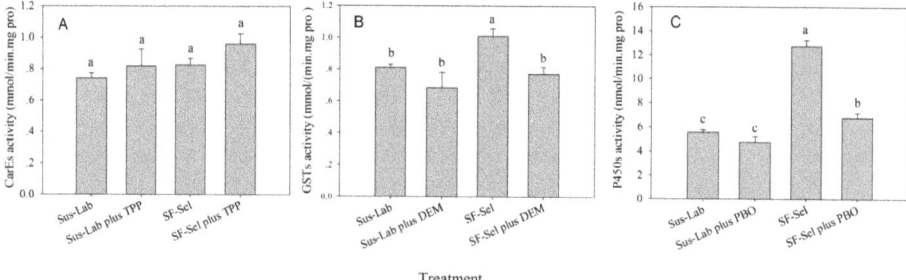

Figure 1. Synergistic effects of TPP, DEM and PBO on activity of detoxification enzymes: CarE (**A**), GST (**B**) and P450 (**C**) in 3rd-instar nymph of *S furcifera*. The activities of CarE, GST and P450 in 3rd-instar nymph of *S furcifera* are presented as the mean of three replications ± SE. Means followed by the same letters did not differ significantly ($p > 0.05$) according to the ANOVA test. The $F_{3,8}$ values of different treatments on CarE, GST and P450 in 3rd-instar nymph of *S furcifera* were 1.818, 5.410, 610.745, and the p values on CarE, GST and P450 in 3rd-instar nymph of *S furcifera* were = 0.222 > 0.05, = 0.025 < 0.05, = < 0.0001, respectively.

2.4. Illumina Sequencing and Read Assembly

The Sus-Lab and SF-Sel strains were each assessed in triplicate. The total numbers of reads (150 bp) obtained were 335,119,142 in the six samples, with over 50,653,904 reads for each sample. The proportion of reads containing duplicate sequences was 0.37% ~ 0.42%, and the proportion of low-quality reads was 1.23%~1.35%, including reads with > 10% Ns and abase number of Q ≤ 10 in > 50% of the total reads. After filtering out the linker sequences or low-quality reads, 329,449, and 482 clean reads were obtained, and the Q20 and Q30 base percentages of clean reads of were over 98.69% and 96.04%, respectively (Supplementary Table S1).

2.5. Transcriptome Data Splicing

The clean reads from the transcriptomes of the Sus-Lab and SF-Sel strains were composed of the mixed pools and were assembled into approximately 74,119 unigenes, with a longest unigene of 34,340 nt and a shortest unigene of 201 nt (Supplementary Figure S1A). The N50 value, at which the cumulative fragment length reaches 50% of the total fragment length, was 2043 nt, and there were 5552 unigenes of over 3000 nt. When the reads were compared with the unigenes, 3971 of the unigenes were found to exhibit more than 10,000 reads, while 34, 847 of unigenes presented only 11~100 reads (Supplementary Figure S1B).

2.6. Transcriptome Annotation

From to the 74,119 assembled unigenes, 24,719 of which were successfully annotated in the Nr database, and the species with the greatest number of homologous sequences included *Zootermopsis nevadensis* (2001 genes), *Bemisia tabaci* (1620 genes), *Cimex lectularius* (1500 genes), and *Halyomorpha halys* (1424 genes) (Supplementary Figure S2A). Additionally, 19,050 unigenes were annotated with the Swissprot database, 17,119 unigenes with the KOG database, and 11,788 unigenes with the KEGG database, and 10,516 unigenes were annotated in all four databases (Supplementary Figure S2B), among the 17,119 unigenes annotated with the KOG database, 6671 unigenes were classified as general function prediction only, accounting for the largest proportion, and 4867 unigenes were classified as being associated with signal transduction mechanisms (Supplementary Figure S2C).

2.7. Analysis of Gene Expression

The results indicated that the correlations of gene expression levels in the three samples (M1, M2 and M3) of the Sus-Lab strain (with correlation index of 0.99 to 1.00) were significantly higher than those in the SF-Sel strain (S1, S2 and S3 samples, with a correlation index of 0.85 to 0.94) (Figure 2A). Principal component analysis (PCA) of the six samples also showed that there was a significant clustering relationship on PC1 between the gene expression levels in the samples of the susceptible strains and the SF-Sel strain. Additionally, the degree of the contribution of PC1 (88.7%) was significantly higher than that of PC2 (7.9%) (Figure 2B).

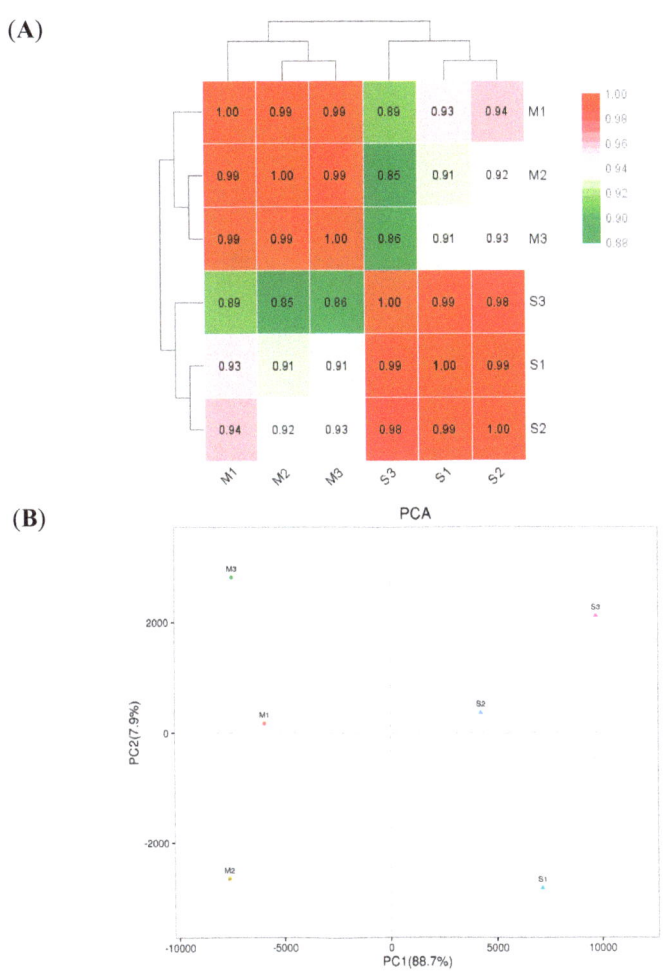

Figure 2. (A) Heat map of gene expression levels in the six samples. The darker the color is, the greater the correlation is. Three samples (M1, M2 and M3) are in the Sus-Lab strain, and other three samples (S1, S2 and S3) are in the SF-Sel strain. The same means as followed. (B) Principal component analysis (PCA) of six samples.

2.8. Cluster Analysis of DEGs

Based on the screening of DEGs with under criteria of an FDR < 0.05 and log2FC| > 1, 786 DEGs were screened from the SF-Sel strain, 557 of which were upregulated, while 229 were downregulated compared with those in the Lab-Sus strain (Figure 3A,B).

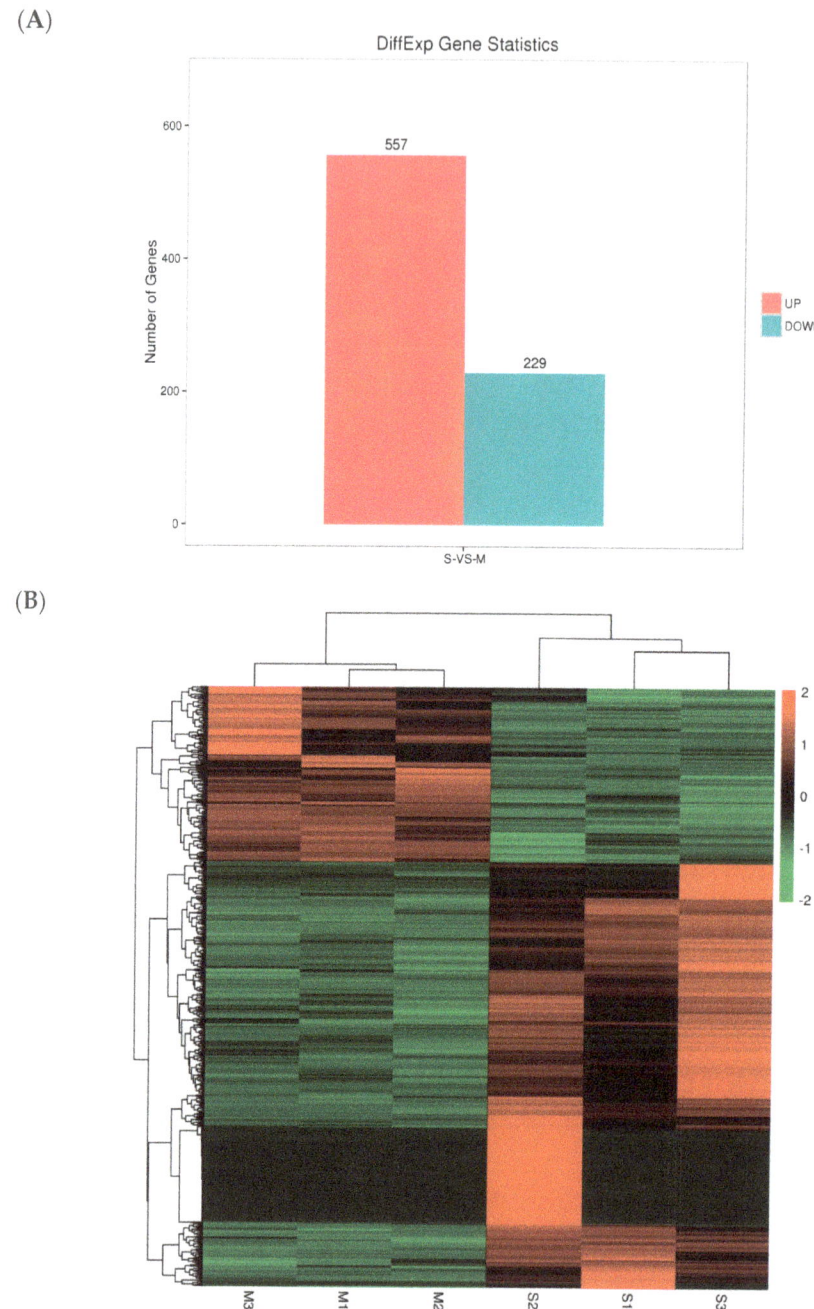

Figure 3. (**A**) The statistical maps of DEGs among S-vs-M. (**B**)The cluster heat maps of DEGs among S-vs-M. The column represents samples, and the row means genes; Color-scaled represents log 2 (fold change) values for resistant lines, the redder the color is, the higher the gene expression is, and on the contrary, the greener color is.

2.8.1. GO Enrich

The up- or downregulated unigenes were enriched and assessed in the three Gene Ontology (GO) categories of biological process, cellular component and molecular function. The enrichment of upregulated unigenes from the SF-Sel strain was found in the Sus-Lab strain. Additionally, the number of unigenes of upregulated unigenes of the binding class in the molecular function category was highest, at 16, and the P450 genes were associated iron ion binding; with up- and downregulated P450 gene were enriched in the binging class (Supplementary Figure S3).

2.8.2. Enrichment of DEGs in the KEGG Database

A total of 84 differentially expressed unigenes were enriched according to the KEGG database, which were related to organismal systems, genetic information processing, and metabolism, etc. The enriched DEGs found in the KEGG database were mainly included in categories such as "microbial metabolism in diverse environments (with a p-value of 0.00337)", "cardiac muscle contraction (p-value of 0.00153), "oxidative phosphorylation (p-value of 0.00114), "metabolic pathways (p-value of 0.000322)", "ribosome (p-value of 0.0000753), and "phototransduction-fly (p-value of 7.27×10^{-8}). Additionally, the greatest number of unigenes enriched in metabolic pathways was 44 (Figure 4).

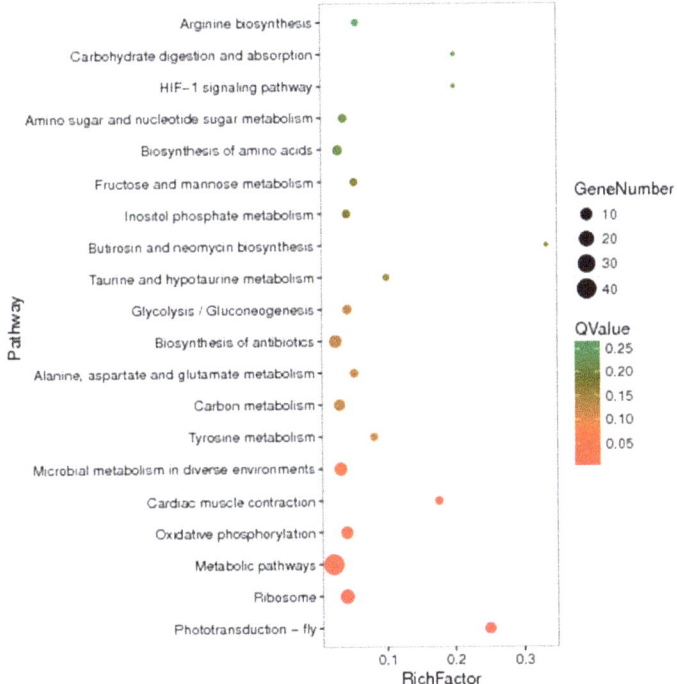

Figure 4. Bubble diagram of pathway enrichment of unigene on the different groups (Top 20). The abscissa means KEGG terms, and the ordinate means rich factor of each term. Red color means the terms of significant enrichment, and the bubble size indicates the number of enriched genes

2.9. Screening the Candidate Genes

The results showed that among the 74,119 unigenes, there were 198 unigenes involved in the detoxification and metabolism of foreign substances, and 138 were labeled cytochrome P450s, enriching the detoxification and metabolism genes of the white-backed planthopper. Compared with the Sus-Lab strain, the SF-Sel strain exhibited 7 significantly upregulated unigenes annotated as P450 genes, and a total of 4 gene types were found by NCBI alignment. One of the significantly upregulated unigenes was annotated as anorganic cation transporter, and 4 of the significantly upregulated unigenes were annotated as transcription factors (Table 2).

Table 2. Candidate P450 gene statistics.

GeneID	log2 Ratio(S/M)	S_vs_M Regulated	Gene Type	Annotation
Unigene0005814	11.17326071	UP	CYP6FD1	P450
Unigene0012458	10.38247993	UP		
Unigene0020537	9.861035196	UP	CYP6FD2	P450
Unigene0020536	9.83315364	UP		
Unigene0069588	8.681589817	UP	CYP4FD1	P450
Unigene0015479	7.934280594	UP		
Unigene0027543	2.07289003	UP	CYP4FD2	P450
Unigene0036498	8.17697521	UP		Transporter
Unigene0042782	1.136503653	UP	C2H2ZF2	Transcription factors
Unigene0051504	1.044721874	UP	NlE78sf	Transcription factors
Unigene0010562	1.411318813	UP	C2H2ZF1	Transcription factors
Unigene0010210	1.036525876	UP	C2H2ZF3	Transcription factors

2.10. P450 Diversity Analysis

Ten motifs (motif 1~motif 10) composed of very conservative amino acid residues were found in 54 of P450 genes (22 from our transcriptome data and other 32 downloaded from NCBI) of white-backed planthopper by meme search (http://meme-suite.org/tools/meme), on the contrary 24 motifs annotated with different functions were achieved by motif search (https://www.genome.jp/tools/motif/) (e.g., P450). Not only highly homologous sequences but also functional domains such as FAD_binding_1, Flavodoxin_1 and NAD_binding_1 domains were found in AHM93009.1 (the Sequences of NADPH-cytochrome P450 reductase in *S. furcifera* downloaded from NCBI) and *unigene0040669* (Figure 5). The results indicated that these sequences were similar in function, and GO annotation showed that both were classified as NADPH-reductases. The other genes all exhibited P450 functional domains and the conserved structural domains of motif1 and motif3. Motif3 presented absolutely conserved EXXR residues, and their general topography and structural folding were highly conserved. The heme-binding loop (with an absolutely conserved cysteine that serves as the 5th ligand for the conserved heme iron core) was composed of a coil known as the 'meander', a four-helix bundle, helices J and K, and two sets of beta-sheets. Additionally, the conserved structural domains of motif7, motif4, motif8, motif6, motif5, motif3, motif1, and motif2 were found in *CYP6FD1* and *CYP4FD2*, but the conserved structural domains of motif9 and AAA (ATPases superfamily, consisted of ATP binding site, Walker A and B) were only found in *CYP4FD2* and *CYP6FD1*, respectively (Supplementary Figure S4).

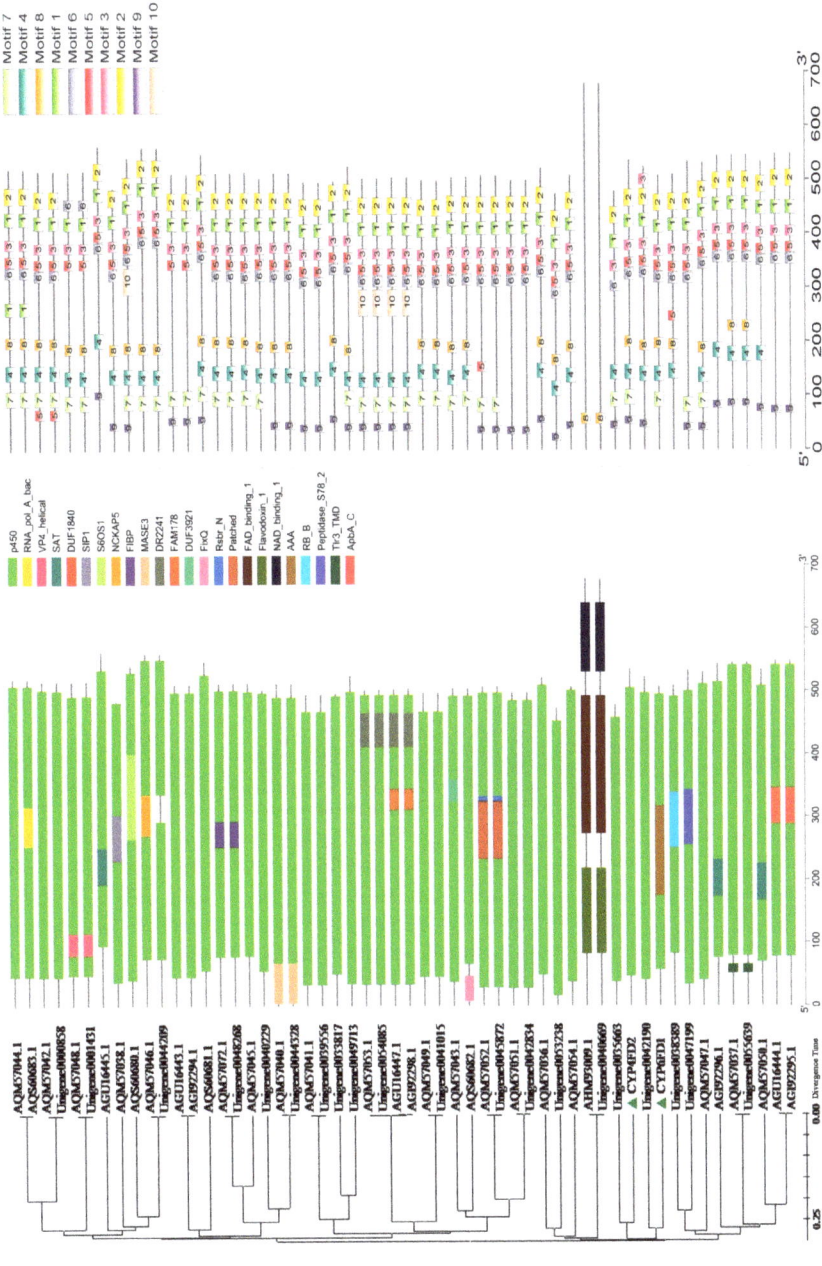

Figure 5. Phylogenetic tree of P450 gene family constructed by NJ method and gene molecular structure map predicted by motif search and meme.

2.11. Quantitative PCR (qRT-PCR)

The results indicated that the relative normalized expressions of *CYP6FD1* and *CYP4FD2* in the SF-Sel strain were 5.22- and 2.99-fold higher, respectively, than in the Sus-Lab strain, which were significantly higher than those of *CYP4FD1* and *CYP6FD2* ($p < 0.05$). The relative normalized expression of the *NlE78sf* transcription factor was highly significantly increased by 1.91-fold ($p < 0.01$), while the transcription factors *C2H2ZF1* and *C2H2ZF3* showed 1.53-and 1.30-fold increases, respectively, compared with the Sus-Lab strain ($p < 0.05$) (Figure 6).

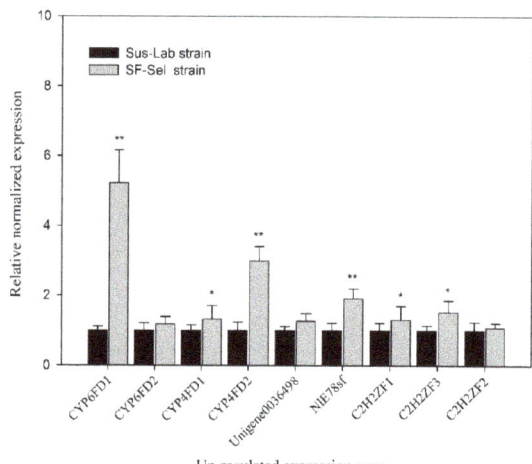

Figure 6. The relative expression of DEGs in Sus-Lab/SF-Sel strains. Each RT-qPCR reaction for each sample was performed in three technical replicates. Asterisks indicate significant differences of up-regulated expression gene in SF-Sel strains compared to the Sus-lab strain (Student's *t*-test, ** $p < 0.01$ and * $p < 0.05$).

2.12. Functional Analysis of CYP6FD1 and CYP4FD2 via RNAi

The results indicated that the relative normalized expression of *CYP6FD1* and *CYP4FD2* was significantly lower than that of dsGFP ($p < 0.05$) at 24 h after treatment, only reaching 0.72- and 0.68-fold, respectively. With prolongation of the interference time (48–72 h), the RNAi efficacy of *CYP6FD1* and *CYP4FD2* dsRNA became greater, with relative normalized expression of 0.42- and 0.55-fold for *CYP6FD1* and *CYP4FD2*, respectively, at 48 h to 72 h after injection, while that of *dsCYP6FD1* was higher than that of *dsCYP4FD2* ($p < 0.05$). However, the RNAi efficacy of *CYP6FD1* and *CYP4FD2* dsRNA gradually decreased, with the relative normalized expression of 0.61- and 0.71-fold for *CYP6FD1* and *CYP4FD2*, respectively, at 96 h after injection (Figure 7A).

The results of the bioassay showed that after 72 h of treatment with sulfoxaflor at the LC_{50}, the mortality of the SF-Sel strain treated with *CYP6FD1* dsRNA was 65.89% ± 5.38, which was significantly higher than the mortality of the same strain treated with *CYP4FD2* dsRNA (48.14% ± 4.33) and that in the control treatment (dsGFP, 38.95% ± 3.07) ($p < 0.05$). With prolongation of the interference time (96 h), the mortality of the *CYP6FD1* and *CYP4FD2* dsRNA treatment groups increased, reaching 62.01% ± 2.12~77.94% ± 5.30, respectively; this difference was also significant ($p < 0.05$), and mortality in these groups was significantly higher than under treatment with dsGFP (48.14% ± 0.93) (Figure 7B).

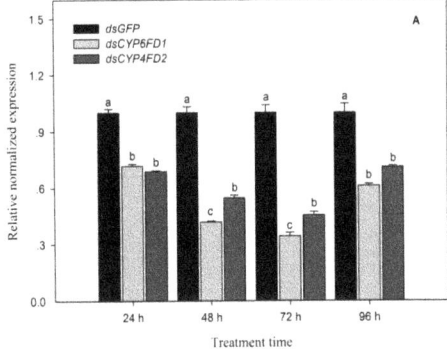 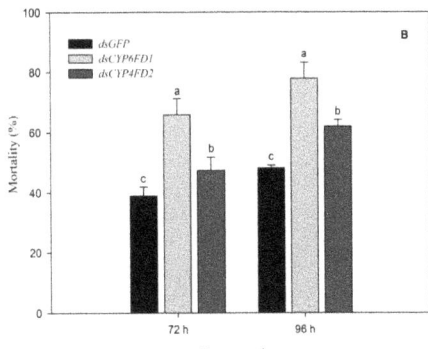

Figure 7. RNA interference and biological activity of two major P450 genes. (**A**)—RNA interference efficiency; (**B**)—Activity of sulfoxaflor against post-RNAi insects. Different letters (a, b, c) above bars indicate significant differences ($p < 0.05$) according to Duncan's multiple range test. The relative normalized expression of *dsGFP*, *dsCYP6FD1* and *dsCYP4FD2* in 3rd-instar nymph of *S furcifera*, and the mortalities for each treatment are presented as the mean of three replications ± SE. Means followed by the same letters did not differ significantly ($p > 0.05$) according to the ANOVA test. The $F_{2,6}$ values of relative normalized expressions at 24 h, 48 h, 72 h, 96 h and mortlities at 72 h, 96 h for different treatments were 158.933, 230.377, 164.477, 51.036 and 9.949, 19.901, respectively. and the corresponding p values were $< 0.0001, < 0.0001, < 0.0001, < 0.0001$, and $= 0.012 < 0.05, = 0.02 < 0.05$, respectively.

2.13. Interaction of the Tertiary Structure of CYP6FD1 and CYP4FD2 with Sulfoxaflor

The *CYP6FD1* domain included these amino acids ILE97, PHE105-GLY109, TYR111, and HIS122-SER126, and the absolute conserved residues of GLU444 and ARG447 were located near the entrance of the active pocket and the N-segment where heme binding occur. The totals core between CYP6FD1 and sulfoxaflor was 5.8954 (crash of −0.7686 and polar of 0.0277). However, the domain of CYP4FD2 was mainly composed of the LEU103, LYS106-LYS108, ALA110-LYS112, and LEU123 amino acids, and the key amino acids PHE317 and ILE459 bind with sulfoxaflor through noncovalent bonds, with a total score of 5.2800 (crash of −1.4457 and polar of 0.5769). Moreover, the absolutely conserved residues of GLU75 and ARG78 were distant from the active pockets and the heme binding region (Figure 8).

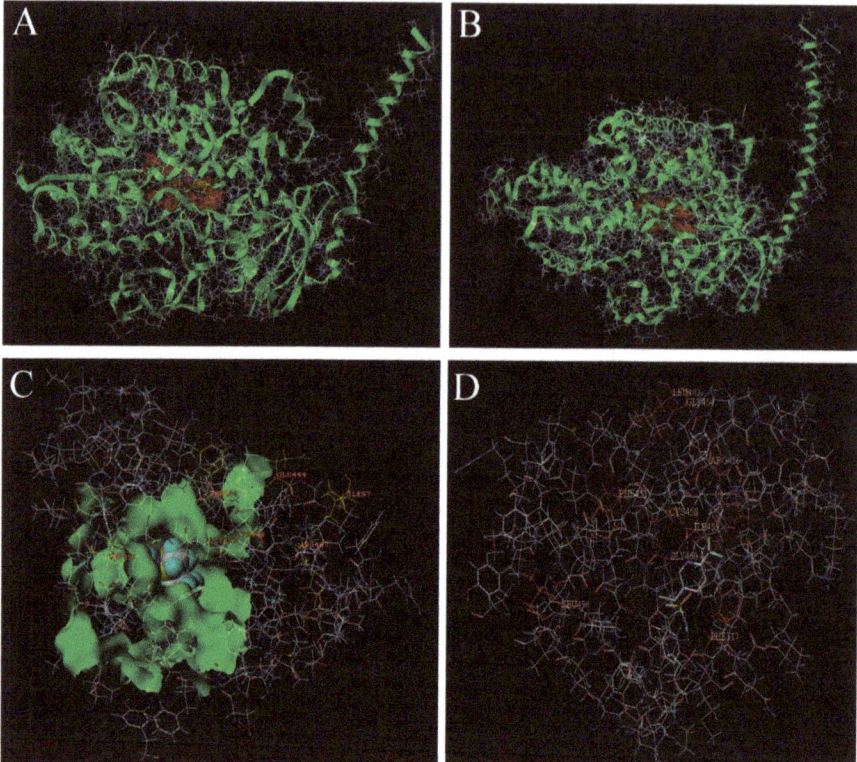

Figure 8. The tertiary structure of *CYP6FD1* and *CYP4FD2* and their docking structure with sulfoxaflor. (**A**) *CYP6FD1*, (**B**) *CYP4FD2*, (**C**) *CYP6FD1* domain and sulfoxaflor, (**D**) *CYP4FD2* domain and sulfoxaflor (molecular docking between sulfoxaflor and the active sites of the target P450 proteins was predicted using the Surflex-Dock program in software Syby lX-2.0 version (Tripos Inc.).

3. Discussion

Although rice planthoppers in China are still sensitive to fluorodinitrile, our results showed that the toxicity of sulfoxaflor to *S. furcifera* decreased when the Sus-Lab strain was successively screened with the sublethal dose of sulfoxaflor, which indicated that *S. furcifera* presented some resistance to sulfoxaflor. Liao et al. [24] monitored the resistance levels of *Nilaparvata lugens* to sulfoxaflor from 2013 to 2016 in China and found that all field-collected populations were still sensitive, with LC_{50} values ranging from 1.63 to 13.20 mg/L (resistance ratio from 0.8 to 6.8-fold). Liao et al. [25] continuously screened *N. lugens* with a sublethal dose of sulfoxaflor in approximately 39 intervals and finally obtained an extremely sulfoxaflor-resistant strain with a resistance ratio of 183.6-fold. Ma et al. [26] performed continuous selection of *Aphis gossypii* gradually increasing LC_{50} concentrations of sulfoxaflor based on bioassays of the parental generations for a total of 27 generations in the laboratory, finally resulting in a 366.4-fold resistance ratio compared with the susceptible strain. Therefore, there is an extremely high risk of insects developing resistance to sulfoxaflor, and it is necessary to perform resistance monitoring in field populations, along with the investigation of resistance mechanisms and cross-resistance to design integrated pest management strategies.

Insecticide resistance is inevitable after the application of insecticides, and the main reasons include a reduced penetration rate, increased detoxification and metabolism of insecticides (MFOs [27], CarEs [28], GSTs [29]) and decreased sensitivity at the target site. Wei et al. [30] found that PBO and

TPP could increase bifenthrin toxicity in resistant *A. gossypii* Glover strains by 2.38- and 4.55-fold, respectively. Liao et al. [25] showed that the toxicity of sulfoxaflor to sulfoxaflor- resistant *N. lugens* (Stål) showed a synergistic effect with PBO resulting in a 2.69-fold relative synergistic ratio, and the P450 enzyme activity of SFX-SEL was increased 3.50-fold compared with that in the unselected strain (UNSEL). Mao et al. [31] also reported that a resistant strain (NR) with a high nitenpyram resistance level (164.18-fold) and cross-resistance to sulfoxaflor (47.24-fold) showed a 3.21-fold increase in P450 activity compared to that in NS, and resistance also showed a synergistic effect (4.03-fold) with the inhibitor PBO, suggesting a role of P450. Our results further demonstrated that the inhibitors PBO, TPP, and DEM showed some synergism with sulfoxaflor regarding the toxicity and inhibition of the activities of three types of metabolic detoxification enzymes in the Sus-Lab and SF-Sel strains; this effect was especially strong for the inhibitor PBO in the SF-Sel strain.

Normally, the contribution of the overexpression of detoxification metabolism genes to an increased detoxification ability, especially which of the P450 genes related to insecticide detoxification metabolism, is the main reason for insect resistance to insecticides [32–34]. Jones et al. [35] reported that the resistance of the ALM07 strain of B-biotype populations of *Bemisia tabaci* adults to imidacloprid reached 180-fold, and the relative normalized expression of the resistance gene *CYP6CM1* in adults and nymphs reached 4.2- and 200-fold in the resistant strain, respectively. The overexpression of *CYP6AY1* contributes to the development of resistance to imidacloprid in *N. lugens* [36]. Liao et al. [25] and Mao et al. [31] also demonstrated that the reducing expression of *CYP6ER1* in sulfoxaflor-resistant strain through RNAi could significantly increase its' susceptibility to sulfoxaflor. Our transcriptome data and qRT-PCR results also indicated that two P450 genes, *CYP6FD1* and *CYP4FD2*, and three transcription factors, *NlE78sf*, *C2H2ZF1* and *C2H2ZF3*, were clearly upregulated in the SF-Sel strain. The RNAi results also showed that when 3rd-instar nymphs were injected with the *CYP6FD1* and *CYP4FD2* dsRNA, the relative expression of *CYP6FD1* and *CYP4FD2* was decreased, causing the insects to be more sensitive and ultimately to show higher mortality compared with negative dsGFP control treatment. However, it is still uncertain which transcription factors are mainly responsible for regulating the overexpression of *CYP6FD1* and *dsCYP4FD2*, and require further study in the future.

The P450s area multi-enzyme complex, and the first step in the metabolism of exogenous toxic substances is recognition by a CYP protein, which binds the toxin; then, electrons are transferred by electron donors to exogenous REDOX substances [37]. At present, the examination of P450 structure in insects generally concentrates on assessing highly conserved sequence motifs, such as the residue pairs WxxxR in helix C, CxxT in helix I, ExxR in helix K, RxxF in the meander region, and FxxGxRxCxG in the canonical heme-binding domain [38]. Our research showed that the molecular structure of sulfoxaflor was surrounded by the active pocket of *CYP6FD1*, while the active pocket was located near the heme-binding region. This protein exhibits a predicted active site structure with an oval shape [39], a large volume, and large substrate channels, allowing sulfoxaflor to fit the active site cavity. The spacious cavity of P450 enzymes enables larger molecules to access the heme-bound oxygen of the reaction center; therefore, we hypothesize that *CYP6FD1* could present a greater metabolic ability than *CYP4FD2* [40,41].

On the basis of our results, the main findings show that it is likely that *S. furcifera* will develop resistance to sulfoxaflor and that upregulation of detoxification enzymes such as P450s is a likely mechanism. However, the authors also show that the toxicity of sulfoxaflor is increased by using synergistic agents, so perhaps this is one possible approach that could be used in the field to prevent rapid development of resistance to this compound. Meanwhile, we also find that two main P450 genes (*CYP6FD1* and *CYP4FD2*) could be related to the development of resistance of *S. furcifera* to sulfoxaflor. Our results should provide a foundation for subsequent efforts to investigate the expression of *CYP6FD1* and *CYP4FD2* in heterologous expression systems, such as baculovirus- infected Sf9 cells, and metabolic processes in vitro and transcriptional regulation of the two genes in the further investigations.

4. Materials and Methods

4.1. Insects and Insecticide

The susceptible laboratory (Sus-Lab) strain of WBPH (*S. furcifera*) established in our laboratory was obtained from the research group of Prof. Li, College of Plant Protection of Hunan Agricultural University (Changsha, China) in 2016, where the strain had been reared in the laboratory without exposure to any insecticide since 2009. All stages were maintained on rice seedlings under standard conditions of a temperature of 27 ± 1 °C, relative humidity (RH) of 70–80% and a light/dark cycle of 16:8 h. Sulfoxaflor (95%, technical grade) was purchased from Dow AgroSciences (Shanghai) Co., Ltd. China (Shanghai, China).

4.2. Selection of the SF-Sel strain with a Sublethal Dose of Sulfoxaflor

The toxicity of sulfoxaflor to *S. furcifera* was performed using the rice seedling dipping method, with some modifications [42]. First, technical grade sulfoxaflor was dissolved in acetone, and a series of suitable concentrations (i.e., 1, 2, 4, 6, and 8 µg/mL) were prepared with 0.1% Triton X-100; the 0.1% Triton X-100 solution alone was used as the blank control. Four to five leaves of rice seedlings were cleaned with water and air-dried at room temperature. Fifteen rice seedlings were bundled together, immersed in the diluted solution for approximately 30 s, and then dried at room temperature. Second, moistened cotton was wrapped around the rice roots, which were immobilized in a 500 mL plastic cup. Then, fifteen 3rd-instar nymphs were transferred to each plastic cup, and all treatments were set up in triplicate. All treatments were performed under standard environmental conditions (26 ± 1 °C, 85 ± 10% R.H., 14:10 L: D), and mortality was recorded after 96 h of treatment. Individual nymphs were considered dead if they did not show movement after being slightly nudged with a #26 soft brush. Probit analyses were conducted using a Statistical Analysis System (SAS) software to calculate the slope, LC50, 95% CI, and χ^2 values of sulfoxaflor or sulfoxaflor plus synergistic agents after 96 h of treatment [16]. Then, continuous selection with the sublethal LC_{25} dose of sulfoxaflor was performed for six generations in the SF-Sel strain.

4.3. Test for Synergism

The synergism bioassays for the Sus-Lab and SF-Sel strains of *S. furcifera* to sulfoxaflor were performed as described by Mu et al. [13] with some modifications. Three synergistic agents, DEM, TPP and PBO, were dissolved with acetone and diluted with Triton X-100 to the highest possible concentrations showing no adverse effect on the tested insects (PBO, 30 µg/mL; TPP, 160 µg/mL; DEM, at 300 µg/mL), after which rice seedlings into the synergistic treatment solutions for 30 s and naturally dried them. Then, approximately 300 of 3rd-instar nymphs were transferred to the rice seedlings treated with each synergistic agent for approximately 2 h. The remaining procedures were similar to the rice seedling dipping method as described above.

4.4. Enzyme Assays

To evaluate the potential role of the detoxification enzymes of *S. furcifera* in resistance to sulfoxaflor, the activities of CarEs, GSTs and P450s in the 3rd-instar nymphs of the Sus-Lab and SF-Sel strains treated with synergistic agents (TPP or DEM or PBO) were determined.

CarE activity was determined according to the method described by van Asperen [43]. Twenty 3rd-instar nymphs were placed in a centrifugal tube and stored in liquid nitrogen as quickly as possible, then homogenized on ice in 2 mL of homogenization buffer (0.04 mol/L phosphate buffer, pH 7.0) using a 5 mL glass homogenizer and centrifuged at 4 °C, 10,000× g for 15 min using a 5417R centrifuge (Eppendorf, Germany). The supernatant was subsequently transferred to a clean Eppendorf tube as the crude enzyme solution. A mixture of 0.45 mL of phosphate buffer (0.04 mol/L, pH 7.0), 1.8 mL of 3×10^{-4} mol/L α-NA solution (containing 3×10^{-4} mol/L physostigmine) and 50 µL of diluted enzyme liquid was added to each tube, followed by mixing and then incubation in a water bath at 30 °C for

15 min, after which the process was stopped with 0.9 mL of staining solution (0.2 g of fast blue-B salt in 20 mL of distilled water plus 50 mL of 5% sodium dodecyl sulfate). The absorbance values were recorded at 600 nm after 5 min in a UV 2000-Spectrophotometer (Unic [Shang Hai] Instruments Incorporated, Shanghai, China).

GST activity was determined using 1-chloro-2, 4-dinitrobenzene (CDNB) as a substrate according to the method of Wang et al. [16] with minor revisions. Twenty 3rd-instar nymphs were homogenized on ice in homogenization buffer (0.1 mol/L phosphate buffer containing 1.0 mmol/L EDTA, pH 6.5) and centrifuged at 10,800 rpm at 4 °C for 10 min, after which the supernatant was used as an enzyme source. A mixture including 2470 µL of phosphate buffer (0.1 mol/L, pH 6.5), 90 µL of CDNB (15 mmol/L), 50 µL of the enzyme source and 90 µL of reduced GSH (30 mmol/L) were added to a 5 mL centrifuge tube, which was promptly shaken. The OD value was recorded at 340 nm for 2 min and calculated as ΔA_{340}/min.

P450 activity was assayed using the method of Rose et al. [44] with some modifications. One hundred and fifty 3rd-instar nymphs were homogenized on ice in 2 mL of homogenization buffer (0.1 mol/L, pH 7.6, containing 20% glycerol, 0.1 mmol/L EDTA, 0.1 mmol/L DTT, and 0.4 mmol/L PMSF). The homogenates were centrifuged at 4 °C at 10,000× g for 10 min using a 5417R centrifuge (Eppendorf, Germany) to obtain the supernatant, which was used as the crude enzyme. Then, 100 µL of 4-nitroanisole (2×10^{-3} mol/L) was added to the cell culture plate and mixed with 90 µL of crude enzyme liquid, followed by incubation for 3 min at 27 °C in a water bath kettle and the addition of 10 µL of NADPH (9.6×10^{-3} mol/L) for reaction. The changes in the OD value were recorded at 405 nm (Model 680 Microplate Reader, Bio-Rad) every 20 s for 2 min. A standard curve was generated using p-nitrophenol, and the specific activity of P450s was finally calculated as nanomoles of p-nitrophenolper minute per milligram of protein.

All treatments were set up with three samples (tubes) as biological repetitions, and each enzyme sample was individually prepared. Each assay of enzymatic activity was replicated three times as mechanical repetitions for each enzyme sample. The total protein content of the enzyme solution was determined by the Bradford method [45] using bovine albumin as a standard. The activities of CarEs, GSTs and P450s were analyzed using unpaired Student's t-tests, and the significance level of the results was set at $p < 0.05$.

4.5. Transcriptome Analysis

4.5.1. Library Construction and Sequencing, Illumina Read Processing, and Assembly and Annotation of Unigenes

According to the manufacturer's protocol for the TRIzol® Reagent (Invitrogen™, ThermoFisher Scientific, USA), approximately 100 nymphs and adults that had either been continuously selected with the LC_{25} dose of sulfoxaflor for six generations (SF-Sel) or not (Sus-Lab) were used for total RNA extraction. cDNA library construction and sequencing, Illumina read processing, assembly and bioinformatics analysis, and the annotation of unigenes, including protein functional annotation, pathway annotation, COG/KOG functional annotation and Gene Ontology (GO) annotation, etc., were performed as described by Wang et al. [46].

4.5.2. Gene Expression and Differential Gene Enrichment

The expression of unigenes was calculated with the of RPKM (reads per kb per million reads) method [47] according to the following formula: RPKM(A) = (1000000 * C)/(N * L/1000)

The RPKM(A) value stands for the expression of gene A; C values stands for the number of reads that uniquely aligned to gene A; N values stands for the total number of reads that uniquely aligned to all genes, and L stands for the number of bases on gene A.

According to the gene expression represented by the RPKM values for each sample, the significant differentially expressed genes (DEGs) among the samples were screened with edge R. The screening criteria were an FDR < 0.05 (*p*-value after calibration by FDR) and |log2FC| > 1, and GO functional analysis and KEGG pathway analysis were performed based on the results for the DEGs.

4.5.3. Diversity and Collinearity of *S. furcifera* P450 Genes

Twenty-two relatively complete P450 amino acid sequences obtained from the transcriptome were compared with thirty-two P450 amino acid sequences of the white-back planthopper downloaded from NCBI and analyzed for the conserved functional domains with the motif (https://www.genome.jp/tools/motif/) and meme (http://meme-suite.org/tools/meme) tools, and their phylogenetic tree was constructed by using MEGA 6.0 software with the default settings and the neighbour-joining method. The results were visualized with TBtools software.

4.6. Quantitative PCR (qRT-PCR)

Total RNA of the Sus-Lab and SF-Sel strains was extracted using TRIzol reagent (Invitrogen™, ThermoFisher Scientific, USA) according to the instructions of the manufacturer's kit, and the reverse transcription reaction was performed with a cDNA Synthesis for qPCR (One-Step gDNA Removal) kit according to the instruction manual. The cDNA was kept at −20 °C for qRT-PCR.

The cDNAs of four P450 genes (*CYP6FD1*, *CYP6FD2*, *CYP4FD1* and *CYP4FD2*), one transporter (*Unigene0036498*), four transcription factors (*NlE78sf*, *C2H2ZF1*, *C2H2ZF3* and *C2H2ZF2*) and one reference gene (*RPL9*) [48] from the Sus-Lab and SF-Sel strains were amplified by PCR with twelve pairs of corresponding primers (Table 3). The qRT-PCR system and procedure were as described by Wang et al. [49]. All experimental results were analyzed in three independent replicates, and the treatment means and variances were analyzed via one-way ANOVA with PROC GLM of the SAS program. All means were compared by least squared difference (LSD) tests at a Type I error = 0.05.

Table 3. The primers of upregulation expression genes used in this study.

Gene Family	Prime	Sequence (5′-3′)	Length
Reference	RPL9-F	TGTGTGACCACCGAGAACAACTCA	131
	RPL9-R	ACGATGAGCTCGTCCTTCTGCTTT	
P450	CYP6FD1-F	CTTCAACATGCGGTTCACGC	187
	CYP6FD1-R	TTCATCCAAGCTCAACGGCT	
	CYP4FD1-F	AACCACTGCATGACTTTGCC	199
	CYP4FD1-R	TCAGCACCCGCAATGAATGT	
	CYP6FD2-F	GAGATGGCACACAAACCGGA	171
	CYP6FD2-R	GCAGAATCGCGCTAGAATGG	
	CYP4FD2-F	CAGCGAATGGTGGCTTCATC	183
	CYP4FD2-R	ATAGCAGCCATGGTCTCACC	
Transporter	Unigene0036498-F	CCCAAACCCTTCAAGACGGA	162
	Unigene0036498-R	GGCTGGATCGGAAATGCTCT	
Transcription factor	NlE78sf-F	GGAGTGTTGGGGTGGTAGTG	181
	NlE78sf-R	GGTGATGAACACTGCTCCGA	
	C2H2ZF1-F	CCATCATCAAGGCGGAACCT	182
	C2H2ZF1-R	ACCAGCGTTTTCAATGGTGC	
	C2H2ZF3-F	GTCGCCTGTGCCTTCTAGTT	165
	C2H2ZF3-R	AGCGGATGCACCTGATACTG	
	C2H2ZF2-F	ACAAGGGCATTCGCAAACAC	159
	C2H2ZF2-R	ATGTGCCGATCCAGATAGCG	

The biological function of *CYP6FD1* and *CYP4FD2* was verified through RNA interference as described by Mao et al. [31] and Wang et al. [49], with some modifications. A 168 bp fragment of *CYP6FD1*, 450 bp of *CYP4FD2* cDNA and a 657 bp green fluorescent protein (gfp) fragment were amplified by PCR using corresponding primer pairs (with the T7 promoter appended). PCR was performed with the primers listed in Table 4. The PCR products were purified for use as templates for dsRNA synthesis using the T7 MEGAscript kit (ThermoFisher, USA) according to the manufacturer's instructions. The dsRNA concentration was measured using a spectrophotometer (Nanodrop) after 1:10 dilution of the dsRNA product in water and adjustment of the ultimate concentration to 4 ng/µL for injection.

Table 4. The RNAi primers of *CYP6FD1* and *CYP4FD2* used in this study.

Prime	Sequence (5'-3')
T7-GFP-F	TAATACGACTCACTATAGGGAAGGGCGAGGAGCTGTTCACCG
T7-GFP-R	TAATACGACTCACTATAGGGCAGCAGGACCATGTGATCGCGC
CYP6FD1dsRNAF	TAATACGACTCACTATAGGGAGAAGTCCCAATTTCACAGACGC
CYP6FD1dsRNAR	TAATACGACTCACTATAGGGAGAGATTCCGGTCTATGCGCTTC
CYP4FD2dsRNAF	TAATACGACTCACTATAGGGAGAAAGGTTTCATCTACAAAGGATTGC
CYP4FD2dsRNAR	TAATACGACTCACTATAGGGAGACATCAGTGAAATCGTGCAGAATC

4.7. Function Analysis of CYP6FD1 and CYP4FD2 via RNAi

Third-instar nymphs were used for dsRNA injection experiments. First, the tested insects were anesthetized with CO_2 for approximately 30 s, and each insect received 120 ng (approximately 30 µL) of the dsRNA for each target gene using an UMP3/Nanoliter2010 microinjection device (World Precision Instruments, Sarasota, Florida), with dsGFP used as a negative control, and 300 3rd-instar nymphs were prepared to check the RNAi efficiency and bioassay for each gene. The relative expression of *CYP6FD1* and *CYP4FD2* was detected at 24, 48, 72 and 96 h after injection. For insecticide bioassays after RNAi, thirty 3rd-instar nymphs were collected 24 h after injection and sixty 3rd-instar nymphs for each treatment were transferred to rice seedlings that had been treated with the LC_{50} of sulfoxaflor in solution. Mortality was calculated at 72 h and 96 h after insecticide treatment. Three biological replicates were performed.

4.8. Prediction the Interaction of Tertiary Structure of CYP6FD1 and CYP4FD2 with Sulfoxaflor

To obtain information about how sulfoxaflor affects P450s, molecular docking between sulfoxaflor and the active sites of the target P450 proteins was carried out using the Surflex-Dock program in SybylX-2.0 version (Tripos Inc.) as previously described [50]. Surflex-Dock scores (total scores) were expressed in kcal/mol units to represent binding affinities [51,52].

4.9. Data Analysis

The relative normalized expression of the upregulated P450 genes in the Sus-Lab and SF-Sel strains, the efficacy of *CYP6FD1* and *CYP4FD2* knockdown in 3rd-instar nymphs of the SF-Sel strain by RNAi, and the mortality of larvae injected with dsRNA with or without the LC_{50} concentration of sulfoxaflor were compared using analysis of variance (ANOVA) followed by Duncan's test for multiple comparisons ($p < 0.05$) with the SPSS version 17.0 software package (IBM).

Supplementary Materials: The following are available online at http://www.mdpi.com/1422-0067/20/18/4573/s1, (Transcriptome data) can be found at https://dataview.ncbi.nlm.nih.gov/object/PRJNA560503.

Author Contributions: Conceived and designed the experiments: X.-G.W. Performed the experiments; X.-G.W., Y.-W.R. and Xiang Xin. Analyzed the data: C.-W.G., Y.-M.Z. Contributed reagents/materials/analysis tools: L.-T.S. and Xiang Xu; drafted and revised the manuscript: X.-G.W., Y.-W.R. and C.-W.G. All authors approved the final version of the article, including the authorship list.

Funding: This research was financially supported by the National Key R&D Program of China (2018YFD0200300).

Conflicts of Interest: The authors declare no conflict of interest.

References

1. Zhang, X.L.; Liao, X.; Mao, K.K.; Yang, P.; Li, D.Y.; Ali, E.; Wan, H.; Li, J.H. Neonicotinoid insecticide resistance in the field populations of *Sogatella furcifera* (Horváth) in Central China from 2011 to 2015. *J. Asia-Pacific Entomol.* **2017**, *20*, 955–958. [CrossRef]
2. Horgan, F.G.; Srinivasan, T.S.; Naik, B.S.; Ramal, A.F.; Bernal, C.C.; Almazan, M.L.P. Effects of nitrogen on egg-laying inhibition and ovicidal response in planthopper-resistant rice varieties. *Crop. Prot.* **2016**, *89*, 223–230. [CrossRef] [PubMed]
3. Wang, Z.; Zhou, C.; Long, G.Y.; Yang, H.; Jin, D.C. Sublethal effects of buprofezin on development, reproduction, and chitin synthase 1 gene (*SfCHS1*) expression in the white-backed planthopper, *Sogatella furcifera*. *J. Asia-Pacific Entomol.* **2018**, *21*, 585–591. [CrossRef]
4. Cheng, Z.N.; Li, S.; Gao, R.Z.; Sun, F.; Liu, W.C.; Zhou, G.H.; Wu, J.X.; Zhou, X.P.; Zhou, Y.J. Distribution and genetic diversity of southern rice black-streaked dwarf virus in China. *Virol. J.* **2013**, *10*, 307. [CrossRef] [PubMed]
5. Matsukura, K.; Towata, T.; Sakai, J.; Onuki, M.; Okuda, M.; Matsumura, M. Dynamics of Southern rice black-streaked dwarf virus in rice and implication for virus acquisition. *Phytopathology* **2013**, *5*, 509–512. [CrossRef] [PubMed]
6. Liu, M.G.; Jiang, C.X.; Mao, M.; Liu, C.; Li, Q.; Wang, X.G.; Yang, Q.F.; Wang, H.J. Effect of the insecticide dinotefuran on the ultrastructure of the flight muscle of female *Sogatella furcifera* (Hemiptera: Delphacidae). *J. Econ. Entomol.* **2017**, *110*, 632–640. [CrossRef] [PubMed]
7. Endo, S.; Tsurumach, I.M. Insecticide susceptibility of the brown planthopper and the white-backed planthopper collected from Southeast Asia. *J. Pestic. Sci.* **2001**, *26*, 82–86. [CrossRef]
8. Jin, J.X.; Jin, D.C.; Li, W.H.; Cheng, Y.; Li, F.L.; Zhou, Y.H.; Zhang, Y.L.; Liu, L. Status of insecticide resistance in *Sogatella furcifera* (Horváth)(Hemiptera: Delphacidae) from Guizhou Province. *J. Nanjing Agric. Univ.* **2017**, *40*, 258–265.
9. Matsumura, M.; Takeuchi, H.; Satoh, M.; Sanada-Morimura, S.; Otuka, A.; Watanabe, T.; Thanh, D.V. Species-specific insecticide resistance to imidacloprid and fipronil in the rice planthoppers *Nilaparvata lugens* and *Sogatella furcifera* in east and southeast Asia. *Pest Manag. Sci.* **2008**, *64*, 1115–1121. [CrossRef]
10. Tang, J.; Li, J.; Shao, Y.; Yang, B.J.; Liu, Z.W. Fipronil resistance in the whitebacked planthopper (*Sogatella furcifera*): possible resistance mechanisms and cross-resistance. *Pest Manag. Sci.* **2010**, *66*, 121–125. [CrossRef]
11. Matsumura, M.; Sanada-Morimura, S.; Otuka, A.; Ohtsu, R.; Sakumoto, S.; Takeuchia, H.; Satoha, M. Insecticide susceptibilities in populations of two rice planthoppers, *Nilaparvata lugens* and *Sogatella furcifera*, immigrating into Japan in the period 2005–2012. *Pest Manag. Sci.* **2014**, *70*, 615–622. [CrossRef] [PubMed]
12. Zhang, K.; Zhang, W.; Zhang, S.; Wu, S.F.; Ban, L.F.; Su, J.Y.; Gao, C.F. Susceptibility of *Sogatella furcifera* and *Laodelphax striatellus* (Hemiptera: Delphacidae) to six insecticides in China. *J. Econ. Entomol.* **2014**, *107*, 1916–1922. [CrossRef] [PubMed]
13. Mu, X.C.; Zhang, W.; Wang, L.X.; Zhang, S.; Zhang, K.; Gao, C.F.; Wu, S.F. Resistance monitoring and cross-resistance patterns of three rice planthoppers, *Nilaparvata lugens*, *Sogatella furcifera* and *Laodelphax striatellus* to dinotefuran in China. *Pestic. Biochem. Physiol.* **2016**, *134*, 8–13. [CrossRef] [PubMed]
14. Mohan, M.; Gujar, G.T. Local variation in susceptibility of the diamondback moth, *Plutella xylostella* (Linnaeus) to insecticides and role of detoxification enzymes. *Crop Prot.* **2009**, *22*, 495–504. [CrossRef]
15. Lai, T.C.; Li, J.; Su, J.Y. Monitoring of beet armyworm *Spodoptera exigua* (Lepidoptera: Noctuidae) resistance to chlorantraniliprole in China. *Pestic. Biochem. Physiol.* **2011**, *101*, 198–205. [CrossRef]
16. Wang, X.G.; Xiang, X.; Yu, H.L.; Liu, S.H.; Yin, Y.; Cui, P.; Wu, Y.Q.; Yang, J.; Jiang, C.X.; Yang, Q.F. Monitoring and biochemical characterization of beta-cypermethrin resistance in *Spodoptera exigua* (Lepidoptera: Noctuidae) in Sichuan Province, China. *Pestic. Biochem. Physiol.* **2018**, *146*, 71–79. [CrossRef] [PubMed]
17. Watson, G.; Loso, M.; Babcock, J.; Hasler, J.; Letherer, T.; Young, C. Novel nicotinic action of the sulfoximine insecticide sulfoxaflor. *Ins. Biochem. Mol. Biol.* **2011**, *41*, 432–439. [CrossRef] [PubMed]
18. Babcock, J.M.; Gerwick, C.B.; Huang, J.X.; Loso, M.R.; Nakamura, G.; Nolting, S.P.; Rogers, R.B.; Sparks, T.C.; Thomas, J.; Watson, G.B.; et al. Biological characterization of sulfoxaflor, a novel insecticide. *Pest Manag. Sci.* **2011**, *67*, 328–334. [CrossRef] [PubMed]

19. Zhu, Y.; Loso, M.R.; Watson, G.B.; Sparks, T.C.; Rogers, R.B.; Huang, J.X.; Gerwick, B.C.; Babcock, J.M.; Kelley, D.; Hegde, V.B.; et al. Discovery and characterization of sulfoxaflor, a novel insecticide targeting sap-feeding pests. *J. Agric. Food Chem.* **2011**, *59*, 2950–2957. [CrossRef]
20. Longhurst, C.; Babcock, J.M.; Denholm, I.; Gorman, K.; Thomas, J.D.; Sparks, T.C. Cross-resistance relationships of the sulfoximine insecticide sulfoxaflor with neonicotinoids and other insecticides in the whiteflies *Bemisia tabaci* and *Trialeurodes vaporariorum*. *Pest Manag. Sci.* **2013**, *69*, 809–813. [CrossRef]
21. Chen, X.; Ma, K.; Li, F.; Liang, P.; Liu, Y.; Guo, T.; Song, D.; Desneux, N.; Gao, X.W. Sublethal and transgenerational effects of sulfoxaflor on the biological traits of the cotton aphid, *Aphis gossypii* Glover (Hemiptera: Aphididae). *Ecotoxicology* **2016**, *25*, 1841–1848. [CrossRef] [PubMed]
22. Cutler, P.; Slater, R.; Edmunds, A.J.; Maienfisch, P.; Hall, R.G.; Earley, F.G.; Pitterna, T.; Pal, S.; Paul, V.L.; Goodchild, J.; et al. Investigating the mode of action of sulfoxaflor: A fourth-generation neonicotinoid. *Pest Manag. Sci.* **2013**, *69*, 607–619. [CrossRef] [PubMed]
23. Guedes, R.N.C.; Smagghe, G.; Stark, J.D.; Desneux, N. Pesticide induced stress in arthropod pests for optimized integrated pest management programs. *Annu. Rev. Entomol.* **2016**, *61*, 43–62. [CrossRef] [PubMed]
24. Liao, X.; Mao, K.K.; Ali, E.; Zhang, X.L.; Li, J.H. Temporal variability and resistance correlation of sulfoxaflor susceptibility among Chinese populations of the brown planthopper *Nilaparvata lugens* (Stål). *Crop Prot.* **2017**, *102*, 141–146. [CrossRef]
25. Liao, X.; Jin, R.H.; Zhang, X.L.; Ali, E.; Mao, K.K.; Xu, P.F.; Li, J.H.; Wan, H.H. Characterization of sulfoxaflor resistance in the brown planthopper, *Nilaparvata lugens* (Stål). *Pest Manag. Sci.* **2019**, *275*, 1646–1654. [CrossRef]
26. Ma, K.S.; Tang, Q.L.; Zhang, B.Z.; Liang, P.; Wang, B.M.; Gao, X.W. Overexpression of multiple cytochrome P450 genes associated with sulfoxaflor resistance in *Aphis gossypii* Glover. *Pestic. Biochem. Physiol.* **2019**, *157*, 204–210. [CrossRef]
27. Reyes, M.; Rocha, K.; Alarcón, L.; Siegwart, M.; Sauphanor, B. Metabolic mechanisms involved in the resistance of field populations of *Tuta absoluta* (Meyrick) (Lepidoptera: Gelechiidae) to spinosad. *Pestic. Biochem. Phys.* **2012**, *102*, 45–50. [CrossRef]
28. Tian, X.R.; Sun, X.X.; Su, J.Y. Biochemical mechanisms for metaflumizone resistance in beet armyworm, *Spodoptera exigua*. *Pestic. Biochem. Phys.* **2014**, *113*, 8–14. [CrossRef]
29. Chen, E.H.; Dou, W.; Hu, F.; Tang, S.; Wang, J.J. Purification and biochemical characterization of glutathione S-transferases in *Bactrocera minax* (Diptera: Tephritidae). *Fla. Entomol.* **2012**, *95*, 593–601. [CrossRef]
30. Wei, X.; Pan, Y.; Xin, X.C.; Zheng, C.; Gao, X.W.; Xi, J.H.; Shang, Q.L. Cross-resistance pattern and basis of resistance in a thiamethoxam-resistant strain of *Aphis gossypii* Glover. *Pestic. Biochem. Phys.* **2017**, *138*, 91–96. [CrossRef]
31. Mao, K.K.; Zhang, X.L.; Ali, E.; Liao, X.; Jin, R.H.; Ren, Z.J.; Wan, H.; Li, J.H. Characterization of nitenpyram resistance in *Nilaparvata lugens* (Stål). *Pestic. Biochem. Phys.* **2019**, *157*, 26–32. [CrossRef] [PubMed]
32. Bao, H.B.; Gao, H.L.; Zhang, Y.X.; Fan, D.; Fang, J.; Liu, Z. The roles of *CYP6AY1* and *CYP6ER1* in imidacloprid resistance in the brown planthopper: Expression levels and detoxification efficiency. *Pestic. Biochem. Phys.* **2016**, *129*, 70–74. [CrossRef] [PubMed]
33. Jin, J.X.; Jin, D.C.; Li, F.L.; Cheng, Y.; Li, W.H.; Ye, Z.C.; Zhou, Y.H. Expression differences of resistance-related genes induced by cycloxaprid using qRT-PCR in the female adult of *Sogatella furcifera* (Hemiptera: Delphacidae). *J. Econ. Entomol.* **2017**, *110*, 1785–1793. [CrossRef] [PubMed]
34. Zimmer, C.T.; Garrood, W.T.; Singh, K.S.; Randall, E.; Lueke, B.; Gutbrod, O.; Matthiesen, S.; Kohler, M.; Nauen, R.; Davies, T.G.E.; et al. Neofunctionalization of duplicated P450 genes drives the evolution of insecticide resistance in the brown planthopper. *Curr. Biol.* **2018**, *28*, 268. [CrossRef] [PubMed]
35. Jones, C.M.; Daniels, M.; Andrews, M.; Lind, R.J.; Gorman, K.; Williamson, M.S. Age-specific expression of a P450 monooxygenase (*CYP6CM1*) correlates with neonicotinoid resistance in *Bemisia tabaci*. *Pestic. Biochem. Phys.* **2011**, *101*, 53–58. [CrossRef]
36. Ding, Z.P.; Wen, Y.C.; Yang, B.J.; Zhang, Y.X.; Liu, S.H.; Liu, Z.W.; Han, Z.J. Biochemical mechanisms of imidacloprid resistance in *Nilaparvata lugens*: Over-expression of cytochrome P450 *CYP6AY1*. *Insect Biochem. Molec.* **2013**, *43*, 1021–1027. [CrossRef]
37. Chen, X.; Zhang, Y. Identification and characterisation of NADPH-dependent cytochrome P450 reductase gene and cytochrome b5 gene from *Plutella xylostella*: Possible involvement in resistance to beta-cypermethrin. *Gene* **2015**, *558*, 208–214. [CrossRef]

38. Denlinger, D.L.; Yocum, G.D.; Rinehart, J.P. Hormonal control of diapause. *Compr. Mol. Insect Sci.* **2005**, 615–650.
39. Stiborová, M.; Indra, R.; Frei, E.; Schmeiser, H.H.; Eckschlager, T.; Adam, V.; Heger, Z.; Arlt, V.M.; Martínek, V. Cytochrome b_5 plays a dual role in the reaction cycle of cytochrome P450 3A4 during oxidation of the anticancer drug ellipticine. *Monatsh. Chem.* **2017**, *148*, 1983–1991. [CrossRef]
40. Cui, S.F.; Wang, L.; Ma, L.; Geng, X.Q. P450-mediated detoxification of botanicals in insects. *Phytoparasitica* **2016**, *44*, 585–599. [CrossRef]
41. Lertkiatmongkol, P.; Jenwitheesuk, E.; Rongnoparut, P. Homology modeling of mosquito cytochrome P450 enzymes involved in pyrethroid metabolism, insights into differences in substrate selectivity. *BMC Res. Notes* **2011**, *4*, 321. [CrossRef] [PubMed]
42. Wang, Y.; Chen, J.; Zhu, Y.C.; Ma, C.; Huang, Y.; Shen, J. Susceptibility to neonicotinoids and risk of resistance development in the brown planthopper, *Nilaparvata lugens* (stål) (Homoptera: Delphacidae). *Pest Manag. Sci.* **2010**, *64*, 1278–1284. [CrossRef] [PubMed]
43. Van, A.K. A study of housefly esterase by means of sensitive colorimetric-method. *J. Insect Physiol.* **1962**, *8*, 401–406.
44. Rose, R.L.; Barbhaiya, L.; Roe, R.M. Cytochrome P450-associated insecticide resistance and the development of biochemical diagnostic assays in *Heliothis virescens*. *Pestic. Biochem. Phys.* **1995**, *51*, 178–191. [CrossRef]
45. Bradford, M.M.A. A rapid and sensitive method for quantitation of microgram quantities of protein utilizing the principle of protein-dye binding. *Anal. Biochem.* **1976**, *25*, 248–256. [CrossRef]
46. Wang, X.G.; Chen, Y.Q.; Gong, C.W.; Yao, X.G.; Jiang, C.X.; Yang, Q.F. Molecular identification of four novel cytochrome P450 genes related to the development of resistance of *Spodoptera exigua* (Lepidoptera: Noctuidae) to chlorantraniliprole. *Pest Manag. Sci.* **2018**, *74*, 1938–1952. [CrossRef] [PubMed]
47. Mortazavi, A.; Williams, B.A.; McCue, K.; Schaeffer, L.; Wold, B. Mapping and quantifying mammalian transcriptomes by RNA-Seq. *Nat. Methods.* **2008**, *5*, 621–628. [CrossRef]
48. An, X.K.; Hou, M.L.; Liu, Y.D. Reference gene selection and evaluation for gene expression studies using qRT-PCR in the white-backed Planthopper, *Sogatella furcifera* (Hemiptera: Delphacidae). *J. Econ. Entomol.* **2016**, *109*, 879. [CrossRef]
49. Wang, L.X.; Niu, C.D.; Zhang, Y.; Jia, Y.L.; Zhang, Y.J.; Zhang, Y.; Zhang, Y.Q.; Gao, C.F.; Wu, S.F. The NompC channel regulates Nilaparvata lugens proprioception and gentletouch response. *Insect Biochem. Mol. Biol.* **2019**, *106*, 55–63. [CrossRef]
50. Revankar, H.M.; Kulkarni, M.V.; Joshi, S.D.; More, U.A. Synthesis, biological evaluation and docking studies of 4-aryloxymethyl coumarins derived from substructures and degradation products of vancomycin. *Eur. J. Med. Chem.* **2013**, *70*, 750–757. [CrossRef]
51. Jain, A.N. Scoring noncovalent protein-ligand interactions: Acontinuous differentiable function tuned to compute binding affinities. *J. Comput. Aid. Mol. Des.* **1996**, *10*, 427–440. [CrossRef]
52. Jain, A.N. Surflex: Fully automatic flexible molecular docking using a molecular similarity-based search engine. *J. Med. Chem.* **2003**, *46*, 499–511. [CrossRef] [PubMed]

© 2019 by the authors. Licensee MDPI, Basel, Switzerland. This article is an open access article distributed under the terms and conditions of the Creative Commons Attribution (CC BY) license (http://creativecommons.org/licenses/by/4.0/).

MDPI
St. Alban-Anlage 66
4052 Basel
Switzerland
Tel. +41 61 683 77 34
Fax +41 61 302 89 18
www.mdpi.com

International Journal of Molecular Sciences Editorial Office
E-mail: ijms@mdpi.com
www.mdpi.com/journal/ijms

www.ingramcontent.com/pod-product-compliance
Lightning Source LLC
LaVergne TN
LVHW070142100526
838202LV00015B/1876